Springer Texts in Statistics

Advisors:
Stephen Fienberg   Ingram Olkin

Springer Texts in Statistics

Alfred	Elements of Statistics for the Life and Social Sciences
Blom	Probability and Statistics: Theory and Applications
Chow and Teicher	Probability Theory: Independence, Interchangeability, Martingales. Second Edition
Christensen	Plane Answers to Complex Questions: The Theory of Linear Models
du Toit, Steyn and Strumpf	Graphical Exploratory Data Analysis
Kalbfleisch	Probability and Statistical Inference: Volume 1: Probability. Second Edition
Kalbfleisch	Probability and Statistical Inference: Volume 2: Statistical Inference. Second Edition
Keyfitz	Applied Mathematical Demography. Second Edition
Kiefer	Introduction to Statistical Inference
Kokoska and Nevison	Statistical Tables and Formulae
Madansky	Prescriptions for Working Statisticians
McPherson	Statistics in Scientific Investigation: Basis, Application and Interpretation
Nguyen and Rogers	Fundamentals of Mathematical Statistics: Volume I: Probability for Statistics
Nguyen and Rogers	Fundamentals of Mathematical Statistics: Volume II: Statistical Inference
Peters	Counting for Something: Statistical Principles and Personalities

(continued after index)

Paul E. Pfeiffer

Probability for Applications

With 77 Illustrations

Springer-Verlag
New York Berlin Heidelberg
London Paris Tokyo Hong Kong

Paul E. Pfeiffer
Department of Mathematical Sciences
Rice University
Houston, Texas 77251-1892

Library of Congress Cataloging-in-Publication Data
Pfeiffer, Paul E.
  Probability for applications / Paul E. Pfeiffer.
    p.   cm.—(Springer texts in statistics)
  ISBN 0-387-97138-6 (alk. paper)
  1. Probabilities.   I. Title.   II. Series.
QA273.P423 1989
519.2—dc20   89-21924

Printed on acid-free paper.

Camera-ready copy formatted using LaTeX.
Printed and bound by R.R. Donnelley and Sons, Harrisonburg, Virginia.
Printed in the United States of America.

9 8 7 6 5 4 3 2 1

ISBN 0-387-97138-6 Springer-Verlag New York Berlin Heidelberg
ISBN 3-540-97138-6 Springer-Verlag Berlin Heidelberg New York

Preface

Objectives. As the title suggests, this book provides an introduction to probability designed to prepare the reader for intelligent and resourceful applications in a variety of fields. Its goal is to provide a careful exposition of those concepts, interpretations, and analytical techniques needed for the study of such topics as statistics, introductory random processes, statistical communications and control, operations research, or various topics in the behavioral and social sciences. Also, the treatment should provide a background for more advanced study of mathematical probability or mathematical statistics.

The level of preparation assumed is indicated by the fact that the book grew out of a first course in probability, taken at the junior or senior level by students in a variety of fields—mathematical sciences, engineering, physics, statistics, operations research, computer science, economics, and various other areas of the social and behavioral sciences. Students are expected to have a working knowledge of single-variable calculus, including some acquaintance with power series. Generally, they are expected to have the experience and mathematical maturity to enable them to learn new concepts and to follow and to carry out sound mathematical arguments. While some experience with multiple integrals is helpful, the essential ideas can be introduced or reviewed rather quickly at points where needed.

Fundamental approach. Modern probability is a sophisticated mathematical discipline, which provides useful models for a wide variety of applications, both physical and behavioral. And significant applications often require highly developed aspects of the theory, for both providing appropriate models and for determining and validating pertinent properties of the solutions. Therefore, it is highly desirable that the practitioner have a sound grasp of the underlying mathematics as well as a reasonable knowledge of its previous and potential uses. This view has shaped the presentation in many ways.

- The central core is the development of those concepts and patterns that constitute essential elements of modern probability theory. Historically, these concepts were not discovered in the clean, tightly organized fashion of a modern treatment. But in the interest of moving efficiently toward a working command of the subject, we seek a logical, step-by-step treatment, with occasional digressions.

- Since modern probability theory is quite abstract and formal it is essential to have concrete examples that give substance to the abstract patterns and help relate them to situations in which they may be applied. Whenever possible, mechanical-graphical representations of concepts and relationships are employed to aid visualization. Not only do these facilitate learning, but they also serve as important aids to application.

- Emphasis on fundamental ideas, interpretations, and use of graphical-mechanical representations makes it possible to treat significant problems with a minimum of mathematical techniques. On the other hand, introduction to a variety of techniques makes possible the solution of more complex problems.

- The examples provided are of various types. Some are quite artificial, intended to serve as illustrations of concepts or as examples of techniques of solution. Some present potential or actual applications but use only illustrative data. Often, the cases are greatly simplified in order to prevent complexities and difficulties of computation from obscuring simple ideas and strategies. Sometimes the method of solution illustrated is more sophisticated than a simplified example really requires. The goal is insight into the structure of the problems and the techniques presented, as well as an appreciation of the possibilities for types of application.

An introductory course. The book contains more material than can be developed in a one-semester course. This allows some flexibility in topical coverage and makes the book more useful as a continuing reference. The following comments may help the prospective teacher plan for use of the book as a text for an introductory course.

Part I. Basic Probability

The six chapters in this part provide a closely knit development. Because of its unfamiliarity, some will be inclined to omit Chapter 5 entirely. Although some selectivity is in order, this treatment clarifies many ambiguities found in the literature and leads to significant applications. The instructor may wish to examine this chapter carefully before making a selection of topics.

Part II. Random Variables and Distributions

The five chapters in this part develop essential features of the model in an integrated way. Section 8.6 on reliability theory is optional; I consider Section 11.5 on coordinate transformations to be secondary, although the topic is considered essential by many who work with probability. The treatment of quantile functions (Section 11.4) can be considerably abbreviated;

the application to simulation, using a random number generator, should probably be included.

Part III. Mathematical Expectation

This part allows for more flexibility of treatment. Chapters 12, 14, 15, 16, and the first three sections of Chapter 18 provide core material. In the usual first course, the treatment of Chapter 13 should probably be somewhat cursory. I generally limit presentation to Section 13.1 and a brief discussion of the Lebesgue integral and the transformation of integrals. Treatment of convergence, in Chapter 17. is usually limited to a few comments. In order to make the tie with statistics, without developing statistical method, I use the last three sections of Chapter 18.

If students intend to pursue studies in mathematical probability or mathematical statistics, more complete coverage of Chapters 13 and 17 may be desirable. Some instructors may wish to expand the treatment of random samples and statistics in Section 18.7. However, they should also consider the introductory treatment to Bayesian statistics in Section 22.2.

Part IV. Conditional Expectation

Because of the importance of conditional expectation in probability theory, the first five sections of Chapter 19 should certainly be used. Also, the concept of conditional independence is finding an growing role in applied probability. The first two sections of Chapter 22 introduce the concept and show its usefulness while sketching some essential ideas of Bayesian statistics. Students find interesting and challenging both the applications and the techniques in Chapter 20; hence, instructors may wish to include some of this material. It is usually not feasible to include the material on Poisson processes and Markov processes in a one-semester first course. However, the development of these very important topics in Chapters 21 and 23 rounds out the introduction to probability by showing how the concepts and techniques established earlier may be used to obtain quite sophisticated results. These topics are useful in several fields of application, and this treatment establishes bridges to subsequent study in those fields.

Supplementary Theoretical Material

Much of the theoretical material, particularly in the chapter appendices, can be omitted in a first course. It is included because some readers will want to know the basis for propositions stated without proof or other justification in the chapters. In particular, those who find that they need extensions or modifications of those propositions may find it useful to examine proofs and theoretical details.

Some features.   Much of the work is quite standard in form and content. Some features that have proved useful in my teaching include the following.

- The use of minterm maps (Chapter 2), popular in switching theory, continues to provide an extremely useful tool for organizing the treatment of logical combinations of events.

- The organization of the treatment of conditioning (Chapter 3) and of independence as lack of conditioning (Chapter 4) is the result of considerable teaching experience. It fosters a grasp of the essential structure of the concepts and clarifies a number of facts that often escape or confuse beginners.

- The concept of conditional independence of events (Chapter 5) has proved increasingly useful in clarifying the treatment of many important applications. Hopefully, this treatment will lead to wider recognition and use of this concept.

- The treatment of independence of random vectors (Chapter 9) presents formal patterns that justify much of the intuitive use of the concept in practice.

- Some of the graphical techniques in the treatment of functions of random variables (Chapter 11) are quite useful, as both teaching tools and as aids to organizing the solution of practical problems.

- After a brief and informal treatment of mathematical expectation in Chapter 12, the sketch of the theory of the Lebesgue integral in Chapter 13 provides an important bridge to more advanced treatments of mathematical probability. Although I ordinarily give only a brief sketch of the ideas (Section 13.1) in an introductory course, I make this treatment available for those who wish to delve deeper or who may want to refer to it in a later course. An awareness of this development makes it possible to put mathematical expectation on a firm footing (Chapter 14), producing and using a powerful list of properties for applications.

- The treatment of conditional expectation, given a random vector (Chapter 19), makes available essential aspects of more advanced (and more abstract) notions that play a central role in modern theory of random processes. Beginning with the usual special cases considered in most texts at this level, the treatment moves to an identification of critical patterns for both theory and application.

- In Chapter 20, transform methods (moment-generating functions and generating functions) are combined with conditional expectation to solve a number of important and challenging problems. The notion of a random number of random variables is examined, including a brief introduction to Markov times (sometimes called stopping times). Certain problems of "random selection" lead to the notion of counting processes, the most notable of which is the Poisson process.

- Treatment of the Poisson process (Chapter 21) is usually delayed until a subsequent applications course. Some of the theoretical material in the appendices to the chapter is included for reference.

- The concept of conditional independence, given a random vector (Chapter 22), has been studied extensively in abstract treatments of Markov processes. Use of conditional expectation, as introduced in Chapter 19, makes available to the undergraduate this essential concept. We show its relevance to certain topics in decision theory and in the Bayesian approach to statistics. In addition, it provides the basis for a modern treatment of Markov sequences or chains (Chapter 23). For lack of time in the introductory course, Markov sequences are usually treated in subsequent applications courses.

Acknowledgment.   With a few exceptions, where my indebtedness is clear, I have not attempted to identify sources or to provide references. The literature is vast and growing. Most of what is provided in this introductory treatment is either in the common fund of knowledge in the field or represents the fruits of my own efforts in teaching this material over many years. My indebtedness to the work of many others—innovators as well as textbook writers—is great and gratefully acknowledged. I only hope that this work can add to the common fund and perhaps aid in introducing others to this fascinating and increasingly important field of intellectual endeavor.

P.E.P.

Contents

III Mathematical Expectation   273

List of Figures

Part I

Basic Probability

1

Trials and Events

A long search for the essence of probability has culminated in a powerful and useful modern theory, which is highly sophisticated yet conceptually simple. As a mathematical discipline, probability theory is increasingly abstract in its formal expression, yet it is remarkably useful as a practical model for an extraordinary range of applied problems. This system has provided the basis for modeling in a growing array of applications in which the central notion is choice and decision under uncertainty.

In this chapter, we begin the task of introducing and elucidating this model in a series of steps or phases. We do not attempt to trace the history nor to guide the beginner through the often tortuous process of discovery that underlies the development of the topic. Rather, we seek to move rather directly to unfold the essence of the modern mathematical discipline, which has so successfully captured the insights of the intellectual pioneers in the field. In the process, we attempt to relate the very abstract concepts and formalisms to the intuitive notions and fundamental interpretations that provide the basis for application. We also seek to develop ways of visualizing the abstract and formal properties in a manner that aids thinking about both the mathematics and the structure of the system or entity represented. While a knowledge of the history of the subject is often important, it seems more fruitful to look at that history from the perspective of modern formulation and development. Such a perspective can often sharpen our perceptions of concepts and relations; frequently, significant advances result from fresh investigation of avenues that once seemed closed or of little promise.

1.1. Trials, Outcomes, and Events

Imprecise but intuitive notions of probability abound in popular thought. Even in such undifferentiated thought, it is clear that *probability* is a number assigned to an *event*, which may or may not occur as a result of performing some trial or experiment. It is natural, therefore, to begin our investigation and exposition by considering how an event can be represented mathematically in a manner that is both intuitively satisfying and sufficiently precise to serve as an element of a sound mathematical system.

We begin by asking several questions: To what kind of situations does probability thinking apply? What are the central features? How can we find a mathematical "handle"? We speak of events and the probability of events. What is an event? How can we describe it mathematically? The answers come quite simply from an examination of a few situations to which probability has been applied.

Example 1.1.1: Games of Chance

The early developments of probability as a mathematical discipline, principally in the work of Pierre de Fermat and Blaise Pascal in the seventeenth century, came as a response to questions about games of chance played repeatedly. A well-defined "play" of the game can result in one of several outcomes. Although the result of any play is not predictable, certain "statistical regularities" of results were observed. In seeking to understand these regularities, the possible results were described in ways that made each result seem equally likely. If there were N such possible "equally likely" results, each was assigned a probability of $1/N$. Our point, at this stage of the analysis, is to see that the game is described in terms of a well-defined trial; the result of any trial is one of a specific set of distinguishable outcomes.

—   □

Example 1.1.2: Statistical Sampling

A second source of stimulus to the development of mathematical probability was the analysis of statistical data. The pioneering work of John Graunt in the seventeenth century concerned the study of "vital statistics" such as records of births, deaths, and various diseases. Graunt gave the fraction of people in London who died from various diseases during a period in the early seventeenth century. Some thirty years later, in 1693, Edmond Halley (for whom the comet is named) published the first life insurance tables. To apply these results, one considers selection of a member of the population on a chance basis. One then assigns the probability that such a person will have a given disease. The trial here is the selection of a person; the outcome is the particular person chosen. We may speak of the event that the person selected will die of a certain disease—say "consumption." Here, we distinguish between the person (the outcome) and certain attributes (death of certain diseases).

—   □

Example 1.1.3: Some Modern Examples

- The temperature at a weather station is recorded each hour during a 24-hour period. A single trial consists of making the set of 24 individual measurements. The outcome of the trial is one of the set of possible "profiles." If the individual temperature readings are taken on a thermometer calibrated from 30° F to 150° F, in one-half degree increments, then there are 241 possible readings each hour so that there are possible 241^{24} distinguishable profiles. Taking a set of readings is tantamount to selecting one of the possible profiles. One could visualize this by supposing each possible profile is printed on a separate card. The trial is equivalent to selecting one of this basic set of cards. Although the number of outcomes is finite, the number is so large that it is frequently useful to assume that the individual temperature readings may be spread over the continuum from 30 to 150°, so that the number of possible profiles is infinite.

- The number of vehicles passing a certain point in a given time period is counted. The trial is the vehicle count; each outcome is characterized by the number of vehicles observed. This number is a nonnegative integer. In actual fact, there must be some upper limit on the number of vehicles passing in a given time interval, so that the set of outcomes must be finite. However, in some situations, it may not be possible to place such an upper limit, so that formally the number is presumed to be without bound. Observation of the count is equivalent to selecting one of the possible outcomes.

- The number of uncorrected errors on a digital data line in a pre-scribed time period is observed. The trial is the error count; the outcome is the number of uncorrected errors observed. Since the number of digits that can arrive during the prescribed period is limited by the data rate, the number of distinguishable outcomes is finite. Again, observation of the count amounts to selecting one of the possible distinguishable outcomes.

- A person who comes to a medical clinic is tested for the presence of symptoms for a certain disease. From the point of view of obtaining statistics on the prevalence of the disease, the trial is the testing of the patient, who is from the population that the clinic serves. The outcome is the result of the test. If the distinguishable results are presence of the symptoms or absence of the symptoms, there are two possible outcomes. From the point of view of the physician who works with the patient, the trial is the selection of a person from the population served by the clinic. The number of outcomes may be quite large. For purposes of treatment, this outcome falls into one of two subclasses of the population determined by the condition of the patient with respect to the symptoms.

□

One could extend such a list indefinitely. These examples illustrate that there are two central features of the various situations in which probability is applied.

- There is a well-defined *trial*, experiment, or operation.

- The result of any trial is an *outcome*, which is one of a specified *set of possibilities*.

The trials may be quite simple or they may be composite. In any case, the result of a trial is a single outcome from a basic set of such potential outcomes. *We may view the performance of the trial as the selection of one of these outcomes.* From this point of view, we can set up a simple physical representation, which then is translated into an equally simple mathematical representation.

If the set of possible outcomes is finite the performance of a trial is analogous to drawing a ball from a jar or selecting a single card from a box. As noted in the description of the composite trial of taking 24 hourly temperature readings, for example, one could think of each distinguishable profile written on a card. Taking a set of readings is equivalent to selecting one such profile card. We find this representation as a jar of balls or box of cards applicable to each of the examples in Example 1.1.3. Even in the infinite case we may imagine a jar with an infinite set of balls.

The desired mathematical notion is that of a *set* of distinguishable elements. A basic set of elements is an abstraction of a jar filled with balls. The process of performing the trial may be represented and visualized as the selection of one element from the basic set. If one's imagination can allow a possible infinity of balls, the correspondence is complete.

Definition 1.1.1. *The* basic space Ω *is the set whose elements* ω *are the possible outcomes of the experiment.*

Remarks

1. The term *sample space* is used by many authors, particularly in the literature on statistics.

2. In most situations, there may be several (perhaps many) possible basic sets. The same experiment may be described equally well in terms of various basic spaces. However, to carry out the analysis, the basic space must be agreed upon.

3. In probabilistic situations, the selection process involves "chance." The preceding formulation allows deliberate selection. One can reach into a candy jar and pick out the piece desired or one can reach in, without looking, and select by chance. The discussion of event, which follows, does not require the chance ingredient in selection, although

the ensuing theory of probability is based on notions of chance and uncertainty.

Having abstracted the notion of a trial as the selection of an element from a set, we have to ask: What do we mean by saying that a given *event* has *occurred*? Consider, again, the trials described in Example 1.1.3.

- In the temperature example, suppose we say the maximum temperature observed today is 90°. This is a shortened expression for "the event that the maximum temperature is 90° has occurred." By this we mean, "The profile selected (observed) is *one of those* in which the maximum number is 90."

- In the vehicle count example, suppose we say the number observed is between 20 and 27. In the language of events, we say, "The event that the count is between 20 and 27 has occurred." This indicates that the actual number of vehicles observed is *one of those integers* n that satisfies the condition $20 \leq n \leq 27$.

- In the medical patient case, suppose the test indicates the symptoms of osteoporosis. From the point of view that it is the patient who is selected, we say, "The patient has the symptoms of osteoporosis." That is, the event that the patient has symptoms of osteoporosis has occurred. Thus, the patient is *one of those* in the population served by the clinic who has these symptoms. In other words, the patient belongs to the subset of patients who are characterized by the presence of the symptoms.

These examples suggest that the occurrence of an event is the selection of an outcome from among the subset of possible outcomes that have a desired property.

A MATHEMATICAL MODEL FOR TRIALS AND EVENTS

Let us summarize the structure of our system in terms of the analogy of a jar with colored balls.

Example 1.1.4: A Jar with Colored Balls

The trial is to chose one of the balls so that each possible outcome corresponds to one of the balls. We are interested in the color of the ball drawn. The trial is performed; a ball is drawn. Did the event "the ball is red" occur?

—   □

MATHEMATICAL REPRESENTATION

- A mathematical jar (basic space) with mathematical balls (elements).

- Pick a mathematical ball—"randomly" or otherwise.

- Did the desired *event* occur? This translates into the question: Does the outcome belong to the *subset* consisting of those outcomes that have the desired characteristic (the color red, in this case)?

EVENTS

- An event A is the subset of those mathematical balls (elements of the basic space) corresponding to outcomes that have the prescribed characteristics (the color red, in this example).

- The event A *occurs* iff the mathematical ball (the element) selected belongs the the prescribed subset (i.e., has the desired characteristic).

We formalize as follows.

Definition 1.1.2. *Suppose the basic space is Ω with elements ω representing outcomes. Let $\pi_A(\cdot)$ be a proposition; $\pi_A(\omega)$ is that proposition made about ω; it may be true or false.*

Event $A = \{\omega : \pi_A(\omega)$ is true$\}$ = the set of outcomes such that the proposition $\pi_A(\omega)$ is true.

Event A occurs iff the element ω selected is an element of A, which is true iff the proposition $\pi_A(\omega)$ is true.

Remark: When the outcome ω is identified completely, many different events may have occurred. This merely means that ω may belong to many subsets of the basic space.

Example 1.1.5: (Example 1.1.3 Continued)

- In the temperature profile case, if we say, "The temperature range today was from a low of 68° to a high of 88°," this means that the temperature profile observed belongs to that subset of profiles in which the minimum reading is 68 and the maximum reading is 88. The event is represented by this subset. Occurrence of this event is represented by the selection of a profile from this subset.

- If we say the number of vehicles observed lies between 20 and 24, we are asserting that the outcome is one of the subset of integers $\{20, 21, 22, 23, 24\}$. The event is this subset of the integers; the fact that this event occurred is the fact that the number observed belongs to this subset.

- A physician in the clinic sees a patient. The tests show the presence of the symptoms. The event "the patient has the symptoms" has occurred. This means that the patient tested belongs to that subset for whom the symptoms are present. From the formal point of view, the event is the subset of afflicted patients. The occurrence of the event is the observation of a patient with those symptoms, who therefore belongs to the subset.

— □

1.2. Combinations of Events and Special Events

In both formal and informal use of probability, we refer not only to events but also to various logical combinations of events. For example, we may say that it rained today *or* turned colder. The person interviewed for a job will have management experience *and* have a business school degree. The combinations may be more complex. In a survey of readers of Time magazine, one may be interested in those who take Time and take a morning newspaper, but do not take Newsweek magazine. A company is trying to meet a deadline in bringing out a new product. It has two development groups, working somewhat independently, attempting to produce a suitable version. One may speak of the event that group one or group two will meet the deadline. In ordinary language, it may not be clear whether or not this includes the possibility that both will meet the deadline. The very ambiguity of common language necessitates a more precise mode of expression for careful, logical analysis.

The representation of events as subsets of the basic space points the way to a simple resolution of the problem. If we let event A be the event that the maximum temperature recorded on a given day by our weather station is no greater than 85° and let B be the event that the minimum temperature is no less than 65°, then event A is understood to be the set of profiles in which the maximum number is less than or at most equal to 85. Similarly, event B consists of that set of profiles in which the minimum number is at least 65°. If we let C be the event that the minimum temperature is at least 65° *or* the maximum temperature is at most 85°, then C is the set of all profiles in which at least one of the conditions holds. Here we use the mathematical convention that A or B means one or the other or both. The set C thus refers to the *union* of sets A and B. Using the conventional mathematical notation, we write

$C = A \cup B$ Event C occurs iff event A occurs or event B occurs.

Since we are representing events by subsets, we may use the traditional *Venn diagram* to aid in visualizing combinations and relations between sets. The *or combination* (union) is represented in Figure 1.1. The points in the

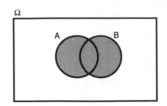

The shaded region represents the union of A, B.

FIGURE 1.1. The "or" combination as set union.

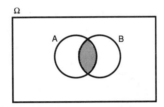

The shaded region represents the intersection of A, B.

FIGURE 1.2. The "and" combination as set intersection.

rectangular figure represent elements of the basic space Ω (i.e., possible outcomes). Elements of the events A and B are represented as points in the indicated circles. The event $C = A \cup B$ is represented by the collection of all points in the shaded region. Thus, any point that is either in A or in B (or in both), is a point in C. This translates into the statement that event C occurs iff either event A occurs or event B occurs or both.

Suppose event D is the event that the minimum temperature is no less than $65°$ *and* the maximum temperature is no greater than $85°$. Since both conditions must be met, the event D consists of those profiles that are in both event A and in event B. Event D occurs iff both event A occurs *and* event B occurs. In set terminology, D is the *intersection* of sets A and B. In the usual notation, we write

$D = A \cap B = AB$ Event D occurs iff event A occurs and event B occurs.

We often use the shorter notation AB rather than $A \cap B$, using the intersection symbol much as we use the multiplication symbol in ordinary algebra. The *and combination* (intersection) is represented in Figure 1.2, which uses the same Venn diagram conventions as Figure 1.1. The points in the common part of sets A and B represent the elements of D. These are the points in the shaded region. Any point that is both in A and in B

is a point in D. This signifies that event D occurs iff both event A occurs and event B occurs.

Often we wish to refer to the fact that an event A did not occur. This means that the outcome was not in set A. In the language of sets, we say the outcome (element) is in the complementary set A^c. This set consists of all elements not in A. In the language of events, we say event A does not occur iff the *complementary event* A^c occurs. On any trial, either event A occurs or event A^c occurs. We may represent this by writing

$$A \cup A^c = \Omega.$$

By definition, the basic set Ω, as an event, always occurs. If a trial is successfully executed, the outcome is in Ω. For this reason, it is often natural to refer to the basic space as the *sure event* or the *certain event*, since it is sure to occur on any trial.

We consider one other special event, which at first may seem quite trivial, but which we find quite useful in a variety of contexts. In set theory, the *empty set*, usually represented by $\emptyset$, is the set without any elements. It is useful, for example, to indicate that two sets have no common elements by saying their intersection is the empty set $\emptyset$. In the language of events, the empty set corresponds to the *impossible event*. It can never occur on any trial, since it contains no outcomes. We can use this to indicate that two events are *mutually exclusive*, which means that no outcome can belong to both events. In symbols

$$A \cap B = \emptyset \quad \text{or} \quad AB = \emptyset \quad \text{iff} \quad A \text{ and } B \text{ are mutually exclusive.}$$

We find it convenient to use a special symbol for the union of mutually exclusive events. If A and B are mutually exclusive, then we may write $A \biguplus B$ rather than $A \cup B$. *This is not a new operation.* It amounts to writing $A \cup B$, with the additional stipulation $AB = \emptyset$. It is not incorrect to indicate the union of mutually exclusive events by $A \cup B$, since this is the basic combination. The crossbar is just a convenient way of asserting that in the case under consideration the events to be combined are mutually exclusive.

In set theory, the *inclusion* relation $A \subset B$ means that every element in A is also in B. This is often indicated by $AB = A$, since the common part of A and B is A. As events, if $A \subset B$, then every occurrence of A requires the occurrence of B also. We therefore say event A *implies* event B. The inclusion relation is represented schematically in the Venn diagram of Figure 1.3, where the representation for A lies inside that for B.

Once the symbolic representation of various combinations of sets is introduced, these conventions can be applied to combinations of two or more sets.

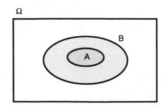

Occurrence of A implies occurrence of B.

FIGURE 1.3. The inclusion relation and implication.

Example 1.2.1:

Suppose a trial consists of checking four items from a production line to determine if they meet specifications. Let

E_i be the event the ith item meets specifications, $1 \le i \le 4$.
A_r be the event exactly r of these meet specifications, $0 \le r \le 4$.
S_r be the event r or more of these meet specifications, $0 \le r \le 4$.

Determine A_2, A_3, A_4 in terms of the E_i and then determine S_2.

Solution

To express A_2, we consider all the patterns of success-failure, where there are exactly two successes (hence two failures). There are six such patterns, representing mutually exclusive events, so that

$$A_2 = E_1^c E_2^c E_3 E_4 \uplus E_1^c E_2 E_3^c E_4 \uplus E_1^c E_2 E_3 E_4^c \uplus E_1 E_2^c E_3^c E_4$$
$$\uplus E_1 E_2^c E_3 E_4^c \uplus E_1 E_2 E_3^c E_4^c.$$

In a similar manner, we may determine A_3 and A_4 to be

$$A_3 = E_1^c E_2 E_3 E_4 \uplus E_1 E_2^c E_3 E_4 \uplus E_1 E_2 E_3^c E_4 \uplus E_1 E_2 E_3 E_4^c$$

and

$$A_4 = E_1 E_2 E_3 E_4.$$

We have two or more successes iff we have exactly two or exactly three or exactly four successes. These are mutually exclusive possibilities, so that

$$S_2 = A_2 \uplus A_3 \uplus A_4 = (A_0 \cup A_1)^c = A_0^c A_1^c.$$

The alternate forms can be obtained by well-known rules on set combinations or they may be derived by noting that two or more successes

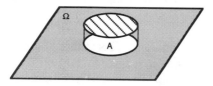

FIGURE 1.4. Schematic graph of indicator function I_A.

requires not zero or one, which means that we have not zero and not one success.

— ☐

It is assumed that the reader is familiar with elementary operations with sets. The primary purpose of the preceding discussion is to relate these facts about sets to their use in representing events and their combinations and relationships.

1.3. Indicator Functions and Combinations of Events

In this section, we introduce the very simple notion of an *indicator function* I_A for a subset A of a basic set Ω. The usefulness of such functions is demonstrated repeatedly throughout this book. As the name suggests, this is a function on the basic space, in the sense that to each ω in Ω we assign a number $I_A(\omega)$. We make the assignment as follows:

$$I_A(\omega) = \begin{cases} 0 & \text{for } \omega \in A^c \\ 1 & \text{for } \omega \in A \end{cases}.$$

The name comes from the fact that I_A has the value one for each element in A and the value 0 for each element not in A. It thus gives an indication whether or not element ω is in set A. It has the nature of a step function. The schematic representation of the graph in Figure 1.4 provides a way of visualizing the concept. The indicator function serves to "lift out" the set from the basic space to make it evident.

If A is an event, the indicator function I_A is a very simple and special case of those functions known as random variables. The notion of random variable, introduced in Chapter 7, is one of the central concepts of mathematical probability. Although the indicator function is a very simple form of random variable, we find it useful in a variety of contexts, both for its intrinsic properties and as a notational device for giving precise expression to other types of random variables.

A part of the usefulness of indicator functions derives from a collection of elementary properties that transform basic combinations and relations among sets into ordinary numeric combinations and relations among the corresponding indicator functions. This makes possible verification of many important facts about combinations of sets by arithmetic and algebraic arguments. We establish and list several very elementary properties. The arguments are as simple as the concept. The first translates set inclusion into inequality of the indicator functions.

IF1) $I_A \leq I_B$ iff $A \subset B$.

Proof

If $A \subset B$, then the condition $\omega \in A$ implies $\omega \in B$. Hence, the condition $I_A(\omega) = 1$ implies $I_B(\omega) = 1$. We may, of course have $I_A(\omega) = 0$ but $I_B(\omega) = 1$ if ω is in B, but not in A. On the other hand, if $I_A(\omega) \leq I_B(\omega)$ for all ω, then $I_A(\omega) = 1$ implies $I_B(\omega) = 1$, so that every ω in A must also be in B.

—   □

From (IF1) and the fact that $A = B$ iff both $A \subset B$ and $B \subset A$, it follows that

IF2) $I_A = I_B$ iff $A = B$.

Since $1 - I_A(\omega) = 1$ iff $I_A(\omega) = 0$ and $1 - I_A(\omega) = 0$ iff $I_A(\omega) = 1$, it follows that

IF3) $I_{A^c} = 1 - I_A$.

The next two properties make the transformation for intersections and unions.

IF4) $I_{AB} = I_A I_B = \min\{I_A, I_B\}$

IF5) $I_{A \cup B} = I_A + I_B - I_A I_B = I_A + I_B - I_{AB} = \max\{I_A, I_B\}$.

Proof

We examine the four cases determined by the four possible combinations of values of I_A, I_B

I_A	I_B	I_{AB}	$I_{A \cup B}$
0	0	0	0
0	1	0	1
1	0	0	1
1	1	1	1

In each case, the value of I_{AB} may be obtained by taking the product of the values of I_A and I_B, which is the minimum of

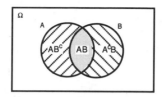

FIGURE 1.5. Representation of the union as a disjoint union.

these values. In each case, the values of $I_{A \cup B}$ may be obtained by taking the sum of the values of I_A and I_B, then subtracting the product of these. As the table shows, this value is also the maximum of the values of I_A and I_B.

—   □

From (IF2) and the facts that $I_\Omega = 1$ and $I_\emptyset = 0$, we may assert

IF6) $I_A = 1$ iff $A = \Omega$ and $I_A = 0$ iff $A = \emptyset$

These fundamental properties lead to a wide variety of useful results, some of which we gather in the following example.

Example 1.3.1:

$$I_{A \bigsqcup B} = I_A + I_B$$
$$I_{A \cup B} = I_A + I_{A^c B} = I_B + I_{AB^c} = I_{AB^c} + I_{AB} + I_{A^c B}.$$

Proof

The first follows from (IF4), (IF5), and the fact that A and B are disjoint. The second makes use of the fundamental properties to write

$$I_{A \cup B} = I_A + I_B(1 - I_A) = I_A + I_{A^c B}$$

or

$$I_{A \cup B} = I_B + I_A(1 - I_B) = I_B + I_{AB^c}$$

or

$$I_{A \cup B} = I_A(1 - I_B) + I_A I_B + I_B(1 - I_A) = I_{AB^c} + I_{AB} + I_{A^c B}.$$

These identities reflect the fact that $A \cup B$ may be expressed as disjoint unions (see Figure 1.5)

$$A \cup B = A \bigsqcup A^c B = B \bigsqcup AB^c = AB^c \bigsqcup AB \bigsqcup A^c B.$$

In the language of events, A or B occurs if A occurs or A does not occur and B does. Similarly, A or B occurs if B occurs or B does not occur and A does. We have occurrence of A or B if one of the mutually exclusive combinations A and not B or both A and B or B and not A occurs.

——   □

The usefulness of the indicator function becomes more apparent when we consider combinations of more than two sets (events). For example, in Section 2.4, the indicator function provides the basis for a simple proof of the important minterm expansion theorem, which we use repeatedly.

INDICATOR FUNCTIONS ON OTHER DOMAINS

The preceding treatment defines indicator functions for subsets of the basic space Ω. It should be apparent that one can define indicator functions for subsets of any basic set of elements in a similar manner. Consider the set $\Re$ of real numbers, represented by points on a line. If M is a subset of the real numbers, then $I_M(t) = 1$ iff $t \in M$ and is zero otherwise. Similarly, we may define indicator functions for the set $\Re^2$ of pairs of real numbers, represented by points on the plane. We find such indicator functions on multidimensional spaces useful when we consider functions of random variables. These ideas are introduced in Chapter 11 and are used extensively throughout subsequent developments.

1.4.   Classes, Partitions, and Boolean Combinations

Combinations of sets produced by the formation of unions, intersections, and complements are known as *Boolean combinations*, after George Boole who pioneered the systematic formal algebraic handling of combinations of sets in the mid nineteenth century. Suppose A, B, and C are sets (these may, for example, represent events). Consider the combination

$$F = A \cup A^c(B \cup C^c).$$

The precise element content of F depends upon the content of the three sets A, B, and C. Thus, it is sometimes useful to look upon the rule for a Boolean combination as determining a *Boolean function* of the "variable" sets A, B, and C. Whatever the choice of these, the pattern of the combination provides a rule for determining the content of F. The Venn diagram of Figure 1.6 indicates three sets represented in such a way that each could have a common part with each of the others. The distribution of elements in Ω could be such that some of these intersections are empty.

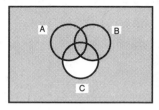

$$F = A \cup A^C (B \cup C^C)$$

FIGURE 1.6. Venn diagram for a Boolean combination.

Careful interpretation of the expression in terms of the Venn diagram representation shows that the set F consists of all elements in the basic set in the shaded region. It is often possible to find equivalent expressions, representing other Boolean combinations, which give the same elements of F for each assignment of elements to A, B, and C.

- We may use the Venn diagram to discover equivalent expressions. For example, the shaded area corresponding to F is seen to be the region for the combination

$$F = \left(C (A \cup B)^c \right)^c.$$

- We may use the rules of "set algebra," perhaps aided by the Venn diagram, to establish equivalent expressions. By the usual rules on set combinations, the preceding expression may modified as follows:

$$F = C^c \cup (A \cup B) = A \cup B \cup C^c = A \cup A^c (B \cup C^c).$$

- We may use indicator functions to show that the new expression is equivalent to the original.

$$
\begin{aligned}
I_F &= 1 - I_C (1 - I_A - I_B + I_{AB}) \\
&= 1 - I_C \big(1 - I_A - I_B (1 - I_A)\big) \\
&= 1 - I_{A^c} (1 - I_{C^c})(1 - I_B) \\
&= I_A + I_{A^c} (1 - 1 + I_B + I_{C^c} - I_B I_{C^c}) \\
&= I_{A \cup A^c (B \cup C^c)}.
\end{aligned}
$$

- The algebraic manipulation of the numerical combinations indicator functions may be tedious. Rather than compare the *forms* of the combinations, one may compare the *possible values* of two expressions to check for equality.

Example 1.4.1: Comparison of Values of Two Indicator Function Expressions

Use indicator functions to determine whether or not the following events F and G are the same:

$$F = A(B \cup C^c) \cup A^c C \qquad G = A^c BC \cup AB.$$

Solution

Use of fundamental patterns shows that

$$I_F = I_A \big(I_B + (1 - I_C) - I_B(1 - I_C) \big) + (1 - I_A) I_C = I_A + I_C + I_A I_B I_C - 2 I_A I_C$$

and

$$I_G = (1 - I_A) I_B I_C + I_A I_B = I_A I_B + I_B I_C - I_A I_B I_C.$$

We compare values of I_F and I_G for all possible combinations of values of I_A, I_B, and I_C. There are $2^3 = 8$ such combinations, as follows:

I_A	I_B	I_C	I_F	I_G	I_A	I_B	I_C	I_F	I_G
0	0	0	0	0	1	0	0	1	0
0	0	1	1	0	1	0	1	0	0
0	1	0	0	0	1	1	0	1	1
0	1	1	1	1	1	1	1	1	1

Examination shows the values of I_F and I_G are different for the second entry and for the fifth entry, so that $F \neq G$. Closer examination shows that $I_G \leq I_F$, so that $G \subset F$.

—   □

Remark: The indicator function analysis illustrates the significant fact that I_F can change value only when one or more of I_A, I_B, or I_C change value. This occurs when moving from at least one of the sets A, B, or C to its complement. A similar statement can, of course, be made for I_G. This fact is used in Section 2.4 to establish the minterm expansion theorem.

CLASSES OF EVENTS

We need to be able to consider classes of sets (events) of various cardinalities. To facilitate writing, we need suitable notation. The following basic scheme, with several variations, is useful.

$\mathcal{A} = \{ A_i : i \in J \}$ indicates there is a set A_i for each i in the index set J.

The class is *finite* if there is a finite number of indices in J, and hence a finite number of sets in the class $\mathcal{F}$.

The class is *countable* if there is a finite or countably infinite number of indices in J.

Sometimes we indicate the index set explicitly. For example

$$\mathcal{A} = \{A_i : 1 \le i \le n\} \qquad \text{or} \qquad \mathcal{A} = \{A_i : 1 \le i\}.$$

For a small finite cardinality, we may use different letters for different sets, with no indices. Whether or not indices are used, we frequently list the sets in the class, as in the following examples.

$$\mathcal{A} = \{A_1, A_2, A_3\} \qquad \text{or} \qquad \mathcal{G} = \{A, B, C, D\}.$$

We extend the notion of union and intersection of pairs of sets (events) to general classes.

Definition 1.4.1. *If $\mathcal{A} = \{A_i : i \in J\}$, then the* union *of the class, designated $\bigcup_{i \in J} A_i$, is the set of all those elements in Ω that are in at least one of the sets A_i in the class. The* intersection *of the class, designated $\bigcap_{i \in J} A_i$, is the set of all those elements that are in every one of the sets A_i in the class. It may well be that there are no such elements, so that the intersection is empty.*

It may seem that we should have little use for an infinite class of events. However, the following classical example illustrates a simple situation that calls for an unlimited class of events.

Example 1.4.2: Events Based on Infinite Classes

Two persons play a game of chance alternately until one is successful. Let E_i be the event of success on the ith play, A be the event the first player wins, and B be the event the second player wins. Since there is no upper limit on the number of plays, we must consider the entire class $\{E_i : 1 \le i\}$. We may write

$$A = E_1 \bigcup E_3 E_1^c E_2^c \bigcup E_5 E_1^c E_2^c E_3^c E_4^c \bigcup \ldots = \bigcup_{k=0}^{\infty} \left(E_{2k+1} \bigcap_{j=1}^{2k} E_j^c \right)$$

and

$$B = E_2 E_1^c \bigcup E_4 E_1^c E_2^c E_3^c \bigcup E_6 E_1^c E_2^c E_3^c E_4^c E_5^c \bigcup \ldots = \bigcup_{k=1}^{\infty} \left(E_{2k} \bigcap_{j=1}^{2k-1} E_j^c \right).$$

The case that neither wins is

$$C = A^c B^c = \bigcap_{i=1}^{\infty} E_i^c.$$

Note that we cannot say that C is impossible, since the game can conceivably continue without limit. Of course, under any reasonable assignment of probability, the probability will be zero.

——   □

PARTITIONS

In the next chapter we see that additivity is a fundamental property of probability. This means that the probability of the union of mutually exclusive events is the sum of the individual probabilities. For this reason, one of the important techniques of analysis is to express an event as an appropriate disjoint union of events whose individual probabilities are known. In Example 1.4.2, we express A and B as such disjoint unions. We speak of the process of decomposing an event into mutually exclusive events as *partitioning* the event. In this regard, we adopt the following terminology.

Definition 1.4.2. *A partition $\{A_i : i \in J\}$ is a class of mutually exclusive events such that exactly one of them occurs on each trial. That is $\Omega = \biguplus_{i \in J} A_i$.*

A partition of event B is a class $\{B_i : i \in J\}$ of mutually exclusive events, exactly one of which occurs whenever B occurs. That is, $B = \biguplus_{i \in J} B_i$.

There are several common ways to partition an event. One the simplest is the decomposition

$$A = AB \biguplus AB^c.$$

Event A can occur in one of two mutually exclusive ways: both A and B occur, or both A and not B occur. A generalization of this pattern is as follows:

Suppose $\{B_i : i \in J\}$ is a mutually exclusive class such that

$$A \subset \biguplus_{i \in J} B_i. \qquad \text{Then } A = \biguplus_{i \in J} AB_i.$$

The event A occurs iff exactly one of the events AB_i occurs. In the event that the class $\{B_i : i \in J\}$ is a partition, A is automatically included in its union. The pattern is illustrated in Figure 1.7.

Another common pattern is illustrated by the analysis of the game in Example 1.4.2. We examine the process in more detail.

Example 1.4.3: Partitioning a Class

Consider the class $\{A, B, C, D\}$. We begin by taking event A. We may then take the part of B that is not in A. The events A and $A^c B$ form a mutually exclusive pair, since we cannot have both A and not A occur.

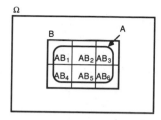

FIGURE 1.7. Partitioning of event A by a disjoint class.

Then we take the part of C that is not in A or B, to get $A^c B^c C$. The class $\{A, A^c B, A^c B^c C\}$ is mutually exclusive, since any two members differ by being in complementary parts of at least one of the original class. Finally, we take $\{A, A^c B, A^c B^c C, A^c B^c C^c D\}$. It should be clear that this process could have been carried out by considering the sets in any order. In each case we get a different partitioning. Also, the example indicates how the process may continue for any number of events. This procedure is the basis of the decomposition in the game problem in Example 1.4.2.

—   □

Remark: In Section 2.4, we consider an alternate decomposition for a finite class of events that produces basic elements for constructing any Boolean combination of the events in the class.

JOINT PARTITIONS

A partition of the basic space may be viewed as a process of *classifying* the elements into mutually exclusive *categories*. If we have several such classification schemes, we may combine these in a multiple classification whose categories form a joint partition of the basic space.

Example 1.4.4: Joint Partitions and Multiple Classification

For purposes of inventory, an automobile dealer categorizes his passenger car stock according to model (two-door, four-door, wagon), transmission control (manual, automatic), and air conditioning (with, without). There are thus three partitions of the stock.

$\mathcal{A} = \{A_1, A_2, A_3\}$ where A_1 is the set of two-door cars, A_2 is the set of four-door models, and A_3 is the set of station wagons.

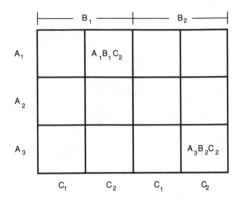

FIGURE 1.8. Venn diagram for the joint partition in Example 1.4.4.

$\mathcal{B} = \{B_1, B_2\}$ where B_1 is the set with manual transmission and $B_2 = B_1^c$ is the set with automatic transmission.

$\mathcal{C} = \{C_1, C_2\}$ where C_1 is the set with air conditioning and $C_2 = C_1^c$ is the set without.

Each car can be assigned to one of $3 \cdot 2 \cdot 2 = 12$ categories, corresponding to one of the sets $A_i B_j C_k$. These sets form the joint partition

$$\{A_i B_j C_k : i \in I, \, j \in J, k \in K\} \text{ where } I = \{1, 2, 3\}, \, J = \{1, 2\}, \, K = \{1, 2\}.$$

This joint partition may be represented by the Venn diagram of Figure 1.8.

—   □

PROBLEMS

1-1. Let Ω be the set of all positive integers. Consider the subsets $A = \{\omega : \omega \le 12\}$, $B = \{\omega : \omega < 8\}$, $C = \{\omega : \omega$ is even$\}$, $D = \{\omega : \omega$ is a multiple of 3$\}$, and $E = \{\omega : \omega$ is a multiple of 4$\}$. Express the following sets in terms of A, B, C, D, and E:

  (a) $\{1, 3, 5, 7\}$ (b) $\{3, 6, 9\}$ (c) $\{8, 10\}$
  (d) The even integers greater than 12.
  (e) The positive integers that are multiples of 6.
  (f) Positive integers even and not greater than 6 or odd and greater than 12.

1-2. The basic space Ω consists of 100 elements ω_1, ω_2, ..., ω_{100}. Use index set notation and describe the index set for each of the following subsets of Ω.

  (a) The set of elements with odd index i.
  (b) The set consisting of ω_{13} and those elements following it in the sequence.
  (c) The set of elements such that the index is a multiple of three.

1-3. Let Ω be the set of integers 0 through 10, A be the set $\{5, 6, 7, 8\}$, B be the set of odd integers in Ω, and C be the integers that are even or less than 3. Describe the following sets by listing their elements:

  (a) AB (b) AC (c) $AB^c \cup C$ (d) ABC^c
  (e) $A \cup B^c$ (f) $A \cup BC^c$ (g) ABC (h) $A^c BC^c$

1-4. Express each of the following events as a logical combination of the events A, B, and C.

  (a) Exactly two of the events   (b) All of the events
  (c) At least two of the events   (d) One or three of the events
  (e) At most two of the events   (f) The first two or the last two
  (g) At least one of the events   (exclusive)

1-5. Which of the following statements are true for any events A, B, and C?

  (a) $A \cup (BC)^c = A \cup B \cup B^c C^c$ (b) $(A \cup B)^c = A^c C \cup B^c C$
  (c) $A \subset AB \cup AC \cup BC$

1-6. Extend Example 1.3.1 to show that $I_C = I_A + I_B$ iff $AB = \emptyset$ iff $C = A \biguplus B$.

1-7. Express the indicator functions for the following events in terms of the indicator functions I_A, I_B, and I_C.

(a) $D = A \cup BC^c$ (b) $E = AB^c \biguplus A^c B$

(c) $F = AB^c C \biguplus A^c B^c C \biguplus ABC$ (d) $G = AC \cup BC$

1-8. Use indicator functions to show that

(a) $AB \cup B^c C \subset A \cup A^c C$,

(b) $A \cup A^c BC = AB \cup BC \cup AC \cup AB^c C^c$.

1-9. Express $A \cup B \cup C \cup D$ in the manner of Example 1.4.3.

1-10. Use complements to express $A \cup B \cup C \cup D$ as an intersection.

1-11. Suppose the class $\{A_1, A_2, A_3, A_4\}$ is a partition. Let B be a subset of the basic set Ω. Draw a Venn diagram that shows the partitioning of B by the class in the following cases.

(a) $A_4 B = \emptyset$ but $A_i B \neq \emptyset$ for $i \neq 4$.

(b) Each intersection $A_i B$ is nonempty.

1-12. Draw a Venn diagram for the joint partition for the partitions $\mathcal{A} = \{A_1, A_2\}$, $\mathcal{B} = \{B_1, B_2, B_3\}$, and $\mathcal{C} = \{C_1, C_2, C_3\}$.

1-13. Among five "identical" instruments, two are defective. A service man selects them one by one, in a random manner. Let E_i be the event the ith unit chosen is defective. Write an expression for the event that he has found both defective units by the third selection?

1-14. Two persons in a game "play" consecutively, until one of them is successful or there are ten unsuccessful plays. Let E_i = event of a success on the ith play of the game. Let

A = event the first player wins. B = event the second player wins.

C = event neither player wins.

Express A, B, and C in terms of the events E_1, E_2, E_3, $\ldots$, E_{10}.

1-15. Suppose a game could, in principle, be played an infinite number of times. Let E_i be the event that a certain result is obtained on the ith play. Let A be the event that only a finite number of plays result in the desired outcome. Express A in terms of the E_i.

2

Probability Systems

In Chapter 1, elementary examples show that probability applies to situations in which there is a well-defined trial whose possible outcomes are found among a given basic set. In that chapter, we begin our modeling of probability by formalizing the notion of an event. Occurrence or nonoccurrence of an event is determined by characteristics or attributes of the outcome observed. The event is taken to be the subset of possible outcomes that have these characteristics. Performing the trial is visualized as selecting an outcome from the basic set. An event occurs whenever the selected outcome is a member of the subset representing the event. As described in Chapter 1, the selection process could be quite deliberate with a prescribed outcome or it could involve the uncertainties associated with "chance." Probability enters the picture only in the latter situation. The probability of an event is a measure of the likelihood that the outcome of the trial will belong to that event. In this chapter, we seek to represent mathematically the likelihood that an event will occur. For this, we introduce the concept of a *probability measure*.

2.1. Probability Measures

Probability is applied to situations in which a well-defined trial or experiment or procedure can issue in one of a set of possible outcomes. Before the trial is performed, there is *uncertainty* about which of these latent possibilities will be realized. In formulating the concept of an event, in Chapter 1 we represent the performance of the trial as the selection of one of the outcomes. The performance of the trial (or act) may be quite deliberate with the outcome predetermined. No probability is involved in such a case. *From this point on, however, we are concerned with trials in which there is uncertainty about which of the possible outcomes will be realized.* The existence of uncertainty due to "chance" or "randomness" does not necessarily suppose the act of performing the trial is haphazard. The trial may be quite carefully planned; the contingency may be the result of factors beyond the control or knowledge of the experimenter.

The mechanism of chance (i.e., the source of the uncertainty) may depend upon the nature of the actual selection process. For example, in taking the temperature profile of Example 1.1.3, the principal variations are not due to experimental error but rather to unknown factors that converge to provide the specific weather pattern experienced. In the case of uncorrected digital transmission errors, the cause of uncertainty lies in the intricacies of the correction mechanisms and the perturbations produced by a very complex environment. The patient at the clinic may be self selected. Before his or her appearance and the result of the test, the physician may not know which patient with which condition will appear. In each case, from the point of view of the experimenter the cause is simply attributed to "chance." Whether one sees this as an "act of the gods" or simply the result of a configuration of physical or behavioral causes too complex to analyze, the situation is one of uncertainty before the trial about which outcome will present itself.

If there were complete uncertainty, the situation would be chaotic. But this is not usually the case. While there is an extremely large number of possible temperature profiles, a substantial subset of these has very little likelihood of occurring. For example, profiles in which successive hourly temperatures alternate between very high then very low values throughout the day constitute an unlikely subset (event). One normally expects trends in temperatures over the 24-hour period. Although the traffic engineer does not know exactly how many vehicles will be observed in a given time period, experience would provide some idea what range of values to expect. While there is uncertainty about which patient with which symptoms will appear, the physician certainly knows approximately what fraction of the clinic's patients have the disease in question. In a game of chance, analyzed into "equally likely" outcomes, the assumption of equal likelihood is based on knowledge of symmetries and structural regularities in the mechanism by which the game is carried out. And the number of outcomes associated with a given event is known or may be determined.

In each case, there is some basis in statistical data on past experience or knowledge of structure, regularity, and symmetry in the system under observation that makes it possible to assign likelihoods to the occurrence of various events. It is this ability to assign likelihoods to the various events that characterizes probability. *Probability* is a number assigned to *events* to indicate their likelihood of occurrence. In the classical probability used to analyze games of chance with N equally likely possible outcomes, the probability assigned to an event A is the fraction of those possible outcomes in the subset. Thus, if N_A of the outcomes result in event A, $P(A) = N_A/N$. Probability so determined is a number between 0 and 1, with 1 assigned to the certain event Ω and 0 assigned to the impossible event $\emptyset$. If A and B are mutually exclusive events, the number of outcomes in the union is the sum of the numbers in the separate events. This implies $P(A \biguplus B) = P(A) + P(B)$.

The original impetus for developing a mathematical theory of probability came principally from the study of games of chance. The classical model of a finite number of equally likely outcomes served well. And this theory achieved a high level of development. Support for this theory came from the fact that when such games were played repeatedly the relative frequency of occurrence of various events tended to the probability values derived from the model. Although many important problems can be analyzed in terms of this classical model, this is often done at the expense of artificially contrived representations. An equally important source of ideas and stimulus for the development of probability is the study of frequencies or fractions of various outcomes in statistical data. If data show that a certain fraction of a population dies from consumption, it is a simple step to supposing that the likelihood or probability that a member of the population will share that fate is just that fraction. If a condition is shared by the entire population, the fraction, hence the probability, is one. Suppose we are interested in the fraction of a population that has an income of at least fifty thousand dollars per year or less than ten thousand dollars per year. Since the conditions are mutually exclusive the fraction of the population in one class or the other is the sum of the fractions in the subclasses. This means that the probability of one condition or the other is the sum of the separate probabilities. If there is a finite population to be dealt with in gathering the data, one may be tempted to suppose that each member is equally likely to be encountered. This may well be a false assumption, as many statistical studies show. If the class of possible outcomes is infinite, the concept of equally likely outcomes loses significance.

A more general model than the classical one is needed. Such a model should be consistent with, and include as a special case, the classical probability model. Which properties of the classical model should we make universal? Certainly, probability should be a number between zero and one. It should assign the value one to the certain event Ω. And the probabilities of mutually exclusive events should add. In fact, it turns out that these are the *essential* patterns for a very satisfactory probability measure. While one can never give logical proof that a model is correct, extensive study of the resulting mathematical system shows that the following model captures the essence of intuitive notions of probability in a remarkable way.

PROBABILITY MEASURE

A probability measure is an assignment of a number $P(A)$ to each event A under the condition that

P1) $P(A) \geq 0$.

P2) $P(\Omega) = 1$.

P3) *Countable additivity.* If $A = \bigcup_{i \in J} A_i$ (J countable), then $P(A) = \sum_{i \in J} P(A_i)$.

A TECHNICAL MATHEMATICAL CONDITION

In order to set up a satisfactory mathematical system, the class of events must be characterized properly. In many elementary problems, *any* subset of the basic space Ω can be an event, particularly if there are only finitely many such subsets. For technical reasons, however, in the general case we must put certain mild restrictions on the class of events, in order to avoid some unusual but highly undesirable mathematical phenomena. We often deal with infinite classes of events. We want to form Boolean combinations of the members of such classes. We need the flexibility to take complements, unions, and intersections of the events. At the same time, we need to avoid the pathological conditions referred to earlier. These requirements are met if the class $\mathcal{F}$ of events has the structure of a *sigma algebra* of subsets of Ω. A class of sets is a sigma algebra if

1. It is closed under complements. That is, if A is in $\mathcal{F}$, so is A^c.

2. It is closed under countable unions. That is, if $\{A_i : i \in J\}$ is a countable class of subsets in $\mathcal{F}$, then the union $A = \bigcup_{i \in J} A_i$ is also in $\mathcal{F}$.

These two conditions imply

3. A sigma algebra is also closed under countable intersections.

A brief discussion of sigma algebras is included in the appendix to this chapter. Fortunately, these technical considerations can be ignored in most applications at the level of this work.

PROBABILITY SYSTEMS

We can now formalize our probability model as a probability system, as follows.

Definition 2.1.1. *A probability system is a triple* $(\Omega, \mathcal{F}, P)$, *where*

Ω *is a basic set, the* basic space.
$\mathcal{F}$ *is a sigma algebra of subsets of* Ω, *the* class of events.
P *is a probability measure on* $\mathcal{F}$.

This system is often referred to as the *Kolmogorov model*, after the Russian mathematician A. N. Kolmogorov (1903–1987), who succeeded in bringing together developments begun at the turn of the century, principally in the work of E. Borel and H. Lebesgue on measure theory. Kolmogorov published his epochal work in German in 1933. It was translated into English and published in 1956 by Chelsea Publishing Company.

At this point, the formal Kolmogorov system seems rather trivial and devoid of any real relevance to applications. In order to make it useful, we proceed along the following lines.

1. We interpret probability as mass assigned to the subsets that are events. This provides a means of visualizing, remembering, and sometimes discovering significant patterns in the mathematical system.

2. We display the mathematical content of the system by deriving additional properties from the three taken as axiomatic. In the process, we are aided by the graphical/mechanical representation of sets on Venn diagrams with associated mass.

3. We interpret the mathematical patterns in terms of the fundamental notion of probability as the likelihood of the occurrence of an event. This interpretation is the key to representing a practical problem in terms of the formalism of the mathematical system as well as to interpreting the mathematical results.

PROBABILITY AS MASS

As a first step, we interpret probability as mass assigned to the elements in a subset. Thus $P(A)$ is viewed as mass assigned to set A. Since the probability of the certain event Ω is one, the total mass is $P(\Omega) = 1$. If A and B are disjoint sets, representing mutually exclusive events, then their probabilities add to give the probability of A or B. Thus $P(A \biguplus B) = P(A) + P(B)$, the combined mass of the two sets. We think of the probability measure as distributing mass over the Venn diagram representation of the basic space. This distribution need not be uniform. If there is a finite number of possible outcomes, then we consider point mass concentrations at the points representing those outcomes. An event is represented by a region on the Venn diagram. The associated probability is proportional to the weight of that region. Thus, if there is a finite set of points in a region corresponding to the event in question, the probability of that event is proportional to the combined weight assigned those points. If the number of elements is uncountably infinite, then the probability mass may be viewed as spread in some smooth, continuous fashion over the Venn diagram. The probability assigned any event is proportional to the weight of that part of the weighted Venn diagram.

This representation of probability as mass is consistent with nonrelativistic views of mass assigned to particles and bodies. This interpretation or representation of probability as mass is no accident. A probability measure is a special case of the mathematical concept of *measure* introduced in the study of certain problems associated with moments of mass about points and lines in space. In fact, the German term for measure, in this sense, is *mass*.

We may derive additional mathematical properties of probability and check them against the mass representation and the likelihood interpretation. Or we may discover the properties by thinking about the mass representation or about the fundamental likelihood interpretation, then verify them mathematically.

2.2.   Some Elementary Properties

The list of three properties taken as axiomatic in defining probability measures does not seem to have much mathematical content. We next establish a number of logical consequences of these axiomatic properties that display much more of the inherent structure. Some of these derived properties are so readily apparent from the mass interpretation that there may seem little point in a mathematical verification. In fact, some of them are so natural that one may wonder why they are not included in the list of defining properties. It is one of the features of an axiomatic approach that we seek the most elementary list of properties from which the others may be derived. In order to determine that an assignment of numbers to events is in fact a probability assignment, only the defining properties need be established. Then all derived properties are automatically satisfied.

If the total probability mass is one and A^c consists of all the outcomes not in A, then we should have

P4) $P(A^c) = 1 - P(A)$.

If the probability is 0.3 that the Dow Jones average will increase by at least 20 points, then the probability is 0.7 that it will not increase by that much. A formal proof is based on the additivity of the probability measure and the fact that

$$\Omega = A \biguplus A^C, \quad \text{so that } P(\Omega) = 1 = P(A) + P(A^c).$$

It is clear that the probability to be assigned to the impossible event must be zero. This follows from (P4) and the fact that $\emptyset = \Omega^c$, so that

P5) $P(\emptyset) = 0$.

Since one can only be promoted to top management in a company if hired by the company, the probability of being hired is greater than the probability of being promoted to top management. In general, if event A implies event B (i.e., $A \subset B$), then we should expect the probability of B to be at least as great as that of A. This is also clear from the mass interpretation. The mass of a subset is at most equal to the mass of the set in which it is included. A formal argument is based on additivity and the fact that

If $A \subset B$, then $B = A \biguplus A^c B$ so that $P(B) = P(A) + P(A^c B) \geq P(A)$.

We thus have the property

P6) $A \subset B$ implies $P(A) \leq P(B)$.

While probabilities for mutually exclusive events add, the same is not true in general for the union of two events. Suppose, for example, $P(A) = 0.6$ and $P(B) = 0.5$. Then the sum is greater than one, so that the probability of the union cannot be the sum of the probabilities. Reference to the Venn diagram for the union $A \cup B$ (see Figure 2.1) shows that the total probability mass in the union can be expressed in several ways, depending upon how

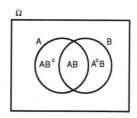

FIGURE 2.1. Partition of the union of the events A and B.

the union is partitioned. If we write

$$A \cup B = A \biguplus A^c B \quad \text{then} \quad P(A \cup B) = P(A) + P(A^c B).$$

For the decomposition

$$A \cup B = B \biguplus A B^c \quad \text{we have} \quad P(A \cup B) = P(B) + P(A B^c).$$

If we express

$$A \cup B = A B^c \biguplus A B \biguplus A^c B$$

then

$$P(A \cup B) = P(A B^c) + P(A B) + P(A^c B).$$

If we add the first two expressions and subtract the third, we get

$$P(A \cup B) = P(A) + P(B) - P(AB).$$

This last formula expresses the fact that if we add the probabilities for A and B, we have counted the mass in the common part twice; subtracting the mass in the common part yields the correct result. We thus have established

$$
\begin{aligned}
\textbf{P7)} \quad P(A \cup B) &= P(A) + P(A^c B) = P(B) + P(A B^c) \\
&= P(A B^c) + P(AB) + P(A^c B) \\
&= P(A) + P(B) - P(AB).
\end{aligned}
$$

If we partition event A with a disjoint class, then by additivity the probability of A is the sum of the probabilities in the parts or components.

P8) Suppose $\{B_i : i \in J\}$ is a countable, mutually exclusive class such that $A \subset \bigcup_{i \in J} B_i$. Then

$$P(A) = \sum_{i \in J} P(AB_i).$$

In the case of two events, property (P7) shows that $P(A \cup B) \leq P(A) + P(B)$. This property, known as *subadditivity*, holds for any countable union of events. To establish this, we note that

$$\bigcup_{i=1}^{\infty} A_i = \bigcup_{i=1}^{\infty} B_i, \quad \text{where} \quad B_i = A_i A_1^c A_2^c \ldots A_{i-1}^c.$$

We therefore have $P(\bigcup_{i=1}^{\infty} A_i) = \sum_{i=1}^{\infty} P(B_i)$. Since each $P(B_i) \leq P(A_i)$, we have

P9) *Subadditivity.* $P(\bigcup_{i=1}^{\infty} A_i) \leq \sum_{i=1}^{\infty} P(A_i)$

The next pair of properties express a limit property that is important in both theory and application. The results may be extended to more general sequences of events (see Theorem 17.1.6). We include a proof to illustrate a type of argument used in dealing with sequences of events.

P10) *Limits of monotone classes.*

(a) If $A_i \subset A_{i+1}$ for all $i \geq 1$ and $A = \bigcup_{i=1}^{\infty} A_i$, then $P(A_i) \to P(A)$.

(b) If $A_i \supset A_{i+1}$ for all $i \geq 1$ and $A = \bigcap_{i=1}^{\infty} A_i$, then $P(A_i) \to P(A)$.

Proof of Property (P10)

(a) Let $B_i = A_i A_{i-1}^c = A_i A_1^c A_2^c \ldots A_{i-1}^c$. Then $A_n = \bigcup_{i=1}^{n} B_i$ and

$$A = \bigcup_{i=1}^{\infty} A_i = \bigcup_{i=1}^{\infty} B_j.$$

$$P(A_n) = \sum_{i=1}^{n} P(B_i) \to \sum_{i=1}^{\infty} P(B_i) = P(A).$$

(b) If $A_i \supset A_{i+1}$, then $A_i^c \subset A_{i+1}^c$. Hence $\bigcap_{i=1}^{\infty} A_i = (\bigcup_{i=1}^{\infty} A_i^c)^c = A$.

By the result of part (a),

$$
\begin{aligned}
P(A) &= 1 - P(A^c) = 1 - \lim_n P(A_n^c) = 1 - \lim_n \left(1 - P(A_n)\right) \\
&= \lim_n P(A_n).
\end{aligned}
$$

$\square$

2.3.   Interpretation and Determination of Probabilities

It should be clear that the formal probability system is a *model* whose usefulness can only be established by examining its structure and determining whether patterns of uncertainty and likelihood in any practical situation can be adequately represented. With the exception of the sure event and the impossible event, the model does not tell us how to assign probability to any given event. The formal system is consistent with many probability assignments, just as the notion of mass is consistent with many different mass assignments to sets in the basic space.

The axioms provide *consistency rules* for making probability assignments. One cannot assign negative probabilities or probabilities greater than one. The sure event is assigned probability one. If two or more events are mutually exclusive, the total probability assigned to the union must equal the sum of the probabilities of the separate events. Any assignment of probability consistent with these conditions is allowed.

One may not know the probability assignment to every event. In the next section, we show that if we start with a class of n events, then knowledge of the probabilities assigned to a class of 2^n generated events, known as *minterms*, suffices to determine the probability of any Boolean combination of the n generating events. Often, it is not necessary to specify the probability assignment to every minterm in order to determine the probability of a specified Boolean combination of the generating events.

On a more formal level, standard techniques of measure theory make it possible to identify certain subclasses that are probability determining, in the sense that if a consistent assignment is made to each event in the subclass, then the probability is determined for all events in the sigma algebra $\mathcal{F}$. We do not deal with this problem, although it has significant consequences for probability distributions for random variables (Chapter 8) and certain questions concerning independence of random variables (Chapter 10).

The formal system allows us to deduce relationships among probabilities assigned to various events when these events have certain logical relationships among themselves. But the system gives no hint about empirical procedures for assigning probabilities. These assignments are influenced by the interpretation given probability. This has been a matter of considerable controversy. Sometimes, significant advances in the development and application of probability theory have been associated with one view or another. However, *each of the dominant points of view and the results discovered by their proponents is consistent with the formal model we are developing.*

RELATIVE FREQUENCIES

By Borel's theorem (Section 17.3), if a trial is performed a large number of times in an independent manner, the *fraction* of times that event A occurs will *approach*, as a limit, *the value* $P(A)$. This is a formal validation of the intuitive notion that lay behind early attempts to formulate probabilities. Inveterate gamblers had noted such long-range statistical regularity and sought explanations from their mathematically gifted friends. From this point of view, probability is meaningful only in repeatable situations. Those who hold these views usually assume that there is something objective about probability. It is a number determined by the nature of things, to be discovered by repeated experiment.

SUBJECTIVE PROBABILITY

A sharply different point of view is expressed by those who insist that probability, as a measure of likelihood, is a condition of the mind of the person making the probability assessment. From this point of view, the laws of probability simply impose rational consistency upon the way one assigns probabilities to events. Various attempts have been made to find objective ways to measure the "strength" of one's belief or certainty that an event will occur. The probability $P(A)$ is the degree of certainty one feels that event A will occur.

> One approach to characterizing an individual's degree of certainty is to equate his assessment of $P(A)$ with the amount a he is willing to pay to play a game that returns one unit of money if A occurs, for a gain of $(1 - a)$, and returns zero if A does not occur, for a gain of $-a$.

Behind this formulation is the notion of a *fair game*, which we examine briefly in Section 14.2.

ONE-TIME TRIALS

There are many applications of probability in which the relative frequency point of view is not feasible. These are unique, nonrepeatable trials. As the popular expression has it, "You only go around once." Sometimes, probabilities in these situations may be quite subjective. The probability that one's favorite football team will win the next Superbowl game may well be only a subjective probability of the bettor. This is certainly not a probability that can be determined by a large number of repeated trials. However, the subjective assessment of probabilities may be based on intimate knowledge of relative strengths and weaknesses of the teams involved, as well as factors such as weather, injuries, and experience. There may be a considerable objective basis for the subjective assignment of probability. There are other nonrepeatable trials in which the probabilities are more than simply a matter of individual opinion or conviction.

Example 2.3.1: The Probability of Rain

Newscasts often report a probability of rain of 20 percent or 60 percent or some other figure. There are several difficulties here.

- To use the formal mathematical model, there must be precision in determining an event. An event either occurs or it does not. How do we determine whether it has rained or not? Must there be a measurable amount? Where must this rain fall to be counted? During what time period?

- What does the statement there is a 30 percent probability of rain mean? Even if it is agreed on the area, the amount, and the time period, there is ambiguity.

  - Does this mean that if the prediction is correct, 30 percent of the area indicated will get rain during the specified time period?

  - Or does it mean that 30 percent of the times when such a prediction is made there will be significant rainfall in the area during the specified time period? The latter alternative may well hide a frequency interpretation. Does the statement mean that 30 percent of the times when conditions are *similar* to current conditions it rains?

Regardless of the interpretation, there is some ambiguity about the event and whether it has occurred. And there is some difficulty with knowing how to interpret the probability figure. While the precise meaning of a 30 percent probability of rain may be difficult to determine, it is generally useful to know whether the conditions lead to a 20 percent or a 40 percent probability assignment. And there is no doubt that as weather forecasting technology and methodology continue to improve the weather probabilities become increasingly useful.

—   ☐

A CHOICE

It is fortunate that we do not have to declare a single position with regard to the "correct" viewpoint and interpretation. As noted earlier, the formal model is consistent with any of the views set forth. *We are free in any situation to make the interpretation most meaningful and natural to the*

problem at hand. It is not necessary to fit all problems into one conceptual mold, nor is it necessary to change mathematical models each time a
different point of view seems appropriate.

There are subtle and difficult *statistical problems* in determining how
probabilities are to be distributed over a class of events of interest. These
problems are certainly an essential aspect of applied probability. While
we make occasional reference to practical considerations in this regard,
we do not attempt to deal systematically with the problem of empirical
determination of probabilities. For one thing, to deal with these problems
in a satisfactory manner requires a considerable background in probability
theory of the kind we are developing. The treatment of empirical probability
distributions is an extensive and sophisticated field with techniques and
approaches of its own.

PROBABILITY AND ODDS

In many experimental situations it is easier to determine and work with
ratios of probabilities or likelihoods than with the probabilities themselves.
Such ratios are known as *odds.*

Definition 2.3.1. *For a given probability measure,*

the odds favoring event A over event B: $O(A : B) = \dfrac{P(A)}{P(B)}$,

the odds favoring event A *(over A^c)*: $O(A) = O(A : A^c) = \dfrac{P(A)}{P(A^c)}$.

There is a one-one relationship between $P(A)$ and $O(A)$. Simple algebra
applied to the defining expression, with $P(A^c)$ replaced by $1 - P(A)$ shows
that

$$O(A) = \frac{P(A)}{1 - P(A)} \quad \text{iff} \quad P(A) = \frac{O(A)}{1 + O(A)}.$$

If one knows the odds, the probabilities may be calculated; conversely, the
probabilities determine the odds.

Example 2.3.2:

If $P(A) = \dfrac{91}{216}$, then $O(A) = \dfrac{91/216}{1 - 91/216} = \dfrac{91}{216 - 91} = \dfrac{91}{125}$.

If $O(A) = \dfrac{91}{125}$, then $P(A) = \dfrac{91/125}{1 + 91/125} = \dfrac{91}{125 + 91} = \dfrac{91}{216}$.

— □

Use of odds is popular among gamblers when referring to the probability
of an event. The notion of a fair game leads to payoffs on bets according
to the odds. If the odds are 1 in 10 favoring event A, then the payoff ratio
should be 10 to 1 should event A occur. This provides an alternate way of

assessing subjective probabilities. The use of odds is also common in certain experimental situations and situations involving assessment of likelihoods. The notion of odds is extended to the case of conditional probabilities in Section 3.2.

2.4.   Minterm Maps and Boolean Combinations

If we partition an event F into component events whose probabilities can be determined, by additivity the probability of F is the sum of these component probabilities. Frequently, the event F is a Boolean combination of members of a finite class—say, $\{A, B, C\}$ or $\{A, B, C, D\}$. For each such finite class, there is a fundamental *partition generated by the class*. The members of the partition are called *minterms*. Any Boolean combination of members of the class can be expressed as the union of a unique subclass of the minterms. If the probability of every minterm in this subclass can be determined, then the probability of the Boolean combination is determined.

The minterms generated by a class may be arranged systematically on a special Venn diagram known as a *minterm map*, which is useful in the analysis of specific problems. We illustrate the fundamental patterns in the case of four events $\{A, B, C, D\}$. We form the *minterms* as intersections of members of the class, with various patterns of complementation. For a class of four events, there are $2^4 = 16$ such patterns, hence 16 minterms. These are, in a systematic arrangement,

$$
\begin{array}{llll}
A^c B^c C^c D^c & A^c B C^c D^c & A B^c C^c D^c & A B C^c D^c \\
A^c B^c C^c D & A^c B C^c D & A B^c C^c D & A B C^c D \\
A^c B^c C D^c & A^c B C D^c & A B^c C D^c & A B C D^c \\
A^c B^c C D & A^c B C D & A B^c C D & A B C D.
\end{array}
$$

These sets form a *partition* (some of the minterms could be empty). Each element ω is assigned to exactly one of the minterms by determining the answers to four questions:

Is it in A?   Is it in B?   Is it in C?   Is it in D?

Suppose, for example, the answers are: Yes, No, No, Yes. Then ω is in the minterm $A B^c C^c D$. In a similar way, we can determine the membership of each ω in the basic space. The result is thus a partition. In terms of events, the minterms are mutually exclusive events, one of which is sure to occur on each trial. The membership of any minterm depends upon the membership of each generating set A, B, C, or D, and the relationships between them. For some classes, one or more of the minterms are empty (impossible). In deriving an expression for a given Boolean combination that holds for any class $\{A, B, C, D\}$, we include all possible minterms. If a minterm is empty for a given class, its presence does not modify the set content or probability

assignment for the Boolean combination. The minterms can be arranged
systematically on a special Venn diagram, as follows:

$A^cB^cC^cD^c$	$A^cBC^cD^c$	$AB^cC^cD^c$	ABC^cD^c
$A^cB^cC^cD$	A^cBC^cD	AB^cC^cD	ABC^cD
$A^cB^cCD^c$	A^cBCD^c	AB^cCD^c	$ABCD^c$
A^cB^cCD	A^cBCD	AB^cCD	$ABCD$

It should be clear that if we begin with a class of n events, there are 2^n
minterms. To aid in systematic handling, we introduce a simple numbering
system for the minterms, which we illustrate by considering again the events
A, B, C, and D, in that order. The answers to the four questions can be
represented numerically by the scheme

$$\text{No} \sim 0 \qquad \text{and} \qquad \text{Yes} \sim 1.$$

Thus, if ω is in $A^cB^cC^cD^c$, the answers are tabulated as 0000. If ω is in
AB^cC^cD, this is designated 1001. With this scheme, the previous minterm
arrangement becomes

$$
\begin{array}{cccc}
0000 & 0100 & 1000 & 1100 \\
0001 & 0101 & 1001 & 1101 \\
0010 & 0110 & 1010 & 1110 \\
0011 & 0111 & 1011 & 1111
\end{array}
$$

Now the combinations of 0s and 1s may be interpreted as binary representations of integers. Thus 0000 is decimal 0; 1001 is decimal 9; 1101 is
decimal 13; etc. The preceding arrangement can be displayed in terms of
the decimal equivalents.

$$
\begin{array}{cccc}
0 & 4 & 8 & 12 \\
1 & 5 & 9 & 13 \\
2 & 6 & 10 & 14 \\
3 & 7 & 11 & 15
\end{array}
$$

We use this numbering scheme on the special Venn diagrams called *minterm
maps*. These are illustrated in Figure 2.2 for the case of three, four, and
five generating events. In the three-variable case, set A is the right half of
the diagram and set C is the lower half; but set B is split, so that it is the
union of the second and fourth columns. Similar splits occur in the other
cases. Since the actual content of any minterm depends upon the sets A,
B, C, and D in the generating class, it is customary to refer to these sets
as *variables*. Frequently, it is useful to refer to the minterms by number. If
the members of the generating class are treated in a fixed order, then each
minterm number specifies a minterm uniquely. Thus, for A, B, C, D, in
that order, we may designate

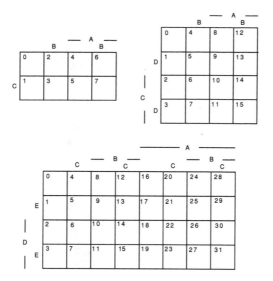

FIGURE 2.2. Minterm maps for three, four, and five variables.

$$A^c B^c C^c D^c = M_0 \quad (\text{minterm } 0), \quad AB^c C^c D = M_9 \quad (\text{minterm } 9), \quad \text{etc.}$$

Remark: Other useful arrangements of minterm maps are employed in the analysis of switching circuits.

Theorem 2.4.1 (Minterm Expansion Theorem) *Each Boolean combination of the elements in a generating class may be expressed as the disjoint union of an appropriate subclass of the minterms. This representation is known as the* minterm expansion *for the combination.*

—   □

We illustrate by two simple examples, then indicate how a general proof may be formulated. To simplify writing, we use a simple notational device

$$M_a \biguplus M_b \biguplus M_c = M(a, b, c), \quad \text{etc.}$$

and

$$P\big(M(a, b, c)\big) = p(a, b, c) = p(a) + p(b) + p(c), \quad \text{etc.}$$

Example 2.4.1: (See Figure 2.3)

Consider $F = A(B^c \cup CD)^c$, a Boolean combination of $\{A, B, C, D\}$. We wish to express F as the union of minterms. We consider members of the generating class in alphabetical order and refer to the minterms by number. Reference to the minterm map in Figure 2.3 shows that

$$A = M(8, 9, 10, 11, 12, 13, 14, 15) = \text{the right half of the minterm map.}$$

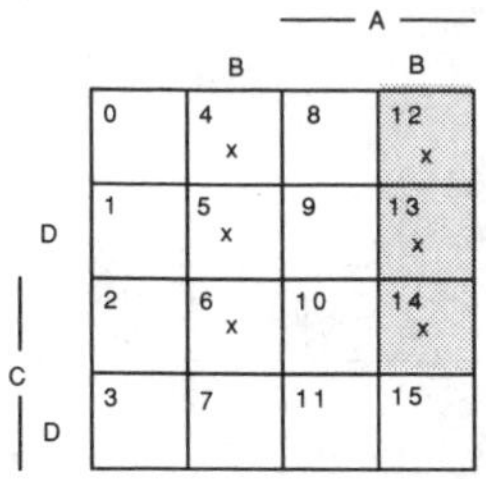

FIGURE 2.3. Minterm map for Example 2.4.1.

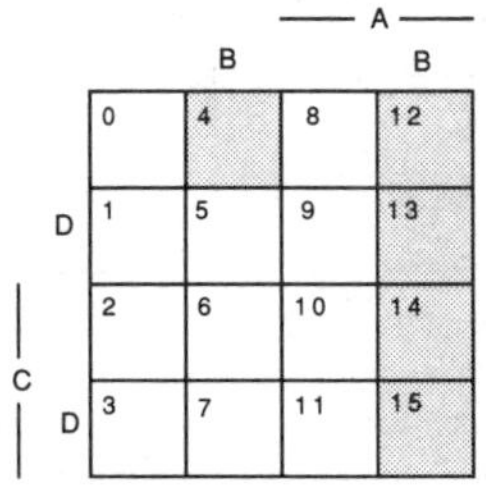

FIGURE 2.4. Minterm map for Example 2.4.2.

The minterms in $B^c \cup CD$ are those either in B or in both C and D. The minterm map shows that

$$B^c \cup CD = M(0, 1, 2, 3, 7, 8, 9, 10, 11, 15)$$

so that

$$(B^c \cup CD)^c = M(4, 5, 6, 12, 13, 14).$$

The intersection $A(B \cup CD)^c$ consists of the minterms common to the two, so that $F = M(12, 13, 14)$.

The minterms comprising F are shown shaded in Figure 2.3. If the probabilities of these minterms are known, then the probability of F is obtained by taking their sum.

—   □

Example 2.4.2: (See Figure 2.4)

Determine the minterm expansion for $F = AB \cup A^c(B^c \cup C \cup D)^c$. We consider the generating sets in alphabetical order. It is clear from examining the minterm map for minterms common to A and B that

$AB = M(12, 13, 14, 15)$. We may handle the other terms by a combination of algebra of sets and the use of the minterm map. By de Morgan's rule

$$(B^c \cup C \cup D)^c = BC^cD^c \quad \text{so that} \quad A^c(B^c \cup C \cup D)^c = A^cBC^cD^c = M(4).$$

Thus, $F = M(4, 12, 13, 14, 15)$. This corresponds to the shaded region on Figure 2.4. If, for example, the probability $p(12, 13, 14, 15)$ of $AB = M(12, 13, 14, 15)$ and the probability $p(4)$ of $M(4)$ were known, the probability of F would be determined.

—   □

In the preceding examples, it is not necessary to know the probabilities of all minterms to calculate the probabilities of the Boolean combinations under consideration. One use of minterm maps and minterm expansions is to determine the probability of a Boolean combination when partial information is given.

Example 2.4.3: Partial Information

In a group of 200 students, 100 take French; 40 of these are men. There are 80 women who take French or a computer science course (or both). A student is picked at random, on an equally likely basis, from this group. What is the probability the student selected is a woman who takes computer science but not French?

Solution

Let A = event the student is a man, B = event the student takes French, and C = event the student takes computer science. The data are taken to mean

$$P(B) = \frac{100}{200} = 0.5 \qquad P(AB) = \frac{40}{200} = 0.2$$

$$P\big(A^c(B \cup C)\big) = \frac{80}{200} = 0.4.$$

Reference to a minterm map for three variables A, B, and C shows that

$$P(B) = p(2, 3, 6, 7), \ \ P(AB) = p(6, 7), \ \text{and} \ P\big(A^c(B \cup C)\big) = p(1, 2, 3).$$

We wish to determine $P(A^cB^cC) = p(1)$. To do this, we note

$$p(2, 3) = p(2, 3, 6, 7) - p(6, 7) = 0.3,$$

so that

$$p(1) = p(1, 2, 3) - p(2, 3) = 0.1.$$

—   □

Remark: For purposes of exposition, we have used algebraic combinations of $p(1, 2, 3)$, $p(2, 3)$, etc. In practice, we usually keep track of the development by writing the probability values in the appropriate regions on the minterm map. Sometimes, however, a combination of algebraic and graphic analysis may be the most effective and least error-prone approach.

The minterm map analysis frequently helps determine whether or not sufficient information is given to obtain the desired probability. The following example illustrates this role.

Example 2.4.4:

In a certain residential section, data on the number of homes with subscriptions to Newsweek, Time, and the morning Post are as follows:

Subscription to	Percentage
Post	70
Newsweek	50
Time	60
Post, but not Newsweek	30
Newsweek and Time	20
All three	10

A home is selected at random, on an equally likely basis. What is the probability that the home selected:

Receives (Post *and* Newsweek, but *not* Time) *or* (Newsweek *and* Time, but *not* the Post).

Solution

Let A = event receives the Post; B = event receives Newsweek; C = event receives Time.

Data are interpreted to mean

$$P(A) = 0.7 \quad P(B) = 0.5 \quad P(C) = 0.6$$
$$P(AB^c) = 0.3 \quad P(BC) = 0.2 \quad P(ABC) = 0.1.$$

The problem is to determine $P(ABC^c \cup A^c BC)$. From a minterm map in three variables, we see that the desired probability is $p(3, 6)$. With the aid of a minterm map, we translate the given probabilities into minterm probabilities as follows:

$$P(A) = p(4, 5, 6, 7) = 0.7 \quad P(AB^c) = p(4, 5) = 0.3$$

$$P(BC) = p(3, 7) = 0.2$$

$$P(C) = p(1, 3, 5, 7) = 0.6 \quad P(ABC) = p(7) = 0.1.$$

We deduce from this that $p(6,7) = P(A) - P(AB^c) = 0.4$, so that $p(6) = 0.3$. Thus

$$p(3,6) = p(3) + p(6) = 0.1 + 0.3 = 0.4.$$

Examination of the minterm map shows that the data are sufficient to determine individual minterm probabilities $p(2)$, $p(3)$, $p(6)$, and $p(7)$. However, we cannot separate the two parts of $p(0,1)$ and of $p(4,5)$. If we had any one of $p(0)$, $p(1)$, $p(4)$, $p(5)$, then the other individual minterm probabilities could be separated, and it would then be possible to calculate the probability of any Boolean combination of the events A, B, and C.

—   □

Verification of the Minterm Expansion Theorem

Although for a specific generating class some minterms may be empty, a general minterm expansion should be obtained as if all minterms were nonempty. Then, if any minterm is empty, it contributes no elements to the union; but if it is not empty, it is needed to complete the set represented. It is clear from the minterm map that there are always basic spaces and generating sets for which none of the minterms is empty. For example, we could simply assign one element to each minterm to obtain such a case.

We do not perform the notational gymnastics necessary to write out a formal general proof. Instead, we illustrate the essential ideas of a general proof by analyzing a special case. Suppose F is a Boolean combination of four events in the class $\{A, B, C, D\}$.

1. Each indicator function I_A, I_B, I_C, and I_D is a constant over each minterm since the respective answers to the questions: Is it in A? Is it in B? Is it in C? Is it in D? are the same for any ω in a given minterm.

2. The rules on indicator functions and set combinations and relations show that I_F is a numerical valued function of I_A, I_B, I_C, and I_D. For example,

   If $F = A(B^c \cup CD)^c$ then $I_F = I_A\Big(1 - \big(1 - I_B + I_C I_D - (1 - I_B)I_C I_D\big)\Big).$

3. The possible values of I_F are 0 and 1. The only way I_F can change value is for one or more of the indicators I_A, I_B, I_C, or I_D to change value. Hence, I_F is constant (at either 0 or 1) over each minterm.

4. Since the set F is the set of ω such that $I_F(\omega) = 1$, the set F must be the union of those minterms for which I_F has the value 1. This is the desired minterm expansion.

2a

The Sigma Algebra of Events

In Chapter 1, we show that events are represented as subsets of the basic space Ω of all possible outcomes of the basic trial or experiment. Also, we see that it is highly desirable to consider logical operations on these subsets (complementation and the formation of unions and intersections). Very simple problems make desirable the treatment of sequences or other countable classes of events. Thus, if E is an event, the complementary set E^c should be an event. If $\{E_i : 1 \leq i\}$ is a sequence of events, then the union $\bigcup_{i=1}^{\infty} E_i$ and the intersection $\bigcap_{i=1}^{\infty} E_i$ should be events. On the other hand, there are technical mathematical reasons for not considering the class of events to be the class of *all* subsets of Ω. Basic spaces with an infinity of elements contain bizarre subsets that put a strain on the fundamental theory. We are thus led to consider some restrictions on the class of events, while still maintaining the desired flexibility and richness of membership. Long experience has shown that the proper choice is a *sigma algebra*. We wish to characterize these classes. Sigma algebras and many other important types of subclasses are characterized by various *closure properties*.

Definition 2a.2. *A class C of subsets of a basic set Ω is said to be* closed *under an operation* iff *application of this operation to appropriate sets or subclasses of sets in this class produces a member of the class.*

The principal closure operations we consider in dealing with sigma algebras are the following.

1. The class C is *closed under complements* iff $E \in C$ implies $E^c \in C$.

2. The class C is *closed under finite intersections* iff $\{E_i : 1 \leq i \leq n\} \subset C$ implies $\bigcap_{i=1}^{n} E_i \in C$.

3. The class C is *closed under finite unions* iff $\{E_i : 1 \leq i \leq n\} \subset C$ implies $\bigcup_{i=1}^{n} E_i \in C$.

4. The class C is *closed under countable intersections* iff $\{E_i : 1 \leq i\} \subset C$ implies $\bigcap_{i=1}^{\infty} E_i \in C$.

5. The class C is *closed under countable unions* iff $\{E_i : 1 \leq i\} \subset C$ implies $\bigcup_{i=1}^{\infty} E_i \in C$.

By the simple device of letting all sets be the same from some point on, we see that closure under countable unions implies closure under finite unions, and closure under countable intersections implies closure under finite intersections.

We now define two classes of sets in terms of these closure properties. We suppose that each class is nonempty, in the sense that the class has at least one member set.

Definition 2a.3. *A nonempty class $\mathcal{F}$ of subsets of a basic set Ω is an* algebra *of sets iff it is closed under complements and finite unions.*

Remark: An algebra must contain $\Omega = E \bigcup E^c$ and $\emptyset = \Omega^c$. It is closed under finite intersections, since

$$\bigcap_{i=1}^{n} A_i = \left(\bigcup_{i=1}^{n} A_i^c \right)^c.$$

Definition 2a.4. *A nonempty class $\mathcal{F}$ of subsets of a basic set Ω is a* sigma algebra *of sets iff it is closed under complements and countable unions.*

Remark: As noted earlier, a sigma algebra is closed under finite unions. Just as in the case of an algebra, a sigma algebra contains the basic set Ω and the empty set $\emptyset$. An argument parallel to that for an algebra shows that a sigma algebra is closed under countable intersections.

Generated Sigma Algebras. The fact that an algebra or a sigma algebra is characterized by closure properties leads to another very important concept. We may begin with a nonempty class $\mathcal{C}$ of subsets of Ω. Then there is a minimal algebra and a minimal sigma algebra that includes $\mathcal{C}$. This comes about for the following reasons.

1. Suppose we have a family or collection of classes of subsets, each of which is closed under some operation. If the intersection class (i.e., the class of all those subsets that are in every one of the classes) is not empty, it also is closed under this operation. Suppose all the classes are closed under complements. Let E be a set common to all the classes, so that it is in the intersection class. Then E^c belongs to each of these classes, and hence is in the intersection class. Similar arguments hold for the other operations.

2. Let $\mathcal{P}$ be the class of *all* subsets of Ω. This class is closed under every operation on sets and is therefore a sigma algebra. And $\mathcal{P}$ must contain any class $\mathcal{C}$.

3. Consider the family of all sigma algebras that include $\mathcal{C}$. Since $\mathcal{P}$ is such a class, the family is not empty. Call the intersection class for this family $\sigma(\mathcal{C})$. Then $\sigma(\mathcal{C})$ must include $\mathcal{C}$ and must be closed under complements and countable unions. Thus $\sigma(\mathcal{C})$ is a sigma algebra that

includes C. It is easy to show that this is the smallest such class and that it is unique.

Definition 2a.5. *The class* $\sigma(C)$ *described earlier is called the* sigma algebra generated by C.

Remark: Although the generated sigma algebra always *exists* and is *unique*, there is no constructive way of defining or describing it (except in some very special cases). Nonetheless, it is extremely important that the sigma algebra exists and is unique, for such generated classes play an essential role in determining the structure of probability theory.

The algebra generated by C is determined similarly. An indication of the significance of generated algebras and sigma algebras is given in Appendix 7a, in the discussion of Borel sets, and in the determination of the distribution induced by a random variable X.

PROBLEMS

2-1. Determine $P(A \cup B \cup C \cup D)$ in terms of probabilities of the events and their intersections.

2-2. Use fundamental properties of probability to establish the following.

(a) $P(AB) \leq P(A) \leq P(A \cup B) \leq P(A) + P(B)$ for any A, B
(b) $P(\bigcap_{i \in J} A_i) \leq P(A_k) \leq P(\bigcup_{i \in J} A_i) \leq \sum_{i \in J} P(A_i)$ for any k.

2-3. The event A is twice as likely as the event C, and B is as likely as the combination of A or C. The events are mutually exclusive and one of them must occur on any trial. Determine the probabilities.

2-4. Suppose $P(A \cup B^c C) = 0.65$, $P(AC) = 0.2$, $P(A^c B) = 0.25$, $P(A^c C^c) = 0.25$, $P(BC^c) = 0.30$. Determine $P((AC^c \cup A^c C)B^c)$.

2-5. Suppose $P((AB^c \cup A^c B)C) = 0.4$, $P(AB) = 0.2$, $P(A^c C^c) = 0.3$, $P(A) = 0.6$, $P(C) = 0.5$, and $P(AB^c C^c) = 0.1$.
Determine $P(A^c C^c \cup AC)$

2-6. Suppose $P(A(B \cup C)) = 0.3$, $P(A^c) = 0.6$, and $P(A^c B^c C^c) = 0.1$. Determine $P(B \cup C)$.

2-7. Suppose $P(A) = 0.6$, $P(C) = 0.4$, $P(AC) = 0.3$, $P(A^c B) = 0.2$, and $P(A^c B^c C^c) = 0.1$. Determine $P((A \cup B)C^c)$.

2-8. Suppose $P(A \cup B^c C) = 0.65$, $P(AC) = 0.2$, $P(A^c B) = 0.25$, $P(A^c C^c) = 0.25$, $P(BC^c) = 0.30$. Determine $P(AC^c \cup A^c B^c C)$.

2-9. Suppose $P(A) = 0.5$, $P(AB) = P(AC) = 0.3$, and $P(ABC^c) = 0.1$. Determine $P(A(BC^c)^c)$.

2-10. Suppose $P(A) = 0.4$, $P(AB) = 0.3$, $P(ABC) = 0.25$, $P(C) = 0.65$, and $P(A^c C^c) = 0.3$. Determine $P(AC^c \cup A^c C)$.

2-11. Consider the class $\{A, B, C, D\}$ of events. Suppose

(a) The probability that at least one of the events A or C occurs is 0.75.
(b) The probability that at least one of the four events occurs is 0.90.

Determine the probability that neither event A nor C occurs, but at least one of the events B or D occurs.

2-12. Consider the class $\{A, B, C, D\}$ of events. Express AC as the union of minterms generated by the class. Express $B \cup D$ in a similar fashion.

2-13. Express the following Boolean functions of $\{A, B, C, D\}$ in canonical form.

(a) $f_1(A, B, C, D) = (A \cup B^c)(C \cup D)^c$

(b) $f_2(A, B, C, D) = A(B^c \cup C^c) \cup CD^c$

(c) $f_3(A, B, C, D) = A \cup B \cup C$

(d) $f_4(A, B, C, D) = AB(B^c \cup C)$

(e) $f_5(A, B, C, D) = A \cup CD^c$.

2-14. Suppose that $F_1 = f(A, B, C) = M_1 \cup M_3 \cup M_4$ and that $F_2 = f(D, E, F, G) = M'_2 \cup M'_3 \cup M'_5 \cup M'_6$, where M_i are minterms generated by $\{A, B, C\}$ and M'_k are minterms generated by $\{D, E, F, G\}$. Express $F_1 F_2$ as the union of minterms generated by $\{A, B, C, D, E, F, G\}$.

2-15. A newspaper is evaluating reader interest in three comic strips it carries as a regular weekday feature. It takes a survey of 1000 of its readers, with the following results:

232 read strip a regularly; 228 read strip b regularly; 230 read strip c regularly; 45 read all three regularly; 62 read both a and c; 197 read at least two regularly; twice as many read both b and c as those who read both a and c.

How many read at least one regularly? How many read only c regularly?

2-16. An air conditioning firm examines its records for 1000 service calls to see how many customers needed new filters, added freon, or thermostat cleaning and adjustment. The records yield the following data:

100 needed all three; 325 needed at least two of the three; 125 needed freon and thermostat maintenance; 550 needed new filters.

How many need new filters only? How many did not need a new filter and not more than one of the other items?

3

Conditional Probability

In this chapter we introduce a measure of conditional probability. Suppose partial information about the outcome of a trial establishes the occurrence of an event C. This removes any uncertainty about the occurrence of event C on that trial. And it may change the uncertainties about the occurrence of other events so that new probabilities or likelihoods should be assigned. We consider a natural revision of the probability assignments, which seems quite plausible. Subsequent theory and the usefulness of the resultant patterns in applications confirm the appropriateness of this new measure. In both theory and application, the conditional probability measures provide an important extension of the probability system.

3.1. Conditioning and the Reassignment of Likelihoods

We begin with an example that shows the need for reassignment of probabilities, as estimates of likelihood are changed by the availability of partial information about the outcome of a trial.

Example 3.1.1: Conditioning Information

Suppose a woman from a middle-class family decides at age 40 years to have regular checkups at a health maintenance clinic. Let A be the event that she develops breast cancer by the time she is 60 years of age. From the point of view of the clinic, she has been selected at random from a certain population. Health statistics indicate a probability $P(A)$ that she will be afflicted. Suppose, however, that her medical history shows that both her mother and one of her mother's two sisters had breast cancer at ages 51 and 58, respectively. This would undoubtedly lead to a new (and higher) assessment of the probability that she may have the cancer. If it also became known that she is a heavy smoker, this additional information would lead to a revision of the assessed likelihood of breast cancer.

The original assessment of probability is based on the assumption that the woman was selected from a certain population. With the additional information about her family history, she is viewed as a member of a smaller subpopulation, with a different likelihood of developing cancer. The further information about her smoking habits identify her as belonging to a still smaller subpopulation, with higher likelihood of breast cancer.

—   □

The situation illustrated may be viewed in terms of our probability model in the following way. The outcome ω is from a given basic space Ω. Likelihoods on this basic space are expressed by a probability measure P, which assigns probability $P(A)$ to event A. Partial information determines that ω is drawn from a subset C of possible outcomes. That is, the partial information establishes that event C has occurred. We refer to such an event C as a *conditioning event*. This raises the question: Given the occurrence of the conditioning event C, how shall we reassign the likelihood that A has occurred? We want to introduce a new probability measure $P(\cdot|C)$ that makes the new assignments. Three considerations lead to a candidate for this probability measure.

- If C has occurred, then ω is an element of C. If A occurs, under this condition, then ω must be in both A and C. That is, A occurs iff AC occurs.

- It seems reasonable to assign a new likelihood to A that is proportional to $P(AC)$. That is, we consider $P(A|C) = aP(AC)$.

- Although $P(AC) \geq 0$ and $P(\bigcup_i A_iC) = \sum_i P(A_iC)$, we have $P(\Omega C) = P(C)$. If we make $a = 1/P(C)$, then the function $P(\cdot|C)$, defined for each event A by

$$P(A|C) = \frac{P(AC)}{P(C)}$$

is a probability measure, for it must satisfy the axiomatic properties (P1), (P2), and (P3).

This line of reasoning leads to the following definition.

Definition 3.1.1. *If C is an event with positive probability, the* conditional probability, given C *is the probability measure $P(\cdot|C)$ defined by*

$$P(A|C) = \frac{P(AC)}{P(C)} \qquad \textit{for any event } A.$$

Remarks

1. The preceding argument does not prove, in a logical sense, that this assignment of conditional probability is correct. "The proof is in the

pudding." Logical consequences and the consistency of results in applications with intuitive notions of conditional probability provide the basis for confidence.

2. Although $P(A|C)$ is a conditional probability, it does not make sense to refer to $A|C$ as a conditional event. No satisfactory formulation of a conditional event has been established.

3. The defining expression is equivalent to the product $P(AC) = P(A) \times P(C|A)$. This form is often useful in applications.

Example 3.1.2:

One thousand persons in a scientific sample in a telephone poll were asked about their views on current presidential policy with regard to Central America. Respondents were asked their political party preference. Their positions were categorized as being (1) generally favorable or (2) opposed or had serious reservations. The data are tabulated here.

Party Affiliation	Favor	Oppose
Republican	175	155
Independent	69	186
Democrat	70	345

Let A be the event the respondent favors current policy, and let R, I, and D be the event the person is Republican, independent, or Democrat, respectively. If the sample can be taken as representative of the voting public and a trial consists of selecting a member of that population at random, then we take the data to mean

$$P(R) = \frac{175 + 155}{1000} = 0.33, \qquad P(RA) = \frac{175}{1000} = 0.175,$$

and

$$P(RA^c) = \frac{155}{1000} = 0.155, \text{ etc.}$$

Then

$$P(A|R) = \frac{P(RA)}{P(R)} = \frac{0.175}{0.330} \approx 0.530$$

$$P(A^c|R) = \frac{P(RA^c)}{P(R)} = \frac{0.155}{0.330} \approx 0.470.$$

From these data we can also calculate $P(R|A)$. We have

$$P(A) = \frac{175 + 69 + 70}{1000} = 0.314$$

so that

$$P(R|A) = \frac{P(RA)}{P(A)} = \frac{0.175}{0.314} \approx 0.557.$$

Various other probabilities and conditional probabilities can be calculated similarly. We note that if, instead of raw data, we had the probabilities $P(R)$, $P(I)$, and $P(D)$ as well as the conditional probabilities $P(A|R)$, $P(A|I)$, and $P(A|D)$, we could calculate $P(A)$ by property (P8) for probability and the product form of the definition for conditional probability, as follows:

$$\begin{aligned} P(A) &= P(AR) + P(AI) + P(AD) \\ &= P(A|R)P(R) + P(A|I)P(I) + P(A|D)P(D). \end{aligned}$$

With $P(A)$ determined, we could obtain $P(R|A)$, $P(I|A)$, and $P(D|A)$.

—   □

Conditional probabilities often provide a natural way to deal with compound trials. In the case of a two-stage trial, events can be determined by the result of the first step and the result of the second step.

Example 3.1.3:

A jet aircraft has two engines. It will fly with only one engine operating. The probability that one engine will fail on a trip is 0.0003. If one engine fails, this puts added load on the other, increasing its probability of failure under this condition to 0.001. What is the probability both will fail on a trip?

Solution

Let F_1 be the event one engine fails and F_2 be the event the second engine fails. The data are taken to mean

$$P(F_1) = 0.0003 \quad \text{and} \quad P(F_2|F_1) = 0.001.$$

According to the product form of the definition

$$P(F_1 F_2) = P(F_1)P(F_2|F_1) = 0.0003 \times 0.001 = 3 \times 10^{-7}.$$

Note that since a second engine can fail only if one has failed previously, we have $F_2 \subset F_1$, so that $P(F_2) = P(F_1 F_2)$.

—   □

A SOURCE OF DIFFICULTY

Sometimes it is difficult to know just which conditioning event C has oc-curred. The source of difficulty usually lies in some ambiguity about the information received.

Example 3.1.4: Ambiguity in Determining the Conditioning Event

A local radio station has a promotional scheme underway. Listeners are encouraged to register for a drawing to win tickets to the baseball World Series games. Three persons are to receive two passes each to the Series. To heighten the suspense (and thus maintain the audience), the station selects, by lot, five finalists and announces their names. Three of these five are selected on an equally likely basis. The names are to be an-nounced, one at a time, on three successive days. To simplify exposition, we number the finalists, 1, 2, 3, 4, and 5, and let A_i be the event that finalist number i is selected among the winning three.

- Upon the announcement that he is one of the five, finalist 1 calcu-lates the probability $P(A_1)$ that he is a winner. In fact, he realizes the probability $P(A_i)$ is the same for each finalist. This can be cal-culated by counting the combinations of three from five in which member number i is included and then dividing by the total num-ber of combinations of three from five. Thus,

$$P(A_i) = \frac{C(4,2)}{C(5,3)} = \frac{6}{10} \qquad \text{for each } i = 1, 2, 3, 4, 5.$$

- The first winner is announced. It is finalist 5. This is information that event A_5 has occurred. Finalist 1 then revises his estimate of the likelihood on the basis of this information to $P(A_1|A_5)$. If two finalists, numbers 1 and 5, are determined, then there are only three ways the third can be selected. Thus

$$P(A_1 A_5) = \frac{3}{10} \qquad \text{so that} \quad P(A_1|A_5) = \frac{P(A_1 A_5)}{P(A_5)} = \frac{3/10}{6/10} = \frac{1}{2}.$$

The knowledge that finalist 5 is a winner has thus made it seem less likely that finalist 1 is a winner.

- Suppose before the first winner is announced finalist 1 calls a friend who works at the station. He knows his friend cannot reveal whether he is a winner. So finalist 1 asks the friend to identify one of the winners other than himself. The friend informs him that finalist 5 is among those selected. Is this the same information as that to be announced shortly? Is the conditioning event A_5? The only information finalist 1 receives is that finalist 5 is one of the winners. It is difficult to escape the conclusion that the information, hence the conditioning event, is the same in both cases. However, many persons are uncomfortable with the assertion that the conditioning event is A_5. They seem to feel that some other information is available. Many would assume some information about how the selection of finalist 5 was made by the friend at the station.

- Suppose finalist 1 asks the friend how he selected finalist 5 from among the three who are winners. The friend explains that he took three cards with the winners names, removed finalist 1, if included, then selected at random from those remaining. This is different information. An event B_5 has occurred. It is clear that the conditioning event B_5 is *not* the same as the conditioning event A_5. In fact (see Example 3.2.4)

$$P(A_1|B_5) = \frac{6}{10} = P(A_1).$$

The information that B_5 has occurred did not change the likelihood that A_1 has occurred.

Regardless of one's position in the second case, the example makes clear that care must be exercised in determining what information is actually available to determine the conditioning event.

$\square$

The following example is taken from the UMAP Module 576, by Paul Mullenix, reprinted in *UMAP Journal*, Vol. 2, no. 4. More extensive treatment of the problem is given there.

Example 3.1.5: Responses to a Sensitive Question on a Survey

In a survey, if answering "yes" to a question may tend to incriminate or otherwise embarrass the subject, the response given may be incorrect or misleading. Nonetheless, it may be desirable to obtain correct responses for purposes of social analysis. The following device for dealing with this problem is attributed to B. G. Greenberg. By a chance process, each subject is instructed to do one of three things:

1. Respond honestly to the sensitive question.

2. Reply "yes" regardless of the truth of the matter.

3. Reply "no" regardless of the true answer.

Let A be the event the subject is told to reply honestly, B be the event the subject is to reply "yes," and C bet the event the subject is to reply "no." The probabilities $P(A)$, $P(B)$, and $P(C)$ arc determined by a chance mechanism. Let E be the event the reply is "yes." On the basis of the survey responses, $P(E)$ is determined. What is desired is the probability that an honest response is "yes." We need to calculate $P(E|A)$. Since $E = EA \bigcup B$, we have

$$P(E) = P(EA) + P(B) = P(E|A)P(A) + P(B).$$

Simple algebra shows that

$$P(E|A) = \frac{P(E) - P(B)}{P(A)}.$$

Since all values on the right-hand side are known, the desired conditional probability is determined.

Suppose there are 250 subjects. The chance mechanism is such that $P(A) = 0.7$, $P(B) = 0.14$, and $P(C) = 0.16$. Of the 250 responses, 62 are "yes." This is interpreted as $P(E) = 62/250 \approx 0.248$. Then

$$P(E|A) = \frac{62/250 - 14/100}{70/100} = \frac{27}{175} \approx 0.154.$$

———   □

Bcfore considering other examples, we develop several general properties of conditional probability that are useful in applications.

3.2. Properties of Conditional Probability

As noted earlier, a conditional probability measure satisfies the axiomatic properties (P1), (P2), and (P3), so that it has all the properties of any probability measure. But it also has properties that are a consequence of the manner in which it is derived from the original probability measure. Some special cases are used in the preceding examples. We wish to identify and establish for ready reference several properties that are encountered in a variety of theoretical and practical situations.

We note in Section 3.1 that the defining expression for conditional probability can be put in product form. The usefulness of this form is suggested in the examples of Section 3.1. This product form can be extended to any number of events, provided their intersections have positive probability. Consider

$$\begin{aligned}
P(A_1 A_2 A_3 A_4) &= P(A_1)\frac{P(A_1 A_2)}{P(A_1)} \cdot \frac{P(A_1 A_2 A_3)}{P(A_1 A_2)} \cdot \frac{P(A_1 A_2 A_3 A_4)}{P(A_1 A_2 A_3)} \\
&= P(A_1)P(A_2|A_1)P(A_3|A_1 A_2)P(A_4|A_1 A_2 A_3).
\end{aligned}$$

It is apparent that this pattern of argument can be extended to any number of events, in any fixed order. We thus have the general

CP1) *Product rule.*

$$P(A_1 A_2 \ldots A_n) = P(A_1)P(A_2|A_1) \ldots P(A_n|A_1 A_2 \ldots A_{n-1}).$$

Example 3.2.1:

There are ten terminals in a computer laboratory, one of which is defective. Four students make a random selection, in succession, on an equally likely basis from among those remaining. Determine the probability that all terminals selected are good.

Solution

Let A_i, $1 \leq i \leq 4$, be the event the ith student to select chooses a good unit. We could use combinatorics to determine $P(A_1 A_2 A_3 A_4)$, but the product rule provides an easy alternative. The first student picks one from 9 out of 10 good units; the second picks one from 8 out of 9 good units; etc. Thus,

$$P(A_1) = \frac{9}{10} \qquad P(A_2|A_1) = \frac{8}{9}$$

$$P(A_3|A_1 A_2) = \frac{7}{8} \qquad P(A_4|A_1 A_2 A_3) = \frac{6}{7}.$$

By the product rule (CP1) for conditional probability, we have

$$P(A_1 A_2 A_3 A_4) = \frac{9}{10} \cdot \frac{8}{9} \cdot \frac{7}{8} \cdot \frac{6}{7} = \frac{6}{10}.$$

—   □

Example 3.2.2:

Determine $P(A_3)$ for the situation in the previous example.

Solution

We may partition A_3 as follows.

$$A_3 = A_1 A_2 A_3 \bigcup A_1 A_2^c A_3 \bigcup A_1^c A_2 A_3 \bigcup A_1^c A_2^c A_3.$$

Arguing as in Example 3.2.1, we have

$$P(A_1 A_2 A_3) = P(A_1)P(A_2|A_1)P(A_3|A_1 A_2) = \frac{9}{10} \cdot \frac{8}{9} \cdot \frac{7}{8} = \frac{7}{10}$$

$$P(A_1 A_2^c A_3) = P(A_1)P(A_2^c|A_1)P(A_3|A_1 A_2^c) = \frac{9}{10}\cdot\frac{1}{9}\cdot 1 = \frac{1}{10}$$

$$P(A_1^c A_2 A_3) = \frac{1}{10}\cdot 1\cdot 1 = \frac{1}{10} \qquad P(A_1^c A_2^c A_3) = 0.$$

Hence

$$P(A_3) = \frac{7}{10} + \frac{1}{10} + \frac{1}{10} + 0 = \frac{9}{10}.$$

We thus have $P(A_3) = P(A_1)$. In like manner, it may be shown that each A_i has the same probability.

—   □

Example 3.2.3:

Five cards, numbered 1 through 5, are put in a container. Three are selected, without replacement, in a random manner. Let A_i be the event card i is among those selected. Determine $P(A_i)$, $P(A_i A_j)$, and $P(A_i|A_j)$.

Solution

This problem is solved in Example 3.1.4 by the use of combinations. We make an alternate approach, using the product rule for conditional probability. Let

E_k be the event card i is picked on the kth selection
F_k be the event card j is picked on the kth selection
G_k be the event that neither i nor j is picked on the kth selection

Then $A_i = E_1 \bigcup E_1^c E_2 \bigcup E_1^c E_2^c E_3$, so that

$$\begin{aligned} P(A_i) &= P(E_1) + P(E_1^c)P(E_2|E_1^c) + P(E_1^c)P(E_2^c|E_1^c)P(E_3|E_1^c E_2^c) \\ &= \frac{1}{5} + \frac{4}{5}\cdot\frac{1}{4} + \frac{4}{5}\cdot\frac{3}{4}\cdot\frac{1}{3} = \frac{6}{10}. \end{aligned}$$

In a similar manner,

$$A_i A_j = E_1 F_2 G_3 \bigcup F_1 E_2 G_3 \bigcup E_1 G_2 F_3 \bigcup F_1 G_2 E_3 \bigcup G_1 E_2 F_3 \bigcup G_1 F_2 E_3.$$

Now

$$P(E_1 F_2 G_3) = P(E_1)P(F_2|E_1)P(G_3|E_1 F_2) = \frac{1}{5}\cdot\frac{1}{4}\cdot 1 = \frac{1}{20}.$$

A similar expansion for each term in the disjoint union gives

$$P(A_i A_j) = \frac{1}{5}\cdot\frac{1}{4}\cdot 1 + \frac{1}{5}\cdot\frac{1}{4}\cdot 1 + \frac{1}{5}\cdot\frac{3}{4}\cdot\frac{1}{3} + \frac{1}{5}\cdot\frac{3}{4}\cdot\frac{1}{3} + \frac{3}{5}\cdot\frac{1}{4}\cdot\frac{1}{3} + \frac{3}{5}\cdot\frac{1}{4}\cdot\frac{1}{3} = \frac{3}{10}.$$

From these it follows easily that $P(A_i|A_j) = \frac{3/10}{6/10} = \frac{1}{2}$.

—   □

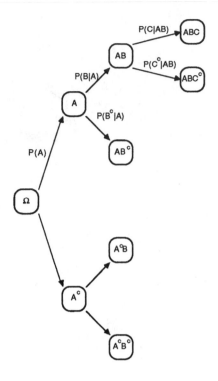

FIGURE 3.1. A probability tree diagram.

Remark: In this example, the combinatoric method may be somewhat simpler to employ. We need the method illustrated, since there are problems for which the combinatoric method is not relevant.

The product rule is often used in conjunction with a tree diagram, illustrated in Figure 3.1. The base node corresponds to the certain event Ω. At the end of the branches emanating from the base node are first-level nodes representing members of a partition (in this case A and A^c). The branches leaving these nodes terminate in second-level nodes representing members of a partitioning of the first-level events. In this example, each first-level event is partitioned by B and B^c. This process can be continued to higher level nodes. Only one extension to the third level is shown on the figure. The branches are assigned weights or values corresponding to the appropriate conditional probabilities, as shown. To determine the probability for the event at any node, simply multiply the conditional probabilities on the path leading from the base node to the node in question. Although this representation is quite popular, we do not make much use of the scheme.

In property (P8) for probability, we show that if a countable, mutually exclusive class $\{B_i : i \in J\}$ partitions event A, then

$$P(A) = P\left(\bigcup_{i \in J} AB_i\right) = \sum_{i \in J} P(AB_i).$$

If each $P(B_i) > 0$, then each of the terms $P(AB_i)$ can be expanded by the product rule to yield an important pattern.

CP2) *Law of total probability.* If $A = \bigcup_{i \in J} AB_i$ with each $P(B_i) > 0$, then

$P(A) = \sum_{i \in J} P(A|B_i)P(B_i)$.

Example 3.2.4: A Compound Experiment (See Example 3.1.4)

Five cards are numbered 1 through 5. A two-step procedure is followed.

1. Three cards are selected without replacement, on an equally likely basis.

   - If card 1 is drawn, the other two are put into a box

   - If card 1 is not drawn, all three are put into the box

2. One of the cards in the box is drawn.

Let A_i be the event card i is drawn on the first selection and B_i be the event card i is drawn on the second selection.

Determine $P(B_5)$, $P(A_1 B_5)$, and $P(A_1|B_5)$.

Solution

From Example 3.2.3, we have $P(A_i) = 6/10$ and $P(A_i A_j) = 3/10$. This implies

$$P(A_i A_j^c) = P(A_i) - P(A_i A_j) = \frac{6}{10} - \frac{3}{10} = \frac{3}{10}.$$

Now event $B_5 \subset A_5$ and $A_5 = A_1 A_5 \cup A_1^c A_5$, so that

$$B_5 = B_5 A_5 = B_5 A_1 A_5 \cup B_5 A_1^c A_5.$$

By the law of total probability (CP2),

$$
\begin{aligned}
P(B_5) &= P(B_5|A_1 A_5)P(A_1 A_5) + P(B_5|A_1^c A_5)P(A_1^c A_5) \\
&= \frac{1}{2} \cdot \frac{3}{10} + \frac{1}{3} \cdot \frac{3}{10} = \frac{1}{4}.
\end{aligned}
$$

Also, $A_1 B_5 = A_1 A_5 B_5$, so that

$$P(A_1 B_5) = P(A_1 A_5 B_5) = P(A_1 A_5) P(B_5 | A_1 A_5) = \frac{3}{10} \cdot \frac{1}{2} = \frac{3}{20}.$$

It follows that

$$P(A_1 | B_5) = \frac{3/20}{5/20} = \frac{6}{10} = P(A_1).$$

Occurrence of the event B_5 has no effect on the likelihood of the occurrence of event A_1. This is a consequence of the manner in which the second selection is made. We examine this "lack of conditioning" in Chapter 4, where the notion of independence of events is introduced.

— $\square$

Frequently, data are available that give conditional probabilities in one "direction," say $P(A|B)$, when it is desired to know $P(B|A)$. That is, we want to reverse the direction of conditioning.

Example 3.2.5: Reversal of Direction of Conditioning

A distributor of electronic parts sells an item that may be obtained from one of three manufacturers. The item is packaged and sold under the distributor's brand, so that the customer does not know which manufacturer made the unit he buys. Let H_i be the event that the item purchased is from manufacturer number i. Experience indicates that $P(H_1) = 0.2$, $P(H_2) = 0.5$, and $P(H_3) = 0.3$. An occasional unit will fail to operate satisfactorily when delivered and must be replaced. Data on the performance of the various manufacturers show that if D is the event that a deficient unit is sold

$$P(D|H_1) = 0.05, \quad P(D|H_2) = 0.01, \quad \text{and} \quad P(D|H_3) = 0.03.$$

A customer receives a satisfactory unit. Determine the probability that it was made by each manufacturer.

Solution

The probabilities to be determined are interpreted to be $P(H_i|D^c)$, for each i. Now

$$P(H_i|D^c) = \frac{P(H_i D^c)}{P(D^c)}.$$

The numerator can be expanded by the product rule to give

$$P(H_i|D^c) = \frac{P(D^c|H_i) P(H_i)}{P(D^c)}.$$

Since the class $\{H_1, H_2, H_3\}$ is a partition, we may use the law of total probability (CP2) to determine

$$
\begin{aligned}
P(D^c) &= P(D^c|H_1)P(H_1) + P(D^c|H_2)P(H_2) + P(D^c|H_3)P(H_3) \\
&= 0.95 \cdot 0.2 + 0.99 \cdot 0.5 + 0.97 \cdot 0.3 = 0.976.
\end{aligned}
$$

The desired conditional probabilities are found to be

$$
P(H_1|D^c) = \frac{0.95 \cdot 0.2}{0.976} = \frac{190}{976} \qquad P(H_2|D^c) = \frac{0.99 \cdot 0.5}{0.976} = \frac{495}{976}
$$

$$
P(H_3|D^c) = \frac{0.97 \cdot 0.3}{0.976} = \frac{291}{976}.
$$

These probabilities sum to one, which provides a check on the results.

—   □

The pattern of reversal in the second expression for $P(H_i|D^c)$ and the use of the law of total probability to obtain $P(D^c)$ are quite general and are expressions of Bayes's rule, after the eighteenth century theologian and mathematician, the Reverend Thomas Bayes.

CP3) *Bayes's rule.* If $A = \bigcup_{i \in J} AB_i$ with each $P(B_i) > 0$, then

$$
P(B_j|A) = \frac{P(A|B_j)P(B_j)}{P(A)}.
$$

Example 3.2.6: An Indirect Test

There are three items in one lot, one of which is defective. In a second lot there are four items, one of which is defective. An item is selected from the first lot, at random on an equally likely basis, and put with the second lot. An item is selected from the augmented second lot, again on an equally likely basis. It is tested and found to be good. What is the probability the item selected from the first lot is good?

Solution

We let G_1 be the event the first item selected is good and G_2 be the event the second item selected is good. Event G_2 has occurred. We know that

$$
P(G_1) = \frac{2}{3}. \qquad \text{Also,} \quad P(G_2|G_1) = \frac{4}{5} \quad \text{and} \quad P(G_2|G_1^c) = \frac{3}{5}.
$$

We want

$$
P(G_1|G_2) = \frac{P(G_2|G_1)P(G_1)}{P(G_2)}.
$$

Now, by the law of total probability (CP2)

$$P(G_2) = P(G_2|G_1)P(G_1) + P(G_2|G_1^c)P(G_1^c) = \frac{4}{5} \cdot \frac{2}{3} + \frac{3}{5} \cdot \frac{1}{3} = \frac{11}{15}.$$

We thus have

$$P(G_1|G_2) = \frac{4/5 \times 2/3}{11/15} = \frac{8}{11}.$$

—   □

Such a reversal of conditional probabilities is needed in a variety of practical situations.

Example 3.2.7: Reversal of Conditional Probabilities

1. *Medical tests.* Suppose T is the event a test is positive and D is the event the patient has the disease. Data are usually of the form $P(T|D)$, $P(T|D^c)$, and $P(D)$.

   The desired probabilities are $P(D|T)$ and $P(D|T^c)$.

2. *Safety device.* If T is the event the safety device is activated and D is the event of a dangerous condition, then data are usually of the form $P(T|D)$, $P(T|D^c)$, and $P(D)$.

   The desired probabilities are $P(D|T)$ and $P(D|T^c)$.

3. *Coding error.* Suppose noise operates on one symbol at a time. Let A be the event that a "1" is sent and B be the event a "1" is received. Probability data may be of the form $P(B^c|A)$, $P(B|A^c)$, and $P(A)$. Since it is desired to know what symbol was transmitted, we wish to know $P(A|B)$ and $P(A|B^c)$.

—   □

Remark: Frequently the data are given in terms of probability of *false positive* and probability of *false negative*. A false positive occurs, when the test is positive but the condition does not exist. Thus the probability of a false positive is $P(T|D^c)$. Similarly, the probability of a false negative is $P(T^c|D)$, the probability that the test is negative when the condition in fact exists.

The notion of odds as ratios of probabilities is introduced in Section 2.3. We may have odds $O(A:B) = P(A)/P(B)$ favoring event A over event B. As a result of a trial, a conditioning event C is observed to have occurred. It may be desirable to determine new odds $P(A|C)/P(B|C)$, in terms of the new likelihood assignments. As in the situations leading to the use of Bayes' rule, we may have data of the form $P(C|A)$ and $P(C|B)$, or, perhaps, in the form $P(C|A)/P(C|B)$. In this case, we find it convenient to use the following form of Bayes' rule.

CP3') *Ratio form of Bayes' rule:*

$$\frac{P(A|C)}{P(B|C)} = \frac{P(C|A)}{P(C|B)} \cdot \frac{P(A)}{P(B)}$$

Posterior odds = likelihood ratio × prior odds.

As in the case of ordinary probability, the *conditional odds* express the ratio of likelihoods for two events, with the same conditioning event. The term *likelihood ratio* is used for the same event, but different probability measures, $P(\cdot|A)$ and $P(\cdot|B)$. The terms *prior* and *posterior* refer to the likelihoods before and after knowledge of the conditioning event is obtained.

Example 3.2.8: Production Testing

Items coming off a production line are subjected to a nondestructive test. It is known that approximately 98 percent of the items meet specifications. The test is known to have probability 0.02 of a false negative (i.e., of rejecting a satisfactory item) and probability 0.05 of a false positive (i.e., of accepting an item that does not meet specifications). If a test is satisfactory, what are the posterior odds that the item meets specifications?

Solution

Let T be the event the item tests satisfactory and S be the event it is actually satisfactory. The data are

$$P(S) = 0.98, \quad P(T^c|S) = 0.02, \quad P(T|S^c) = 0.05.$$

By the ratio form of Bayes' rule

$$\frac{P(S|T)}{P(S^c|T)} = \frac{P(T|S)}{P(T|S^c)} \cdot \frac{P(S)}{P(S^c)} - \frac{0.98}{0.05} \cdot \frac{0.98}{0.02} = \frac{9604}{10}.$$

We may convert from odds to probability to obtain

$$P(S|T) = \frac{9604/10}{1 + 9604/10} = \frac{9604}{9614} = 1 - \frac{10}{9614} \approx 1 - 0.00104$$

so that $P(S^c|T) \approx 0.00104$. A positive test result is incorrect only about one time in a thousand. The calculation of conditional odds first then conditional probabilities is often easier than direct calculation of the probabilities.

—— $\square$

Example 3.2.9: An Inference Problem

A subscriber to a cable television service notices that the likelihood of a program interruption during the evening is eight times as great if the weather is stormy. At this time of the year, the likelihood of storms in the area is about 0.1. This evening there is a program interruption. What is the probability of a storm in the area?

Solution

Let S be the event of a storm in the area and A be the event of a program interruption. The data are taken to mean

$$P(S) = 0.1 \qquad \text{and} \qquad \frac{P(A|S)}{P(A|S^c)} = 8.$$

We wish to obtain $P(S|A)$. By Bayes' rule

$$\frac{P(S|A)}{P(S^c|A)} = \frac{P(A|S)}{P(A|S^c)} \cdot \frac{P(S)}{P(S^c)} = 8 \cdot \frac{1}{9} = \frac{8}{9}.$$

Thus

$$P(S|A) = \frac{8/9}{1 + 8/9} = \frac{8}{17} \approx 0.47.$$

□

Remark: In this example, as in others to follow, we speak of events that are not sharply defined. What is meant by a program interruption? What is meant by a storm in the area? Strictly speaking, events must be delineated with complete precision. On any trial an event either occurs or it does not. In spite of some ambiguity, one can have general agreement on whether or not the conditions "a storm in the area" or "a program interruption" prevail, so that it is clear whether or not the corresponding events occur. A probability analysis may thus be useful in situations in which it is difficult to define the events in question with the precision the mathematical model assumes.

The next property is a set of equivalent conditions expressed in a form that allows a large number of cases. Not only are these patterns useful in analysis of conditional probabilities, but they provide essential patterns for characterizing independence, introduced in the next chapter. We use a simple notational scheme to write compactly and efficiently a variety of cases. We use $\top$ to indicate $<$, $\leq$, $=$, $\geq$, or $>$, with the understanding that the same meaning is to be used in each case. Consider the pair of relations $P(A|B) \top P(B)$ iff $P(AB) \top P(A)P(B)$. If we interpret the first $\top$ to mean $<$, then we must interpret the second in the same way. The same equivalent pair in terms of $\top$ corresponds to five equivalent pairs in terms of the ordinary relations, according to the meaning assigned $\top$. In a similar way, we use the symbol $\perp$ to indicate the inverse of $\top$. If $\top$ is interpreted as $>$, then $\perp$ is interpreted as $\leq$, etc.

CP4) If $0 < P(A) < 1$ and $0 < P(B) < 1$, then

$$P(A|B) \top P(A) \text{ iff } P(AB) \top P(A)P(B) \text{ iff } P(AB^c) \perp P(A)P(B^c)$$

where $\top$ is $<, \leq, =, \geq$, or $>$ and $\perp$ is $>, \geq, =, \leq$, or $<$, respectively.

Remarks

1. Because of the symmetry of the treatment of A and B in the second condition, the two events may be interchanged to give an equivalent expression. For example

$$P(A|B) \top P(A) \quad \text{iff} \quad P(B|A) \top P(B).$$

2. If we complement one of the events A or B in an expression (using $(A^c)^c = A$, etc.) and take the inverse relation, we obtain an equivalent expression. For example

$$P(B|A) \top P(B) \quad \text{iff} \quad P(B^c|A) \perp P(B^c).$$

3. If we complement both of the events A and B in an expression, we obtain an equivalent expression.

4. Since $=$ is its own inverse, in an equality we may interchange A and B or complement consistently either or both and still have an equality.

Proof

We need establish only the basic equivalences and the second remark.

1. Since multiplying both sides of an expression by a positive quantity does not change the $\top$ relation

$$P(A|B) \top P(A) \quad \text{iff} \quad P(A|B)P(B) = P(AB) \top P(A)P(B).$$

2. We use the facts that $P(B) = 1 - P(B^c)$ and $P(AB) = P(A) - P(AB^c)$ to assert

$$P(AB) \top P(A)P(B)$$
$$\text{iff} \quad P(A) - P(AB^c) \top P(A)(1 - P(B))$$
$$\text{iff} \quad -P(AB^c) \top -P(A)P(B^c).$$

Since multiplying both sides of a $\top$ relation by a negative quantity yields the inverse relation, we have

$$-P(AB^c) \top -P(A)P(B^c) \quad \text{iff} \quad P(AB^c) \perp P(A)P(B^c).$$

3. The validity of the second remark follows from the facts that A and B may be interchanged and that $\top$ and $\perp$ are inverse relations, so that

$$P(AB^c) \top P(A)P(B^c) \quad \text{iff} \quad P(AB) \perp P(A)P(B)$$

and

$$P(A^cB) \top P(A^c)P(B) \quad \text{iff} \quad P(AB) \perp P(A)P(B).$$

$\square$

Example 3.2.10: Further Discussion of Results in Example 3.1.4 and 3.2.4

In Example 3.1.4, we let A_i be the event that person number i among five finalists is selected in the final drawing. We show that $P(A_i) = 0.6$ for each i and that

$$P(A_1|A_5) = 0.5 < P(A_1).$$

Because of (CP4), we can assert immediately that $P(A_5|A_1) < P(A_5)$. Also, we define an event B_5 and show in Example 3.2.4 that $P(A_1|B_5) = P(A_1)$. That is, knowledge that B_5 has occurred does not change the likelihood that A_1 has occurred. Again, we can assert immediately that

$$P(B_5|A_1) = P(B_5).$$

Thus, knowledge that A_1 has occurred does not change the likelihood that B_5 has occurred also.

$\square$

3.3.   Repeated Conditioning

When we have information that conditioning event C has occurred, we reassign probabilities to events according to the rule

$$P(A|C) = P_C(A) = \frac{P(AC)}{P(C)}.$$

The symbols $P(A|C)$ and $P_C(A)$ are alternate expressions for the same quantity. Sometimes one notation will be more convenient than the other. In the discussion leading up to the definition of conditional probability, in Section 3.1, we show that $P(\cdot|C) = P_C(\cdot)$ is a probability measure, defined for every event A.

Suppose after receiving information that C has occurred, we receive additional information that event D has occurred. Since both C and D have occurred, we have a new conditioning event CD. We suppose that $P(CD) > 0$. How shall we reassign probabilities? Two possibilities suggest themselves. We show that they are equivalent.

- *Reassign the conditional probabilities.* Conditional probability $P_C(A)$ becomes

$$P_C(A|D) = \frac{P_C(AD)}{P_C(D)}.$$

  Here we use the hybrid notation $P_C(\cdot|D)$ to indicate that we make the same kind of change from $P_C(\cdot)$ to $P_C(\cdot|D)$ as we do in going from $P(\cdot)$ to $P(\cdot|C)$.

- *Reassign the total probabilities.* $P(A)$ becomes

$$P_{CD}(A) = P(A|CD) = \frac{P(ACD)}{P(CD)}.$$

The following simple argument shows that the two approaches yield the same assignments.

$$P_C(A|D) = \frac{P_C(AD)}{P_C(D)} = \frac{P(AD|C)}{P(D|C)} \cdot \frac{P(C)}{P(C)} = \frac{P(ACD)}{P(CD)} = P(A|CD).$$

Since the role of C and D can be interchanged in the preceding argument, we have the basic result

$$P_C(A|D) = P(A|CD) = P_D(A|C).$$

The result shows that conditioning with respect to two events can be done in either order or it may be done in one step. It is not difficult to show that this extends to conditioning by any finite number of events such that their intersection has positive probability. This notion of repeated conditioning aids in the analysis of conditional independence, introduced in Chapter 5.

PROBLEMS

3-1. A radar computer system has a primary unit and a backup unit. Let A be the event the primary system operates successfully and B be the event the backup system operates successfully when called on. Suppose $P(A) = 0.98$ and $P(B|A^c) = 0.90$. Determine the probability the computer system will operate successfully.

3-2. Four students are to be selected from a group of twenty, of whom 13 are women and 7 are men. The selections are made sequentially, with each choice made on an equally likely basis from those not already chosen. Determine the probability of choosing a woman, man, man, woman, in that order.

3-3. Three persons are selected from a group of twenty, eight of whom are men. What is the probability that the first person selected is a man and that at least one of the others is a woman? Assume that the selection at each stage is on an equally likely basis from those remaining.

3-4. Four candidates for a job consist of two men and two women. They are selected randomly, one by one, for interview. What is the probability the second woman to be interviewed is the second person selected? The fourth?

3-5. Among five "identical" instruments, two are defective. A service man selects them one by one, on an equally likely basis, for test and repair. What is the probability he has found both defective units by the third selection?

3-6. In a College "room draw" there are two unassigned single rooms on the third floor. Eight persons have requested these rooms. Eight cards are put into a box. Two of them have designations for the single rooms. The interested persons draw in succession; each draws on an equally likely basis from the remaining cards. What is the probability that both rooms will have been assigned by the fourth draw?

3-7. The cultural committee for a residential college has two tickets to the Ungrateful Live Ones concert, to be given to interested college members. Ten people request them. It is decided to have a drawing. Ten cards are placed in a hat. Two of them indicate "winner." The interested individuals draw at random from the hat. What is the probability that both winning cards will be drawn by the fourth selection?

3-8. A car rental agency has twenty cars on the lot, nine of which are Chevrolets and eleven of which are Fords. Three are selected in sequence, with each choice made on an equally likely basis from those remaining. What is the probability the first and third selected are Fords? *Hint*: Note that the second selected may or may not be a Ford.

3–9. Two persons are to be selected from a group of 12. Individual names are put on identical cards, and two cards are selected at random, without replacement. The group consists of five men and seven women. The second person selected is a woman. Given this fact, what is the probability the first person chosen is also a woman?

3–10. An electronics repair shop receives five terminals to be repaired. In two of them a certain module has failed. Experience leads the technicians to check this module early in the testing procedure. The terminals are tested in random order. Determine the probability that both of the defective modules will be discovered by the third test.

3–11. In a survey, 85 percent of the employees say they favor a certain company policy. Previous experience indicates that 20 percent of those who do not favor the policy say that they do out of fear of reprisal. What is the probability that an employee picked at random really does favor the company policy?

3–12. A computer has a component that has experienced intermittent failure. A critical project is to be run during a certain eight-hour shift. Let R be the event the component operates successfully throughout the period and H be the event the project is completed during the period. Success or failure of the component does not completely determine success or failure of the project. Experience indicates

$$P(H|R) = 0.98 \quad \frac{P(H|R^c)}{P(H^c|R^c)} = 0.3 \quad \text{and} \quad P(R) = 0.75.$$

Determine $P(H)$.

3–13. Five percent of the patients seen in a clinic have a certain disease. Experience shows that 93 percent of the patients afflicted with the disease had previously exhibited a certain symptom, while only two percent of those who do not develop the disease display the symptom. A patient is examined and found to have the symptom. What is the probability, given this fact, that the patient will have the disease?

3–14. Two manufacturers, Excello Corp and Cutrate Supply, produce a certain replacement part. Because the items must be interchangeable, the external appearance of the units is the same. However, Excello's product has a probability 0.95 of operating for one year, while the competitor's part has a probability of 0.75 of successful operation. A retail store normally stocks the more reliable part; by mistake, however, a batch of inferior parts is received and put on the shelf. The stock arrangement was such that it was three times as likely to select an Excello part as to select an inferior part. A sale is made; the part fails in three months. What is the probability (conditional) that the unit was made by Excello?

3–15. Fred, Mary, and Eric take an examination. Let A, B, and C be the events that Fred, Mary, and Eric, respectively, make a grade of 90 or better. Suppose

$$P(A) = P(B) = 0.9, \quad P(C) = 0.85, \quad P(A|B^c) = 0.75,$$

$$P(C|B^c) = 0.7, \quad \text{and} \quad P(A|BC^c) = 0.8.$$

If Fred makes 90 or better, what is the probability (conditional) that both Mary and Eric make 90 or better?

3–16. A quality control group is designing an automatic test procedure for compact disk players coming from a production line. Experience shows that one percent of the units produced are defective. The automatic test procedure has a probability 0.05 of giving a false positive indication and a probability 0.02 of giving a false negative. That is, if D is the event a unit tested is defective, and T is the event that it tests satisfactory, then $P(T|D) = 0.05$ and $P(T^c|D^c) = 0.02$. Determine the probability $P(D^c|T)$ that a unit that tests good is, in fact, free of defects.

3–17. Two percent of the units received at a warehouse are defective. A nondestructive test procedure gives two percent false positive indications and five percent false negative. Units that fail to pass the inspection are sold to a salvage firm. This firm applies a corrective procedure, which does not affect any good unit and that corrects 90 percent of the defective units. A customer buys a unit from the salvage firm. It is good. What is the probability the unit was originally defective? Let

D be the event the unit was originally defective.
T be the event the first test indicates a defective unit.
G be the event the unit purchased is good.

The data may be taken to mean

$$P(D) = 0.02, \quad P(T^c|D) = 0.05, \quad P(T|D^c) = 0.02,$$

$$P(GT^c) = 0, \quad P(G|DT) = 0.90, \quad P(G|D^cT) = 1.$$

The problem is to determine $P(D|G)$.

3–18. You are a member of a jury trying a felony case. At this point in the proceedings, you think the odds are eight to one favoring innocence. A new item of evidence is presented. You think this evidence is four times as likely if the defendant is guilty than if he is innocent. In the light of this evidence, what should be your estimate of the probability of guilt.

3–19. Suppose $P(A)P(B) > 0$. Which of the following statements are generally true?

(a) $P(A|B) + P(A|B^c) = 1$ (b) $P(A|B) + P(A^c|B) = 1$

(c) $P(A|B) + P(A^c|B^c) = 1$.

3–20. Establish the following important propositions.

(a) If $0 < P(A) < 1$, $0 < P(B) < 1$, and $P(A|B) = P(A|B^c)$, then A and B cannot be mutually exclusive.

(b) If $P(B) > 0$ and $A \subset B$, then $P(A|B) = \dfrac{P(A)}{P(B)}$.

(c) If $P(B) > 0$ and $B \subset C$, then $P(A|B) = \dfrac{P(AB|C)}{P(B|C)}$.

3–21. An individual is to select from among n alternatives in an attempt to obtain a particular one. This might be selection from answers on a multiple-choice question, when only one is correct. Let A be the event he makes a correct selection and B be the event he knows which is correct before making the selection. We suppose $P(B) = p$ and $P(A|B^c) = 1/n$. Determine $P(B|A)$; show that $P(B|A) \geq P(B)$ and $P(B|A)$ increases with n for fixed p.

3–22. *Polya's urn scheme for a contagious disease.* An urn contains initially b black balls and r red balls ($r + b = n$). A ball is drawn on an equally likely basis from among those in the urn, then replaced along with c additional balls of the same color. The process is repeated. There are n balls on the first choice, $n + c$ balls on the second choice, etc. Let B_k be the event of a black ball on the kth draw and R_k be the event of a red ball on the kth draw. Determine (a) $P(B_2|R_1)$, (b) $P(B_1 B_2)$, (c) $P(R_2)$, and (d) $P(B_1|R_2)$.

4

Independence of Events

The concept of independence in probability, called *stochastic independence*, was introduced early in the history of the topic. If events form an independent class, then the calculation of probabilities of Boolean combinations is highly simplified. In many applications, events of interest may be the result of processes that are *operationally independent*. That is, these processes are related in a such a manner that one process does not affect, nor is it affected by, the other. Thus the occurrence of an event associated with one of the processes does not affect the likelihood of the occurrence of an event associated with the other process. It would seem reasonable to suppose these events are independent in a probabilistic sense. The concept of stochastic independence can be represented mathematically by a simple *product rule*. In spite of the mathematical simplicity, logical consequences of this definition show that it captures much of the intuitive notion of independence and hence provides a useful model for applications. We introduce stochastic independence for a pair of events, examine some basic patterns, then extend the notion to arbitrary classes of events. We then identify a variety of patterns that indicate why the concept is both appropriate and useful.

4.1. Independence as a Lack of Conditioning

In probability, we suppose two events form an independent pair iff knowledge of the occurrence of either of the events does not affect the uncertainty about the occurrence of the other. In terms of conditional probability, it seems appropriate to model this condition by the requirement

$$P(A|B) = P(A) \quad \text{or} \quad P(B|A) = P(B).$$

The following example exhibits an operational independence of two processes that would seem to justify an assumption of stochastic independence.

Example 4.1.1: Two Contractors Working Independently

Suppose two contractors, Smith and Jones, are working on separate jobs in different parts of the city. The performance of one has no effect on, nor is it affected by, the performance of the other. In fact, neither may be aware of the activities of the other. If A is the event Smith completes his job on schedule and B is the event Jones completes his job as planned, then it seems reasonable to suppose

$$P(A|B) = P(A) \quad \text{or} \quad P(B|A) = P(B).$$

—   □

In Example 3.2.4, we encounter a similar pattern of conditioning in the compound experiment in which card 5 is drawn on the second selection.

Example 4.1.2: The Compound Experiment of Example 3.2.4

Recall that in Example 3.2.4, five cards are numbered 1 through 5. A two-step procedure is followed.

  1. Three cards are selected without replacement, on an equally likely basis.

   - If card 1 is drawn, the other two are put into a box.

   - If card 1 is not drawn, all three are put into the box.

  2. One of the cards in the box is drawn.

A_i is the event card i is drawn on the first selection and B_i is the event card i is drawn on the second selection. We show in Example 3.2.4 that $P(A_1|B_5) = P(A_1)$. Use of property (CP4) for conditional probability enables us to claim also that $P(B_5|A_1) = P(B_5)$. The latter fact could also be verified by direct calculation. Since knowledge of the occurrence of B_5 does not alter the uncertainty about A_1 and knowledge of the occurrence of A_1 does not give any information about the likelihood of B_5, it would seem that this pair should be considered independent in the probability sense.

—   □

If we make direct use of property (CP4) for conditional probability, with both ⊤ and ⊥ interpreted as =, we obtain the first twelve of the sixteen equivalent properties in the following table. In the case of equality, we can interchange A and B in any of the equivalent expressions in (CP4) to get a new equivalent expression. And we may complement either A or B, or both, to get an equivalent expression.

Sixteen equivalent conditions

$P(A\|B) = P(A)$	$P(B\|A) = P(B)$	$P(AB) = P(A)P(B)$
$P(A\|B^c) = P(A)$	$P(B^c\|A) = P(B^c)$	$P(AB^c) = P(A)P(B^c)$
$P(A^c\|B) = P(A^c)$	$P(B\|A^c) = P(B)$	$P(A^cB) = P(A^c)P(B)$
$P(A^c\|B^c) = P(A^c)$	$P(B^c\|A^c) = P(B^c)$	$P(A^cB^c) = P(A^c)P(B^c)$

$P(A\|B) = P(A\|B^c)$	$P(A^c\|B) = P(A^c\|B^c)$
$P(B\|A) = P(B\|A^c)$	$P(B^c\|A) = P(B^c\|A^c)$

The last four equalities, in the lower section of the table, are derived easily from the equivalence of the first twelve. Using appropriate pairs from among the twelve equalities in the upper part of the table, we have

$$P(A|B) = P(A) = P(A|B^c) \qquad P(A^c|B) = P(A^c) = P(A^c|B^c)$$
$$P(B|A) = P(B) = P(B|A^c) \qquad P(B^c|A) = P(B^c) = P(B^c|A^c).$$

This establishes the last four pairs as consequences of the first twelve. Suppose $P(A|B) = P(A|B^c)$. Then by the law of total probability (CP2),

$$\begin{aligned} P(A) &= P(A|B)P(B) + P(A|B^c)P(B^c) = P(A|B)\big(P(B) + P(B^c)\big) \\ &= P(A|B) \end{aligned}$$

so that this condition implies all the others. The other cases follow similarly. For the next condition, replace A by A^c. For the third, interchange A and B in the argument. Finally, in the argument for the third case, replace B by B^c.

We may choose any one of these conditions as the defining condition and use the others as equivalents for the defining condition. Because of its simplicity and symmetry with respect to A and B, we select the product rule in the upper right-hand corner of the table of equivalents.

Definition 4.1.1. *The pair $\{A, B\}$ of events is said to be* independent *iff the following* product rule *holds.*

$$P(AB) = P(A)P(B).$$

Remark: Although the product rule is adopted as the basis for definition, in many applications the fundamental assumptions may be stated more naturally in terms of one of the equivalent equalities. We are free to do this, for the effect of assuming one condition is to assume them all.

As an immediate consequence of the equivalences, particularly the last column in the upper part of the table, we have the following result.

Theorem 4.1.2. *If any of the pairs $\{A, B\}$, $\{A, B^c\}$, $\{A^c, B\}$, and $\{A^c, B^c\}$ is independent, so are the others.*

— $\square$

A second simple, but important result, should be noted.

Theorem 4.1.3. *If N is an event with $P(N) = 0$ (a null event), and S is an event with $P(S) = 1$ (an almost sure event), then for any event A, each of the pairs $\{A, N\}$ and $\{A, S\}$ is independent.*

Proof

$0 \leq P(AN) \leq P(N) = 0$ implies $P(AN) = 0 = P(A)P(N)$.
Since $P(S^c) = 0$, $\{A, S^c\}$ is independent; hence, by Theorem 4.1.2, $\{A, S\}$ is independent.

—   □

CAUTION AND INTERPRETATION

- Independence and mutual exclusiveness are *not* the same. Suppose $P(A) > 0$ and $P(B) > 0$. Then $P(AB) = P(A)P(B) > 0$ implies $AB \neq \emptyset$. On the other hand, $AB = \emptyset$ implies $P(AB) = 0 \neq P(A)P(B)$.

  If $AB = \emptyset$, then the occurrence of one guarantees that the other did *not* occur. Thus, knowledge of the occurrence of one gives considerable information about the occurrence of the other. On the other hand, if knowledge of the occurrence of A leaves the likelihood of the occurrence of B unchanged, then A and B are not mutually exclusive.

- With the exception of the certain event Ω and the impossible event $\emptyset$, which are always assigned probability one and zero, respectively, or events that are mutually exclusive, *independence is a property of the probability measure rather than of the events*. For this reason, we cannot give a a structural characterization of independent events. Thus there is nothing that can be shown on the Venn diagram (except nonmutual exclusiveness) to characterize independent events. We see the importance of this in Chapter 5 for the concept of conditional independence.

If the pair $\{A, B\}$ is independent, then knowledge of $P(A)$ and $P(B)$ determines the probability of each minterm generated by the pair and hence determines the probability of any Boolean combination of the pair.

Example 4.1.3:

Suppose the pair $\{A, B\}$ is independent, with $P(A) = 0.6$ and $P(B) = 0.3$. Determine $P(A \cup B^c)$.

Solution

Examination of a minterm map shows that the minterm expansion is

$$A \cup B^c = AB \bigcup AB^c \bigcup A^cB^c.$$

Since independence of $\{A, B\}$ implies that of $\{A, B^c\}$ and $\{A^c, B^c\}$, by additivity and the product rule we have

$$\begin{aligned} P(A \cup B^c) &= P(A)P(B) + P(A)P(B^c) + P(A^c)P(B^c) \\ &= 0.6 \cdot 0.3 + 0.6 \cdot 0.7 + 0.4 \cdot 0.7 = 0.88. \end{aligned}$$

This is not the most efficient way to calculate the probability. The event $A \cup B^c$ may be expressed alternately as

$$A \cup B^c = A \bigcup A^cB^c = (A^cB)^c.$$

The probability may be calculated more efficiently according to either of the two expansions, as follows

$$P(A \cup B^c) = 0.6 + 0.4 \cdot 0.7 = 1 - 0.4 \cdot 0.3 = 0.88.$$

In the independent case, the probability of each minterm can be calculated at the beginning. Then when the minterm expansion for any Boolean combination is determined, the probability is immediately obtained as the sum of the probabilities of the appropriate minterms.

— □

4.2. Independent Classes

We need to extend the concept of stochastic independence to arbitrary classes of two or more events. As in the case of a pair of events, we make the *product rule* fundamental. While an independent class may be infinite, we merely require the product rule to apply to any *finite* subclass.

Definition 4.2.1. *A class* $\{A_i : i \in J\}$ *of events is said to be* independent *iff for every finite subclass* $\{A_{i_1}, A_{i_2}, \ldots, A_{i_m}\}$ *of two or more members the following* product rule *holds.*

$$P\left(\bigcap_{k=1}^m A_{i_k}\right) = \prod_{k=1}^m P(A_{i_k}).$$

Remarks and examples

- We may have independence of each pair without having independence of the class.

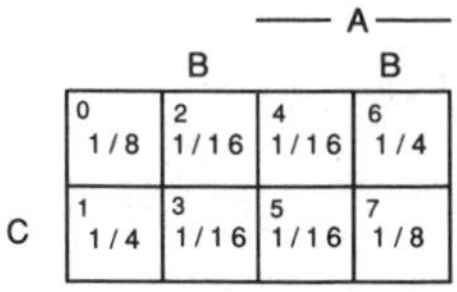

$$\begin{array}{c}\text{—— A ——}\\[2pt]\quad\text{B}\qquad\qquad\text{B}\end{array}$$

	B		B	
	0 1/8	2 1/16	4 1/16	6 1/4
C	1 1/4	3 1/16	5 1/16	7 1/8

FIGURE 4.1. The probability distribution for Example 4.2.2.

Example 4.2.1:

Consider the partition $\{A_1, A_2, A_3, A_4\}$ with $P(A_i) = 1/4$ for each i. Define the events

$$A = A_1 \cup A_4 \qquad B = A_2 \cup A_4 \qquad C = A_3 \cup A_4.$$

We have

$$P(A) = P(B) = P(C) = \frac{1}{2}$$

and

$$P(AB) = P(AC) = P(BC) = P(A_4) = \frac{1}{4}.$$

The product rule holds for each pair, so that each pair is independent. But

$$P(ABC) = P(A_4) = \frac{1}{4} \neq P(A)P(B)P(C) = \frac{1}{8}.$$

Thus, although the class is *pairwise independent*, it is *not* independent.

— $\square$

- The product rule may hold for the entire class, but may hold for no pair.

Example 4.2.2:

Consider the class $\{A, B, C\}$. Probabilities 1/8, 1/4, 1/16, 1/16, 1/16, 1/16, 1/4, 1/8 are assigned to minterms M_0 through M_7 (see Figure 4.1). Then we may calculate directly from the minterm map

$$P(A) = P(B) = P(C) = \frac{1}{2}.$$

We also find

$$P(ABC) = \frac{1}{8} = P(A)P(B)P(C) \qquad P(AB) = \frac{3}{8}$$

$$P(AC) = P(BC) = \frac{3}{16}.$$

Thus the product rule holds for the entire class, but holds for no pair.

—

□

- The examples illustrate that whether or not the product rule is satisfied depends upon the distribution of probability mass among the minterms. Thus, as in the case of a pair of events, independence of a class is not a property of the events.

In Section 4.1, we note that if a pair of events is independent, we may replace either, or both, by its complement, by a null event, or by an almost sure event, and still have an independent pair. We have such a theorem for arbitrary independent classes.

Theorem 4.2.2 (Replacement Theorem) *Suppose $\mathcal{A} = \{\, A_i : i \in J \,\}$ is an independent class of events. Consider the class $\mathcal{A}^* = \{\, A_i^* : i \in J \,\}$ in which each A_i^* is either A_i, A_i^c, N_i (a null event), or S_i (an almost sure event). Then the class $\mathcal{A}^*$ is also independent.*

Proof

By definition, it is sufficient to show the theorem is true for the replacement of any finite subclass of members of $\mathcal{A}$. We form a proof by induction on the size n of the subclass that may be replaced.

1. Consider any finite subclass $\mathcal{A}_m$ of $m > 1$ events in $\mathcal{A}$. Let A be any event in $\mathcal{A}_m$ and let M be the intersection of the remaining $m - 1$ events in $\mathcal{A}_m$. Then the basic product rule shows $\{A, M\}$ is independent. By Theorems 4.1.2 and 4.1.3, we may make a replacement A^* for A and have $\{A^*, M\}$ independent. This implies the product rule for $\mathcal{A}_m$, with A replaced. Since the number $m \geq 2$, the membership of $\mathcal{A}_m$, and the choice of A in $\mathcal{A}_m$ are all arbitrary, the product rule must hold for $\mathcal{A}$ with any one member replaced.

2. Suppose the theorem is true for $n = k$. That means we can replace as many as k of the events and still have an independent class $\mathcal{A}^*$. Consider any finite subclass $\mathcal{A}_m$ of $m > k$ events from $\mathcal{A}$. Let A be any event in $\mathcal{A}_m$ and let M be the intersection of the remaining events, with at most k of them replaced. Then the basic product rule shows $\{A, M\}$ is independent. By Theorems 4.1.2 and 4.1.3, we can replace A by its complement, or a null event, or an almost sure event and still have an independent pair. This means that the product rule holds for the replacement of $n = k+1$ or fewer events in $\mathcal{A}_m$. Since the choice of $m > k$, $\mathcal{A}_m$, and A in $\mathcal{A}_m$ are all arbitrary, we can replace any $k + 1$ events. Thus if the theorem is true for $n = k$ replacements, it is true for $n = k + 1$ replacements.

3. By the principle of mathematical induction, the theorem is true for any finite number n of replacements.

—   □

The following result on Boolean combinations of a finite class is an easy consequence of Theorem 4.2.2 and the minterm expansion theorem.

Theorem 4.2.3. *If a finite class of events is independent and the probability of each event is known, then the probability of any Boolean combination is determined.*

Proof

The replacement theorem implies that the probability of any minterm is obtained by applying the product rule to the class with the appropriate members complemented. By the minterm expansion theorem, the Boolean combination is the disjoint union of an appropriate subclass of minterms. The probability of this disjoint union is the sum of the probabilities of the minterms in the union.

—   □

Example 4.2.3:

Four persons attempt a game of skill. Let E_i be the event player i succeeds. If their attempts are independent (in an operational sense), we assume the class $\{E_1, E_2, E_3, E_4\}$ is independent (in the probability sense). Suppose $P(E_1) = 0.8$, $P(E_2) = 0.7$, $P(E_3) = 0.6$, and $P(E_4) = 0.9$. What is the probability that at least three will be successful?

Solution

The event of three or more successes is

$$A = E_1 E_2 E_3 E_4^c \cup E_1 E_2 E_3^c E_4 \cup E_1 E_2^c E_3 E_4 \cup E_1^c E_2 E_3 E_4 \cup E_1 E_2 E_3 E_4.$$

This is the minterm expansion. By the replacement theorem, each of the classes with one member complemented is independent, so that the product rule holds for each of the minterms in the expansion. If we set $P(E_i) = p_i$ and $P(E_i^c) = q_i$, we have

$$P(A) = p_1 p_2 p_3 q_4 + p_1 p_2 q_3 p_4 + p_1 q_2 p_3 p_4 + q_1 p_2 p_3 p_4 + p_1 p_2 p_3 p_4.$$

Substitution and straightforward calculation give the result $P(A) = 0.7428$. We can determine the probability of any minterm so that we can calculate the probability of any Boolean combination of the E_i.

—   □

We now consider a sequence of theorems leading to a very general result on independence of Boolean combinations of events. This result provides a significant level of freedom and flexibility in working with independent classes of events. Although familiarity with the proofs is not important for specific applications, the arguments display some of the structure of the independence concept as embodied in the product rule.

Theorem 4.2.4. *Suppose $\mathcal{A} = \{A_i; 1 \leq i \leq m\}$ is a finite class of events. Let M_r be any minterm generated by $\mathcal{A}$ and let F be any Boolean combination of the events in $\mathcal{A}$. If G is any event such that each pair $\{M_r, G\}$ is independent, then $\{F, G\}$ is independent.*

Proof

By the minterm expansion theorem, F is the disjoint union of a subclass of the minterms generated by $\mathcal{A}$. Thus

$$F = \biguplus_{r \in J_F} M_r \qquad \text{and} \qquad FG = \biguplus_{r \in J_F} M_r G.$$

By additivity and the product rule

$$P(FG) = \sum_{r \in J_F} P(M_r G) = \sum_{r \in J_F} P(M_r)P(G) = P(F)P(G).$$

We thus have the desired independence of $\{F, G\}$.

— $\qquad\qquad\qquad\qquad\qquad\qquad\qquad\qquad\qquad\qquad\qquad\qquad$ $\square$

Theorem 4.2.5. *Consider any two finite classes of events, $\mathcal{A} = \{A_i : 1 \leq i \leq n\}$ and $\mathcal{B} = \{B_j : 1 \leq j \leq m\}$. Let M_r be any minterm generated by $\mathcal{A}$ and N_s be any minterm generated by $\mathcal{B}$. Also, let F be any Boolean combination of events in $\mathcal{A}$ and G be any Boolean combination of events in $\mathcal{B}$. If every pair $\{M_r, N_s\}$ is independent, then $\{F, G\}$ is independent.*

Proof

By Theorem 4.2.4, each pair $\{M_r, G\}$ is independent. Again, by Theorem 4.2.4, $\{F, G\}$ is independent.

— $\qquad\qquad\qquad\qquad\qquad\qquad\qquad\qquad\qquad\qquad\qquad\qquad$ $\square$

Theorem 4.2.6. *Suppose $\mathcal{A} = \{A_i : 1 \leq i \leq m\}$ is an independent class of events. If G is any event such that $\{F, G\}$ is independent for each Boolean combination F of events in $\mathcal{A}$, then the augmented class $\mathcal{A}^* = \{A_1, A_2, \ldots, A_m, G\}$ is independent.*

Proof

Let J_F be any subclass of the class $J = \{1, 2, \ldots, m\}$, and set $F = \bigcap_{i \in J_F} A_i$. Then F is a Boolean combination of the events in $\mathcal{A}$. By hypothesis, we have

$$P(FG) = P(F)P(G) = \prod_{i \in J_F} P(A_i)P(G).$$

This and the fact the product rule holds for every subclass of $\mathcal{A}$ guarantees that the product rule holds for every subclass of the augmented class $\mathcal{A}^*$.

— $\square$

These simple results and their proofs are interesting and useful in themselves. They also provide the tools for obtaining a proof of the following very general result.

Theorem 4.2.7 (Independence of Boolean Combinations of Events)
For each $k = 1, 2, \ldots, n$, let $\mathcal{A}_k$ be a finite class of events, such that no event is in more than one of the classes; let M_{r_k} be any minterm generated by $\mathcal{A}_k$; and let F_k be any Boolean combination of the events in $\mathcal{A}_k$. If each class $\{M_{r_k} : 1 \leq k \leq n\}$ is independent, so is the class $\{F_k : 1 \leq k \leq n\}$.

Proof

We sketch a proof by induction on the number n of classes involved.

1. Theorem 4.2.5 establishes the theorem for $n = 2$.

2. We suppose the theorem is true for $n = q$ and show that this implies the theorem for $n = q + 1$. Let $\mathcal{B}_q = \bigcup_{k=1}^{q} \mathcal{A}_k$. The minterms for $\mathcal{B}_q$ are intersections of minterms for the various $\mathcal{A}_k$. Let $M_{r_{q+1}}$ be any minterm generated by $\mathcal{A}_{q+1}$ and N_s be any minterm generated by $\mathcal{B}_q$. By hypothesis, $\{M_{r_{q+1}}, N_s\}$ is an independent pair. By Theorem 4.2.5, if G is any Boolean combination of events in $\mathcal{B}_q$, then $\{F_{q+1}, G\}$ is independent. By hypothesis, $\{F_1, F_2, \ldots, F_q\}$ is an independent class and if F is any Boolean combination of this class (and hence a Boolean combination of events in $\mathcal{B}_q$), then $\{F_{q+1}, F\}$ is independent. By Theorem 4.2.6, the augmented class $\{F_1, F_2, \ldots, F_q, F_{q+1}\}$ is independent. Thus the theorem is true for $n = q + 1$.

3. By the principle of mathematical induction, the theorem must be true for any $n \geq 2$.

— $\square$

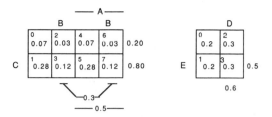

FIGURE 4.2. Minterm maps for Example 4.2.4.

Corollary 4.2.8. *For each $k = 1, 2, \ldots, n$, let $\mathcal{A}_k$ be a finite class of events, such that no event is contained in more than one of the classes; and let F_k be any Boolean combination of the events in $\mathcal{A}_k$. Let $\mathcal{B}_n = \bigcup_{k=1}^{n} \mathcal{A}_k$ be the class consisting of all the sets in any of the $\mathcal{A}_k$. If $\mathcal{B}_n$ is an independent class, then so is the class $\{F_k : 1 \le k \le n\}$.*

Proof

Let M_{r_k} be any minterm generated by $\mathcal{A}_k$, $1 \le k \le n$. Independence of $\mathcal{B}_n$ implies independence of the class $\{M_{r_k} : 1 \le k \le n\}$. By Theorem 4.2.7, we have independence of each class $\{F_k : 1 \le k \le n\}$.

—   □

Remark: Each $\mathcal{A}_k$ is a subclass of $\mathcal{B}_n$, with the property that no $\mathcal{A}_k$ shares any event with another. Independence of any class that contains $\mathcal{B}_n$ implies independence of the class $\{F_k : 1 \le k \le n\}$ of Boolean combinations of subsets of the $\mathcal{A}_k$.

Example 4.2.4:

Suppose the class $\{A, B, C, D, E\}$ is independent, with

$$P(A) = 0.5 \quad P(B) = 0.3 \quad P(C) = 0.8 \quad P(D) = 0.6 \quad P(E) = 0.5.$$

Consider the Boolean combinations $F = A(B \cup C^c)$ and $G = D^c E \cup DE^c$. Determine $P(FG)$.

Solution

We have F a Boolean combination of events in the class $\mathcal{A}_1 = \{A, B, C\}$ and G is a Boolean combination of the class $\mathcal{A}_2 = \{D, E\}$. The class $\mathcal{B}_2 = \mathcal{A}_1 \cup \mathcal{A}_2$ is independent. Therefore, the class $\{F, G\}$ is independent. If the minterms for class $\mathcal{A}_1$ are M_r and those for class $\mathcal{A}_2$ are N_s, then (see minterm maps in Figure 4.2)

$$F = M_4 \cup M_6 \cup M_7 \quad \text{and} \quad G = N_1 \cup N_2.$$

The minterm probabilities, obtained by use of the product rule, are shown on the minterm maps in Figure 4.2. From these we see

$$P(F) = p(4, 6, 7) = 0.07 + 0.03 + 0.12 = 0.22$$

and

$$P(G) = p^*(1, 2) = 0.2 + 0.3 = 0.5.$$

From the independence of the pair $\{F, G\}$, we have $P(FG) = 0.22 \cdot 0.5 = 0.11$.

—   □

In spite of its simplicity, the concept of stochastic independence plays a significant role in many applications. Whenever independence can be assumed, a great simplification of the analytical treatment of a problem usually follows. In the next chapter, we consider a modification of the independence condition that has been found appropriate and useful in many applications.

PROBLEMS

4-1. The class $\{A, B, C\}$ is independent, with $P(A) = 0.3$, $P(B^cC) = 0.32$, and $P(AC) = 0.12$. Determine the probability of each minterm generated by the class, and use these to determine $P(AB^c \cup BC^c)$.

4-2. Suppose the class $\{A, B, C\}$ is independent, with $P(A) = 0.5$, $P(AB^c) = 0.1$, and $P(A \cup B^cC) = 0.55$. Determine $P\big(A(B \cup C^c)\big)$.

4-3. The class $\{A, B, C\}$ is independent, with $P(A \cup B) = 0.6$, $P(A \cup C) = 0.7$, and $P(C) = 0.4$. Determine the probability of each minterm generated by the class.

4-4. The class $\{A, B, C\}$ is independent, with $P(A) = 0.6$, $P(A \cup B) = 0.76$, and $P(BC) = 0.08$. Calculate the probability of each minterm generated by the class.

4-5. Suppose $P(A) = 0.3$, $P(B) = 0.4$, and $P(C) = 0.2$. Calculate the probability of each minterm generated by the class $\{A, B, C\}$.

   (a) Under the assumption the class $\{A, B, C\}$ is independent.

   (b) Under the condition the class $\{A, B, C\}$ is mutually exclusive.

4-6. Suppose $P(A) = 0.4$, $P(A^cB) = 0.18$, and $P(C) = 0.2$. Let $E = (A^c \cup B)C^c$

   (a) Determine $P(E)$ under the condition $\{A, B, C\}$ is independent.

   (b) Determine $P(E)$ under the condition $\{A, B, C\}$ is a mutually exclusive class.

4-7. Mary receives a subscription advertisement for *Exquisite Home* magazine at $18 per year; Joan receives a mailing for *Elite* at $24 per year; and Bill receives a subscription solicitation for *Sports and Games* at $36 per year. They make their decisions whether or not to subscribe independently, with respective probabilities of 0.6, 0.9, and 0.8. What is the probability that the total amount of their annual subscription will be at least $42?

4-8. An investor seeks advice concerning an investment opportunity from three consultants, who arrive at their conclusions independently. He follows the recommendation of the majority. The probabilities that the advisers are wrong are 0.1, 0.15, and 0.05, respectively. What is the probability the man takes the wrong course of action?

4-9. Mary is wondering about her prospects for the weekend. There is a probability of 0.5 that she will get a letter from home with money. There is a probability of 0.8 that she will get a call from Bill at UT

Austin. There is also a probability of 0.4 that Jim will ask her for a date. These events are independent. What is the probability she will get money and Bill will not call or that both Bill will call and Jim will ask for the date?

4-10. The minterms generated by the class $\{A, B, C\}$ have probabilities $p(0)$ through $p(7)$, 0.15, 0.05, 0.02, 0.18, 0.25 0.05, 0.18, 0.12, respectively. Show that $P(ABC) = P(A)P(B)P(C)$, but the class is not independent.

4-11. The class $\{A, B, C\}$ is pairwise independent and the pair $\{AB, C\}$ is independent. Show whether or not the class $\{A, B, C\}$ is independent.

4-12. Four students are taking a competitive exam. Let A, B, C, D be the events that Alice, Bob, Carl, and Debby, respectively, score 95 or better on the exam. Suppose the class $\{A, B, C, D\}$ is independent, with $P(A) = 0.70$, $P(B) = 0.50$, $P(C) = 0.60$, and $P(D) = 0.60$. Calculate $P(F)$, where $F = A(B \cup C^c) \cup CD$.

4-13. Suppose problem 4-12 is modified so that A is the event that Alice is the *first* to finish with a score of 95 or better, B is the event Bob is the *first* to finish with a score of 95 or better, etc. Assume there are no ties; then $\{A, B, C, D\}$ is a mutually exclusive, but not necessarily exhaustive class. Suppose $P(A) = 0.25$, $P(B) = 0.15$, $P(C) = 0.25$, and $P(D) = 0.20$.

   (a) Calculate the probability of each minterm generated by the class.
   (b) Determine $P(F)$, where $F = A(B \cup C^c) \cup CD$.

4-14. A family is having a garage sale. They are particularly interested in selling four unrelated items. Let

   E_1 be the event they sell a man's sport coat for $20.
   E_2 be the event they sell a small radio for $10.
   E_3 be the event they sell a lawnmower for $30.
   E_4 be the event they sell a television set for $25.

Suppose the class $\{E_1, E_2, E_3, E_4\}$ may be considered independent, with

$$P(E_1) = 0.60, \ P(E_2) = 0.40, \ P(E_3) = 0.8, \text{ and } P(E_4) = 0.3.$$

What is the probability they realize $50 or more from the sale of these items?

4-15. Two persons in a game play consecutively, until one of them is successful or there are ten unsuccessful plays. Let E_i be the event of a success on the ith play of the game. We suppose $\{E_i : 1 \le i\}$ is an independent class with $P(E_i) = p_1$ for i odd, and $P(E_i) = p_2$ for i even.

Let A be the event the first player wins, B be the event the second player wins, and C be the event neither player wins.

(a) Express A, B, and C in terms of the component events E_1, E_2, ..., E_{10}.

(b) Determine $P(A)$, $P(B)$, and $P(C)$ in terms of p_1 and p_2 or $q_1 = 1 - p_1$ and $q_2 = 1 - p_2$. Then obtain numerical values in the case $p_1 = 1/4$ and $p_2 = 1/3$.

(c) Use appropriate facts about the geometric series to show that $P(A) = P(B)$ iff $p_1 = p_2/1 + p_2$ and that $P(C) = (q_1 q_2)^5$.

(d) Suppose $p_2 = 0.5$. Use the result of part (c) to find the value of p_1 to make $P(A) = P(B)$ and then determine $P(C)$, $P(A)$, and $P(B)$.

4–16. Suppose $\mathcal{A} = \{A_i : i \in I\}$, $\mathcal{B} = \{B_j : j \in J\}$, and $\mathcal{C} = \{C_k : k \in K\}$ are disjoint classes with $A = \bigcup_i A_i$, $B = \bigcup_j B_j$, and $C = \bigcup_k C_k$. Use arguments similar to those in the proof of Theorem 4.2.4 to show that if $\{A_i, B_j, C_k\}$ is independent for each permissible triple $\{i, j, K\}$, then $\{A, B, C\}$ is independent.

5

Conditional Independence of Events

In Chapter 4, we introduced a mathematically simple characterization of stochastic independence to model the lack of conditioning between two events. Independence is assumed for events that are the result of operationally independent actions or processes. In some cases, however, closer examination of the operational independence shows a "conditioning" by some common chance factor. If this common condition exists, the processes affecting the events may then be operationally independent. This suggests the need for some modification of the concept of stochastic independence.

Two mathematical facts point to a possible approach to the problem. For one thing, independence is a property of the probability measure and not of the events themselves. Also, a conditional probability measure is a probability measure. This suggests that independence with respect to a conditional probability measure might be useful. We show that this is indeed the case in many important applications.

5.1. Operational Independence and a Common Condition

Frequently we have two events that are the result of operationally independent actions. That is, one action does not affect, nor is it affected by, the other. In most situations, the natural probabilistic assumption is that the events constitute a stochastically independent pair. However, in some apparently independent situations there is a common chance factor that affects both operations. Under any given state of the chance factor, the actions may be operationally independent and yet the results are not stochastically independent. We consider some simple examples.

Example 5.1.1:

An electronic system has two modules that operate "independently," in the sense that failure of one is not affected by and does not affect the failure of the other. Let

A be the event the first unit operates one year without failure.

B be the event the second unit operates one year without failure.

Ordinarily, it would seem reasonable to assume that the pair $\{A, B\}$ is independent. However, line voltage fluctuations may occur in a manner that affects the likelihood of failure. Let

C be the event the line voltage is 20 for 10

Suppose

$$P(A|C) = 0.60, \quad P(A|C^c) = 0.90, \quad P(B|C) = 0.75,$$

$$P(B|C^c) = 0.95, \quad \text{and} \quad P(C) = 0.3.$$

What can be said about independence? Since the units operate independently under any given line condition, it seems reasonable to suppose

$$P(A|BC) = P(A|C) \quad \text{and} \quad P(A|BC^c) = P(A|C^c).$$

If the line voltage is high, success or failure of the first unit is not affected by success or failure of the second unit. The same is true if the line voltage is normal. These conditions are equivalent to

$$P(AB|C) = P(A|C)P(B|C) \quad \text{and} \quad P(AB|C^c) = P(A|C^c)P(B|C^c).$$

This means independence of the pair $\{A, B\}$ with respect to the conditional probability measures $P(\cdot|C)$ and $P(\cdot|C^c)$. However, the following simple calculations show that the pair $\{A, B\}$ is not independent (with respect to the total probability measure P). We have

$$
\begin{aligned}
P(A) &= 0.60 \cdot 0.3 + 0.90 \cdot 0.7 = 0.81 \\
P(B) &= 0.75 \cdot 0.3 + 0.95 \cdot 0.7 = 0.89 \\
P(AB) &= P(A|C)P(B|C)P(C) + P(A|C^c)P(B|C^c)P(C^c) \\
&= 0.7335 \neq 0.7209 = P(A)P(B).
\end{aligned}
$$

Since the product rule does not hold, $\{A, B\}$ cannot be independent. The tie comes through the common condition. If the first unit fails, this indicates that the line voltage is probably high so that the likelihood of failure of the second unit is increased. If the first unit does not fail, this indicates the line voltage is probably normal, so that the likelihood of failure of the second unit is probably the lower value.

——— $\square$

Example 5.1.2:

As a second example, consider an electronic game. The system has two states. In the first state, the probability of success on any trial is 0.15; in the second state, the probability of success on any trial is 0.20. Let

  A be the event of success on first play.
  B be the event of success on second play.
  C be the event the system is in state 1.
  C^c be the event the system is in state 2.

The trials are "operationally independent" so that it seems reasonable to assume

$$P(A|BC) = P(A|B^cC) \quad \text{and} \quad P(A|BC^c) = P(A|B^cC^c).$$

This may be shown to be equivalent to

$$P(AB|C) = P(A|C)P(B|C)$$

and

$$P(AB|C^c) = P(A|C^c)P(B|C^c).$$

Now, assuming $P(C) = 0.3$,

$$P(B) = P(B|C)P(C) + P(B|C^c)P(C^c) = 0.15 \cdot 0.3 + 0.20 \cdot 0.7 = 0.1850$$

and

$$
\begin{aligned}
P(AB) &= P(AB|C)P(C) + P(AB|C^c)P(C^c) \\
&= P(A|C)P(B|C)P(C) + P(A|C^c)P(B|C^c)P(C^c) \\
&= 0.0348.
\end{aligned}
$$

By symmetry, $P(A) = P(B)$. We have

$$P(A|B) = \frac{P(AB)}{P(B)} = \frac{0.0348}{0.1850} = 0.1878 \neq 0.1850 = P(A).$$

This last condition shows that the pair $\{A, B\}$ is *not* independent, although we have independence with respect to the conditional probability measures $P(\cdot|C)$ and $P(\cdot|C^c)$.

—   □

Example 5.1.3:

Two students are working to complete a term paper on time. They are working independently, each unaware of the other's predicament. The library has at least one copy of a book they both need to complete the project. Let

A be the event the first student completes his work on time.
B be the event the second student completes his work on time.
C be the event the library has a second copy available.

If both are able to obtain copies of the book, then their work is effectively independent. If only one copy is available, then the student fortunate enough to obtain that copy has a better chance to complete his work on time than does the other. Thus, we may have

$$P(A|BC) = P(A|C) \qquad \text{but} \quad P(A|BC^c) < P(A|B^cC^c)$$

so that

$$P(A|BC^c) < P(A|C^c).$$

Occurrence of BC^c indicates that the second student probably got the single copy, to the detriment of the first student. Fundamental properties of conditional probability may be used to show that the preceding condition also requires $P(B|AC^c) < P(B|C^c)$. In this example, we have an asymmetry with respect to the conditioning events. The pair $\{A, B\}$ is independent with respect to the conditional probability measure $P(\cdot|C)$. But the pair is not independent with respect to the conditional probability measure $P(\cdot|C^c)$.

—   □

It is appropriate to ask how "natural" these examples are. Is "conditional independence" in the sense of independence with respect to a conditional probability measure an academic curiosity, or is it a condition likely to be encountered in practical modeling situations? The following illustrative examples indicate further the likely usefulness of the concept.

Example 5.1.4: Contractors and Weather (See Example 4.1.1)

Two contractors, Smith and Jones, are working on different construction jobs in the same city. Let

A be the event Smith completes his job on schedule.
B be the event Jones completes his job on time.

The two contractors may work in such a way that the performance of one has no effect on the performance of the other. We speak of operational independence. In this case, it would ordinarily seem reasonable to suppose $P(A|B) = P(A|B^c)$, so that the events $\{A, B\}$ form an independent pair, as in Example 4.1.1. However, if the jobs are outside, performance may be affected by the weather. Let C be the event the weather is good. It may be reasonable to suppose that

$$P(A|BC) = P(A|B^cC) \quad \text{and} \quad P(A|BC^c) = P(A|B^cC^c).$$

Although the contractors work in an operationally independent manner, there is a "stochastic tie" provided by the weather. There is conditional independence if the weather is good; similarly, there is conditional independence if the weather is bad. But there may not be total independence, since the two jobs are both affected by the weather. If B occurs (Jones completes on time), then the weather was probably good and the likelihood of A is higher than if B^c occurs.

—   □

Example 5.1.5: Independent Medical Tests

A doctor may order "independent" laboratory tests on a patient to determine the existence of some disease or disorder. The tests are operationally independent, in that the result of any one is not affected by, nor does it affect, the results of any other. Yet the tests, if meaningful, will not be independent, for all are affected by the presence or absence of the disease. Thus, the operational independence results in conditional independence, given the presence of the disease, or given its absence.

—   □

Example 5.1.6: Polls and Public Opinion

Two professional polling organizations may take simultaneous political polls to determine the popularity of a given candidate or the general views on a controversial question. The organizations will operate in an independent manner, in that the results of one poll will not directly affect the results of the other. The results of either one will not depend on whether the other is conducted. Clearly, if the polls are at all effective in determining public opinion, the results will not be independent. They will be independent, given the condition of the public sector polled.

—   □

Example 5.1.7: Survey Response and Categories

A well-designed survey consists of a group of "independent" questions. That is, for any individual the answers to the questions in the list are independent in the sense that the answer to any question does not depend on the answers to any other questions, the order in which they are asked, or whether or not the others are even asked. The respondents will belong to one of several categories or groups. The answers by an individual respondent to the various questions may be affected by his or her group membership, although the questions are for that individual quite "independent." Again, the proper modeling condition seems to be conditional independence of the answers, given the group membership of the respondent.

—   □

Example 5.1.8: Memoryless Situations

Although probability does not inherently exhibit time as a variable, in the study of many systems, events may be determined by the condition of the system at various times. Suppose a system is observed at a given instant, considered the "present." Let C be the event the system is in a given condition at the present. Suppose A is an event affected by the system in the future, and B is an event reflecting the condition of the system in the past. If the system is memoryless, in the sense that the future is influenced by the present condition, but not by the manner in which the present is reached, then the appropriate assumption may be $\{A, B\}$ is conditionally independent, given the event C. This condition is examined more carefully under the topic of the Markov property, in Chapter 23, in which the notion of conditional independence is given a broader formulation.

—   □

In Example 5.1.1, 5.1.2, and 5.1.4–5.1.8, the pairs of events considered are conditionally independent given either C or C^c. However, the condition illustrated in Example 5.1.3 is asymmetrical with regard to conditional independence. The following provide some further examples of this asymmetry of conditional independence.

Example 5.1.9: Contractors and Scarce Material

Consider again the situation of the contractors in Example 5.1.4. It may be that both jobs require a certain material. If the supply is good, then the jobs proceed without interaction. On the other hand, if the supply is short, there may well be interaction of the following sort. Let D be the event the common material is in good supply. It may be reasonable to suppose $\{A, B\}$ is conditionally independent, given that D occurs. However, if the material is in short supply (event D^c occurs), then we may well have $P(A|BD^c) < P(A|B^cD^c)$, since if the second contractor completes on time, it will likely be because he has obtained the scarce material to the detriment of the first contractor. This condition precludes the possibility that $\{A, B\}$ is conditionally independent, given D^c.

—   □

Example 5.1.10: Computer Jobs and File Space

Two personal computers share a common hard disk storage. Let

A be the event the first computer operates successfully on a job.
B be the event the second computer operates successfully on another job.
C be the event the disk has sufficient space for both jobs.

Unless there is some other unforeseen cause for interaction, it seems reasonable to suppose that $\{A, B\}$ is conditionally independent, given C. However, if there is not enough file space for both (C^c occurs), then the performance of one unit may affect that of the other. For example, if the first job is completed successfully, there would not be enough disk space for the second job. Thus, we might well have the condition $P(B|AC^c) < P(B|A^cC^c)$, with the consequent conditional nonindependence, given C^c.

— $\square$

Example 5.1.11: Medical Tests that may be Affected by Unknown Conditions

Although operationally independent medical tests, as described in Example 5.1.5, are usually modeled properly as conditionally independent tests, one must be aware of possible significant interactions. One test may affect the patient physically in some unsuspected manner that influences the results of the second test. Or the interaction could be the result of some subtle psychological response of the patient. This interaction may take place if a person has the disease, and not take place otherwise, leading to an asymmetry with regard to conditional independence.

— $\square$

Example 5.1.12: Course Performance and Joint Preparation

If students have prepared independently for an examination, then their performance on a given question may be quite independent; however, if they studied together, they are likely to give the same answers. Thus if we let

A be the event the first student answers question correctly.
B be the event the second student answers correctly.
C be the event they studied together.

then it is reasonable to assume $P(A|BC^c) = P(A|B^cC^c)$ but $P(A|BC) > P(A|B^cC)$. In this case, we have conditional independence of the pair $\{A, B\}$, given C^c, but not given C.

— $\square$

5.2. Equivalent Conditions and Definition

The examples of Section 5.1 illustrate a variety of situations in which independence of a pair of events is "conditioned" by an event that reflects
a common chance factor. In some of these cases, we find it appropriate to
assume independence with respect to a conditional probability measure. It
is natural to refer to such independence as "conditional independence." We
want to formalize this concept and to develop some general patterns that
facilitate modeling and analysis in the many practical situations for which
it is appropriate.

We begin by looking at the formal pattern representing independence
with respect to a conditional probability measure $P(\cdot|C) = P_C(\cdot)$. In the
hybrid notation we use for repeated conditioning in Section 3.3, we write

$$P_C(A|B) = P_C(A) \qquad \text{or} \qquad P_C(AB) = P_C(A)P_C(B).$$

This translates into

$$P(A|BC) = P(A|C) \qquad \text{or} \qquad P(AB|C) = P(A|C)P(B|C).$$

If it is known that C has occurred, then additional knowledge of the occurrence of B does not change the likelihood of A.

If we write the sixteen equivalent conditions for independence in terms
of the conditional probability measure $P_C(\cdot)$, then translate as herein, we
have the following equivalent conditions.

Sixteen equivalent conditions

$P(A|BC) = P(A|C)$	$P(AB|C) = P(A|C)P(B|C)$
$P(A|B^cC) = P(A|C)$	$P(AB^c|C) = P(A|C)P(B^c|C)$
$P(A^c|BC) = P(A^c|C)$	$P(A^cB|C) = P(A^c|C)P(B|C)$
$P(A^c|B^cC) = P(A^c|C)$	$P(A^cB^c|C) = P(A^c|C)P(B^c|C)$
$P(B|AC) = P(B|C)$	$P(A|BC) = P(A|B^cC)$
$P(B^c|AC) = P(B^c|C)$	$P(A^c|BC) = P(A^c|B^cC)$
$P(B|A^cC) = P(B|C)$	$P(B|AC) = P(B|A^cC)$
$P(B^c|A^cC) = P(B^c|C)$	$P(B^c|AC) = P(B^c|A^cC)$

The patterns of conditioning in the examples of the previous section
belong to this set. In a given problem, one or the other of these conditions may seem a reasonable assumption. As soon as *one* of these patterns
is recognized, then *all* are equally valid assumptions. Because of its simplicity and symmetry, we take as the defining condition the *product rule*
$P(AB|C) = P(A|C)P(B|C)$.

Definition 5.2.1. *A pair of events* $\{A, B\}$ *is said to be* conditionally in-
dependent, *given* C, *designated* $\{A, B\}$ ci| C, *iff the following product rule
holds*

$$P(AB|C) = P(A|C)P(B|C).$$

Because of the equivalence of the first four entries in the right-hand column of the table, we may assert

Theorem 5.2.2. *If any of the pairs $\{A, B\}$, $\{A, B^c\}$, $\{A^c, B\}$, or $\{A^c, B^c\}$ is conditionally independent, given C, then so are the others.*

—   □

Examples 5.1.1, 5.1.2, and 5.1.4–5.1.8 show that we may have $\{A, B\}$ ci| C and $\{A, B\}$ ci| C^c, and yet not have independence. Examples 5.1.3 and 5.1.9–5.1.12 show that we may have conditional independence, given C, but not conditional independence, given C^c. Such situations arise when some chance factor has an asymmetrical effect on the two events in question.

An examination of the relations involved show that only in very special cases is it true that a pair $\{A, B\}$ is conditionally independent with respect to both C and C^c and is independent. The next example displays such a case. One might suppose this would imply that $\{A, B, C\}$ must be independent. Calculations show that it is not so.

Example 5.2.1:

In a large group of students, there is the same number of males and females. Three fourths of the men and one fourth of the women like to play basketball. One half of each group likes to play computer games. A student is selected at random and questioned concerning his or her preferences. Let

   A be the event the student selected likes to play basketball.
   B be the event the student selected likes to play computer games.
   C be the event the student selected is male.

The data are assumed to mean that

$$P(A|C) = \frac{3}{4}, \qquad P(A|C^c) = \frac{1}{4},$$

$$P(B|C) = P(B|C^c) = \frac{1}{2}, \qquad P(C) = P(C^c) = \frac{1}{2}.$$

Since there seems to be no connection, among either men or women, between interest in playing basketball and playing computer games, it is reasonable to suppose that $\{A, B\}$ ci| C and $\{A, B\}$ ci| C^c.

Show that $\{A, B\}$ is independent, but $\{A, B, C\}$ is not.

Solution

By the law of total probability

$$P(A) = P(A|C)P(C) + P(A|C^c)P(C^c) = \left(\frac{3}{4} + \frac{1}{4}\right)\frac{1}{2} = \frac{1}{2}.$$

Similarly, we have $P(B) = 1/2$. Use of the law of total probability and the assumed conditional independence gives

$$
\begin{aligned}
P(AB) &= P(AB|C)P(C) + P(AB|C^c)P(C^c) \\
&= P(A|C)P(B|C)P(C) + P(A|C^c)P(B|C^c)P(C^c) \\
&= \frac{3}{4} \cdot \frac{1}{2} \cdot \frac{1}{2} + \frac{1}{4} \cdot \frac{1}{2} \cdot \frac{1}{2} = \frac{1}{4} = P(A)P(B).
\end{aligned}
$$

which establishes the independence of $\{A, B\}$. By the product rule on conditional probability and conditional independence, we have

$$
\begin{aligned}
P(ABC) &= P(AB|C)P(C) = P(A|C)P(B|C)P(C) = \frac{3}{4} \cdot \frac{1}{2} \cdot \frac{1}{2} = \frac{3}{16} \\
&\neq P(A)P(B)P(C) = \frac{1}{8}.
\end{aligned}
$$

Thus, the product rule fails for the class $\{A, B, C\}$.

— $\square$

Example 5.2.2:

An investor consults three advisers about a certain stock issue. He plans to follow the advice of the majority. Let A, B, C be the events that the first, second, and third adviser, respectively, recommends the stock. Let H be the event that the stock is a good buy and G the event the investor makes the right choice (i.e., buys if the stock is good, does not if it is a poor risk). Suppose he thinks $P(H) = 0.8$. The reliability of the advisers is expressed in the following probabilities:

$$P(A|H) = 0.9, \quad P(A|H^c) = 0.3, \quad P(B|H) = 0.85,$$

$$P(B|H^c) = 0.2, \quad P(C|H) = 0.8, \quad P(C|H^c) = 0.25.$$

Since the advisers work independently and are influenced only by the indicators they use to estimate the worth of the stock, it seems reasonable to suppose $\{A, B, C\}$ ci| H and $\{A, B, C\}$ ci| H^c. Under these assumptions, determine the probability $P(G)$ that the investor makes the right choice.

Solution

By the law of total probability

$$P(G) = P(G|H)P(H) + P(G|H^c)P(H^c).$$

The event the investor makes the right choice is the event that the majority of his advisers give him correct advice. Thus

$$P(G|H) = P(ABC|H) + P(ABC^c|H) + P(AB^cC|H) + P(A^cBC|H)$$

and

$$P(G|H^c) = P(A^c B^c C^c|H^c) + P(A^c B^c C|H^c)$$
$$+ P(A^c B C^c|H^c) + P(A B^c C^c|H^c).$$

Because of conditional independence

$$P(ABC|H) = P(A|H)P(B|H)P(C|H)$$
$$P(ABC|H^c) = P(A|H^c)P(B|H^c)P(C|H^c)$$

with similar expressions for the various patterns of complementation of ABC. We therefore have

$$P(G|H) = 0.9 \cdot 0.85 \cdot 0.8 + 0.9 \cdot 0.85 \cdot 0.2$$
$$+ 0.9 \cdot 0.15 \cdot 0.8 + 0.1 \cdot 0.85 \cdot 0.8 = 0.941$$
$$P(G|H^c) = 0.7 \cdot 0.8 \cdot 0.75 + 0.7 \cdot 0.8 \cdot 0.25$$
$$+ 0.7 \cdot 0.2 \cdot 0.75 \cdot 0.3 \cdot 0.8 \cdot 0.75 = 0.845.$$

The final result is

$$P(G) = 0.941 \cdot 0.8 + 0.845 \cdot 0.2 \approx 0.922.$$

—   □

EXTENSION TO LARGER CLASSES

Since conditional independence, given event C, is independence with respect to the conditional probability measure P_C, the extension to arbitrary classes of events follows the same pattern as for the extension of the concept of independence from a pair to an arbitrary class.

Definition 5.2.3. *A class $\{A_i : I \subset J\}$ of events is said to be conditionally independent, given C, iff for every finite subclass $\{A_{i_1}, A_{i_2}, \ldots, A_{i_m}\}$ of two or more members the following product rule holds.*

$$P\left(\bigcap_{k=1}^{m} A_{i_k}|C\right) = \prod_{k=1}^{m} P(A_{i_k}|C).$$

Remarks

1. Most of the theorems on independence carry over to conditional independence with appropriate adjustment of notation and terminology. This is true of Theorem 4.2.2 (the replacement theorem) and of Theorem 4.2.7 on independence of Boolean combinations.

2. It is not necessary to develop formally the concept of conditional independence. We could simply identify the proper pattern of conditioning in any specific problem and work out the implications of that pattern for the case at hand. However, one or another of the sixteen conditions is an appropriate assumption in so many situations that it is useful to formalize the notion and identify some of its properties. In so doing, we facilitate analysis of any problem in which the pattern is encountered.

5.3.   Some Problems in Probable Inference

In this section, we examine some common problems in probable inference for which conditional independence assumptions are quite natural and useful.

EVIDENCE CONCERNING A HYPOTHESIZED CONDITION

A fundamental problem of probable inference arises when there is an attempt to establish the presence or absence of a *hypothesized condition* by means of one or more appropriate *tests*. The test results may be less than completely reliable. In some instances, a test may consist of obtaining a report or affirmation from a source that is not completely reliable. The results of such tests do not directly establish the truth or falsity of the hypothesized condition; instead, they provide statistically related *evidence* concerning the hypothesized condition. Some typical situations are tabulated here.

Hypothesis	Evidence
Job success	Personal traits
Presence of oil	Geological structure
Win a game	Condition of players
Operation of a device	Physical conditions
Market condition	Test market condition
Presence of disease	Diagnostic test results

In the following analysis, we identify for ready reference a number of important cases commonly encountered in applications.

Case 1. Hypothesis : Evidence

We let H be the event the hypothesized condition exists and E be the event that the item of evidence in question is observed. Usually, we have data of the form $P(E|H)$, $P(E|H^c)$, with a prior estimate of the odds $P(H)/P(H^c)$. A straightforward use of Bayes' rule yields the posterior odds $P(H|E)/P(H^c|E)$. No conditional independence is involved.

Questions of conditional independence come into play when there are several items of evidence obtained in an operationally independent manner. Examples 5.1.5 and 5.1.6 illustrate such situations. In Example 5.1.5, the hypothesized condition is the presence of a disease. The evidence is provided by the results of "independent" diagnostic tests. In Example 5.1.6, the hypothesized condition is the degree of popularity of a candidate for political office. The evidence is provided by the result of "independent" opinion polls.

Case 2. Hypothesis : Independent items of evidence

Often there are two or more "operationally independent" items of evidence, with data concerning reliability of the evidence in the form $P(E_i|H)$ and $P(E_i|H^c)$, where E_i is the event the ith item of evidence is observed. Suppose there are two such items of evidence. Under the condition of operational independence, we find it reasonable to assume

$$P(E_1|E_2 H) = P(E_1|E_2^c H) \quad \text{and} \quad P(E_1|E_2 H^c) = P(E_1|E_2^c H^c).$$

This is equivalent to the pair of conditions

$$\{E_1, E_2\} \text{ ci}| H \qquad \text{and} \qquad \{E_1, E_2\} \text{ ci}| H^c.$$

Direct use of this conditional independence shows that

$$\frac{P(H|E_1 E_2)}{P(H^c|E_1 E_2)} = \frac{P(H E_1 E_2)}{P(H^c E_1 E_2)} = \frac{P(H)}{P(H^c)} \cdot \frac{P(E_1|H) P(E_2|H)}{P(E_1|H^c) P(E_2|H^c)}.$$

It should be apparent that this analysis may be extended to any finite number of operationally independent items of evidence.

Example 5.3.1: Medical Tests

As a result of a routine physical examination, a physician suspects his patient is suffering from a liver disorder. On the basis of the examination, the doctor thinks the odds are about two to one that the patient is so afflicted. He orders two tests. Both are positive (i.e., indicate the presence of the disease). The first test has probability 0.1 of a false positive and 0.05 of a false negative. The second test has probability 0.05 of a false positive and 0.08 of a false negative. If the tests are carried out independently, what are the posterior odds that the patient has the suspected liver condition?

Solution

We let H be the event the patient has the disease, A be the event the first test is positive, and B be the event the second test is positive. The prior odds are $P(H)/P(H^c) = 2$. The data on the tests are taken to mean

$$P(A|H^c) = 0.1, \qquad P(A^c|H) = 0.05,$$
$$P(B|H^c) = 0.05, \qquad P(B^c|H) = 0.08.$$

The operational independence of the tests leads to the assumption

$$\{A, B\} \text{ ci} | H \qquad \text{and} \qquad \{A, B\} \text{ ci} | H^c.$$

According to the pattern of solution, we have

$$\frac{P(H|AB)}{P(H^c|AB)} = \frac{P(H)}{P(H^c)} \cdot \frac{P(A|H)P(B|H)}{P(A|H^c)P(B|H^c)} = 2 \cdot \frac{0.95 \cdot 0.92}{0.10 \cdot 0.05} = \frac{1748}{5}$$
$$= 349.6.$$

The posterior probability is $P(H|AB) = 1 - 5/1753 \approx 1 - 0.0029$. Because of their individual reliability and the fact that conditional independence allows combining their evidential value in a simple way, the results are highly diagnostic. The prior odds of two to one are increased to posterior odds of nearly 350 to 1, which means that the posterior likelihood of not having the disease is about 3 in 1000.

——   □

EVIDENCE FOR A SYMPTOM

In many practical situations, not only is evidence less than completely reliable, but it does not point directly to the hypothesized condition. Rather, the evidence concerns some associated condition, which we call a *symptom*. There are thus two stages of uncertainty:

- The evidence provides partially reliable information about the symptom.

- The symptom provides partially reliable information about the hypothesized condition.

In such situations, illustrated here, conditional independence assumptions are often appropriate.

Hypothesis	Symptom	Evidence for the symptom
Job success	Personal traits	Diagnostic test results
Presence of oil	Geological structure	Geophysical survey results
Win a game	Condition of players	Insider's report
Operation of device	Physical conditions	Monitoring report
Market condition	Test market condition	Market survey report
Presence of disease	Physical symptoms	Report on test for symptom

We let

H be the event a certain *hypothesized condition* is present.
S be the event a *symptom* is present.
E be the event a test provides *evidence* the symptom is present.

We consider, first, the case of one item of evidence, with four subcases reflecting various ways the stochastic tie between H and S is described.

Case 3. Hypothesis : Symptom : Evidence about the symptom.

Usually we have data concerning the stochastic tie between evidence and symptom in terms of the conditional probabilities $P(E|S)$ and $P(E|S^c)$. If the test is for the presence of the symptom, the result is not directly affected by the presence or absence of the hypothesized condition. In terms of conditional probabilities, we take this to mean that $P(E|SH) = P(E|SH^c)$ and $P(E|S^cH) = P(E|S^cH^c)$. These assumptions are equivalent to the conditions

$$\{H, E\} \text{ ci} | \ S \qquad \text{and} \qquad \{H, E\} \text{ ci} | \ S^c.$$

The posterior odds favoring H may be written

$$\frac{P(H|E)}{P(H^c|E)} = \frac{P(HE)}{P(H^cE)} = \frac{P(HSE) + P(HS^cE)}{P(H^cSE) + P(H^cS^cE)}.$$

Now

$$P(HSE) = P(HS)P(E|HS) = P(HS)P(E|S)$$

(and similarly for the other cases), so that

$$\frac{P(H|E)}{P(H^c|E)} = \frac{P(HS)P(E|S) + P(HS^c)P(E|S^c)}{P(H^cS)P(E|S) + P(H^cS^c)P(E|S^c)}.$$

There are four subcases for relating H and S stochastically, deriving from

$$P(HS) = P(H)P(S|H) = P(S)P(H|S)$$

(and similarly for the other cases).

Case 3a.

Data are $P(S|H)$ and $P(S|H^c)$, along with an estimate of $P(H)$.

Example 5.3.2: Anticipating a Competitor's Actions

Let

H be the event a competitor will announce a new product (hypothesis).
S be the event he will undertake a new expensive advertising campaign (symptom).
E be the event of a significant increase in advertising budget (evidence).

The data are as follows:

$$P(H)/P(H^c) = 4, \quad P(S|H) = 0.95, \quad P(S|H^c) = 0.25,$$

$$P(E|S) = 0.99, \quad P(E|S^c) = 0.30.$$

Use of the data in the expression obtained gives

$$\frac{P(H|E)}{P(H^c|E)} = \frac{P(H)}{P(H^c)} \cdot \frac{P(S|H)P(E|S) + P(S^c|H)P(E|S^c)}{P(S|H^c)P(E|S) + P(S^c|H^c)P(E|S^c)}$$

$$= 4 \cdot \frac{0.95 \cdot 0.99 + 0.05 \cdot 0.30}{0.25 \cdot 0.99 + 0.75 \cdot 0.30} \approx 8.09.$$

While far from conclusive, the data improve the prior odds of four to one to the posterior odds of slightly more than eight to one.

—   □

Case 3b.

If we have $P(S|H)$, $P(S|H^c)$ and a prior estimate of $P(S)$, we may obtain the corresponding estimate of $P(H)$, and proceed as in Case 3a. Since $P(S) = P(S|H)P(H) + P(S|H^c)(1 - P(H))$, it follows that

$$P(H) = \frac{P(S) - P(S|H^c)}{P(S|H) - P(S|H^c)}.$$

Note that for consistency we must have one of the following conditions

$$P(S|H^c) < P(S) < P(S|H) \quad \text{or} \quad P(S|H^c) > P(S) > P(S|H).$$

The condition $P(S|H^c) = P(S) = P(S|H)$ means $\{H, S\}$ is independent, which makes the symptom useless.

Example 5.3.3:

A professional basketball team in the playoff games depends heavily on the scoring of its star forward. Team statistics indicate that in 80 percent of the games that the team wins, he scores 30 or more points, while in only 30 percent of the games lost does he score 30 or more points. The season records indicate he scores 30 or more points in 70 percent of the games played. An assistant coach notices that in 85 percent of his high-scoring games he has a good pregame warmup, but has a good pregame warmup in only 60 percent of the games in which he scores less than 30 points. He has a good pregame warmup, what is the probability the team will win?

Solution

Let H be the event the team wins the game, S be the event the star scores 30 or more points, and E be the event of a good pregame warmup. Winning or losing depends upon how the player performs during the game, not how he warms up. Thus, it seems reasonable to assume $\{H, E\}$ ci$|$ S and $\{H, E\}$ ci$|$ S^c. The data are interpreted to mean

$$P(S|H) = 0.80, \qquad P(S|H^c) = 0.30, \qquad P(S) = 0.70,$$

$$P(E|S) = 0.85, \qquad P(E|S^c) = 0.60.$$

Now

$$P(H) = \frac{P(S) - P(S|H^c)}{P(S|H) - P(S|H^c)} = \frac{0.70 - 0.30}{0.80 - 0.30} = 0.80.$$

As in Case 3a (see Example 5.3.2), we can now write

$$\begin{aligned}
\frac{P(H|E)}{P(H^c|E)} &= \frac{P(H)}{P(H^c)} \cdot \frac{P(S|H)P(E|S) + P(S^c|H)P(E|S^c)}{P(S|H^c)P(E|S) + P(S^c|H^c)P(E|S^c)} \\
&= \frac{0.8}{0.2} \cdot \frac{0.8 \cdot .85 + 0.2 \cdot 0.60}{0.3 \cdot 0.85 + 0.7 \cdot 0.6} = \frac{128}{27} \approx 4.74.
\end{aligned}$$

Although the posterior odds move in the right direction, the change from prior to posterior odds is so slight that the pregame warmup has little evidential value for predicting the outcome of the game.

—   □

Case 3c.

Data are $P(H|S)$ and $P(H|S^c)$, along with an estimate of $P(S)$.

Example 5.3.4: Meeting a Product Deadline

A firm is attempting to have a new product ready to announce on the first of July. A prototype is available six weeks in advance. A series of tests is run on the prototype. Let

H be the event the product is ready on time.
S be the event the prototype is satisfactory.
E be the event the test shows the prototype to be satisfactory.

Experience indicates that $P(S) = 0.9$ and $P(H|S) = 0.95$, while $P(H|S^c) = 0.30$. Since the test results depend only on the condition of the prototype and not on whether the product will be ready on time, we suppose $\{E, H\}$ ci$|$ S and $\{E, H\}$ ci$|$ S^c. The reliability of the testing is such that $P(E|S) = 0.9$ and $P(E|S^c) = 0.05$. The tests indicate the prototype is satisfactory. Determine the posterior odds favoring the event of meeting the deadline for the product.

Solution

According to the patterns previously developed,

$$\frac{P(H|E)}{P(H^c|E)} = \frac{P(S)P(H|S)P(E|S) + P(S^c)P(H|S^c)P(E|S^c)}{P(S)P(H^c|S)P(E|S) + P(S^c)P(H^c|S^c)P(E|S^c)}$$

$$= \frac{0.9 \cdot 0.95 \cdot 0.90 + 0.1 \cdot 0.30 \cdot 0.05}{0.9 \cdot 0.05 \cdot 0.90 + 0.1 \cdot 0.70 \cdot 0.05} = \frac{771}{44} \approx 17.5.$$

The data imply the prior probability

$$P(H) = P(H|S)P(S) + P(H|S^c)P(S^c) = 0.95 \cdot 0.9 + 0.30 \cdot 0.1$$
$$= 0.885$$

so that

$$\frac{P(H)}{P(H^c)} \approx 7.7.$$

The modest increase of posterior odds over prior odds reflects, in part, the fact that meeting the deadline is not too tightly coupled to having a satisfactory prototype. The data indicate a 30 percent probability of meeting deadline even if the prototype is unsatisfactory.

—   □

Case 3d.

If we have $P(H|S)$, $P(H|S^c)$ and a prior estimate of $P(H)$, we may obtain the corresponding estimate of $P(S)$ and proceed as in Case 3c.

$$P(S) = \frac{P(H) - P(H|S^c)}{P(H|S) - P(H|S^c)}.$$

For consistency we must have one of the following conditions

$$P(H|S^c) < P(H) < P(H|S) \quad \text{or} \quad P(H|S^c) > P(H) > P(H|S).$$

Example 5.3.5:

Let H be the event the annual profit of a company for the current year exceeds that of the previous year by at least ten percent. This will be influenced strongly by the success or failure of a new product to be introduced. A market analysis of the new product is made. Let S be the event the product is successful and E be the event the report of the analyst is favorable. Management assumes the following:

$$P(H) = 0.65 \qquad P(H|S) = 0.90 \qquad P(H|S^c) = 0.15.$$

The reliability of the analyst is expressed in terms of the conditional probabilities

$$P(E|S) = 0.95 \qquad P(E|S^c) = 0.10.$$

Determine the posterior odds favoring increased profitability if the report of the analyst is favorable.

Solution

On the one hand the analyst studies the product and cannot know directly the profitability of the company; on the other hand, the profitability of the company depends upon the status of the product, regardless of the analysis. We therefore assume $\{H, E\}$ ci$|$ S and $\{H, E\}$ ci$|$ S^c. According to the general analysis for Case 3, we need to express $P(HS)$ as $P(S)P(H|S)$ and similarly for $P(H^cS)$. Now

$$P(S) = \frac{P(H) - P(H|S^c)}{P(H|S) - P(H|S^c)} = \frac{0.65 - 0.15}{0.90 - 0.15} = \frac{2}{3}.$$

We use the same pattern as in Case 3c (see Example 5.3.4) to obtain

$$\frac{P(H|E)}{P(H^c|E)} = \frac{P(S)P(H|S)P(E|S) + P(S^c)P(H|S^c)P(E|S^c)}{P(S)P(H^c|S)P(E|S) + P(S^c)P(H^c|S^c)P(E|S^c)}$$

$$= \frac{2 \cdot 0.90 \cdot 0.95 + 1 \cdot 0.15 \cdot 0.10}{2 \cdot 0.10 \cdot 0.95 + 1 \cdot 0.85 \cdot 0.10} = \frac{69}{11} \approx 6.27.$$

— $\square$

Remark: Our discussion of conditional probability shows that one must be careful with respect to conditioning events. In a similar vein, the question of independence is seen to be a subtle one. Because of the simplifications resulting from an assumption of independence or conditional independence, it is easy to make an unwarranted assumption. In Example 5.3.5, we find an assumption of conditional independence quite appropriate. The problem in the following example has a very similar formal structure, yet it does not admit of such an assumption.

Example 5.3.6: Conditional Nonindependence

Let H be the event that a business will be sold to a prospective buyer. The buyer is interested in the prospects of the company for profitability. This will be influenced strongly by the success or failure of a new product to be introduced. A market analysis of the new product is made. Let S be the event the product is successful and E be the event the report of the analyst is favorable.

Discussion

In Example 5.3.5, we assume that $\{E, H\}$ is conditionally independent, given S and given S^c. Is that a good assumption in this problem? Almost surely the buyer is going to want to know the outcome of the product analysis. If the company does not tell him, he will make his own analysis. Thus, it is not true that $P(H|SE) = P(H|SE^c)$, so that we cannot assert $\{E, H\}$ ci$|$ S.

— $\square$

INDEPENDENT TESTS FOR A SYMPTOM

A further stage of complexity occurs when there are independent tests for the symptom. Again we have four subcases, as in Case 3.

Case 4. Hypothesis : Symptom : Independent evidence about the symptom.

Suppose, for example, we have two "operationally independent" items of evidence. Since the results of observations depend upon the symptom and not directly on the hypothesized condition it seems reasonable to suppose

$$
\begin{aligned}
P(E_1|E_2 S) = P(E_1|E_2^c S) && P(E_1|E_2 S^c) = P(E_1|E_2^c S^c) \\
P(E_1|HS) = P(E_1|H^c S) && P(E_1|HS^c) = P(E_1|H^c S^c) \\
P(E_2|HS) = P(E_2|H^c S) && P(E_2|HS^c) = P(E_2|H^c S^c) \\
P(E_1 E_2|HS) = P(E_1 E_2|H^c S) && P(E_1 E_2|HS^c) = P(E_1 E_2|H^c S^c)
\end{aligned}
$$

These conditions imply

$$\{E_1, E_2, H\} \text{ ci} | \, S \qquad \text{and} \qquad \{E_1, E_2, H\} \text{ ci} | \, S^c.$$

Now, elementary operations show that

$$\frac{P(H|E_1 E_2)}{P(H^c|E_1 E_2)} = \frac{P(HE_1 E_2)}{P(H^c E_1 E_2)} = \frac{P(HSE_1 E_2) + P(HS^c E_1 E_2)}{P(H^c S E_1 E_2) + P(H^c S^c E_1 E_2)}.$$

With the aid of the conditional independence, we may expand

$$P(HSE_1 E_2) = P(HS)P(E_1|HS)P(E_2|HSE_1) = P(HS)P(E_1|S)P(E_2|S)$$

with similar expansions for the other terms, to give the posterior odds

$$\frac{P(H|E_1 E_2)}{P(H^c|E_1 E_2)} = \frac{P(HS)P(E_1|S)P(E_2|S) + P(HS^c)P(E_1|S^c)P(E_2|S^c)}{P(H^c S)P(E_1|S)P(E_2|S) + P(H^c S^c)P(E_1|S^c)P(E_2|S^c)}.$$

As in Case 3, we have the four subcases for handling the stochastic tie between H and S. The following example is illustrative of Case 4a.

Example 5.3.7: A Market Survey Problem

The Sparkle Bottling Company is planning to market nationally a new diet soft drink. Its executives feel confident that the odds are at least 3 to 1 the product would be successful. Before launching the new product, the company decides to investigate a test market. Previous experience indicates that the reliability of the test market is such that if the national market is favorable, there is a probability of 0.9 that the test market is also. On the other hand, if the national market is unfavorable, there is a probability of only 0.2 that the test market will be favorable. These facts lead to the following analysis. Let

H be the event the national market is favorable (hypothesis).

S be the event the test market is favorable (symptom).

The initial data are the following probabilities, based on past experience:

(a) Prior odds: $P(H)/P(H^c) = 3$

(b) Reliability of the test market: $P(S|H) = 0.9$, $P(S|H^c) = 0.2$.

If it were known that the test market is favorable, we should have

$$\frac{P(H|S)}{P(H^c|S)} = \frac{P(S|H)P(H)}{P(S|H^c)P(H^c)} = \frac{0.9}{0.2} \cdot 3 = 13.5.$$

Unfortunately, it is not feasible to know with certainty the state of the test market. The company decision makers engage two market survey companies to make *independent surveys* of the test market. The reliability of the companies may be expressed as follows. Let

A be the event the first company reports a favorable test market (first item of evidence).

B be the event the second company reports a favorable test market (second item of evidence).

On the basis of previous experience, the reliability of the evidence about the test market (the symptom) is expressed in the following conditional probabilities.

$$P(A|S) = 0.9 \quad P(A|S^c) = 0.3 \quad P(B|S) = 0.8 \quad P(B|S^c) = 0.2.$$

Both survey companies report that the test market is favorable. What is the probability the national market is favorable, given this result?

Solution

In order to use this data, we need to examine the independence conditions. The two survey firms work in an "operationally independent" manner. The report of either company is unaffected by the work of the other. Also, each report is affected only by the condition of the test market—regardless of what the national market may be. These facts make the following assumptions reasonable:

$$
\begin{aligned}
P(A|BS) &= P(A|B^cS) & P(A|BS^c) &= P(A|B^cS^c) \\
P(A|HS) &= P(A|H^cS) & P(A|HS^c) &= P(A|H^cS^c) \\
P(B|HS) &= P(B|H^cS) & P(B|HS^c) &= P(B|H^cS^c) \\
P(AB|HS) &= P(AB|H^cS) & P(AB|HS^c) &= P(AB|H^cS^c)
\end{aligned}
$$

This is evidently the pattern listed earlier, with $A \sim E_1$, and $B \sim E_2$. We can therefore assert

$$\{A, B, H\} \text{ ci} | \ S \quad \text{and} \quad \{A, B, H\} \text{ ci} | \ S^c.$$

We may use the pattern for the solution developed earlier to obtain

$$\frac{P(H|AB)}{P(H^c|AB)} = \frac{P(H)}{P(H^c)} \cdot \frac{\left[\begin{array}{l} P(S|H)P(A|S)P(B|S) \\ + P(S^c|H)P(A|S^c)P(B|S^c) \end{array}\right]}{\left[\begin{array}{l} P(S|H^c)P(A|S)P(B|S) \\ + P(S^c|H^c)P(A|S^c)P(B|S^c) \end{array}\right]}$$

$$= 3 \cdot \frac{0.9 \cdot 0.9 \cdot 0.8 + 0.1 \cdot 0.3 \cdot 0.2}{0.2 \cdot 0.9 \cdot 0.8 + 0.8 \cdot 0.3 \cdot 0.2} = \frac{327}{32} \approx 10.22.$$

In terms of the posterior probability, we have

$$P(H|AB) = \frac{327/32}{1 + 327/32} = \frac{327}{359} = 1 - \frac{32}{359} \approx 0.91.$$

We note that the odds favoring H, given positive indications from both survey companies, is 10.2 as compared with the odds favoring H, given a favorable test market, of 13.5. The difference reflects the residual uncertainty about the test market after the market surveys. Nevertheless, the results of the market surveys increase the odds favoring a satisfactory market from the prior 3 to 1 to a posterior 10.2 to 1. In terms of probabilities, the market surveys increase the likelihood of a favorable market from the original $P(H) = 0.75$ to the posterior $P(H|AB) = 0.91$. The conditional independence of the results of the survey make possible direct use of those results.

— $\square$

The next example illustrates Case 4d, in which the joint data on $\{H, S\}$ are given in terms of $P(H)/P(H^c)$, $P(H|S)$, and $P(H|S^c)$.

Example 5.3.8: Weather Forecasts and Successful Outdoor Activities

A community civic club is planning a flea market and auction on a weekend in a shopping mall parking lot. Its goal is to raise $5,000 for childrens' playground equipment in a public park. It is assumed that the odds favoring a successful sale are about three to one. The big source of uncertainty about success is the weekend weather. Let

H be the event of a successful sale.
S be the event of good weather.
E_1 be the event of a favorable forecast from the national weather service.
E_2 be the event of a favorable forecast from a private meteorological service.

The planners assume $P(H|S) = 0.9$ and $P(H|S^c) = 0.2$. Experience with the reliability of the forecasters indicates

$$P(E_1|S) = 0.9 \quad P(E_1|S^c) = 0.2 \quad P(E_2|S) = 0.8 \quad P(E_2|S^c) = 0.1.$$

The forecasters operate independently and their forecasts depend only upon factors influencing the weather. Neither forecast is affected by the other or by whether or not the sale is successful. The analysis of Case 4 shows that it is reasonable to assume

$$\{E_1, E_2, H\} \text{ ci}|\ S \quad \text{and} \quad \{E_1, E_2, H\} \text{ ci}|\ S^c.$$

Suppose both forecasters predict good weather. What are the posterior odds favoring a successful sale?

Solution

We wish to determine

$$\frac{P(H|E_1 E_2)}{P(H^c|E_1 E_2)} = \frac{\begin{matrix}P(S)P(H|S)P(E_1|S)P(E_2|S) \\ + P(S^c)P(H|S^c)P(E_1|S^c)P(E_2|S^c)\end{matrix}}{\begin{matrix}P(S)P(H^c|S)P(E_1|S)P(E_2|S) \\ + P(S^c)P(H^c|S^c)P(E_1|S^c)P(E_2|S^c)\end{matrix}}.$$

Now

$$P(S) = \frac{P(H) - P(H|S^c)}{P(H|S) - P(H|S^c)} = \frac{0.75 - 0.20}{0.90 - 0.20} = \frac{11}{14} \quad \text{and} \quad P(S^c) = \frac{3}{14}.$$

Substitution of the various values gives

$$\frac{P(H|E_1 E_2)}{P(H^c|E_1 E_2)} = \frac{11 \cdot 9 \cdot 9 \cdot 8 + 3 \cdot 2 \cdot 2 \cdot 1}{11 \cdot 1 \cdot 9 \cdot 8 + 3 \cdot 8 \cdot 2 \cdot 1} = \frac{714}{84} = 8.5.$$

This means the posterior probability $P(H|E_1 E_2) = 17/19$ as compared with the prior probability $P(H) = 3/4$

— □

5.4. Classification Problems

A population consists of members of two subgroups. It is desired to formulate a battery of questions to aid in identifying the subclass membership of randomly selected individuals in the population. The questions are designed so that for each individual the answers are independent, in the sense that the answers to any subset of these questions is not affected by and does not affect the answers to any other subset of the questions. Consider the following

Example 5.4.1: Classification on the Basis of a Profile

A population consists of two mutually exclusive subgroups. A battery of six "independent" questions is devised. The answer to each question is either yes, no, or uncertain. The set of answers by an individual constitutes a *profile*. On the basis of this profile, the individual is classified into one of the two groups. In order to provide a statistical basis, one hundred members of the population, of known membership in the two groups, are interviewed to provide basic data as follows.

	Group 1 (57 members)			Group 2 (43 members)		
	Yes	No	Unc		Yes	No	Unc
$i = 1$	18	27	12	$i = 1$	29	10	4
2	8	31	18	2	26	15	2
3	8	41	8	3	29	11	3
4	14	32	11	4	24	17	2
5	15	27	15	5	12	18	13
6	9	34	14	6	30	6	7

If a member is selected at random from the entire population, let

A_i be the event answer to question i is "yes".
B_i be the event answer to question i is "no".
C_i be the event answer to question i is "uncertain".
G_k be the event member is from group k, $k = 1$ or 2.

The data are taken to mean

$$\frac{P(G_1)}{P(G_2)} = \frac{57}{43}, \quad P(A_1|G_1) = \frac{18}{57}, \quad P(B_3|G_2) = \frac{11}{43}, \quad \text{etc.}$$

A subject is selected at random. His answers to the six questions provide a profile, as follows:

Yes, Yes, No, Uncertain, No, Yes   (in that order)

This corresponds to the event $E = A_1 A_2 B_3 C_4 B_5 A_6$. To make the classification, we calculate $P(G_1|E)/P(G_2|E)$. If this quantity is greater than one, we classify the subject in group 1; if the quantity is less than one, we assume the subject is from group 2. For this case, by the ratio form of Bayes' theorem, we have

$$\frac{P(G_1|E)}{P(G_2|E)} = \frac{P(G_1)}{P(G_2)} \cdot \frac{P(E|G_1)}{P(E|G_2)}.$$

Now

$$P(E|G_1) = P(A_1 A_2 B_3 C_4 B_5 A_1|G_1)$$

with a similar expression for $P(E|G_2)$. At this point we use the conditional independence ideas. Because of the operational independence of the questions for each individual and the data for members of each group, we assume that

$$\{A_1, A_2, B_3, C_4, B_5, A_6\} \text{ ci}| \ G_1 \quad \text{and} \quad \text{ci}| \ G_2.$$

Thus

$$P(E|G_1) \;=\; P(A_1|G_1)P(A_2|G_1)$$
$$\times P(B_3|G_1)P(C_4|G_1)P(B_5|G_1)P(A_6|G_1)$$

with a similar expression for $P(E|G_2)$. Using the numbers in the table, we obtain

$$\frac{P(G_1|E)}{P(G_2|E)} = \frac{57}{43} \cdot \frac{18}{29} \cdot \frac{8}{26} \cdot \frac{41}{11} \cdot \frac{11}{2} \cdot \frac{27}{18} \cdot \frac{9}{30} \cdot \frac{43^6}{57^6} \approx 0.43.$$

We classify the subject in group 2. The subject's answers in this case are mixed. The answer to question 3 points strongly to group 1; the answer to question 5 points mildly to group 1. The answer to question 6 points strongly to group 2; the answers to questions 1 and 2 point somewhat less strongly to group 2. In the aggregate, group 2 is the more likely.

——   □

Remarks

1. Without the conditional independence, the estimation of the various $P(E|G_1)$ and $P(E|G_2)$ would require immense amounts of statistical data. In this case, which is typical of a well-designed battery of questions, the conditional independence assumption is both appropriate and extremely useful.

2. It is possible to use logarithms of the ratios, appropriate random variables, and the central limit theorem to analyze this approach for probabilities of erroneous classification. The point of interest in the preceding discussion and example is the role of the conditional independence assumption in making the problem tractable.

PROBLEMS

5-1. Show that if $P(A|BC) = P(A|BC^c)$, then the common value is $P(A|B)$.

5-2. Prove the equivalence of at least four of the sixteen conditions for $\{A, B\}$ ci| C.

5-3. Suppose $\{E_1, E_2, E_3\}$ ci| A and ci| A^c. Experience indicates that $P(A) = 0.3$, with

$$P(E_i|A) = 0.9,\ 0.8,\ 0.85 \quad \text{and} \quad P(E_i|A^c) = 0.15,\ 0.1,\ 0.2$$

for $i = 1, 2, 3$, respectively. Let $C = E_1 E_2^c E_3$. Determine the posterior odds $P(A|C)/P(A^c|C)$.

5-4. Suppose $\{A, B\}$ ci| C and $\{A, B\}$ ci| C^c, with

$$P(A|C) = 0.4, \quad P(B|C) = 0.6, \quad P(A|C^c) = 0.2,$$

$$P(B|C^c) = 0.1, \quad \text{and} \quad P(C) = 0.3.$$

Show whether or not $\{A, B\}$ is independent by showing whether or not the product rule holds.

5-5. A group of 100 students consists of 50 males and 50 females. Three fourths of the males and one fourth of the females play intramural sports. One half of each group has taken at least one computer course. A student is chosen at random from the group. Let A be the event the student plays intramural sports, B be the event the student has taken a computer course, and C be the event the student is male. Suppose $\{A, B\}$ ci| C and $\{A, B\}$ ci| C^c. Show that $\{A, B\}$ is independent, $\{B, C\}$ is independent, but $\{A, B, C\}$ is *not* independent.

5-6. Fifteen percent of the patients seen in a clinic have a certain disorder. They are routinely given two tests, which are operationally independent. That is, neither is affected by whether or not the other is given, nor by the result of the other if it is given. Let A and B be the respective events that the two tests indicate the presence of the disorder, and let D be the event the patient has the disease. The reliability of the tests is expressed in terms of the following probabilities:

$$P(A|D) = 0.90 \quad P(A|D^c) = 0.05$$

$$P(B|D) = 0.95 \quad P(B|D^c) = 0.15$$

A patient is examined. The results of both tests are positive (i.e., indicate the presence of the disorder). What is the probability the patient is afflicted, given these results?

5–7. The Excello Co. is planning to market a new product. Top management thinks the odds favoring successful sales is four to one. Two market analysis firms are consulted for their evaluations. Both predict a favorable market. Let

S be the event of successful sales.
A be the event the first consultant predicts favorable sales.
B be the event the second consultant gives a favorable report

Previous performance of these companies indicate

$$P(A|S) = 0.9 \qquad P(A|S^c) = 0.15$$

$$P(B|S) = 0.8 \qquad P(B|S^c) = 0.05.$$

The companies operate in an independent manner. Thus, it seems reasonable to suppose

$$P(A|BS) = P(A|B^cS) \qquad \text{and} \qquad P(A|BS^c) = P(A|B^cS^c)$$

so that $\{A, B\}$ ci| S and $\{A, B\}$ ci| S^c. Determine the posterior odds $P(S|AB)/P(S^c|AB)$

5–8. The campaign managers for a political candidate feel that the odds favoring their candidate's election are two to one. They hire two opinion poll firms to sample the electorate. The pollsters operate in an independent manner, with the work of neither affecting the other. Let

E be the event the candidate will win.
A be the event the first poll shows the candidate ahead.
B be the event the second poll shows the candidate ahead.

The previous record of the pollsters indicates

$$P(A|E) = 0.85, \qquad P(A|E^c) = 0.20,$$

$$P(B|E) = 0.75, \qquad P(B|E^c) = 0.10.$$

Because of the operational independence, we may suppose that we have $\{A, B\}$ ci| E and $\{A, B\}$ ci| E^c. Both polls show the candidate is ahead. Determine the posterior odds favoring the candidate, given this result.

5–9. The EZ-Comp software company is planning to introduce a new line of educational/recreational software aimed at the national market of home users. Its business analysts see the odds favoring a successful market as 4 to 1. However, because of the expense of a national advertising campaign, inducements to dealers, etc., company executives decide to survey a test market. They analyze the situation as follows. Let

M be the event the national market is favorable.

T be the event the test market is favorable.

S be the event the survey indicates the test market is favorable.

The initial assumption is that $P(M)/P(M^c) = 4$. Past experience provides the following data.

Reliability of the test market: $P(T|M) = 0.90$ $P(T|M^c) = 0.20$

Reliability of the survey: $P(S|T) = 0.85$ $P(S|T^c) = 0.10$.

If the result of the survey is affected only by the test market, regardless of the state of the national market, we may suppose that $\{M, S\}$ ci$|$ T and $\{M, S\}$ ci$|$ T^c. Suppose the survey indicates the test market is favorable. Determine the posterior odds $P(M|S)/P(M^c|S)$.

5–10. An oil company is deciding whether to drill a test well in a certain location. On the basis of general experience in the area, the decision makers feel that the odds favoring success are about five to one. The situation is most favorable if a certain geological formation is present. A geophysical survey is initiated to determine whether or not the formation exists. On the basis of past experience, various probabilities may be assigned as follows. Let

H be the event that a well at the proposed site would be successful.

C be the event that the geological formation is present.

T be the event the geological survey indicates the desired formation.

The initial odds are $P(H)/P(H^c) = 5$. Previous field data indicate

$$P(C|H) = 0.9 \qquad P(C|H^c) = 0.25$$

$$P(T|C) = 0.95 \qquad P(T|C^c) = 0.15.$$

The result of the geological survey depends upon the geological structure and is not affected by the presence or absence of oil. Thus, it is reasonable to assume that

$$P(T|CH) = P(T|CH^c) \qquad \text{and} \qquad P(T|C^cH) = P(T|C^cH^c).$$

Reference to the equivalent conditions shows that this implies

$$\{H, T\} \text{ ci} | \ C \qquad \text{and} \qquad \{H, T\} \text{ ci} | \ C^c.$$

The geological survey indicates the presence of the desired structure. Determine the posterior odds favoring a successful well.

5–11. Jim and some friends are planning a deer hunting trip in a favorite locale. In order to assess the likelihood of finding game, Jim phones the local game warden about the presence of deer. He also calls the operator of a camp where they plan to stay. Past experience indicates the game warden has a five percent probability of a false positive prediction and a ten percent probability of a false negative. The camp owner has a fifteen percent chance of a false positive and a five percent chance of a false negative. Previous experience indicates a seventy percent chance of good hunting in the area. Let

H be the event of good hunting conditions.
W be the event the warden reports good hunting conditions.
C be the event the camp owner reports good conditions.

The warden and camp owner operate independently in assessing conditions to arrive at their respective reports. Thus, it seems reasonable to suppose

$$P(C|HW) = P(C|HW^c) \quad \text{and} \quad P(C|H^cW) = P(C|H^cW^c).$$

The warden indicates poor hunting conditions and the camp owner reports good hunting conditions. On the basis of these reports, what is the likelihood of good hunting?

5–12. A company plans to hire students for a job that requires both logical skills and the ability to interact well with people. Let H be the event a student hired will be successful, S be the event the student has the required skills and abilities, and E be the event the student has good grades in mathematically oriented subjects and has held positions of leadership in campus organizations. On the basis of experience, the company believes that a student hired from a certain university has a probability 0.50 of succeeding. Furthermore, experience indicates

$$P(S|H) = 0.95 \qquad P(S|H^c) = 0.20$$

$$P(E|S) = 0.85 \qquad P(E|S^c) = 0.20.$$

Since the students' grades and extracurricular activities have nothing directly to do with success in the job, it seems reasonable to suppose that $\{H, E\}$ ci$|$ S and $\{H, E\}$ ci$|$ S^c.

A student with good grades in mathematical subjects and an impressive record of extracurricular activities is offered a job. On this basis, what are the odds favoring success as an employee?

5–13. Show the conditions (a) $P(A|BD) = P(A|B^cD)$, (b) $P(A|HD) = P(A|H^cD)$, (c) $P(B|HD) = P(B|H^cD)$, and (d) $P(AB|HD) = P(AB|H^cD)$ together imply that $\{A, B, H\}$ ci$|$ D.

5–14. A student decides to determine the odds on the forthcoming basketball game with State University. The odds depend on whether State's
star center will be able to play. He contacts a couple of sources. Each
report depends upon the facts related to the condition of the player
and not on the outcome of the game (which, of course, is not known).
The two contacts arrive at their opinions quite independently. The
student proceeds as follows. He lets

> W be the event his team wins.
> C be the event the star center will play for State.
> A be the event the first source says he will play.
> B be the event the second contact says he will play.

The student, having been introduced to conditional independence
in his probability course, decides to assume $\{A, B, W\}$ ci| C and
$\{A, B, W\}$ ci| C^c. On the basis of past experience, he assesses the
reliability of his advisers and assumes the following probabilities:

$$P(A|C) = P(A^c|C^c) = 0.8, \quad P(B|C) = 0.6, \text{ and } P(B^c|C^c) = 0.7.$$

Initially, he could only assume $P(C) = 0.5$. Expert opinion assigns
the odds:

$$\frac{P(W|C)}{P(W^c|C)} = \frac{1}{3} \quad \text{and} \quad \frac{P(W|C^c)}{P(W^c|C^c)} = \frac{3}{2}.$$

On these assumptions, determine

$$\frac{P(W|AB^c)}{P(W^c|AB^c)} \quad \text{and} \quad P(W|AB^c).$$

5–15. A computer has a component that has experienced intermittent failure. A critical project is to be run during a certain eight-hour shift.
Let

> R be the event the component operates successfully throughout
> the period.
> H be the event the project is completed during the period.

Success or failure of the component does not completely determine
success or failure of the project. Experience indicates

$$P(H|R) = 0.98 \frac{P(H|R^c)}{P(H^c|R^c)} = 0.3 \quad \text{and} \quad P(R) = 0.75.$$

Two independent diagnostic tests are run to determine the condition
of the component. The tests are not completely reliable. For a given
condition of the component, neither test affects the other and the

tests are not affected by the performance of the project. Thus, if A is the event the first test is positive and B is the event the second test is positive, we assume $\{A, B, H\}$ ci$|$ R and ci$|$ R^c. Also, experience indicates

$$P(A|R) = 0.95 \qquad P(A^c|R^c) = 0.90$$

$$P(B|R) = 0.90 \quad \text{and} \quad P(B^c|R^c) = 0.85.$$

Determine $P(H|AB)$.

5-16. An investor is debating whether to buy stock that he thinks has a good chance of increasing in value in the near future. The odds favoring this depend upon whether or not a new product under development will be marketed successfully. He consults two analysts familiar with the company. Each reports on the prospects for successful marketing of the new product on the basis of facts within the company. The analysts reach their conclusions quite independently. The investor proceeds as follows. He lets

> H be the event the stock will escalate in value.
> M be the event the new product will be marketed successfully.
> A be the event the first analyst reports favorably.
> B be the event the second analyst reports favorably.

Initially, the investor assumes $P(M)/P(M^c) = 2$, and on the basis of expert opinion assumes the follow posterior odds on success, given market performance,

$$\frac{P(H|M)}{P(H^c|M)} = 5 \quad \text{and} \quad \frac{P(H|M^c)}{P(H^c|M^c)} = 1.$$

It seems reasonable to suppose $\{A, B, H\}$ ci$|$ M and $\{A, B, H\}$ ci$|$ M^c. Past experience leads to the following assessment of the reliability of the analysts:

$$P(A|M) = 0.8, \qquad P(B|M) = 0.85,$$

$$P(A^c|M^c) = 0.75, \quad \text{and} \quad P(B^c|M^c) = 0.8.$$

Using these assumptions, determine $P(H|AB)/P(H^c|AB)$.

5-17. A student is planning to buy a new word processor for her personal computer. The program Mighty Fine has the desired features, including speed of use. Although the company has a good reputation, software does have bugs, even after extensive testing. She analyzes the probability the program will perform satisfactorily for her intended use, as follows. Let

H be the event the program performs satisfactorily.

S be the event the program is free of significant "bugs."

E_1 be the event the program operates satisfactorily on one test document.

E_2 be the event the program operates satisfactorily on a second test document.

The best information available indicates the following probabilities are in effect:

$$P(H|S) = 0.95 \qquad P(H|S^c) = 0.40 \qquad P(E_1|S) = 0.90$$

$$P(E_1|S^c) = 0.20 \qquad P(E_2|S) = 0.95 \qquad P(E_2|S^c) = 0.30.$$

Experience with the software produced by the company suggests $P(S) = 0.8$. Since the tests are operationally independent of each other and of the event H, it seems reasonable to suppose that we have $\{E_1, E_2, H\}$ ci| S and $\{E_1, E_2, H\}$ ci| S^c. Determine the posterior odds favoring H, given that both tests are satisfactory.

5–18. Suppose whether or not it rains on any day is influenced primarily by whether or not it rained on the previous day, but earlier weather has negligible influence. If R_n is the event of rain on the nth day, we suppose $P(R_{n+1}|R_n) = P(R_{n+1}^c|R_n^c) = p$, regardless of previous weather patterns. Let $P_1 = P(R_1)$ be the probability of rain on the first day of the year. Show that

$$P_n = P(R_n) = \frac{1}{2}\big(1 + (2P_1 - 1)(2p - 1)^{n-1}\big).$$

Remark: This is a special case of a *Markov sequence* (see Chapter 23), in which the past and future of the system are conditionally independent, given the present.

5–19. Persons selected at random from two groups are interviewed. Three questions are asked, each having the possible answers: yes, no, and uncertain (designated 1, 2, and 3, respectively). Let

G_i be the event the person is from group $i(i = 1, 2)$.
A_j be the event the answer to first question is j.
B_j be the event the answer to second question is j.
C_j be the event the answer to third question is j.

Assume $P(G_1) = 0.4$ and $P(G_2) = 0.6$. If for each individual, no matter which group, the questions are independent, we may assume

$\{A_i, B_j, C_k\}$ ci| G_1 and ci| G_2 for each triple (i, j, k). Previous statistical studies have established the following conditional probabilities.

$j =$	1	2	3
$P(A_j|G_1)$	0.3	0.5	0.2
$P(B_j|G_1)$	0.1	0.7	0.2
$P(C_j|G_1)$	0.2	0.7	0.1
$P(A_j|G_2)$	0.6	0.2	0.2
$P(B_j|G_2)$	0.4	0.5	0.1
$P(C_j|G_2)$	0.7	0.1	0.2

If a subject answers the three questions: yes, no, uncertain, in that order, into which group should he be classified?

5–20. A sample of 100 persons from a certain population were asked five questions. The answer to each question is either no, yes, or uncertain. The population consists of two subgroups. The group membership of each person in the sample was ascertained, as were the answers to the questions. The data obtained are tabulated here.

Q	Group 1 (53 Members)			Group 2 (47 Members)		
	Yes	No	Uncertain	Yes	No	Uncertain
1	23	17	13	7	35	5
2	28	20	5	12	21	14
3	35	7	11	14	28	5
4	17	21	15	30	12	5
5	20	30	3	24	11	12

The questions were designed so that there is no interaction among questions for each individual. The answers are affected by group membership, but given group membership, the answers are independent. A person is selected at random from the population and asked the five questions. Let E be the event the answers are No, Yes, No, Uncertain, Yes, in that order. Determine the posterior odds $P(G_1|E)/P(G_2|E)$. In which group should the individual be classified?

5–21. A certain population consists of two subgroups. A sample of 100 persons is selected, and each person in the sample is asked five questions. The answer to each question is either no, yes, or uncertain. In addition to the answers to the questions, the subgroup membership of each person in the sample is ascertained. The data obtained are tabulated here.

Q	Group 1 (61 Members)			Group 2 (39 Members)		
	Yes	No	Uncertain	Yes	No	Uncertain
1	11	32	18	22	7	10
2	38	18	5	12	25	2
3	30	25	6	17	18	4
4	18	35	8	21	12	6
5	13	22	26	18	11	10

The questions are designed so that there is no interaction among questions for each individual. The answers are affected by group membership, but given group membership, the answers are independent (i.e., the answers are conditionally independent, given group membership).

A person is selected at random from the population and asked the five questions. Let E be the event the answers are No, Yes, No, No, Uncertain, in that order. Determine the posterior odds $P(G_1|E)/P(G_2|E)$. In which group should the individual be classified?

6

Composite Trials

Often a trial is a composite one. That is, the fundamental trial is completed by performing several steps. In some cases, the steps are carried out sequentially in time. In other situations, the order of performance plays no significant role. Some of the examples in Chapter 3, using conditional probability, involve multistep trials. In this chapter we examine more systematically how to model composite trials in terms of events determined by the components of the trials. We illustrate this approach in some important special cases, notably Bernoulli trials, in which each outcome results in a success or failure to achieve a specified condition.

6.1. Events and Component Events

The basic trial in a probability model is often a composite one. That is, to carry out the trial, we must perform several (perhaps a whole sequence of) steps that we call *component trials* Performance of a single trial consists of performing all of the component trials. While there may be intermediate results or *component outcomes*, an element of the basic space is determined by the whole set of component outcomes.

Example 6.1.1: Some Experiments with Composite Trials

- Flip a coin n times. The experiment could be carried out sequentially, flipping the same coin n times. Or the experiment could be carried out by flipping ten coins at the same time. In either case, each flip of a coin constitutes the performance of one of the component trials. The outcome is the n-tuple of the component outcomes, indexed in some suitable way to distinguish between the various components. In order to carry out an analysis, we consider the component outcomes sequentially, whether or not the component trials were performed sequentially.

- Select n members from a population of size N. If the selection is with replacement, the component selections are usually done in sequence. However, one could have N identical populations and select one member from each component population. If the selection is made without replacement, all n selections could be made in one operation from a single population. However, for purposes of analysis, we frequently need to impose an order on the selections and keep track of the order by some type of indexing.

- Receive stock market reports on n stocks. The report on each stock is a component trial. The component trials could be sequential in time or simultaneous or some mixture thereof.

—   □

BASIC SPACE AND EVENTS

Some authors give considerable attention to the structure of the basic space Ω. Since the complete outcome, in the case of n component trials, is an n-tuple of component outcomes, the standard approach is to model the basic space as a Cartesian product space. A component space is set up for each component trial; the *basic space* for the complete trial is the Cartesian product of the component spaces. Although what we do in the following analysis amounts to making such a construction, we find it expedient to bypass explicit consideration of the structure of the basic space. Instead, we concentrate on identifying appropriate events and their relationships. In doing so, we suppose the underlying basic space has sufficient elements to distinguish between the needed events.

EVENTS AND COMPONENT EVENTS

We begin by identifying appropriate *component events*. A component event is an event determined by properties of, or propositions about, the outcomes of the corresponding component trial. In the next example, we illustrate by describing some events determined by outcomes of the composite trials in Example 6.1.1.

Example 6.1.2: Component Events

- Consider the event H_i that a head turns up on the ith toss of the coin. The ith component trial is the ith toss. The event H_i occurs whenever the result of the complete trial is one of those in which the result of the ith toss in the sequence is a head. Thus if the results of five tosses, in order, is

  Head   Head   Tail   Head   Tail

  then the outcome is in event H_2, for it is true of the sequence of component outcomes that the result of the second toss is a head. It is, of course, also true that this outcome belongs to events H_4 and H_5^c, among others.

- Let B_3 be the event the third person chosen from a group of N students is an English major. This event occurs on any trial (sequence of choices) whenever the third person chosen is an English major.

- The owner of a personal computer has access by modem and telephone line to a data base that includes late quotations on ten stocks in which she is interested. Let C_7 be the event the seventh stock reported on posts a gain for the day of at least five points. This event occurs whenever the list of ten current quotations shows that stock number seven has gained at least five points during the trading day.

— $\square$

With appropriate component events defined, more complex events may be formed as Boolean combinations of the component events.

Example 6.1.3: Boolean Combinations of Component Events

- Let A be the event of exactly two heads in four tosses. The trial consists of four component trials—each being one of the four tosses of the coin that constitute a complete trial. If H_i is the event of a head on the ith toss (ith component trial), then

$$A \;=\; H_1 H_2 H_3^c H_4^c \biguplus H_1 H_2^c H_3 H_4^c \biguplus H_1 H_2^c H_3^c H_4 \biguplus H_1^c H_2 H_3 H_4^c$$
$$\biguplus H_1^c H_2 H_3^c H_4 \biguplus H_1^c H_2^c H_3 H_4.$$

- Let B be the event of two or more English majors among three students selected from N. As in the previous example, we let B_i be the event the ith student selected is an English major. Then,

$$B = B_1 B_2 B_3^c \biguplus B_1 B_2^c B_3 \biguplus B_1^c B_2 B_3 \biguplus B_1 B_2 B_3.$$

- Let C be the event that either the first and third or the first and fourth of the stocks post a gain of at least five points for the day. If we let C_i be the event the ith stock posts a gain of at least five points, then
$$C = C_1 C_3 \cup C_1 C_4.$$

— $\square$

The previous example illustrates formulation of events from component events. The following considers the assignment of probabilities.

Example 6.1.4: Bias in Jury Selection

A jury of six persons is to be selected from a panel of 20 persons. Seven of these are black. A civil rights group thinks there is bias against the choice of blacks. Calculate the probability that no black will be chosen under each of the following assumptions.

1. The members of the jury are selected on an equally likely basis from the panel.

2. A nonblack member of the panel is twice as likely to be chosen on any component selection as a black.

Solution

Let A_i be the event the ith person selected is nonblack. The event of no blacks selected is $A = A_1 A_2 A_3 A_4 A_5 A_6$. Using the product rule for conditional probability, we have

$$P(A) = P(A_1)P(A_2|A_1)P(A_3|A_1 A_2)P(A_4|A_1 A_2 A_3)$$
$$\times P(A_5|A_1 A_2 A_3 A_4)P(A_6|A_1 A_2 A_3 A_4 A_5).$$

1. For equally likely choices from among those remaining at each step (i.e., no bias), we have

$$P(A) = \frac{13}{20} \cdot \frac{12}{19} \cdot \frac{11}{18} \cdot \frac{10}{17} \cdot \frac{9}{16} \cdot \frac{8}{15} \approx 0.0443.$$

2. To model the "twice as likely" condition, we double the number of nonblacks in the selection population at each stage. If a nonblack is chosen, then the effective number in the "selection pool" is reduced by two. For the event that no blacks are chosen, we have a constant seven blacks in the pool and successively 26, 24, 22, 20, 18, and 16 nonblacks in the selection pool. Thus

$$P(A_1) = \frac{26}{33}, \qquad P(A_2|A_1) = \frac{24}{31}, \qquad \text{etc.}$$

so that

$$P(A) = \frac{26}{33} \cdot \frac{24}{31} \cdot \frac{22}{29} \cdot \frac{20}{27} \cdot \frac{18}{25} \cdot \frac{16}{23} \approx 0.172.$$

Thus, if the bias is operating as modeled, it is nearly four times as likely that no blacks will be selected as in the no-bias situation.

$\square$

The calculations are much easier if the results of the component trials are independent.

Example 6.1.5: Computer Systems Manager

The manager of a computer system has routine duties, but also duties that are due to unpredictable occurrences. Suppose A is the event that a memory disk is down for emergency service in a given week, B is the event the system is down because of other component or power failure during the week, and C is the event a laser printer is inoperative during the week. A trial consists of observing these three elements during the week. Suppose

$\{A, B, C\}$ is independent with $P(A) = 0.2$, $P(B) = 0.1$, $P(C) = 0.1$.

Determine the probability that at least one of the three events will occur.

Solution

We want the probability of the event $E = A \cup B \cup C = A \biguplus A^c B \biguplus A^c B^c C$. Use of independence gives

$$
\begin{aligned}
P(E) &= P(A) + P(A^c)P(B) + P(A^c)P(B^c)P(C) \\
&= 0.2 + 0.8 \cdot 0.1 + 0.8 \cdot 0.9 \cdot 0.1 = 0.352.
\end{aligned}
$$

—   □

6.2.   Multiple Success-Failure Trials

In this section, we consider a special case of composite trials. We consider a sequence or group of trials such that each component trial results in either a "success" or a "failure" to meet some specified condition.

Example 6.2.1: Success-Failure Trials

- The sequence of component trials may be repeated sampling from a production line to determine whether or not each unit selected meets manufacturing specifications. If the unit selected on the ith component trial meets specifications, this is usually considered a "success." However, since we are interested in determining the number of units in the sample that do not meet specifications, it may be useful to consider a "success" to be the selection of a unit that does not meet specifications.

- An electronic device has five "identical" modules, which fail independently. The units are in service for a given period of time.

Some may fail. We may not be interested in, or may not be able to ascertain, the order of failure. But, for purposes of analysis, it is usually desirable to assign the units a number, and consider them in sequence to determine survival or failure. The ith component trial is the observation or testing of the unit numbered i. In the case under consideration, the outcome is a sequence of five component results, each a success or failure. Thus, if the first, third, and fourth units survive, the outcome could be represented by S, F, S, S, F. There are $2^5 = 32$ distinguishable outcomes.

- An investor buys stock in ten companies. At the end of each quarter, he receives a report on his dividend. Since he can invest in a bank certificate of deposit at eight percent per annum interest, he considers each investment a "success" iff his earning for the quarter is better than two percent of the amount he paid for the stock. The reports on dividends are not naturally sequenced. However, he can assign each stock a number, in some suitable fashion, and consider each stock in that order. He can then think in terms of success or failure of the ith stock, so that one observation or trial, at the end of the quarter, consists of ten component trials.

—   □

A FINITE NUMBER OF SUCCESS-FAILURE TRIALS

Each situation described represents a sequence of a predetermined number n of component trials, each of which results in a "success" or a "failure." The events associated with such trials can be modeled in terms of component events as follows. Let

E_i be the event of a success on the ith component trial, $1 \le i \le n$.

Then we may express any event concerning successes and failures as a Boolean combination of the component events $\{E_i : 1 \le i \le n\}$.

Example 6.2.2:

The owner of a small shop is trying to obtain a special item for a customer. He calls, in succession, three suppliers who handle the item. If any of these have the item, he calls no further. Let E_i be the event the ith supplier has the item in stock. Suppose $\{E_1, E_2, E_3\}$ is independent, with $P(E_1) = 0.7$, $P(E_2) = 0.6$, and $P(E_3) = 0.4$. What is the probability he locates the item among these three suppliers?

Solution

Let E be the event the shopkeeper is successful. Then we may express

$$E = E_1 \bigcup E_1^c E_2 \bigcup E_1^c E_2^c E_3.$$

Use of the stated probabilities and the product rule for independent classes gives

$$P(E) = 0.7 + 0.3 \cdot 0.6 + 0.3 \cdot 0.4 \cdot 0.4 = 0.928.$$

—   □

Remark: In this example, the number of component trials is not known before the trial, but the maximum number is fixed.

One of the most common and important questions asked of a success-failure sequence is, "How many successes are there in the n trials?" To illustrate the analysis, we consider the case of four component trials. It should be clear how the essential arguments and patterns extend to any number of trials. We find it useful to express the events as a union of minterms generated by the class $\{E_i : 1 \leq i \leq n\}$.

Example 6.2.3: Analysis of a Sequence of Four Component Trials

If E_i is the event of a success on the ith component trial, then the minterms are

$$
\begin{array}{llll}
E_1^c E_2^c E_3^c E_4^c & E_1^c E_2 E_3^c E_4^c & E_1 E_2^c E_3^c E_4^c & E_1 E_2 E_3^c E_4^c \\
E_1^c E_2^c E_3^c E_4 & E_1^c E_2 E_3^c E_4 & E_1 E_2^c E_3^c E_4 & E_1 E_2 E_3^c E_4 \\
E_1^c E_2^c E_3 E_4^c & E_1^c E_2 E_3 E_4^c & E_1 E_2^c E_3 E_4^c & E_1 E_2 E_3 E_4^c \\
E_1^c E_2^c E_3 E_4 & E_1^c E_2 E_3 E_4 & E_1 E_2^c E_3 E_4 & E_1 E_2 E_3 E_4.
\end{array}
$$

Consider the event A_{r4} that exactly r successes occur in the four trials. Any of the minterms that has exactly r places uncomplemented and $4 - r$ places complemented is in this event. Using the numbering scheme introduced in Section 2.4, we have

$$A_{04} = M_0 = E_1^c E_2^c E_3^c E_4^c, \quad A_{14} = M(1,2,4,8),$$

$$A_{24} = M(3,5,6,9,10,12), \quad A_{34} = M(7,11,13,14),$$

$$A_{44} = M_{15}.$$

The number of minterms in the union for A_{rn} is thus the number $C(4,r)$ of ways r places may be selected from 4 to leave uncomplemented.

If we let B_{r4} be the event of r or more successes in four trials, then, for example

$$B_{24} = A_{24} \bigcup A_{34} \bigcup A_{44} = \bigcup_{k=2}^{4} A_{k4}.$$

That is, we get r or more successes if we get exactly r, or exactly $r + 1$, or ..., or exactly 4. These events are mutually exclusive, as indicated by the symbol $\bigsqcup$ for the union.

— □

GENERAL RESULT FOR A FINITE NUMBER OF TRIALS

The pattern of analysis in the preceding example may be extended to the general case to give the following result.

Theorem 6.2.1 (Number of Successes in n Success-Failure Trials)
If A_{rn} is the event of r successes in n component trials and B_{rn} is the event of r or more successes, then

1. A_{rn} *is the union of the $C(n,r)$ minterms generated by the class of component events $\{E_i : 1 \leq i \leq n\}$ that have r places uncomplemented and $n - r$ places complemented.*

2. $B_{rn} = \bigsqcup_{k=r}^{n} A_{kn}.$

— □

UNLIMITED NUMBER OF TRIALS

There are situations in which no limit may be placed on the number of component trials before the trial is actually performed. In Example 1.4.2, we consider a game that is played alternately until one of the players is successful. In analyzing this game, we must consider the possibility the game will continue indefinitely. Other situations exhibit similar patterns.

Example 6.2.4: Unlimited Success-Failure Trials

- In sampling from a population in which the probability of a success is small, it may be desirable to sample until a certain number of successes are realized, in order that the sample be representative of the population. There is no way of knowing, before the trial is performed, how many component trials will be needed. In principle, we must allow for an unlimited number of possibilities.

- Some processes by nature tend to continue without limit. Consider, for example, the Dow-Jones average for the stock market. At the end of each day of trading, the net gain (or loss) is posted. A success could be the event of a net gain; a failure would then be the event of either a net loss or no gain. The result on each trading day would be a component result. Since there is no way of knowing how long this institution will endure, a satisfactory model must allow for an unlimited sequence of component trials.

- At the end of each day, the net output of a production line is tabulated. If the output reaches a specified goal, this constitutes a success; otherwise, the result is a failure. The observation on each work day is a component trial.

- A personal computer is connected to a power line that experiences occasional voltage surges, due to load switching, atmospherics, etc. The magnitudes of successive surges vary in a random manner. It is important to have surge protection equipment to guard against surges that exceed a safe level. The trial consists of observing or experiencing the successive voltage surges. A "success" on the ith component trial could consist of the condition that the magnitude of the surge exceeded the critical level.

—   □

In situations such as these, the analysis does not differ significantly from the case of a specified number of component trials. We consider an infinite class $\{E_i : 1 \leq i\}$ where E_i is the event of a success on the ith component trial.

- If we let A_{rn} be the event of r successes in the first n trials, then we express A_{rn} in terms of the minterms generated by the finite subclass $\{E_i : 1 \leq i \leq n\}$, just as in the case of n predetermined trials.

- On the other hand, suppose C_k is the event that there are k failures before the first success. Then,

$$C_k = \bigcap_{i=1}^{k} E_i^c E_{k+1}.$$

- As these two results suggest, analysis is usually broken down into steps such that only a finite number of the E_i are considered at each step.

Before the analysis can go beyond mere formulation of the events, appropriate information about probabilities must be given or assumed. One simple pattern is illustrated in Example 6.2.2. In the next section, we consider one of the simplest yet most important cases that can be analyzed quite extensively.

6.3.   Bernoulli Trials

In this section, we consider a pattern that serves as a useful model in a wide variety of applications. Calculation of the probabilities associated with this pattern are so common that extensive tables have been published.

Definition 6.3.1. *We say a sequence of success-failure trials is a* Bernoulli sequence *iff two conditions hold.*

1. The class $\{E_i : 1 \leq i\}$ is independent. That is, the result of any component trial is not affected by, nor does it affect, the result of any other component trial.

2. The probability $P(E_i) = p$, invariant with i. The probability of success is the same on any component trial. We let $q = 1 - p = P(E_i^c)$.

Remark: We often refer informally to Bernoulli trials when speaking of either a Bernoulli sequence or a trial in a Bernoulli sequence. The context serves to make it evident which is the case.

For Bernoulli trials, we can obtain quite readily the probability of the event of r successes in n trials, of the event of r or more successes in n trials, and of the event of k failures before the first success. The results are as follows.

Theorem 6.3.2. *Consider a Bernoulli sequence in which $P(E_i) = p$ is the probability of success on any component trial. Let A_{rn} be the event of exactly r successes in n trials, B_{rn} be the event of r or more successes in n trials, and C_k be the event of k failures before the first success. Then*

$$P(A_{rn}) = p_r = C(n,r)p^r q^{n-r} \qquad P(B_{rn}) = \sum_{k=r}^{n} p_k$$

$$P(C_k) = pq^k \quad \text{where } q = 1 - p.$$

Proof

1. By Theorem 6.2.1, A_{rn} consists of the union of $C(n,r)$ minterms generated by $\{E_i : 1 \leq i \leq n\}$. Each of these minterms is characterized by r uncomplemented and $n - r$ complemented component sets E_i. The probability of the minterm is the product of the probabilities of the intersected sets. Each uncomplemented event contributes a factor p; each complemented event contributes a factor q. Thus, each minterm in the union has probability $p^r q^{n-r}$. The minterms are mutually exclusive, so that their probabilities add. The result is as asserted.

2. There are r or more successes iff there are r or $r + 1$ or ... or n successes. Since these conditions are mutually exclusive, we have

$$P(B_{rn}) = p_r + p_{r+1} + \cdots + p_n = \sum_{k=r}^{n} p_k.$$

3. As noted earlier, $C_k = \bigcap_{i=1}^{k} E_i^c E_{k+1}$. By the product rule for independent classes, $P(C_k) = pq^k$.

— $\qquad\qquad\qquad\qquad\qquad\qquad\qquad\qquad\qquad\qquad\qquad\qquad$ $\square$

Corollary 6.3.3 **(The Number of Successes in Any n Trials)** *For a Bernoulli sequence with probability p of success on any component trial, let $\{E_{i_k} : 1 \leq k \leq n\}$ be any subclass of n component events. If A_{rn} is the event of r successes among the indicated component trials and B_{rn} is the event of r or more successes, then*

$$P(A_{rn}) = p_r = C(n,r)p^r q^{n-r} \qquad and \qquad P(B_{rn}) = \sum_{k=r}^{n} p_k.$$

Example 6.3.1: A Reliability Problem

A system has five similar modules that fail independently. The system operates iff three or more of these modules operate. Suppose each module has probability 0.8 of operating successfully for one year. What is the probability the system will operate successfully for one year?

Solution

If E_i is the event the ith unit survives, we assume the class of component events $\{E_i : 1 \leq i \leq n\}$ is independent, with $P(E_i) = 0.8$ for $i \leq 1 \leq n$. By the general result in Theorem 6.3.2,

$$P(B_{35}) = C(5,3)0.8^3 \cdot 0.2^2 + C(5,4)0.8^4 \cdot 0.2 + C(5,5)0.8^5 \approx 0.9421.$$

— $\qquad\qquad\qquad\qquad\qquad\qquad\qquad\qquad\qquad\qquad\qquad$ $\square$

USE OF TABLES

The calculations for small n and r are quite simple to perform with modern calculators. However, for larger n and r, the work can become rather tedious. Because the Bernoulli trial model is so frequently useful, extensive tables are available. If S_n represents the number of successes in n trials, then

$$P(A_{kn}) = P(S_n = k) = p_k = C(n,k)p^k q^{n-k}$$

and

$$P(B_{rn}) = P(S_n \geq r) = \sum_{k=r}^{n} p_k.$$

The set of probabilities $\{p_k : 0 \leq k \leq n\}$ form the *binomial distribution*, with parameters n and p. The name binomial distribution is associated with the fact that by the Binomial Theorem of algebra

$$1 = (p+q)^n = \sum_{k=0}^{n} C(n,k)p^k q^{n-k} = \sum_{k=0}^{n} p_k.$$

Published tables take a variety of forms. The most common are the following two.

- *Binomial distribution, individual terms*: There are columns for various values of the probability parameter p, listed at the top of the columns. At the left of the table are indicated values of the parameter n. In a second column are listed various values of r (sometimes different letters are used). The numbers in the body of the table are values of $P(S_n = k)$ for various values of n, p, and k. If, for example, $p = 0.20$ and $n = 10$, then $P(S_{10} = 6)$ is found to be 0.0055. For values of p between those in the table, one may interpolate, although linear interpolation is not very accurate.

- *Binomial distribution, cumulative terms*: The parameter designations are the same as in tables for individual terms. However, the numerical entries in the body of the table are

$$P(S_n \geq r) = p_r + p_{r+1} + \cdots + p_n.$$

If $n = 10$ and $p = 0.20$, then $P(S_{10} \geq 6) = 0.0064$.

This result could have been obtained from the table of individual terms by adding the appropriate individual terms (five, in this case). On the other hand, the entries in the cumulative table may be used to get individual terms. For example, since

$$P(S_{10} \geq 6) = p_6 + p_7 + p_8 + p_9 + p_{10} \quad \text{and} \quad P(S_{10} \geq 7) = p_7 + p_8 + p_9 + p_{10}$$

the difference $P(S_{10} \geq 6) - P(S_n \geq 7) = P(S_{10} = 6) = 0.0064 - 0.0009 = 0.0055$. It should be clear that this pattern holds in any case.

$$P(S_n = k) = P(S_n \geq k) - P(S_n \geq k + 1).$$

Most tables have values of p no greater than $1/2$. For values of p greater than $1/2$, we may work with the number of failures. Let F_n be the number of failures in n trials. Then

$$F_n + S_n = n \quad \text{or} \quad S_n = n - F_n.$$

We therefore have

$$S_n \geq r \quad \text{iff} \quad n - F_n \geq r \quad \text{iff} \quad F_n \leq n - r \quad \text{iff} \quad F_n \not\geq n - r + 1.$$

Hence,

$$P(S_n \geq r) = 1 - P(F_n \geq n - r + 1).$$

Now $P(F_n = k) = C(n,k) q^k p^{n-k} = q_k$ and $P(F_n \geq n - r + 1) = \sum_{k=n-r+1}^{n} q_k$. This validates the procedure:

To find $P(S_n \geq r)$ when $p > 0.5$, set $n' = n$, $p' = q$, $r' = n - r + 1$. Enter the table for cumulative terms at (n', p', r'). Subtract the value found in the table from one.

Example 6.3.2:

Suppose $n = 10$, $p = 0.75$, and $r = 4$. Then $n' = 10$, $p' = 0.25$, and $r' = 10 - 4 + 1 = 7$. The entry in the table of cumulative terms at (n', p', r') is 0.0035. Thus,

$$P(S_{10} \geq 4) = 1 - 0.0035 = 0.9965.$$

—   □

Example 6.3.3: Quality Control

At a quality control station on a production line, units are selected for testing according to a specified schedule. A success is interpreted to mean the unit tested meets specifications. If the line is operating satisfactorily, the results of the sequence may be modeled as a Bernoulli sequence with $p = 0.95$. What is the probability of 19 or more successes in 20 trials?

Solution

We work with the number of failures. The probability of a failure on any trial is 0.05.

Nineteen or more successes iff one or fewer failures iff *not* two or more failures.

We enter the cumulative table at $n' = 20$, $p' = 0.05$, and $r' = 2$ to find probability 0.2642. Thus

$$P(S_{20} \geq 19) = 1 - 0.2642 = 0.7358.$$

—   □

COMPOUND BERNOULLI TRIALS

Frequently, we have two or more Bernoulli sequences operating independently. We consider the case of two such sequences, which we model as follows.

- *First sequence:* E_i is the event of a success on the ith trial in the first sequence. As a Bernoulli trial, we suppose

  $$\{E_i : 1 \leq i\} \quad \text{is independent and} \quad P(E_i) = p_1.$$

- *Second sequence:* F_j is the event of a success on the jth trial in the second sequence. We suppose

  $$\{F_j : 1 \leq j\} \quad \text{is independent and} \quad p(F_j) = p_2.$$

- As *independent sequences*, we suppose $\{E_i, F_j : 1 \leq i, 1 \leq j\}$ is an independent class.

If A_{rn} is the event of r successes in n trials in the first sequence and B_{km} is the event of k successes in m trials in the second sequence, then properties of independent classes ensures that

$$\{A_{rn}, B_{km}\} \quad \text{is independent for each permissible } r, n, k, \text{ and } m.$$

Example 6.3.4: Battle of the Sexes

Bill and Mary take ten basketball "free throws" each. Let E_i be the event Bill is successful on his ith try and F_j be the probability Mary is successful on her jth try. We assume the two sequences of trials may be modeled as a compound Bernoulli sequence, with $P(E_i) = 0.85$ and $P(F_j) = 0.80$. What is the probability Mary makes more free throws than Bill?

Discussion of the assumptions

The assumption that either sequence of trials is Bernoulli assumes that both players are experienced and have "warmed up," so that there is no appreciable learning or adjustment as the trial proceeds. It also assumes that there is no appreciable effect of "pressure" felt by the players as they concentrate on winning. Whether or not these are good assumptions depends upon the experience of the players.

The assumption of independence of the sequences assumes no appreciable effect of pressure from the other player. If one player goes first and does exceedingly well, the pressure might affect adversely the performance of the other if he or she were aware of the earlier results.

We assume the independence and constant probability conditions hold.

Solution

Since the probabilities of success on any trial are greater than $1/2$, we find it expedient to work with failures or "misses." Let M_k be the event Mary misses k shots in 10 and let B_j be the event Bill misses j or more shots. If M is the event Mary wins, we may express this as

$$M = \bigcup_{k=0}^{9} M_k B_{k+1} \quad \text{so that} \quad P(M) = \sum_{k=0}^{9} P(M_k) P(B_{k+1}).$$

From tables of the Binomial distribution we obtain the values

k	$P(M_k)$	$P(B_{k+1})$	k	$P(M_k)$	$P(B_{k+1})$
0	0.1074	0.8031	5	0.0264	0.0014
1	0.2684	0.4557	6	0.0055	0.0001
2	0.3020	0.1798	7	0.0008	0.0000
3	0.2013	0.0500	8	0.0001	0.0000
4	0.0881	0.0099	9	0.0000	0.0000

The sum of the products give $P(M) \approx 0.2738 \approx 1/3.65$, so that Mary has slightly better than one chance in four of winning, even though her probability of making any shot is less than Bill's.

— $\square$

Example 6.3.5: Compound Bernoulli Sequences, with no Limit on the Number of Trials

Two persons in a game play alternately until one wins. If E_i is the event that there is a success on the ith component trial, we suppose $\{E_i : 1 \le i\}$ is independent, with

$$P(E_i) = \begin{cases} p_1 & \text{for } i \text{ odd} \\ p_2 & \text{for } i \text{ even} \end{cases}.$$

We set $q_1 = 1 - p_1$ and $q_2 = 1 - p_2$. If A_i is the event the first success is in the ith component trial, then

$$A_i = E_i \bigcap_{j=1}^{i-1} E_j^c.$$

Use of the product rule for independent classes and the assumed probabilities shows that

$$P(A_{2k+1}) = p_1(q_1 q_2)^k \quad \text{and} \quad P(A_{2k}) = p_2 q_1 (q_1 q_1)^{k-1}.$$

If A is the event the first player wins and B is the event the second player wins, then

$$A = \bigcup_{k=0}^{\infty} A_{2k+1} \quad \text{and} \quad B = \bigcup_{k=1}^{\infty} A_{2k}.$$

We may use the geometric series (see the appendix on mathematical aids) to determine the probabilities.

$$P(A) = p_1 \sum_{k=0}^{\infty} (q_1 q_2)^k = \frac{p_1}{1 - q_1 q_2}$$

$$P(B) = p_2 q_1 \sum_{k=1}^{\infty} (q_1 q_2)^{k-1} = \frac{p_2 q_1}{1 - q_1 q_2}.$$

To make the probability of winning the same for each player we need the condition

$$p_1 = p_2 q_1 \qquad \text{so that} \qquad p_2 = \frac{p_1}{q_1} = \frac{p_1}{1 - p_1} = \frac{1}{1/p_1 - 1}.$$

If $p_1 > 1/2$, then we would need $p_2 > 1$, which is impossible. In every case, we must have $p_2 > p_1$ in order to balance out the advantage of playing first. Some typical values of the pair (p_1, p_2) to make $P(A) = P(B)$ are

p_1	p_2
1/5	1/4
1/4	1/3
1/3	1/2
2/5	2/3

Simple algebra shows that $P(A \cup B) = 1$, so that $P(A^c B^c) = 0$. However, $A^c B^c = \bigcap_{i=1}^{\infty} E_i^c \neq \emptyset$.

This is a simple example of a case in which zero probability does not ensure that the event is impossible. But with probability one, we can operate as if it will never occur.

—   □

PROBLEMS

6–1. Twenty folders are placed in a file cabinet in random order. What is the probability that the first four are in the first four positions (not necessarily in order in those four positions)? Treat as a composite trial, and compare results with the classical approach of counting permutations.

6–2. Among five "identical" instruments, two are defective. A service man selects them one by one, on an equally likely basis, for test and repair. What is the probability he has found both defective units by the third selection?

6–3. Five units are put into operation simultaneously. Let F_i be the event the ith unit fails within the first 100 hours. Suppose the units fail independently, with $P(F_i) = 0.4$. What is the probability that at least three are still operating at the end of 100 hours?

6–4. Three marksmen shoot at a target simultaneously, with respective probabilities of 0.8, 0.75, and 0.9 of hitting the target. If their performance is independent, what is the probability the target is hit at least twice?

6–5. Three recent graduates of a law school take the state bar exam. Their respective probabilities for passing the first time are 0.9, 0.6, and 0.8. Their performances are independent. What is the probability that at least two will pass the exam the first time?

6–6. Ten customers come into a video store. If the probability is 1/5 that each customer will buy a TV set, and they decide independently, what is the probability the store will sell three or more sets?

6–7. Under extreme noise conditions, the probability that a given message will be transmitted correctly is 0.1. Successive messages are acted on independently by the noise. Suppose the same message is sent ten times. What is the probability that it gets through correctly at least once?

   Hint: The event of at least one success is the event of not all failures.

6–8. Five "identical" devices are put into service at time $t = 0$. The units fail independently. The probability of failure of any unit in the first 500 hours is 0.1. What is the probability that three or more of the units are still in operation at the end of 500 hours.

6–9. A computer system has ten "identical" modules. The circuit has redundancy features that ensure that the system operates if any eight or more of the modules are operative. Suppose the units fail independently, with probability 0.9 that any unit will survive between preventive maintenance periods. What is the probability of no system failure due to these modules?

6–10. In order to increase reliability, a system has built in redundancy. It contains five modules of a certain type, at least three of which must be operative for successful performance of the larger system. The modules fail independently, and each has a probability of 0.9 of surviving for one year. What is the probability the system will operate for one year?

6–11. Items from a production line are subjected to a nondestructive test for compliance with specifications. The probability of failure is 0.02, and the failures can be considered independent. What is the probability of a run of twenty five or more good units?

6–12. In a "shootout," each contestant makes a fixed number of attempts. Assume conditions of a compound Bernoulli trial, with probability p_1 of success on each trial by the first contestant and probability p_2 of success on each trial by the second. Jim and Ralph make five trials each. Let

A be the event Jim wins, B be the event Ralph wins,

and

N be the event neither wins.

Suppose $p_1 = p_2 = 0.6$. Determine $P(N)$ and $P(A)$.

Hint: Use the symmetries of the problem.

6–13. Suppose in problem 6-12 the probabilities are $p_1 = 0.85$ and $p_2 = 0.80$. Determine $P(N)$.

6–14. Fred and John play a game in which a wheel is spun, resulting in one of the integers 0 through 9. The ten possible digits are equally likely to turn up on any trial. Each person spins the wheel five times. What is the probability Fred spins a seven more times than does John?

6–15. Alice and Robert play a game with a wheel that turns up one of the decimal digits on an equally likely basis. Alice spins the wheel four times. She scores a point each time she gets an even digit, 0, 2, 4, 6, or 8. Robert spins five times. He scores a point each time he turns up a 1, 2, 5, or 7. What is the probability Robert scores more points than Alice?

6–16. A box contains 100 balls: 30 red, 40 blue, and 30 green. Martha and Alex select at random, with replacement and mixing after each selection. Alex has a success iff he selects a red ball; Martha has a success iff she selects a blue ball. If Alex selects six times and Martha selects five times, what is the probability Martha has more successes than Alex?

6–17. In a reliability study, two alternatives are under consideration.

- *First alternative*: four units, each with probability $p_1 = 0.5$ of survival

- *Second alternative*: five units, each with probability $p_2 = 0.4$ of survival

If the units fail independently, what is the probability the second alternative will yield at least as many survivals as the first?

6–18. A device has N similar components, which may fail independently with probability p of failure of any component. The device fails if any one or more of the components fail. In the event of failure of the device, the components are tested sequentially.

   (a) What is the probability the first defective unit tested is the nth, given that one or more units have failed?

   (b) What is the probability the defective unit is the nth, given that exactly one unit has failed?

   (c) What is the probability that more than one unit is defective, given that the first defective unit is the nth $(n < N)$?

6–19. A coin is flipped repeatedly, until a head appears. Under the usual assumptions on coin flipping, show that with a probability of one the game will terminate. Note that if E_i is the event of a head on the ith trial, the event of eventually obtaining a head and thus terminating the game is $H = \bigcup_{i=1}^{\infty} E_i$. This may be expressed as a disjoint union

$$H = \bigcup_{k=1}^{\infty} H_k \qquad \text{where} \quad H_k = E_1^c E_2^c \ldots E_{k-1}^c E_k.$$

Calculate $P(H_k)$ and use the geometric series to sum them.

6–20. Two players each roll a single fair die n times. Assuming independent Bernoulli trials, what is the probability that both roll the same number of sixes?

6–21. A warehouse has a stock of a stock of n items of the same kind, r of which are defective. Two of the items are chosen at random, without replacement. What is the probability that at least one is defective? Show that for large n, the number is very close to that for selection with replacement, so that this corresponds to two Bernoulli trials with probability $p = r/n$ of success on any trial.

6–22. What is the probability of a success on the ith trial in a Bernoulli sequence of n trials, given that there are r successes?

6–23. Three persons in a game "play" consecutively, until one achieves his objective. Let E_i be the event of a success on the ith play of the game. We suppose $\{E_i : 1 \leq i\}$ is an independent class with

$$P(E_i) = p_1 \quad \text{for } i = 1, 4, 7, \ldots \qquad P(E_i) = p_2 \quad \text{for } i = 2, 5, 8, \ldots$$

$$P(E_i) = p_3 \quad \text{for } i = 3, 6, 9, \ldots.$$

Let

A be the event the first player wins.
B be the event the second player wins.
C be the event the third player wins.

(a) Express A, B, and C in terms of the component events E_1, E_2, E_3, ...

(b) Determine $P(A)$, $P(B)$, and $P(C)$ in terms of p_1, p_2, and p_3. Then obtain numerical values in the case

$$p_1 = 1/4, \qquad p_2 = 1/3, \qquad \text{and} \qquad p_3 = 1/2.$$

Hint: Use the geometric series.

Part II

Random Variables and Distributions

7

Random Variables and Probabilities

The previous treatment deals with trials, outcomes, and events. The basic space is the set Ω of all possible outcomes. An event is determined by a set of properties or attributes of the outcomes; it consists of those outcomes that have the prescribed properties or attributes. Probability is assigned to events and to logical combinations of events. In order to ensure that such combinations are also events, the class of events is assumed to have the closure properties of a sigma algebra. That is, complements, countable unions, and countable intersections of events are themselves events.

Frequently the outcome of a basic experiment is characterized by one or more numbers. This is perfectly natural if the experiment involves counting or measuring some physical quantity. Observations of such quantities as temperature, length, cost, value of an economic indicator, speed, price of a share of stock, life duration of a physical device, or production level are common. In many situations, results that are not inherently numerical may be represented usefully by assigning numbers. In Bernoulli trials, for example, a success may be assigned the number 1 and a failure may be assigned the number 0. From these, it is easy to derive significant numbers associated with a sequence of trials. For example, we may want the number of successes in the first ten trials, the number of the trial on which the first success is observed, or the fraction of successes in a long sequence of trials. In this chapter, we introduce a formalism for dealing with such quantities associated with the outcome of an experiment.

7.1. Random Variables as Functions—Mapping Concepts

We wish to deal with numbers associated with the individual outcomes of a basic trial or experiment. These quantities are random in the sense that it is not known before the trial which outcome, hence which value, will be observed. The first step in formulation is to note that:

For each possible outcome ω, there is a number.

Such an assignment of a number to each $\omega \in \Omega$ is precisely the notion of defining a function, whose domain is the basic space Ω. We represent a function of this sort by a letter, say X or Y. The value of the function for a given outcome ω is designated by $X(\omega)$ or $Y(\omega)$, as the case may be.

Example 7.1.1: Successes in a Bernoulli Sequence

Consider the successes in a Bernoulli sequence. For each i, let X_i be the function defined by

$$X_i(\omega) = \begin{cases} 1 & \text{iff there is a success on trial } i \\ 0 & \text{otherwise} \end{cases}.$$

If there are five trials and the event $E_1 E_2^c E_3^c E_4 E_5$ occurs, then for this outcome (i.e., this ω).

$$X_1(\omega) = 1, \quad X_2(\omega) = 0, \quad X_3(\omega) = 0, \quad X_4(\omega) = 1, \quad X_5(\omega) = 1.$$

If X is the function that counts the number of successes in the first five trials, then for this outcome, $X(\omega) = 3$. As a matter of fact, we can relate the functions as follows:

$$X = X_1 + X_2 + X_3 + X_4 + X_5.$$

For any outcome ω, the $X_i(\omega)$ tally the individual successes, and $X(\omega)$ represents the sum of these and hence gives the number of successes.

— $\square$

Example 7.1.2: Random Numbers

A random-number generator produces numbers that are decimal fractions, to four places, between 0 and 0.9999. A trial consists of causing the generator to produce one of the possible numbers. The value on any trial is unknown until observed. The output of the generator can be represented by a function X whose possible values constitute the set $\{0.0000, 0.0001, 0.0002, \ldots, 0.9998, 0.9999\}$, so that $X(\omega)$ is one of this set of numbers.

— $\square$

Example 7.1.3: Spots Turned up on Two Dice

Consider two ordinary dice, which are distinguished (say by color, red and green). There are 36 distinct ways these can turn up on a throw. Thus, a model for the throwing of these dice has a basic space Ω with 36 elements. There are several readily identified functions that may be useful. For a given outcome ω, let

$X(\omega) =$ the number of spots on the upturned face of the red die.

$Y(\omega) =$ the number of spots on the upturned face of the green die.
$V(\omega) =$ the smaller of the numbers of spots turned up.
$W(\omega) =$ the larger of the numbers of spots turned up.
$Z(\omega) =$ the total number of spots turned up.

Now the latter three functions are related to the first two. Thus

$$Z = X + Y \qquad V = \min\{X, Y\} \qquad W = \max\{X, Y\}.$$

—   □

We have avoided the use of the term random variable in referring to the functions just described. In order for one of these functions to be a random variable, it must be possible to make appropriate probability statements about its values. Thus, we would like to assign a probability that the random variable X, which counts the number of successes in ten trials, is greater than three, or that its value lies between 2 and 4, inclusive. We would like to assign the probability that the value of the number X produced by the random-number generator lies in the interval $[0.4950, 0.5050]$. Or we would like to assign the probability that the total number of spots Z turned up on two dice is one of the numbers 2, 4, or 6. Probability is assigned to events. In order to assign probabilities to the conditions described, these conditions must correspond to events.

We formalize the matter as follows. Once a random variable is defined, the value $X(\omega)$ assigned to an outcome ω becomes an attribute of that outcome. If we make the assertion $X(\omega) \leq 4$, this is a proposition that may be true for some ω and false for others. Thus, we may speak of the subset of the basic space

$$\{\omega : X(\omega) \leq 4\} \qquad \text{which we abbreviate } \{X \leq 4\}.$$

If this set is an event, we may assign probability. This is an essential formal condition. In most practical situations of the type we deal with, this condition can be assumed without any real examination. However, to understand the fundamental notion of the probability distribution induced by a random variable, we must look at some aspects of the problem. To consider another example, suppose we want to deal with the set (which we suppose to be an event)

$$\{\omega : 2 < X(\omega) \leq 3\} = \{2 < X \leq 3\} = \{X \in M\} \qquad \text{where } M = (2, 3].$$

These examples illustrate the second major characteristic of a random variable.

We want to assign probabilities to sets of the form
$$\{X \in M\}$$
for suitable sets M.

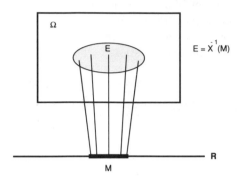

FIGURE 7.1. Basic mapping diagram showing set M and its inverse image.

We view this problem with the aid of a graphical representation in a fundamental *mapping diagram* as in Figure 7.1. In this diagram, the basic space Ω is represented by the rectangle. An outcome ω is represented by a point in the rectangle. The function X assigns to each ω a point $t = X(\omega)$ on the real line. We say the function X *maps* ω into t. If M is a set of points on the line, then the set of all ω, which are mapped into M by X, is called the *inverse image* of M under the mapping X. This is the set

$$E = \{\omega : X(\omega) \in M\} = \{X \in M\} = X^{-1}(M).$$

We want to be able to assign probabilities that the values of X lie in a set M, where M is any interval or any set that can be made up by logical combinations (unions, intersections, complements) of intervals. That is, we want the inverse images of such sets to be events. In the appendix to this chapter, we discuss briefly a suitable class of sets of numbers (viewed as subsets of the real line $\Re$), known as the class of *Borel sets*. This is the smallest sigma algebra that includes all intervals and logical combinations of intervals.

Definition 7.1.1. *A real-valued function X defined on the basic space Ω is called a (real) random variable iff*

$$X^{-1}(M) \qquad \textit{is an event for each Borel set } M.$$

Remark: The condition on the inverse images is known as a *measurability condition*. Technically, a random variable is a *measurable function* on the basic space. This measurability has far reaching consequences for the structure of the probability model. Fortunately, for most applications it is not necessary to examine the details. On the other hand, many practical aspects of the probability model are based on the measurability of random variables. A brief survey of some of the fundamental facts is presented in the appendix to this chapter.

In the next section, we introduce the essential notion of a probability distribution induced by, or associated with, a random variable. It is only through such distributions that practical probability calculations can be made for most random variables. Fundamental to such distributions is the following theorem on inverse images. Although it is stated in terms of real random variables, it is a quite general property of any function considered as a mapping from its domain to its codomain (i.e., the set in which it takes its values).

Theorem 7.1.2   (Preservation of Set Operations and Relations by the Inverse Mapping) *Suppose X is a function from its domain Ω to its codomain $\Re$. Let M, N, and M_i be subsets of $\Re$. Let the inverse images of these sets be*

$$E = X^{-1}(M), \qquad F = X^{-1}(N), \qquad and \qquad E_i = X^{-1}(M_i).$$

That is, $E = X^{-1}(M)$ is the set of all elements of Ω mapped into M by X, etc. Then the following conditions hold.

(1) $X^{-1}(\emptyset) = \emptyset$, $X^{-1}(\Re) = \Omega$ (2) $X^{-1}(M^c) = E^c$
(3) $M \subset N$ implies $E \subset F$

$$(4)\ \ X^{-1}\left(\bigcup_i M_i\right) = \bigcup_i E_i \ \ and$$
$$X^{-1}\left(\bigcap_i M_i\right) = \bigcap_i E_i$$

(5) If $MN = \emptyset$, then $EF = \emptyset$ (6) $X^{-1}\left(\bigcup_i M_i\right) = \bigcup_i E_i$

Proof

Proofs of these propositions amount to careful reading of the notation. For example, (3) is valid, since if every point in E is mapped into M and $M \subset N$, then every point in E is mapped into N. This means that every point in E is also a point in F.

— □

Much of the structure of measurable functions, which includes random variables and Borel functions, stems from these mapping properties. This structure is developed in that branch of mathematics known as *measure theory*, of which probability theory is a significant part. A brief discussion of some of these ideas is included in the appendix to this chapter. We note herein several facts that have immediate implications for our development.

Our definition of a random variable, as a measurable function, requires that the inverse image of *every* Borel set must be an event. It turns out that this condition is ensured if the inverse images of the members of certain subclasses of the Borel sets are all events. In particular

Theorem 7.1.3. *A function X from the basic space Ω to the real line is a random variable iff for every real t the inverse image of the interval $(-\infty, t]$ is an event.*

— □

THE SIGMA ALGEBRA OF EVENTS DETERMINED BY A RANDOM VARIABLE

While the inverse image of every Borel set is an event, it does not follow that the class of inverse images under the mapping X for a random variable exhausts the class of events.

Example 7.1.4:

Suppose random variable X takes on one of the three values 0, 1, or 2, and no others. Let A_0 be the event that X takes on the value 0, A_1 be the event that X takes on the value 1, and A_2 be the event that X takes on the value 2. For each ω, the value $X(\omega)$ is one of these three numbers. That means that each ω must belong to one, and only one, of the A_i. Thus

$$\{A_1, A_2, A_3\} \quad \text{is a partition.}$$

Now consider any Borel set M.

1. If M contains none of the possible values 1, 2, or 3, then there is no ω that is mapped into M. That is, $X^{-1}(M) = \emptyset$.

2. If M contains all of the possible values 1, 2, and 3, then every ω is mapped into M, so that in this case $X^{-1}(M) = \Omega$.

3. If M contains the values 1 and 3, but not 2, then every point in either A_1 or A_2 is mapped into M. We have $X^{-1}(M) = A_1 \bigcup A_2$.

Arguing in this way to consider all cases, we see that the class of inverse images of Borel sets consists of the empty set, the basic space, the individual A_i, and the unions of pairs of these. For most probability systems, these sets will *not* exhaust the class of events.

—   □

As noted in the appendix, the class of Borel sets is a sigma algebra of sets. That means that unions, intersections, and complements of Borel sets are Borel sets. Because of the preservation of set operations by inverse mappings, it is not difficult to establish the following theorem.

Theorem 7.1.4. *The class of all those events that are inverse images of Borel sets under the mapping by a function X is a sigma algebra of sets. If X is a random variable, then this class is a subsigma algebra of the class of events.*

—   □

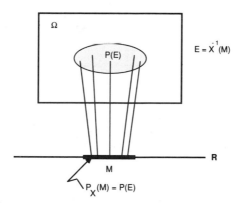

FIGURE 7.2. Mass assignment for induced probability distribution.

We find it useful to introduce a symbol and name for the class of inverse images of Borel sets under the mapping of a random variable X.

Definition 7.1.5. *The class of events determined by a random variable X is called the* sigma algebra determined by X *and is denoted $\sigma(X)$, or sometimes $\mathcal{F}(X)$.*

Remark: It should be clear that $\sigma(X)$ is a subsigma algebra of the sigma algebra $\mathcal{F}$ of events.

7.2.   Mass Transfer and Probability Distributions

Much of the introduction to random variables, thus far, is rather formal and abstract. Even the graphical notion of mapping tells little about a particular random variable. While the treatment lays the groundwork for assigning probabilities to events determined by a random variable, it provides little concrete, practical grasp of this task. However, the preceding development permits us to take a significant step to rectify that deficiency.

Consider the mapping diagram in Figure 7.2. For a Borel set M on the real line, there is the event $E = X^{-1}(M)$, consisting of all those ω that are mapped into M by X. To any such Borel set M, we assign a probability mass $P_X(M)$ equal to the probability mass $P(E)$ in the inverse image. If we make this assignment for each Borel set M, the theorem on preservation of set operations ensures a consistent and unique assignment of probability mass to the Borel sets. We may view this as a point-by-point transfer of mass from the basic space to the real line. The result is to define a new set function P_X on the Borel sets on the real line $\Re$. We note that this set function satisfies the defining properties for a probability measure

P1) $P_X(M) = P(X^{-1}(M)) = P(E) \geq 0.$

P2) $P_X(\Re) = P(X^{-1}(\Re)) = P(\Omega) = 1.$

P3) $P_X(\bigcup_i M_i) = P(X^{-1}(\bigcup_i M_i)) = P(\bigcup_i X^{-1}(M_i))$
$$= \sum_i P(X^{-1}(M_i)) = \sum_i P_X(M_i).$$

Thus, P_X is a *probability measure* on the class of Borel sets on the real line $\Re$. We have transformed the original abstract probability system $(\Omega, \mathcal{F}, P)$ into a new probability system $(\Re, \mathcal{B}, P_X)$. This new system has the probability mass distributed along the real line $\Re$. The Borel sets on the line play the role of events. P_X is the probability measure induced by X. The particular arrangement of the probability mass is referred to as the probability distribution for X.

This new, induced probability system is much more concrete in nature than the original system on the subsets of the basic space. We may visualize the probability mass spread along the line. The probability that X takes on a value in a given set of numbers is just the probability mass assigned to that set. This is particularly easy to visualize in the case of discrete random variables. The probability that any particular value is taken on is just the probability mass concentrated at that point. All probability questions may be answered in terms of this probability distribution. In the next chapter, we introduce analytical methods for describing a distribution, which facilitate visualization and make possible systematic calculations.

While the formal transfer procedure calls for specification of the probability mass assigned to every Borel set, some results of measure theory show that the distribution is determined completely and uniquely by the assignment of probabilities to intervals. In particular, the probability distribution is determined by the assignment of probability mass to each of the semi-infinite intervals of the form $(-\infty, t]$. We accept this fact without proof. Although the result is highly intuitive, a careful proof requires considerable theoretical development.

If two random variables have the same distribution, then in terms of probabilities of taking on various sets of values, they are indistinguishable. However, one must not make the mistake of supposing that if two random variables have the same distribution they are identical. The following simple examples illustrate this fact. In Chapter 10, we see that random variables may be "independent" yet have the same distribution.

Example 7.2.1: Different Random Variables with Identical Distributions

1. Suppose X is a random variable that takes on only the values 0 or 1. Let Y be the random variable defined by $Y = 1 - X$. Thus, Y has the value 1 when X has the value 0, and Y has the value 0 when X has the value 1. If $P(X = 0) = P(X = 1) = 1/2$, then it

follows that $P(Y = 0) = P(Y = 1) = 1/2$. Both random variables have the same distribution. They both have mass $1/2$ at 0 and at 1. Yet these random variables are never equal.

2. Suppose U has a distribution that spreads probability mass uniformly on the unit interval $[0, 1]$. Then the probability assigned to the semi-infinite interval $(-\infty, t]$ is $P(X \leq t)$. Graphically, we see that this probability is

$$0 \quad \text{for} \quad t \leq 0, \quad t \quad \text{for} \quad 0 \leq t \leq 1, \quad \text{and} \quad 1 \quad \text{for} \quad t \geq 1.$$

Now consider $V = 1 - U$. Then

$$\{V \leq t\} = \{1 - U \leq t\} = \{U \geq 1 - t\} = \{U < 1 - t\}^c.$$

Hence,

$$P(V \leq t) = 1 - P(U < 1 - t) = 1 - P(U \leq 1 - t).$$

The last equality is due to the fact that no probability mass is concentrated at point t, so that the mass in $(-\infty, t]$ is the same as the mass in $(-\infty, t)$. It makes no difference whether or not the endpoint is included. We have

$$P(V \leq t) = 1 - (1 - t) = t = P(U \leq t) \qquad \text{for } 0 \leq t \leq 1.$$

The proper values are obtained for $t < 0$ and $t > 1$, so that the distributions for the two random variables are the same. Yet these two random variables are equal only when the common value is $1/2$, an event that has zero probability.

—   □

7.3.   Simple Random Variables

Many random variables have only a finite set of possible values (i.e., finite range). Others may have an infinite range, but have values that are discrete (i.e., the range is an infinite sequence of numbers). The distributions for such random variables consist of point mass concentrations at the various possible values. As such, the distributions are easy to visualize. We consider next some useful methods of representing such random variables, which make possible precise formulation and handling of a variety of important problems. We restrict our treatment principally to the finite case. Extensions to discrete random variables with infinite range are usually obvious.

Definition 7.3.1. *A function X with finite range $\{t_i : 1 \leq i \leq n\}$ is called a* simple *function.*

Remark: If simple function X is a random variable, it is called a simple random variable.

The most elementary simple random variable, other than a constant function, is the indicator function for an event, as introduced in Section 2.4.

Example 7.3.1: Indicator Function for an Event

If E is any event, then its indicator function I_E is defined by

$$I_E(\omega) = \begin{cases} 1 & \text{for } \omega \in E \\ 0 & \text{for } \omega \in E^c \end{cases}.$$

The sigma algebra $\sigma(X)$ is the class $\{\emptyset, \Omega, E, E^c\}$. The distribution for X places mass $P(E)$ at 1 and mass $P(E^c) = 1 - P(E)$ at 0.

—   □

Remark: The first random variable in Example 7.2.1 is an indicator function. In fact, any random variable that takes on only the values 0 and 1 must be the indicator function for the event $E = \{X = 1\}$.

Simple Random Variables as Linear Combinations of Indicator Functions

We first express a simple random variable X in an elementary form. Suppose

X has range $\{t_1, t_2, \ldots, t_n\}$ with the t_i distinct.

For each i, $1 \le i \le n$, we let $A_i = \{X = t_i\}$. Since each $X(\omega)$ is one of these values, each ω belongs to one (and only one) of the A_i. That means the A_i form a partition. Now consider the expression

$$X = \sum_{i=1}^{n} t_i I_{A_i}.$$

Any ω belongs to one and only one A_i, so that $I_{A_i}(\omega) = 1$ and $I_{A_j}(\omega) = 0$ for $i \ne j$. For this ω, the sum has the value t_i, which is exactly the value of the random variable X. The sum thus gives the correct value for any ω. We call this expression the *canonical form* of the simple random variable X.

It is apparent that any linear combination of indicator functions of events must be a simple function. But not every such combination is in canonical form.

Example 7.3.2: The Number of Successes in n Trials

Suppose E_i is the event of a success on the ith trial in a Bernoulli sequence. Then, the random variable

$$X = \sum_{i=1}^{n} I_{E_i}$$

is the simple random variable whose value is the number of successes in the first n trials. The indicator functions I_{E_i} tally the successes as they occur. The expression is not the canonical form, however. We obtain that form as follows. Let A_{kn} be the event of exactly k successes in the first n trials. See the analysis of Bernoulli trials in Section 6.3 for the relation of the A_{kn} to the E_i. Put

$$X = \sum_{k=0}^{n} k I_{A_{kn}} \qquad \text{(cf)}.$$

This expression is in canonical form, which we have indicated by the symbol (cf) following the sum. For some purposes, the canonical form is highly desirable; for other purposes, the original sum, not in canonical form, is more useful. We see this, for example, in considering mathematical expectations.

— ◻

THE CLASS OF EVENTS DETERMINED BY A SIMPLE RANDOM VARIABLE

It is useful to determine the sigma algebra of events $\sigma(X)$ determined by a simple random variable X. It is easy to see, with the aid of the mapping diagram for the canonical form, what this class must be. To illustrate, suppose X has five possible values, t_1, t_2, t_3, t_4, and t_5. Then, as in the mapping diagram in Figure 7.3, every ω in A_1 is mapped into the point t_1, every point in A_2 is mapped into t_2, etc. To determine the inverse image of any Borel set M, simply determine which of the t_i lie in M. Then $X^{-1}(M)$ is the union of those A_i such that t_i is in M. If no t_i lies in M, then $X^{-1}(M) = \emptyset$. If all t_i lie in M, then $X^{-1}(M) = \Omega$. It is apparent that this pattern is quite general. We may thus assert

Theorem 7.3.2 (Events Determined by a Simple Random Variable) *Suppose X is a simple random variable whose canonical form is*

$$X = \sum_{i=1}^{n} t_i I_{A_i} \qquad \text{(cf)}.$$

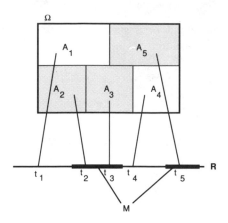

inverse image of M is shaded

FIGURE 7.3. Mapping produced by simple random variable.

Then the sigma algebra of events $\sigma(X)$ determined by X consists of the empty set, the basic space, and all unions of subclasses of the partition $\{A_i : 1 \leq i \leq n\}$.

—   □

DETERMINATION OF CANONICAL FORM

Any linear combination of indicator functions or such a combination plus a constant determines a simple random variable. We wish to derive the canonical form for such a combination. To keep notation simple, we illustrate with an example that displays the essential procedure.

Example 7.3.3: Determination of Canonical Form for a Simple Random Variable

Suppose random variable X is defined by

$$X = 3I_A - I_B - 2I_C + 2.$$

The key to our approach is the fact that the value of X is constant over each minterm generated by $\{A, B, C\}$.

If we evaluate $X(\omega)$ for one ω in each minterm, the possible values of X are displayed. Since the same value may be encountered on more than one minterm, the union of these minterms is the event for that value. It is possible that some minterm may be empty. This does not affect the formal expression, since an empty minterm contributes no elements to the union. Also, a minterm may not be empty but may be assigned zero probability mass. For our function, we may proceed systematically in tabular form.

Minterm	$3I_A$	$-I_B$	$-2I_C$	X
M_0	0	0	0	2
M_1	0	0	-2	0
M_2	0	-1	0	1
M_3	0	-1	-2	-1
M_4	3	0	0	5
M_5	3	0	-2	3
M_6	3	-1	0	4
M_7	3	-1	-2	2

Examination of the value column shows that the range consists of seven values

$$\{-1, 0, 1, 2, 3, 4, 5\}.$$

If we let $A_i = \{X = i\}$, then each A_i consists of one minterm, except $A_2 = M_0 \biguplus M_7$. Again, we have the possibility that some minterms are empty. Also, some minterms may have zero probability, even though they are not empty. These facts depend upon the relationships between the sets A, B, and C and the particular probability distribution.

— $\square$

APPROXIMATION OF A RANDOM VARIABLE BY A SIMPLE RANDOM VARIABLE

Some random variables have a continuum of possible values, so that their distributions do not produce any point mass concentrations. It is frequently desirable to be able to approximate such random variables by simple random variables (or at least discrete random variables). This is sometimes important in practical applications. It plays a very important role in developing the concept of mathematical expectation in Chapters 12, 13, and 14.

To simplify exposition, we consider the case that random variable X is bounded. That means there is some interval $[a, b]$ that contains all the values of X. We partition such an interval into n subintervals by use of the subdivision points

$$a = b_0 < b_1 < \cdots < b_{n-1} < b_n = b.$$

- For each i, $1 \leq i \leq n - 1$, we have the subintervals $M_i = [b_{i-1}, b_i)$. We set $M_n = [b_{n-1}, b_n]$.

- For each i, put $E_i = X^{-1}(M_i)$. Then $\{E_i : 1 \leq i \leq n\}$ is a partition.

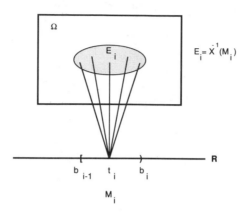

FIGURE 7.4. Mapping for approximating simple random variable.

Consider any of the M_i and the corresponding inverse image E_i (see Figure 7.4). Random variable X maps each point of E_i somewhere in the subinterval M_i. Pick an arbitrary point t_i in each subinterval M_i. Form a new random variable Y by assigning to each ω in E_i the value $Y(\omega) = t_i$. When this is done for each i, the result is that Y is the simple random variable

$$Y = \sum_{i=1}^{n} t_i I_{E_i} = \sum_{i=1}^{n} t_i I_{M_i}(X).$$

The last expression uses the fact $I_{E_i} = I_{M_i}(X)$, since $\omega \in E_i$ iff $X(\omega) \in M_i$. In any set E_i, the difference between the value $Y(\omega) = t_i$ and the value of $X(\omega)$, a point in M_i, cannot be any greater than the length of subinterval M_i. Since this is true for each i, we can assert

$$|X - Y| \leq \text{the length of the longest subinterval.}$$

- If we make the subdivision fine enough, by forming enough subintervals with sufficiently small maximum length, we are able to force the simple random variable Y to approximate X as closely as desired.

- If in each subinterval we take $t_i = b_{i-1}$, the smallest value in the interval, then we have

$$Y(\omega) \leq X(\omega) \qquad \text{for every } \omega.$$

NONNEGATIVE RANDOM VARIABLES AND NONDECREASING SEQUENCES OF SIMPLE RANDOM VARIABLES

Again, we suppose X is bounded, so that the range is included in an interval $[0, b]$. We use the approximation procedure with the convention of assigning

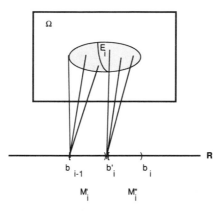

FIGURE 7.5. Approximating random variable for refined partition.

the smallest value in each subinterval. It should be clear from the preceding discussion that if we take a sequence of increasingly finer subdivisions, the sequence of approximating simple random variables Y_n should converge to the random variable X. The remaining fact to be established is that the sequence of approximating random variables can be made nondecreasing. That is, $Y_n(\omega) \leq Y_{n+1}(\omega)$ for every ω. We ensure this by requiring that each new subdivision should be a refinement of the previous. That is, the next subdivision has all the old subdivision points plus some new ones. This means that at least some of the old subintervals are subdivided. Suppose subinterval $M_i = [b_{i-1}, b_i)$ is divided into two subintervals $M_i' = [b_{i-1}, b_i')$ and $M_i'' = [b_i', b_i)$ (see Figure 7.5). For any ω in E_i, $Y_n(\omega) = b_{i-1}$, while $Y_{n+1}(\omega)$ has value either b_{i-1} or b_i'. Thus $Y_{n+1}(\omega) \geq Y_n(\omega)$ on E_i. Since this is true on each E_i, the desired inequality holds for all ω.

If X is unbounded but finite for every ω, one approximates increasingly closely over increasingly larger intervals on the real line. In the limit, the convergence occurs for all ω. The details are somewhat cumbersome to express precisely, but there are no essentially new ideas.

7a

Borel Sets, Random Variables, and Borel Functions

In Chapter 3, we note that the class of events is assumed to be a sigma algebra of subsets of the basic space Ω. In Appendix 2a, we characterize a sigma algebra of events as a class closed under complements and countable unions, and show that these conditions imply closure under countable intersections. Also, we introduce the notion of a sigma algebra generated by a class as the smallest sigma algebra that contains that class.

The definition of a random variable is motivated by the need to be able to ascribe probabilities to events of the form $\{X \in M\}$, where M is an interval or a logical combination of intervals. Considerable mathematical development has shown the desirability of considering the sigma algebra generated by the class of intervals. This is the class of Borel sets on the real line. It turns out that we can start with a class of intervals of any type and produce the same generated sigma algebra. It is customary to begin with the class of semi-infinite intervals of the form $(-\infty, t]$, for all real t.

Definition 7a.3. *The class of* Borel sets *on the real line is the sigma algebra generated by the class of semi-infinite intervals of the form* $(-\infty, t]$ *for all real t.*

It is not possible to give a constructive characterization of the class of Borel sets. Yet the class exists and is determined uniquely. Many important properties can be established. The class includes any sets likely to be needed in applications.

In Section 7.1, we define the class $\sigma(X)$ of events determined by X as the class of inverse images of Borel sets. Because of the preservation of set operations by inverse images (Theorem 7.1.2) and the closure of the class of Borel sets under complements and countable unions, it follows that $\sigma(X)$ is closed under complements and countable unions. Thus, as claimed after the definition of $\sigma(X)$ in Section 7.1, that class must be a sigma algebra.

The condition that the inverse images of Borel sets are events is called *measurability*. The term refers to the ability to assign probability measure to the inverse images. Another important class of measurable functions is the class of Borel functions.

Definition 7a.4. *If g is a real valued function of a single real variable, it is a* Borel function *iff the inverse image of every Borel set is a Borel set.*

A detailed theoretical examination shows that the class of Borel functions includes any continuous function and any function that is continuous except at discrete points where it has a jump discontinuity. If a sequence of Borel functions converges, the limit function is Borel. We accept these statements without proof, but note that they ensure that most any function encountered in ordinary applications will be a Borel function.

The class of Borel sets may be defined on Euclidean spaces of any dimension as the sigma algebra generated by multi-dimensional intervals. As a result, the notion of Borel functions is not limited to real valued functions of a single real variable. The functions may be vectore-valued, and the domain can be Euclidean spaces of any dimension.

One reason for the importance of Borel functions in probability theory is that a Borel function of a random variable is a new random variable. This fact is used repeatedly in Chapter 11. Some of the more pertinent facts are summarized in the discussion of mappings and measurability in Section 11.1 (see especially Theorems 11.1.1 and 11.1.2).

Determination of probabilities. The proper domain of a probability measure is a sigma algebra of sets. Sometimes, however, it is known that a probability exists on a sigma algebra, but the probabilities can be specified explicitly for a subclass. Our primary example is the probability measure P_X induced by a random variable. It is defined for all Borel sets. But it is only necessary to assign values in a consistent manner for all semi-infinite intervals of the type $(-\infty, t]$. This is sufficient to determine the probability uniquely for all Borel sets M. We do not attempt to develop these ideas. They are found under the general topic of extension and uniqueness of measures in theoretical treatments of probability or more general measure theory.

PROBLEMS

7–1. For the simple random variables described here,

(a) Determine the canonical form.

(b) Determine the induced probability distribution on the line.

1. $X = -3I_A + 2I_B + I_C$ with $\{A, B, C\}$ independent and $P(A) = 0.3$, $P(B) = 0.6$, and $P(C) = 0.4$.

2. $X = I_A + I_B + I_C$ with $\{A, B, C\}$ independent and $P(A) = P(B) = P(C) = 0.2$.

3. $X = -I_A + 3I_B + I_C$, where the minterms generated by $\{A, B, C\}$ have probabilities $p(0)$ through $p(7)$ of 0.04, 0.16, 0.13, 0.12, 0.22, 0.18, 0.07, and 0.08.

7–2. For the simple random variables described here in canonical form,

(a) List all of the events determined by X.

(b) Determine $P\big(X^{-1}(M_i)\big)$, where $M_1 = (1, 3]$, $M_2 = [0, 2]$, and $M_3 = (-\infty, -1]$.

1. $X = -4I_{A_1} - I_{A_2} + 3I_{A_3}$, with $P(A_1) = 0.4$, $P(A_2) = 0.5$, and $P(A_3) = 0.1$

2. $X = I_{A_1} + 2I_{A_2} + 3I_{A_3}$, with $P(A_1) = P(A_2) = 0.3$, and $P(A_3) = 0.4$

3. $X = -I_{A_1} + 2I_{A_2} + 4I_{A_3}$, with $P(A_1) = 0.2$, $P(A_2) = 0.3$, and $P(A_3) = 0.5$

7–3. In a Bernoulli trial, random variable X designates the number of the trial on which the first *failure* occurs. Suppose the probability p of success on any trial is 0.7. Determine

(a) $P(X = 7)$ (b) $P(X \geq 8)$.

Hint: Consider the events represented by each case.

7–4. A marker is placed on a straight line at a reference position, which we take to be the origin. A coin is tossed repeatedly. If a head appears, the marker is moved one unit to the right. If a tail appears, the marker is moved one unit to the left.

(a) After ten tosses, what are the possible positions and what are the probabilities of being in these positions?

(b) Show that the position at the end of ten tosses is given by the random variable

$$X = \sum_{i=1}^{10} I_{E_i} - \sum_{i=1}^{10} I_{E_i^c} = 2 \sum_{i=1}^{10} I_{E_i} - 10$$

where E_i is the event of a head on the ith trial.

7–5. A sequence of n trials (not necessarily independent) is performed. Let E_i be the event of success on the ith component trial. We associate with each trial a "payoff" function $X_i = aI_{E_i} + bI_{E_i^c}$. That is, an amount a is earned if there is a success on the ith trial, and an amount b (usually negative) is earned if there is a failure on that trial. Let N be the number of successes in the n trials and W be the net payoff. Show that $W = (a - b)N + bn$.

7–6. A gambler places three bets. He puts down one dollar for each bet. He picks up \$2.00 (his original wager plus one dollar) if he wins the first bet. Similarly, he picks up \$3.00 for a win on the second and \$5.00 for a win on the third. His net winning (in dollars) may be expressed by the random variable

$$X = 2I_A + 3I_B + 5I_C - 3$$

with $P(A) = 0.5$, $P(B) = 0.3$, and $P(C) = 0.2$. What are the possible values of X and the probabilities of taking each value?

8

Distribution and Density Functions

In Chapter 7, mapping properties of a random variable X are used to effect a transfer of probability mass from the basic space Ω to the real line $\Re$. The resultant probability distribution induced by X assigns to each Borel set M a mass equal to the probability that X takes a value in that set. The distribution induced by a discrete random variable X is easy to visualize since it concentrates probability mass at each value in the range of X.

In this chapter, we introduce the probability distribution function as a means of representing analytically any distribution induced by a real random variable. If the probability mass is spread smoothly on the real line, then there is also a probability density function. For many purposes this density function, when it exists, is easier to work with. After introducing these concepts, we illustrate by describing some standard distributions, which we use repeatedly in subsequent developments and applications.

8.1.   The Distribution Function

The primary analytical tool for representing the probability distribution induced by a real random variable is as simple as it is useful. For each real t, we set the value $F_X(t)$ to be the amount of the probability mass located at or to the left of point t on the real line.

Definition 8.1.1. *The (probability) distribution function F_X for a real random variable X is given by*

$$F_X(t) = P(X \le t) = P\big(X \in (-\infty, t]\big) = P_X\big((-\infty, t]\big) \qquad \text{for each real } t.$$

A number of important properties of the distribution function are simple consequences of the nature of the probability mass distribution on the line. Among these are

F1) For $a < b$, $F_X(b) - F_X(a) = P_X\big((a, b]\big) = P(a < X \le b) \ge 0$, so that F_X is nondecreasing.

F2) F_X is continuous from the right, with a jump in the amount $p_0 > 0$ at $t = t_0$ iff $P(X = t_0) = p_0$.

F3) Except in very special cases, requiring extended-valued random variables,

$$F_X(t) \to \begin{cases} 0 & \text{as } t \to -\infty \\ 1 & \text{as } t \to \infty \end{cases}.$$

We usually abbreviate the limits by writing

$$F_X(-\infty) = 0 \qquad \text{and } F_X(\infty) = 1.$$

It is quite possible to verify these properties by purely formal arguments; on the other hand, they are readily visualized in terms of the mass distribution representation. For example, by additivity of the probability measure P_X,

$$P_X\big((-\infty, b]\big) = P_X\big((-\infty, a]\big) + P_X\big((a, b]\big).$$

This is equivalent to

$$F_X(b) = F_X(a) + P_X\big((a, b]\big)$$

from which the equation in (F1) follows immediately. The nondecreasing character of F_X follows from the fact that $P(a < X \le b) = P_X\big((a, b]\big) \ge 0$. This result could be obtained geometrically by noting that the probability mass at or to the left of b is at least as great as the probability mass at or to the left of a.

Property (F2) may be visualized as follows. If there is a mass concentration of p_0 at t_0, then for any $t \ge t_0$ this mass is included in the determination of $F_X(t)$. As t is decreased to t_0, this mass is "seen" by F_X. If t approaches t_0 from the left, the point mass at t_0 is not seen. At exactly $t = t_0$ this mass is picked up, resulting in a jump in the amount p_0 between the value $F_X(t - 0)$ and $F_X(t)$. It is not difficult to write a formal proof of this fact, but the essential idea is included in the graphical-mechanical visualization.

Property (F3) simply states that the limiting values of the distribution function correspond to *excluding* all probability mass as $t \to -\infty$ or *including* all probability mass as $t \to \infty$. Exceptions occur only when the random quantity takes on "infinite values," so that some of the probability mass "escapes to infinity." This somewhat unusual behavior may be encountered in certain problems involving the limit of an infinite class of random variables in a random sequence or a continuous-parameter random process. For an example, see the discussion of counting random variables in Section 20.1. We note such exceptional cases when they occur, since they must be handled appropriately.

In the next section, we consider the distribution functions for several commonly used discrete distributions.

8.2.   Some Discrete Distributions

Distributions for discrete random variables (including simple random variables) are characterized by probability mass concentrations at the points in the range of the random variable. Suppose the range is $\{t_1, t_2, \ldots, t_n\}$, with $P(X = t_i) = p_i$ for each t_i in the range. Then $\sum_i p_i = 1$. The distribution function F_X is a step function, right-continuous, with jump of p_i at $t = t_i$. To obtain simple analytical expressions, we may use the right-continuous unit step function

$$u_+(t) = \begin{cases} 0 & \text{for } t < 0 \\ 1 & \text{for } t \geq 0 \end{cases} \qquad u_+(t - a) = \begin{cases} 0 & \text{for } t < a \\ 1 & \text{for } t \geq a \end{cases}.$$

Then

$$F_X(t) = \sum_i p_i u_+(t - t_i).$$

1. *Indicator function for event E.*

   The range is $\{0, 1\}$ $P(X = 0) = q = 1 - p$ $P(X = 1) = p$

   $$F_X(t) = q u_+(t) + p u_+(t - 1).$$

2. *Uniform on $\{t_1, t_2, \ldots, t_n\}$.*

   There are n point masses, each $1/n$, so that $P(X = t_i) = 1/n$, $1 \leq i \leq n$. Thus,

   $$F_X(t) = \frac{1}{n} \sum_{i=1}^{n} u_+(t - t_i).$$

Example 8.2.1: Random Numbers

Consider the random-number generator of Example 7.1.2. These are usually designed so that each of the numbers generated is equally likely. In practice, this is so nearly the case that the assumption is reasonable. If the random numbers consist of 0.0000 through 0.9999 in equal steps, then $n = 10,000$, so that

$$p_i = \frac{1}{n} = 10^{-4} \qquad \text{and} \qquad t_i = \frac{i - 1}{10^4} \quad 1 \leq i \leq 10^4.$$

— $\square$

3. *Binomial (n, p) with n a positive integer and $0 < p < 1$.*

   Our primary model for this distribution is the random variable that counts the number of successes in the first n trials of a Bernoulli sequence with probability p of success on any trial. As our analysis of Bernoulli sequences in Section 6.3 shows

$$P(X = k) = p_k = C(n, k)p^k q^{n-k} \quad \text{so that } F_X(t) = \sum_{k=0}^{n} p_k u_+(t - k).$$

For any specific choice of n and p, the values of p_k can be calculated or found in a table of the binomial distribution (see Section 6.3).

A related random variable $Y = n - X$ counts the number of failures in the first n trials and is binomial (n, q). This follows from the fact that a "success" for Y is a failure for X.

4. *Geometric* (p) with $0 < p < 1$.

There are two related distributions called the geometric distribution by various authors. We distinguish them later by indicating a primary model or example for the distribution.

(a) Let X count the number of failures in a Bernoulli sequence before the first success. This is called the waiting time to the first success. The event $\{X = k\}$ is the event of k consecutive failures followed by a success. For a Bernoulli sequence, with probability p of success on any trial, the event of k failures then a success has probability

$$P(X = k) = pq^k \quad 0 \le k.$$

We say X has the *geometric* distribution with parameter (p). We usually shorten this to the statement X is *geometric* (p).

(b) Let Y be the number of the trial on which the first success occurs. This is the *time* of the first success. If the first success comes on trial k, then it is preceded by $k - 1$ failures, so that

$$P(Y = k) = pq^{k-1} \quad 1 \le k.$$

The argument also shows that $Y = X + 1$ or $X = Y - 1$, so that we designate the distribution for Y by saying $Y - 1$ is *geometric* (p).

If X is geometric (p), then the graph of F_X has jumps in the amounts pq^k at $t = 0, 1, 2, \ldots$, so that

$$F_X(t) = \sum_{k=0}^{\infty} pq^k u_+(t - k).$$

For integer values of the argument, the value of the distribution function can be expressed simply as

$$F_X(n) = P(X \le n) = \sum_{k=0}^{n} pq^k = p\frac{1 - q^{n+1}}{1 - q} = 1 - q^{n+1}.$$

Thus, $P(X \geq n) = 1 - F_X(n-1) = q^n$. In terms of the Bernoulli sequence, the waiting time is n or more iff there are n successive failures.

AN IMPORTANT PROPERTY OF THE GEOMETRIC DISTRIBUTION

$$P(X \geq n + k | X \geq n) = \frac{P(X \geq n + k)}{P(X \geq n)} = \frac{q^{n+k}}{q^n} = q^k = P(X \geq k).$$

This is an expression of the independence of the successive events in a Bernoulli sequence. If no success has been observed by trial n, then the probability (conditional) of waiting at least k more trials is the same as the probability of waiting for k or more trials at the beginning of the sequence. The sequence essentially "starts over" at each trial.

Example 8.2.2:

In a Bernoulli sequence with probability $p = 0.3$ of success on any trial, what is the probability that the waiting time to the first success is no greater than three?

Solution

According to the preceding discussion, $P(X \leq 3) = 1 - q^4 \approx 0.760$. This is the probability of not four successive failures.

— $\square$

5. *Negative binomial (or Pascal) (m, p).*

This distribution is a generalization of the geometric distribution. We take as our basic example a Bernoulli sequence and consider two related random variables. Let

$X_m =$ the number of failures before the mth success.
$Y_m =$ the number of the trial on which the mth success occurs.

We note that X_1 is geometric (p). Also $X_m = Y_m - m$ or $Y_m = X_m + m$.

Now $Y_m = k \geq m$ iff there are $m - 1$ successes in the first $k - 1$ trials and a success on trial number k. Hence

$$\begin{aligned} P(Y_m = k) &= pC(k-1, m-1)p^{m-1}q^{k-m} \\ &= C(k-1, m-1)p^m q^{k-m} \quad k \geq m. \end{aligned}$$

It follows that

$$P(X_m = k) = P(Y_m = k + m) = C(m + k - 1, m - 1)p^m q^k \, 0 \le k.$$

By using extended binomial coefficients (see the appendix on mathematical aids), we may express

$$
\begin{aligned}
C(m + k - 1, m - 1) \\
&= \frac{(m + k - 1)(m + k - 2)\ldots(m + 1)m}{k!} \\
&= (-1)^k \frac{(-m)(-m - 1)(m - 2)\ldots(-m - k + 1)}{k!} \\
&= (-1)^k C(-m, k)
\end{aligned}
$$

so that

$$P(X_m = k) = C(-m, k)p^m(-q)^k.$$

Use of some facts on the power series expansion of $(1 - q)^{-m}$ shows that the sum of the probabilities is one, as required. Hence, we say $X_m = Y_m - m$ has a negative binomial distribution.

Example 8.2.3:

In a Bernoulli sequence with probability $p = 0.3$ of success on any trial, what is the probability that the second success comes no later than the 5th trial?

Solution

$$
\begin{aligned}
P(Y_2 \le 5) &= \sum_{k=2}^{5} P(Y_2 = k) \\
&= p^2 \left(C(1,1)q^0 + C(2,1)q^1 + C(3,1)q^2 + C(4,1)q^3 \right) \\
&= 0.3^2 \left(1 + 2 \cdot 0.7 + 3 \cdot 0.7^2 + 4 \cdot 0.7^3 \right) \approx 0.472.
\end{aligned}
$$

—   □

6. *Poisson* (μ) with $\mu > 0$

The Poisson distribution is used in a wide variety of applications. In Section 11.4, the quantile function is used to obtain several important approximation theorems involving the Poisson distribution. The discussion in Section 21.1 shows that the Poisson distribution arises naturally in a significant class of counting processes that are used as models in many important applications.

A random variable X with the Poisson distribution takes on nonnegative integer values with

$$p_k = P(X = k) = e^{-\mu}\frac{\mu^k}{k!} 0 \le k.$$

Use of the power series expansion for e^x about the origin shows that the probabilities sum to one.

Because of the extensive use of the Poisson distribution in applications, tables of values of the probabilities for various values of the parameter μ are readily available. Generally, there are two types of tables:

$$\begin{array}{ll} \text{Individual terms:} & p_k = P(X = k) \\ \text{Cumulative terms:} & P(X \ge n) = \sum_{k=n}^{\infty} p_k. \end{array}$$

As in the case of tables of the binomial distribution (see Section 6.3), a table of cumulative terms may be used to obtain individual terms.

POISSON APPROXIMATION TO THE BINOMIAL DISTRIBUTION

In many important problems, the Poisson distribution can be used to approximate the binomial distribution. Suppose X is binomial (n, p). If n is large, but p is small enough to keep the product np moderate in value, then X is approximately Poisson $(\mu = np)$. That is,

$$P(X = n) = C(n, k)p^k(1 - p)^{n-k} \approx e^{-np}\frac{(np)^k}{k!}.$$

Suppose $p = \mu/n$, with n large, but $\mu/n < 1$. Then we have

$$\begin{aligned} P(X = k) &= C(n, k)(\mu/n)^k(1 - \mu/n)^{n-k} \\ &= \frac{n(n-1)\dots(n-k+1)}{n^k}\left(1 - \frac{\mu}{n}\right)^{-k}\left(1 - \frac{\mu}{n}\right)^n \frac{\mu^k}{k!}. \end{aligned}$$

The first factor is the ratio of polynomials in n with the same degree k, hence goes to 1 as n becomes large. The second factor approaches 1 as n becomes large. A well-known property of the exponential shows

$$\left(1 - \frac{\mu}{n}\right)^n \to e^{-\mu} \qquad \text{as} \quad n \to \infty.$$

As a result, for large n, $P(X = k) \approx e^{-\mu}\frac{\mu^k}{k!}$. An alternate derivation in Example 11.4.2 gives a somewhat stronger result, with an indication of how rapidly the convergence takes place.

8.3.   Absolutely Continuous Random Variables and Density Functions.

If a random variable X assigns zero mass to every Borel set of zero length, then X is said to be *absolutely continuous*. This means that the probability mass is distributed smoothly, with no concentration of mass at any point. In this case, there is a (probability) *density function* f_X such that

$$P(X \in M) = P_X(M) = \int_M f_X .$$

The density function has three characteristic properties:

(f1) $f_X \geq 0$ (f2) $\int_{\Re} f_X = 1$ (f3) $F_X(t) = \int_{-\infty}^{t} f_X .$

Remarks

1. Technically, the condition of "zero length" is the condition of "zero Lebesgue measure." The existence of the density function with the desired properties is a special case of the Radon-Nikodym theorem, discussed briefly in Section 13.8. Practically, if the distribution for X has no point mass concentrations, the existence of the density function may be assumed.

2. Strictly speaking, the integral in the definition is a Lebesgue integral, as discussed in Section 13.6. However, for practical densities whose ordinary Riemann integral exists, the Lebesgue integral coincides with the Riemann integral.

3. It is customary in the literature on probability to omit the indication of the region when integrating over the whole space. Thus we write $\int f_X$ for $\int_{\Re} f_X$. Note that $\int f_X$ is *not* an indefinite integral, but a definite integral over the whole space.

4. In any practical case, the density function will be integrable in the ordinary Riemann sense. That means we can appeal to the fundamental theorem of calculus to assert

$$f_X(t) = F_X'(t) \qquad \text{at every point of continuity of } f_X .$$

Any integrable, nonnegative function f on $\Re$ is a candidate for a density function iff $\int f = 1$. If the latter equation does not hold, a constant factor may be employed to provide the necessary "normalization." A density function determines a distribution function, which determines a probability distribution on $\Re$. This ensures the existence of a random variable having this distribution (see Section 11.4).

Example 8.3.1:

Suppose $f_X(t) = 2t \ 0 \le t \le 1$ (and zero elsewhere). Then

$$F_X(t) = \int_{-\infty}^{t} f_X = \begin{cases} 0 & \text{for } t < 0 \\ t^2 & \text{for } 0 \le t \le 1 \\ 1 & \text{for } 1 < t \end{cases}.$$

We note that $f_X(t) = F_X'(t)$ for $t \ne 0, 1$.

— $\qquad\qquad\qquad\qquad\qquad\qquad\qquad\qquad\qquad\qquad$ $\square$

Remark: The notion of a density function, as described earlier, precludes any point mass concentrations in the distribution. However, in some engineering treatments of density, the notion of "impulse function" or "delta function" is used to extend densities to describe mixed distributions. This provides a convenient formalism in many cases, provided proper care is exercised. We do not use this convention in this book.

8.4. Some Absolutely Continuous Distributions

In this section, we describe several absolutely continuous distributions and their density functions. These will be used in a variety of examples and applications in subsequent chapters.

1. *Uniform on (a, b).*

    The probability mass is spread uniformly on the interval $[a, b]$. It is immaterial whether or not the endpoints are included, since zero probability mass is assigned to any point. The probability that X takes a value on any subinterval of $[a, b]$ is proportional to the length of the subinterval. The constant of proportionality must be the reciprocal of the length $b - a$ of the interval on which the mass is spread. Thus,

    $$f_X(t) = \frac{1}{b - a} \quad \text{for } a < t < b \qquad \text{and is zero elsewhere.}$$

    There is no probability mass at or to the left of a, and all probability mass is at or to the left of b. Between a and b, the amount of mass at or to the left of t increases linearly with t. Thus, the distribution function is

    $$F_X(t) = \frac{t - a}{b - a} \qquad \text{for } a < t < b.$$

    This distribution is used when it is known that the value of X is between a and b, but there is no reason to suppose that an observed value is more likely to lie in one part of the interval than any other.

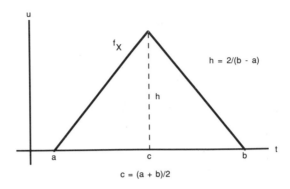

FIGURE 8.1. Density function for symmetric triangular distribution.

2. *Symmetric triangular* $[a, b]$.

The density f_X has a graph as shown in Figure 8.1. This is an isosceles triangle with base on the interval $[a, b]$. In order for the area to be one, the height is $2/(b - a)$ at $t = c = (a + b)/2$. From these facts, expressions for f_X and F_X may be obtained. The expression for f_X is

$$f_X(t) \quad = \quad \frac{4}{(b - a)^2}(t - a) \qquad \text{for } a \leq t \leq c$$

$$f_X(t) \quad = \quad \frac{2}{b - a} - \frac{4}{(b - a)^2}(t - c) \qquad \text{for } c \leq t \leq b.$$

We have used a convention in which f_X is described over the range of X and is understood to be zero elsewhere. The expression for F_X is obtained by elementary integration.

$$F_X(t) \quad = \quad 2\frac{(t - a)^2}{(b - a)^2} \qquad \text{for } a \leq t \leq c$$

$$F_X(t) \quad = \quad 1 - 2\frac{(t - b)^2}{(b - a)^2} \qquad \text{for } c \leq t \leq b.$$

The symmetric triangular distribution appears as the distribution of the sum or difference of independent random variables with uniform distributions over intervals with a common length (see Section 18.2).

3. *Exponential* (λ) $\lambda > 0$.

This distribution is attractive because of some analytical properties of the exponential function. In addition, it displays some probabilistic properties that make it a reasonable and useful model for "length of life" or "time to failure" of a variety of devices and systems. Reasons for this are developed in the discussion of reliability in Section 8.6. The distribution can be defined directly in terms of its distribution function F_X or of its density function f_X. A nonnegative random variable X has the exponential distribution with parameter (λ) iff

$$F_X(t) = P(X \le t) = u(t)\left(1 - e^{-\lambda t}\right) \qquad \lambda > 0.$$

The unit step function u is zero for negative values of argument and one for positive values of argument. The density function is obtained as the derivative of the distribution function

$$f_X(t) = u(t)\lambda e^{-\lambda t}.$$

A CHARACTERISTIC PROPERTY OF THE EXPONENTIAL DISTRIBUTION

The exponential distribution is the only absolutely continuous distribution that has the property that

$$P(X > t + h \mid X > t) = P(X > h), \qquad h > 0, \quad t \ge 0.$$

Suppose a device has exponentially distributed time to failure. If it survives to time t, then the probability (conditional) of surviving h units into the future is the same as the original probability of surviving for h units of time. In reliability analysis (see Section 8.6), this suggests a device that does not "age" or "wear out." Failure is due to conditions that do not change with time or use. This is typical of certain electronic devices, in which failure is due to the impact of external perturbations (say supply voltage fluctuations or ambient temperature variations) rather than degeneration of the device itself.

Proof of the Exponential Property

Since $X > t + h$, $h > 0$ implies $X > t$, we have $\{X > t + h, X > t\} = \{X > t + h\}$ so that

$$P(X > t + h \mid X > t) = \frac{P(X > t + h)}{P(X > t)}.$$

(a) If X is exponential (λ), then $P(X > t + h \mid X > t) = e^{-\lambda(t+h)}/e^{-\lambda t} = e^{-\lambda h} = P(X > h)$.

(b) If the equality holds, then $P(X > t)P(X > h) = P(X > t + h)$. Put $\phi(t) = P(X > t)$, $t \ge 0$.

Then ϕ satisfies $\phi(t + h) = \phi(t)\phi(h)$.

By the product form of Cauchy's equation (see the appendix on mathematical aids), there is a number $a \ne 0$

such that $\phi(t) = e^{at}$, $t > 0$. Since $\phi(t) \to 0$, as $t \to \infty$, the number a must be negative. Let $a = -\lambda$, with $\lambda > 0$; then $\phi(t) = 1 - F_X(t) = e^{-\lambda t}$, $t > 0$.

$\square$

4. *Gamma* (α, λ) $\qquad \alpha > 0$, $\lambda > 0$.

A nonnegative random variable X has the gamma distribution with parameters α, λ iff

$$f_X(t) = u(t)\frac{\lambda^{\alpha} t^{\alpha-1} e^{-\lambda t}}{\Gamma(\alpha)}.$$

The gamma function Γ is given by (see the appendix on mathematical aids)

$$\Gamma(\alpha) = \int_0^{\infty} u^{\alpha-1} e^{-u} \, du \qquad \alpha > 0.$$

For $\alpha = n$, a positive integer, $\Gamma(n) = (n-1)!$. It is apparent that $\int_0^{\infty} f_X = 1$, so that f_X is a probability density function.

Remark: For $\alpha = 1$, $f_X(t) = u(t)\lambda e^{-\lambda t}$ so that X is exponential (λ). In Section 18.2, we show that the sum of n independent random variables, each exponential (λ), is gamma (n, λ).

Theorem 8.4.1 (Relationship Between the Gamma and the Poisson Distributions)

(a) *If X is gamma (n, λ), then $P(X \leq t) = P(Y \geq n)$, where Y is Poisson (λt), $t > 0$.*

(b) *If Y is Poisson (λt), then $P(Y \geq n) = P(X \leq t)$, where X is gamma (n, λ).*

$\square$

Proof

Use of one of the integrals in the appendix on mathematical aids shows that

$$\frac{\lambda^n}{(n-1)!} \int_0^t u^{n-1} e^{-\lambda u} \, du = 1 - e^{-\lambda t} \sum_{k=0}^{n-1} \frac{(\lambda t)^k}{k!} = e^{-\lambda t} \sum_{k=n}^{\infty} \frac{(\lambda t)^k}{k!}.$$

The left-hand member is $P(X \leq t)$ for X gamma (n, λ); the right-hand member is $P(Y \geq n)$ for Y Poisson (λt).

$\square$

Example 8.4.1:

The times (in minutes) between emergency calls to the police dispatch unit in a small city are independent random quantities, each distributed exponentially with parameter ($\lambda = 1$). What is the probability of eight or more calls in five minutes?

Solution

Let X be the time of arrival of the eighth call. There are eight or more calls in five minutes iff $X \leq 5$. According to the second remark, X is gamma $(8, 1)$. Using Theorem 8.4.1, with $n = 8$, $\lambda = 1$, $t = 5$, we have

$$P(X \leq 5) = 1 - e^{-5} \sum_{k=0}^{7} \frac{5^k}{k!} = e^{-5} \sum_{k=8}^{\infty} \frac{5^k}{k!} \approx 0.133.$$

This value can be calculated directly from the second expression, or obtained from a table of cumulative terms for the Poisson (5) distribution.

—   □

5. *Beta* (r, s) $r > 0$, $s > 0$.

In many investigations we consider absolutely continuous random variables whose range is the unit interval $(0, 1)$ (see, for example, Section 22.2 on Bayesian statistics). Often the exact form of the density is not known, but there is some information that indicates the density peaks near a certain point. The Beta distribution, which has two positive parameters, has a density that takes on a variety of forms according to the choice of the parameters (see Figure 8.2). Thus it is possible, in many situations to fit a theoretical density to empirical data.

Many of the mathematical properties of the Beta distribution are based on the definite integral

$$B(r, s) = \frac{\Gamma(r)\Gamma(s)}{\Gamma(r + s)} = \int_0^1 t^{r-1}(1 - t)^{s-1} \, dt \qquad r > 0, \ s > 0.$$

Random variable X, with range $(0, 1)$, has the Beta distribution with positive parameters (r, s) iff its density function is

$$f_X(t) = \frac{\Gamma(r + s)}{\Gamma(r)\Gamma(s)} t^{r-1}(1 - t)^{s-1} \qquad 0 < t < 1, \ r > 0, \ s > 0.$$

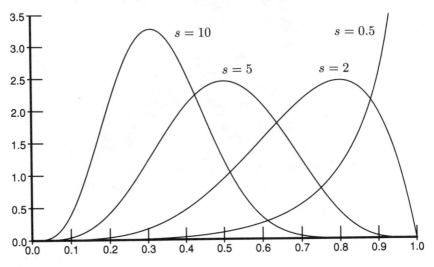

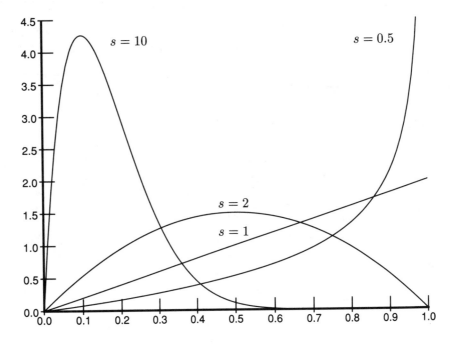

$$f_X(t) = \frac{\Gamma(r+s)}{\Gamma(r)\Gamma(s)} t^{r-1}(1-t)^{s-1} \qquad 0 < t < 1, \ r > 0, \ s > 0$$

FIGURE 8.2. Beta density function for various parameter values.

If $r \geq 1$, $s \geq 1$, the endpoints are included in the range.

If $r \geq 2$, $s \geq 2$, f_X has a maximum at

$$t = t_0 = \frac{r-1}{(r-1)+(s-1)} = \frac{r-1}{r+s-2}.$$

For r, s positive integers, $f_X(t)$ is given by a polynomial on $[0, 1]$, and is zero elsewhere. For example

r	s	$f_X(t)$ for $0 \leq t \leq 1$	Max at
1	1	1	
1	2	$2(1-t)$	
1	5	$5(1-t)^4$	
2	1	$2t$	
2	2	$6t(1-t)$	$1/2$
2	5	$30t(1-t)^4$	$1/5$

The distribution function F_X is given by

$$F_X(t) = \frac{1}{B(r,s)} \int_0^t u^{r-1}(1-u)^{s-1}\, du \qquad 0 < t < 1.$$

Since the antiderivative of a polynomial is a polynomial, for r, s positive integers, $F_X(t)$ is given by a polynomial in t for $0 < t < 1$. Thus, for $r = 2$, $s = 2$,

$$F_X(t) = \int_0^t (6u - 6u^2)\, du = 3t^2 - 2t^3 \qquad 0 \leq t \leq 1.$$

For the general case, the integral defines the *incomplete beta function*. This function has been studied intensively, and a variety of special forms, approximations, and relations to other distributions have been found. These are summarized in several handbooks of mathematical functions, which also include tables of values of the incomplete beta function.

Example 8.4.2:

Suppose random variable X has a Beta distribution and that its density function peaks at $t_0 = 2/3$. Determine the probability $0.6 \leq X \leq 0.7$.

Solution

The condition $t_0 = 2/3$ is met if $r - 1 = 2$ and $r + s - 2 = 3$. This gives $r = 3$, $s = 2$, so that

$$f_X(t) = 12(t^2 - t^3) \qquad \text{and} \qquad F_X(t) = 4t^3 - 3t^4 \qquad 0 \leq t \leq 1.$$

We thus have

$$P(0.6 \le X \le 0.7) = F_X(0.7) - F_X(0.6) \approx 0.1765.$$

———   □

6. *Weibull* (α, λ, ν) $\alpha > 0, \ \lambda > 0, \ \nu \ge 0$

A nonnegative random variable X has the Weibull distribution with parameters α, λ, and ν iff

$$F_X(t) = u(t - \nu)\left(1 - e^{-\lambda(t-\nu)^\alpha}\right)$$

and

$$f_X(t) = u(t - \nu)\alpha\lambda(t - \nu)^{\alpha-1}e^{-\lambda(t-\nu)^\alpha}.$$

The parameter $\nu \ge 0$ serves to shift the distribution so that the range starts at $t = \nu$. We generally take $\nu = 0$, so that

$$F_X(t) = u(t)\left(1 - e^{-\lambda t^\alpha}\right) \quad \text{and} \quad f_X(t) = u(t)\alpha\lambda t^{\alpha-1}e^{-\lambda t^\alpha}.$$

By proper choice of the parameters $\alpha > 0$ and $\lambda > 0$, a variety of density function shapes may be realized (see Figure 8.3).

The Weibull distribution appears as a generalization of the exponential distribution.

- For $\nu = 0$, $\alpha = 1$, $F_X(t) = u(t)(1 - e^{-\lambda t})$, so that X is exponential (λ).

- Suppose Y is exponential (1). Put $X = (Y/\lambda)^{1/\alpha}$. Then

$$F_X(t) = P\left(\left(\frac{1}{\lambda}Y\right)^{1/\alpha} \le t\right) = P(Y \le \lambda t^\alpha)$$
$$= 1 - e^{-\lambda t^\alpha} \qquad \text{for } t \ge 0.$$

Hence, X is Weibull $(\alpha, \lambda, 0)$. This modification of the exponential distribution makes the Weibull distribution useful as a model in many reliability problems (see Section 8.6).

Remark: The Weibull distribution is often expressed in terms of parameters (α', β', ν), as follows:

$$F_X(t) = u(t - \nu)\left(1 - e^{-((t-\nu)/\alpha')^{\beta'}}\right)$$

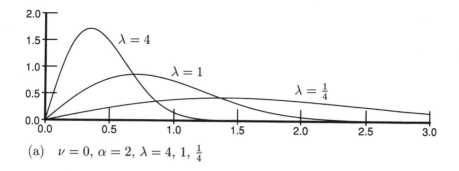

(a) $\nu = 0$, $\alpha = 2$, $\lambda = 4$, 1, $\frac{1}{4}$

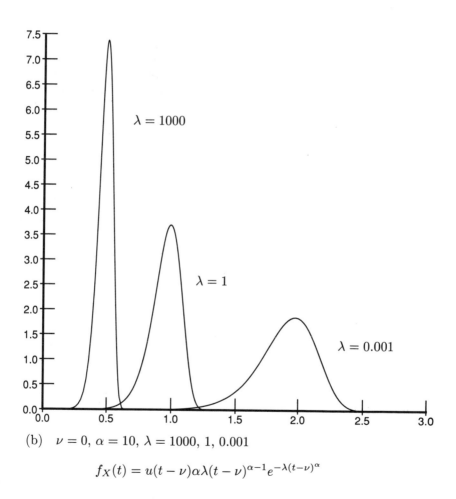

(b) $\nu = 0$, $\alpha = 10$, $\lambda = 1000$, 1, 0.001

$$f_X(t) = u(t - \nu)\alpha\lambda(t - \nu)^{\alpha-1}e^{-\lambda(t-\nu)^{\alpha}}$$

FIGURE 8.3. Weibull density function for various parameter values.

with corresponding expression for f_X obtained by differentiation. Now ν plays the same role in both forms (providing a shift of the distribution). The other parameters are related as follows

$$\beta' = \alpha, \qquad \left(\frac{1}{\alpha'}\right)^{\beta'} = \lambda.$$

We use the first form, since this displays the relationship with the exponential more directly.

8.5.   The Normal Distribution

One of the most frequently used distributions in both applied and theoretical probability is the *normal* or *Gaussian* distribution. The central limit theorem (see Section 18.6) shows that the sum of a sufficiently large number of independent random variables is approximately normal. Many random quantities (e.g., electrical noise or measurement errors) can be considered the result of adding a large number of random quantities. A normally distributed random variable is absolutely continuous and can take on any real value (positive or negative). The distribution is characterized by two parameters, traditionally designated μ and σ. The probability mass is distributed symmetrically about the point $t = \mu$. The parameter μ, which can be any real number, thus determines the center of the distribution. The parameter σ gives an indication of how the probability is spread around the center of the distribution.

It is not possible to express the distribution function in terms of any of the ordinary functions of mathematics. On the other hand, the density function is expressed directly, as follows

$$f_X(t) = \frac{1}{\sigma\sqrt{2\pi}} \exp\left(-\frac{1}{2}\left(\frac{t-\mu}{\sigma}\right)^2\right) \qquad \text{for all real } t.$$

This means that the distribution function is

$$F_X(t) = \int_{-\infty}^{t} f_X.$$

Notation

It is often convenient to designate the fact that X is normal with parameters μ and σ (or σ^2) by the notation $X \sim N(\mu, \sigma^2)$. Thus, if $X \sim N(7,4)$, then X has the normal distribution, with parameter $\mu = 7$ and parameter $\sigma = \sqrt{4} = 2$.

The integral for the distribution function must be evaluated by numerical methods, since it cannot be expressed in closed form in terms of ordinary

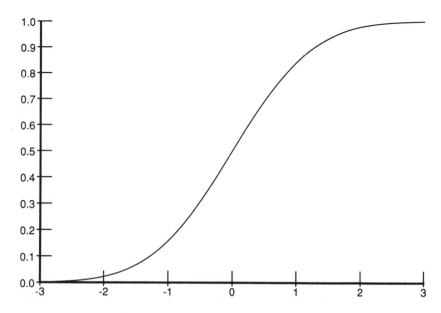

FIGURE 8.4. Density and distribution functions for the normal distribution.

functions of mathematics. In fact, this integral expression defines a new function. Extensive tables for the special case $\mu = 0$ and $\sigma = 1$ are readily available. We see later that some simple properties of the distribution make it possible to use these special tables to evaluate the distribution for any choice of the parameters μ, σ.

Figure 8.4 shows graphs of the density function f_X and the distribution function F_X. A study of either the graph or the analytical expression for f_X shows that the density is symmetric about the point $t = \mu$. Since e^{u^2} has a maximum at $u = 0$ and is symmetrical about $u = 0$, it follows that f_X has a maximum at $t - \mu = 0$ or $t = \mu$. Symmetry about $t - \mu = 0$ is equivalent to symmetry about $t = \mu$. This means that half of the probability mass lies to the left and half to the right of this point. Also, the density is greatest near $t = \mu$, so that probability mass tends to be concentrated about that point. If parameter σ is made larger, then the maximum value of f_X is smaller and values drop off less rapidly with distance of t from μ. Hence, larger σ means greater spread and smaller σ means greater concentration of probability mass about $t = \mu$.

STANDARDIZED NORMAL DISTRIBUTION

If $X \sim N(0, 1)$, we say X is *standardized normal*. In this case

$$\phi(t) = f_X(t) = \frac{1}{\sqrt{2\pi}} e^{-t^2/2} \quad \text{and} \quad \Phi(t) = F_X(t) = \int_{-\infty}^{t} \phi.$$

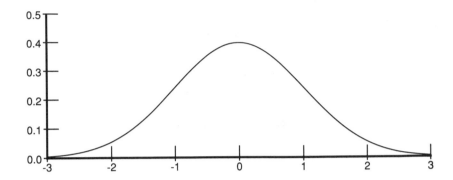

FIGURE 8.5. Density function for the standardized normal distribution.

The graph for the density ϕ is shown in Figure 8.5. Reference to either the graph or the analytical expression shows the curve is symmetric about the origin (i.e., f_X is an *even* function). The area to the left of the origin is one half. At any point $t = a$, the area under the density curve to the left of a is $\Phi(a) = P(X \leq a)$. Extensive tables for the standardized normal distribution are available in most mathematical handbooks or books of tables. Some care is needed in using such tables since there are several variations in common use. Usually, these tables are of one of the two following types.

- Values of $\Phi(a) = P(X \leq a)$ are given for positive values of a. These values are equal to the area under the density curve to the left of point a.

- Values of $\Phi(a) - 0.5 = P(0 \leq X \leq a)$ are given for positive values of a. These values are equal to the area under the density curve between point 0 and point a.

Such tables are equally convenient to use. Suppose we have the first type of table with values $\Phi(a)$ for a positive. We may use the symmetry of the density curve to obtain values for negative arguments. Let a be a positive number; we wish to evaluate $\Phi(-a)$. Now this is the area to the left of point $-a$. By symmetry (see Figure 8.5), the area to the left of point $-a$ is equal to the area to the right of point a. The latter area is $1 - \Phi(a)$.

Example 8.5.1:

Suppose $X \sim N(0, 1)$.
Determine $P(X \leq 1.5)$, $P(X \leq -1)$, and $P(-1 \leq X \leq 2)$.

Solution

We use a table of values of $\Phi(a)$ for a positive.

1. Read $P(X \leq 1.5) = \Phi(1.5) = 0.9332$ directly from the table.

2. $P(X \leq -1)$ is the area under the density curve to the left of -1. By symmetry, this is the same as the area to the right of point 1. This is

$$1 - P(X \leq 1) = 1 - \Phi(1) = 1 - 0.8413 = 0.1587.$$

3. $P(-1 \leq X \leq 2)$ is the area under the density curve between points -1 and 2. This is the area to the left of 2 minus the area to the left of -1. Thus

$$
\begin{aligned}
P(-1 \leq X \leq 2) &= \Phi(2) - \big(1 - \Phi(1)\big) = \Phi(2) + \Phi(1) - 1 \\
&= 0.9772 + 0.8413 - 1 = 0.8185.
\end{aligned}
$$

$\square$

Remark: One could reduce the procedures in these and other cases to a set of rules, so that it is only necessary to find the required values in the table and combine them according to the appropriate rule. However, it is quite simple to deal with each case as it arises by using the geometric interpretation and the symmetry of the density curve. A sketch is usually helpful.

STANDARDIZED TABLES AND THE GENERAL CASE

We show next how to express $F_X(t)$ for the general case in terms of values of the distribution Φ for the standardized case. See Section 11.1 for an alternate approach.

Suppose $X \sim N(\mu, \sigma^2)$. An examination of the defining expressions shows that

$$f_X(t) = \frac{1}{\sigma} \phi \left(\frac{t - \mu}{\sigma} \right).$$

Hence

$$F_X(t) = \int_{-\infty}^{t} f_X = \int_{-\infty}^{t} \phi \left(\frac{x - \mu}{\sigma} \right) \frac{1}{\sigma} \, dx.$$

Make the change of variable $u = (x - \mu)/\sigma$, so that $du = 1/\sigma \, dx$. The limits of integration become $u = -\infty$ and $u = (t - \mu)/\sigma$. As a consequence, we have

$$F_X(t) = \int_{-\infty}^{\frac{t-\mu}{\sigma}} \phi(u) \, du = \Phi \left(\frac{t - \mu}{\sigma} \right).$$

Thus, if we have the standardized normal distribution function Φ for $N(0, 1)$, we can determine values of the distributions function F_X for any random variable $X \sim N(\mu, \sigma^2)$.

TABLE 8.1. Standardized Normal Distribution Function

$$\Phi(t) = \frac{1}{\sqrt{2\pi}} \int_{-\infty}^{t} e^{-u^2/2}\, du \qquad \Phi(-t) = 1 - \Phi(t)$$

	0.00	0.01	0.02	0.03	0.04	0.05	0.06	0.07	0.08	0.09
0.0	0.5000	0.5040	0.5080	0.5120	0.5160	0.5199	0.5239	0.5279	0.5319	0.5359
0.1	0.5398	0.5438	0.5478	0.5517	0.5557	0.5596	0.5636	0.5675	0.5714	0.5753
0.2	0.5793	0.5832	0.5871	0.5910	0.5948	0.5987	0.6026	0.6064	0.6103	0.6141
0.3	0.6179	0.6217	0.6255	0.6293	0.6331	0.6368	0.6406	0.6443	0.6480	0.6517
0.4	0.6554	0.6591	0.6628	0.6664	0.6700	0.6736	0.6772	0.6808	0.6844	0.6879
0.5	0.6915	0.6950	0.6985	0.7019	0.7054	0.7088	0.7123	0.7157	0.7190	0.7224
0.6	0.7257	0.7291	0.7324	0.7357	0.7389	0.7422	0.7454	0.7486	0.7517	0.7549
0.7	0.7580	0.7611	0.7643	0.7673	0.7704	0.7734	0.7764	0.7794	0.7823	0.7852
0.8	0.7881	0.7910	0.7939	0.7967	0.7995	0.8023	0.8051	0.8078	0.8106	0.8133
0.9	0.8159	0.8186	0.8212	0.8238	0.8264	0.8289	0.8315	0.8340	0.8365	0.8389
1.0	0.8413	0.8438	0.8461	0.8485	0.8508	0.8531	0.8554	0.8577	0.8599	0.8621
1.1	0.8643	0.8665	0.8686	0.8708	0.8729	0.8749	0.8770	0.8790	0.8810	0.8830
1.2	0.8849	0.8869	0.8888	0.8907	0.8925	0.8944	0.8962	0.8980	0.8997	0.9015
1.3	0.9032	0.9049	0.9066	0.9082	0.9099	0.9115	0.9131	0.9147	0.9162	0.9177
1.4	0.9192	0.9207	0.9222	0.9236	0.9251	0.9265	0.9279	0.9292	0.9306	0.9319
1.5	0.9332	0.9345	0.9357	0.9370	0.9382	0.9394	9.9406	0.9418	0.9429	0.9441
1.6	0.9452	0.9463	0.9474	0.9484	0.9495	0.9505	0.9515	0.9525	0.9535	0.9545
1.7	0.9554	0.9564	0.9573	0.9582	0.9591	0.9599	0.9608	0.9616	0.9625	0.9633
1.8	0.9641	0.9649	0.9656	0.9664	0.9671	0.9678	0.9686	0.9693	0.9699	0.9706
1.9	0.9713	0.9719	0.9726	0.9732	0.9738	0.9744	0.9750	0.9756	0.9761	0.9767
2.0	0.9772	0.9778	0.9783	0.9788	0.9793	0.9798	0.9803	0.9808	0.9812	0.9817
2.1	0.9821	0.9826	0.9830	0.9834	0.9838	0.9842	0.9846	0.9850	0.9854	0.9857
2.2	0.9861	0.9864	0.9868	0.9871	0.9875	0.9878	0.9881	0.9884	0.9887	0.9890
2.3	0.9893	0.9896	0.9898	0.9901	0.9904	0.9906	0.9909	0.9911	0.9913	0.9916
2.4	0.9918	0.9920	0.9922	0.9925	0.9927	0.9929	0.9931	0.9932	0.9934	0.9936
2.5	0.9938	0.9940	0.9941	0.9943	0.9945	0.9946	0.9948	0.9949	0.9951	0.9952
2.6	0.9953	0.9955	0.9956	0.9957	0.9959	0.9960	0.9961	0.9962	0.9963	0.9964
2.7	0.9965	0.9966	0.9967	0.9968	0.9969	0.9970	0.9971	0.9972	0.9973	0.9974
2.8	0.9974	0.9975	0.9976	0.9977	0.9977	0.9978	0.9979	0.9979	0.9980	0.9981
2.9	0.9981	0.9982	0.9982	0.9983	0.9984	0.9984	0.9985	0.9985	0.9986	0.9986
3.0	0.9987	0.9987	0.9987	0.9988	0.9988	0.9989	0.9989	0.9989	0.9990	0.9990

Example 8.5.2:

Suppose $X \sim N(10, 4)$. Determine $P(9 \le X \le 11)$.

Solution

$P(9 \le X \le 11) = F_X(11) - F_X(9) = \Phi((11 - 10)/2) - \Phi((9 - 10)/2) = \Phi(0.5) - \Phi(-0.5)$.

Use of tables, as in Example 8.5.1, gives

$$P(9 \le X \le 11) = 2\Phi(0.5) - 1 = 2 \cdot 0.6915 - 1 = 0.3830.$$

—   □

We interpret the parameters μ and σ in the chapters on mathematical expectation and variance. Some distributions related to the normal are discussed in Example 11.1.6 and 11.2.7. The central limit theorem and the role of the normal distribution in statistics are discussed briefly in Section 18.6 and 18.7.

8.6.   Life Distributions in Reliability Theory

An analysis of the reliability of a system can be quite complex. Some devices may be treated as individual units. Others may be analyzed as systems of components. When possible, the components or subunits are chosen in such a manner that they "fail independently" in some appropriate sense, so that the assumption of stochastic independence is reasonable. Overall system reliability is then determined by component reliability and the manner in which the failure of the various components affects system behavior.

In this section, we look at the problem of describing in useful ways the reliability of a single component—possibly a subsystem or a complete system viewed as a unit. The underlying idea is that the *reliability* of the unit for a given period of time is measured by the probability that the unit will operate successfully throughout that period. To set up our model, we represent the operating lifetime as the value of a random variable X. We set a reference time $t = 0$ (i.e., we start our clock) at the time the unit is put into service. Then $X(\omega) = t$ means that the device operates over the time interval $[0, t)$ and fails at the instant corresponding to time t. The event the unit survives at time t is the event $\{X > t\}$. The distribution function F for X is called the *life distribution function* (or, sometimes, the *failure distribution function*). Since X is nonnegative, $F(t) = 0$ for $t < 0$. We describe the reliability of the unit in terms of the probability that it is operating at any time t, as follows:

Definition 8.6.1. *The* reliability function R *is defined by*

$$R(t) = P(X > t) = 1 - F(t).$$

Remark: The term *survival function* is used by some authors.

The study of R is, in effect, the study of the life distribution. This study is aided by the introduction of an auxiliary function h, known variously as the *hazard rate function*, the *hazard function*, the *failure rate function*, or the *conditional failure rate function*. Before giving a formal definition, we develop some intuitive background, which motivates the definition and provides the principal interpretation.

1. The probability of failure in the time interval $(t, t + u]$ is

$$P(t < X \le t + u) = F(t + u) - F(t) = R(t) - R(t + u) = f(t)u + o(u).$$

2. The conditional probability of failure in $(t, t + u]$, given survival to time t, is
$$\frac{P(t < X \le t + u)}{P(t < X)} = -\frac{R(t + u) - R(t)}{R(t)}.$$

3. If we divide the quantity in (2) by u, we get a conditional (time) rate of failure, given survival to t

$$-\frac{R(t + u) - R(t)}{u} \cdot \frac{1}{R(t)}.$$

4. Upon taking the limit as u decreases to zero, we obtain an instantaneous conditional rate of failure, given survival to t.

$$-\frac{R'(t)}{R(t)} = \frac{F'(t)}{R(t)} = \frac{f(t)}{R(t)}.$$

Definition 8.6.2. *The* hazard rate function h *is defined by* $h(t) = f(t)/R(t)$.

We trace some important relationships between $f(t)$ and $h(t)$

- $h(t) \ge f(t)$, since $1/R(t) \ge 1$

- $f(t)u \approx f(t)u + o(u) =$ the probability (unconditional) that a failure occurs in time interval $(t, t + u]$. On the other hand $h(t)u \approx h(t)u + o(u) =$ the conditional probability of a failure in $(t, t + u]$, given survival to t. These may be quite different. Consider an individual who has reached 99 years of age. Consider the probability of death in the next year. If $t = 99$ and $t + u = 100$, $f(t)u$ will be quite small, since very few people survive to age 99. However, $h(t)u$ may be considerably larger, since a person who survives to 99 has a relatively high conditional probability of dying during the next year.

- Since $h(t) = -R'(t)/R(t)$, $R(0) = 1$, and $h(t) = 0$ for $t < 0$, elementary calculus shows that

$$-\ln R(t) = \int_0^t h(u)\,du = H(t)$$

so that

$$R(t) = e^{-H(t)} \quad \text{and} \quad f(t) = R(t)h(t) = h(t)e^{-H(t)} = H'(t)e^{-H(t)}.$$

The function defined by $H(t) = \int_0^t h(u)\,du$ is often referred to as the *cumulative hazard function*.

Failure rates depend upon both the condition of the component and the nature of external conditions causing stresses. Typically, the following stages are encountered in the life of a component or unit.

1. During early life, failures are usually due to inherent defects that may be attributed to faulty design, manufacturing, assembly, or adjustment. This period is known variously as the *burn-in* period, the *break-in* period, the *debugging* phase, or the *infant mortality* phase. Usually, this phase is characterized by a decreasing failure rate as it is survived. For many components, this phase is not seen by the customer—at least after the product has been available for some time.

2. During a middle period, referred to as the *useful life*, the failures are due primarily to stresses from the working environment. These are usually random (i.e., unpredictable except in a statistical sense). For example, there may be electrical power line voltage surges, unusual temperature extremes, a road hazard, etc. This period is characterized by an approximately *constant failure rate*.

3. When components approach the end of the useful life period, they begin to fail because of changes of characteristics due to wear or other types of deterioration. This period is generally characterized by an increasing failure rate. The rated life of the unit is placed somewhere in the transition between the middle period and the wear-out phase.

One useful way to exhibit the life characteristics of a unit or component is to plot the hazard function against time to obtain a *mortality curve*, or *life characteristic*, or *hazard curve*. A typical shape is shown in Figure 8.6. Such a graph is often referred to as a *bathtub curve*, because of this suggestive shape. The region A corresponds to the burn-in period, with decreasing failure rate $h(t)$; the region B represents the useful life, with approximately constant failure rate; and finally region C represents the wear-out phase.

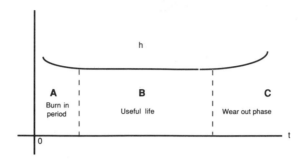

FIGURE 8.6. Typical life characteristic curve: $h(t)$ versus t.

SOME COMMON LIFE DISTRIBUTIONS

We consider some probability distributions commonly assumed as life distributions. The usual approach in detailed reliability studies is to attempt to fit one of a certain class of distributions to empirical data by adjustment of the parameters for the distribution or density function. We do not attempt to deal with the statistical problem of finding the best fit, although this is a major task in the practical analysis of reliability.

1. *Exponential* (λ).

   As we show in Section 8.4, $F(t) = 1 - e^{-\lambda t}$ and $f(t) = \lambda e^{-\lambda t}$ for $t \geq 0$. From this we see that

   $$R(t) = e^{-\lambda t} \quad \text{and} \quad h(t) = \frac{f(t)}{R(t)} = \lambda \quad \text{for } t \geq 0.$$

   This distribution is probably the most widely used in theoretical studies of reliability. On the one hand, it is mathematically tractable. On the other, with suitable choice of λ, it provides a reasonable model of practical life distributions for many types of components. Since $h(t) = \lambda$, constant for $t \geq 0$, the life characteristic shows no burn-in or wear-out period. A unit that survives to time t is "as good as new." Many components (e.g., solid-state electronics devices, if put into service after appropriate selection, adjustment, testing, etc.) have such a long useful life that this distribution provides a good model. Failure is due to the effect of external conditions rather than internal defects. This characteristic could also be approximated if surviving units were replaced before the end of the useful-life period; the composite would give a good approximation to a single unit with constant hazard rate function.

2. *Weibull* $(\alpha, \lambda, 0)$

For this distribution, we have $F(t) = 1 - e^{-\lambda t^\alpha}$ and $f(t) = \alpha \lambda t^{\alpha-1} e^{-\lambda t^\alpha}$ for $t \geq 0$. The reliability function and the hazard rate function are given by

$$R(t) = e^{-\lambda t^\alpha} \qquad \text{and} \qquad h(t) = \alpha \lambda t^{\alpha-1} \qquad \text{for } t \geq 0.$$

By adjusting the parameters α and λ, one can shape the life characteristic in useful ways.

For $\alpha > 1$, the curve exhibits increasing failure rate for all $t > 0$. This does not provide a very good model for the burn-in period, but this may not be important, since burn-in is effectively taken care of before the unit is put into service.

For $\alpha < 1$, the curve exhibits decreasing failure rate. If it is necessary to model a burn-in period with only secondary interest in a wear-out period, this case can be quite useful.

In spite of the limitations noted, the Weibull distribution is used in a wide variety of applications. Much experimental and theoretical work has been done to achieve good fits to actual data. A combination of mathematical tractability and empirical suitability make this an extremely useful class of distributions.

Example 8.6.1:

A plastic seal subject to radiation damage has hazard rate function

$$h(t) = 3 \cdot 10^{-6} t \qquad t \text{ in hours.}$$

What is the probability the unit will survive 500 hours?

Solution

$h(t) = at$ implies the distribution is Weibull with $\alpha = 2$, in which case $h(t) = 2\lambda t$. Thus $\lambda = a/2 = 1.5 \cdot 10^{-6}$. Now $R(t) = e^{-\lambda t^2}$. For $t = 500$, we have $\lambda t^2 = 1.5 \cdot 10^{-6} 500^2 = 0.375$. Hence, $R(500) = e^{-0.375} \approx 0.69$

—

$\square$

3. *The gamma* (n, λ) *distribution.*

The density function is given by

$$f(t) = \frac{\lambda^n}{(n-1)!} t^{n-1} e^{-\lambda t} = \frac{\lambda}{(n-1)!} (\lambda t)^{n-1} e^{-\lambda t}.$$

Use of one of the integrals in the appendix on mathematical aids gives

$$R(t) = \int_t^\infty f(u)\,du = e^{-\lambda t}\left(1 + \lambda t + \frac{(\lambda t)^2}{2!} + \cdots + \frac{(\lambda t)^{n-1}}{(n-1)!}\right).$$

Thus

$$h(t) = \frac{f(t)}{R(t)} = \frac{\lambda(\lambda t)^{n-1}}{(n-1)!\left(1 + \lambda t + \frac{(\lambda t)^2}{2!} + \cdots + \frac{(\lambda t)^{n-1}}{(n-1)!}\right)}.$$

We see that h is a rational function in λt that converges to λ for large t.

4. *The truncated normal distribution*

The density function is of the form

$$f(t) = \frac{1}{K}u(t)f^*(t)$$

where f^* is the density function for a random variable $N(\mu, \sigma^2)$ and $K = \int_0^\infty f^*$ provides the normalizing factor to make $\int_0^\infty f = 1$. The truncated normal distribution is ordinarily used when the parameter μ is large enough that most of the normal probability mass is to the right of the origin. In such a case, $K \approx 1$, and the correction may be omitted. Use of the usual standardizing procedures makes it possible to use tables of the standardized normal distribution to determine K and $f^*(t)$. No simple expression is available for $h(t)$, though, of course, it can be evaluated for any t with the aid of tables.

PROBLEMS

8–1. Suppose $\{A, B, C\}$ is an independent class with $P(A) = 0.2$, $P(B) = 0.5$, $P(C) = 0.3$. Let $X = -I_A + 2I_B + 3I_C$.

Determine the distribution function for X.

8–2. Suppose X is uniform on $(0, 1)$ and Y is the output of a random generator with $n = 10{,}000$ (see Example 8.2.1). Show that

$$|F_X(t) - F_Y(t)| \leq 10^{-4} \qquad \text{for all } t.$$

8–3. In a Bernoulli sequence with probability 0.2 of success on any trial

(a) What is the probability of three or more successes in the first ten trials?

(b) Determine the probability the third success will occur no sooner than the fifth and no later than the tenth trial.

8–4. In a Bernoulli trial, random variable X designates the number of the trial on which the first *failure* occurs. Suppose the probability p of success on any trial is 0.85.

Determine $P(X = 7)$ and $P(X \geq 8)$.

8–5. In a Bernoulli trial, random variable X designates the number of the trial on which the first *failure* occurs. Suppose the probability p of success on any trial is 0.7.

Determine $P(X = 4)$ and $P(X \geq 3)$.

8–6. For the density function $f_X(t) = 2t$, $0 \leq t \leq 1$, calculate the value of $P(0.25 \leq X \leq 0.75)$.

Hint: Sketch the graph and determine the appropriate area.

8–7. Random variable X is Poisson (2.5). Use a table of cumulative terms, Poisson distribution to obtain $P(X = 4)$ and $P(3 \leq X \leq 6)$.

8–8. The number of incoming calls to a switchboard in one hour is Poisson (10). What is the probability of no more than four calls in one hour?

8–9. The number of noise pulses arriving on a power circuit in an hour is a random quantity that may be represented by random variable X with Poisson distribution having parameter $\mu = 5$.

Determine $P(3 \leq X \leq 6)$.

8–10. The number of customers arriving in a specialty store in an hour is a random quantity that may be represented by random variable X with Poisson distribution having parameter $\mu = 3$.

Determine $P(2 \leq X \leq 5)$.

8–11. For X exponential (λ), determine $P(X \geq 1/\lambda)$, $P(X \geq 2/\lambda)$.

8–12. The time to failure, in hours of operating time, of a television set on a line subject to random voltage surges is exponential, with parameter $\lambda = 0.002$. Suppose the unit has "survived" 500 hours of use. What is the probability (conditional) that it will operate for another 500 hours?

8–13. Suppose the life duration, in hours, of a device is exponential $(1/1000)$. That is, the time to failure may be represented by a random variable X having the exponential distribution with parameter $\lambda = 1/1000$.

Determine the probability $P(X > 800)$ that the device is still operating at 800 hours and the value of the conditional probability $P(X > 1000|X > 800)$ that the device will last for at least 1000 hours given that it has survived for 800 hours.

8–14. For X gamma $(2, 1/2)$, determine $P(X \geq 2)$.

8–15. For X Beta $(3, 2)$, determine $P(0.2 < X \leq 0.6)$

8–16. Random variable $X \sim N(9, 16)$. That is, the distribution is normal, with parameter $\mu = 9$ and parameter $\sigma = \sqrt{16} = 4$. Use a table of the standardized normal distribution function to determine the value of $P(7 \leq X \leq 10)$ and $P(|X - 9| \geq 1.6)$.

8–17. Random variable $X \sim N(12, 9)$. That is, the distribution is normal, with parameter $\mu = 12$ and parameter $\sigma = \sqrt{9} = 3$.

Determine $P(11 \leq X \leq 13.5)$ and $P(|X - 12| \geq 1.5)$.

8–18. For $X \sim N(2, 4)$, determine the following values: $P(2 < X \leq 4)$, $P(X > 0)$, and $P(-2 < X < 0)$.

8–19. For $X \sim N(\mu, \sigma^2)$, determine $P(|X - \mu| < a\sigma)$ for $a = 0.25$, 0.50, 0.75, 1, 2, and 3.

8–20. The result of extensive quality control sampling shows that digital watches coming off a production line have accuracy, in seconds per month, that is normally distributed with $\mu = 5$ and $\sigma^2 = 400$. The watches are tested individually and graded. To achieve top grade, a watch must have an accuracy within the range -5 to $+10$ seconds per month. What is the probability that a watch taken from the production line to be tested will achieve top grade?

8–21. Random variable X is $N(10, 4)$. That is, X has the normal distribution with parameters $\mu = 10$ and $\sigma = \sqrt{4} = 2$. Use a table of the standardized normal distribution to determine $P(9 \leq X \leq 13)$.

8–22. The life (in miles) of a certain brand of radial tires may be represented by a random variable X with density function f_X given by

$$f_X(t) = \begin{cases} t^2/a^3 & \text{for } 0 \leq t < a \\ \frac{b}{a}e^{-k(t-a)} & \text{for } t > a \\ 0 & \text{elsewhere} \end{cases}$$

where $a = 40{,}000$, $b/a = 1/6000$, and $k = 1/4000$.

Determine $P(X \geq 45{,}000)$.

8–23. The distribution functions listed here are for mixed distributions. For each function,

(a) Sketch the graph of the function.

(b) Determine the point mass distribution for the discrete part.

(c) Determine the density function for the absolutely continuous part.

1. $F_X(t) = \begin{cases} 0.4(t+1) & \text{for } -1 \leq t < 0 \\ 0.6 + 0.4t & \text{for } 0 \leq t < 1 \end{cases}$

2. $F_X(t) = \begin{cases} e^{2t}/4 & \text{for } -\infty < t < 0 \\ 1 - e^{-2t}/4 & \text{for } 0 \leq t < \infty \end{cases}$.

9

Random Vectors and Joint Distributions

In Chapters 7 and 8, we formulate the concept of real random variables and describe the distributions that they induce on the real line. In that development, we direct attention to a single number associated with each outcome of the basic experiment. Real random variables are considered singly. In this chapter, we consider the case of several numbers associated with the outcome of the experiment. These numbers are viewed separately as values of real random variables; but they are considered jointly as the coordinates of a random vector. We use fundamental mapping ideas in the same manner as for real random variables to produce a joint distribution that describes the probabilistic, or stochastic, behavior of the random vector as well as the probabilistic relations between the coordinate random variables.

9.1. The Joint Distribution Determined by a Random Vector

In many experiments, there are several—perhaps infinitely many—numbers associated with each outcome. Each of these numbers may be viewed as the value of a random variable; each induces its own probability distribution. These individual distributions describe the probabilistic behavior of the various random quantities, considered separately, without revealing anything about relationships between them. The quantities may be related by some functional rule, so that if one or more is known, the others are determined. More typically, the association is partial and probabilistic. The relationships are stochastic. Values of one or more of the variables may affect the probabilities for the others. A simple, classical example serves to illustrate the new element introduced when the random variables are considered jointly.

Example 9.1.1: Separate and Joint Probabilities

Suppose we have a set of 100 cards of identical size. On each card there are two numbers. These are distinguished as a "first number" and a "second number." Each number is represented by one of the decimal digits, 0 through 9. Thus a card may have the pair $(0, 3)$ or $(7, 4)$ or any other of the 100 possible pairs. We suppose that exactly one-tenth of the cards have 0 as the first number, one-tenth have 1 as the first number, etc. Also, one-tenth have 0 as the second number, one-tenth have 1 as the second number, etc. The cards are shuffled thoroughly, and one card is drawn at random (i.e., on an equally likely basis). There are 100 possible outcomes of this experiment. Let X be the random variable whose value is the first number on the card drawn and Y be the random variable whose value is the second number on the card drawn. We have

$$P(X = k) = \frac{1}{10} \quad 0 \le k \le 9 \quad \text{and} \quad P(Y = j) = \frac{1}{10} \quad 0 \le j \le 9.$$

But can we assign a value to $P(X = 2, Y = 7)$? This is the probability that both $X = 2$ and $Y = 7$. What is $P(X = Y)$?

Discussion

Without more information about the assignment of the numbers, we cannot answer questions about joint probabilities, such as those posed here. Consider several ways of assigning the numbers to the cards.

1. Suppose the first and second number is the same on each card. That is, there are ten cards with $(0, 0)$, ten with $(1, 1)$, etc. Then we have
$$P(X = k) = P(Y = j) = \frac{1}{10}$$
$$P(X = 2, Y = 7) = 0 \qquad P(X = Y) = 1.$$

2. Suppose there are ten cards for each of the following pairs:
$$(0, 9), \ (1, 8), \ (2, 7), (3, 6), \ (4, 5), \ (5, 4), \ (6, 3), \ (7, 2), \ (8, 1), \ (9, 0).$$
Then
$$P(X = k) = P(Y = j) = \frac{1}{10} \qquad P(X = 2, Y = 7) = \frac{1}{10}$$
$$P(X = Y) = 0.$$

3. For fifty of the cards, let five each have the pairs $(0, 0)$, $(1, 1)$, $(2, 2)$, etc. For the other fifty cards let five each have the pairs in case (2). Then
$$P(X = k) = P(Y = j) = \frac{1}{10} \qquad P(X = 2, Y = 7) = \frac{1}{20}$$
$$P(X = Y) = \frac{1}{2}.$$

4. Let each card have one of the 100 possible pairs, with no two cards alike.

$$P(X = k) = p(Y = j) = \frac{1}{10} \qquad P(X = 2, Y = 7) = \frac{1}{100}$$

$$P(X = Y) = \frac{1}{10}.$$

In each case, the random variables X and Y have the same distribution. Both are uniformly distributed over the integers 0, 1, ..., 9. Yet the joint probabilities are markedly different.

—   □

This example illustrates why we must consider more than the distributions for the individual random variables. We proceed to introduce the notion of a *joint probability distribution*. To do so, we extend the fundamental mapping ideas employed in Chapters 7 and 8 to the case of random vectors. To keep notation simple and to make possible graphical representations, we consider the case of two random variables treated jointly. The ideas extend to any number of random variables, at the expense of some notational complication and increasing difficulty with geometric visualization.

Suppose we have a pair $\{X, Y\}$ of real random variables. Form the vector function $W = (X, Y)$ whose value is $W(\omega) = (t, u)$ when $X(\omega) = t$ and $Y(\omega) = u$. A value $W(\omega)$ corresponds to the pair of values $(X(\omega), Y(\omega))$. Now a pair of real numbers (t, u) is represented by a point on the plane $\Re^2$. The function $W = (X, Y)$ produces a mapping from the basic space Ω to the plane $\Re^2$, just as individually the random variables X and Y produce mappings from Ω to the real line $\Re$. The notion of inverse images is as meaningful in the multidimensional case as in the one-dimensional case. Thus the inverse image of a set of points Q on the plane under the mapping $W = (X, Y)$ is the set of all those ω such that $(X, Y)(\omega) = (t, u)$ is a point in Q (see Figure 9.1). In symbols

$$W^{-1}(Q) = (X, Y)^{-1}(Q) = \{\omega : (X, Y)(\omega) \in Q\} = \{(X, Y) \in Q\}.$$

We have *measurability* considerations parallel to those for real random variables. The essential facts are these.

- The notion of *Borel sets* can be extended to the plane $\Re^2$ and to higher dimensional Euclidean spaces $\Re^n$ whose elements are n-tuples $(t_1, t_2, \ldots, t_n)$ of real numbers.

- A mapping W from the basic space Ω to $\Re^n$ is a *random vector* iff $W^{-1}(Q)$ is an event for each Borel set Q in $\Re^n$.

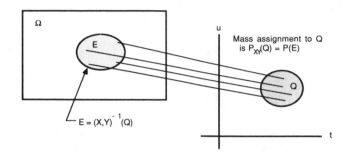

FIGURE 9.1. Mapping and mass transfer induced by $W = (X, Y)$.

- A detailed examination of the measurability problem shows that $W = (X_1, X_2, \ldots, X_n)$ is a random vector iff each of the coordinate functions X_i is a random variable. We call X_i the ith *coordinate random variable* for random vector W.

As in the one-dimensional case, we may map the probability from the events on the basic space to the Borel subsets on $\Re^n$. For example, in the two-dimensional case, we assign to region Q (a Borel set) on the plane the probability mass in event $E = (X, Y)^{-1}(Q)$. We define a set function P_{XY} on the Borel sets by the assignment

$$P_{XY}(Q) = P(E), \qquad \text{where } E = (X, Y)^{-1}(Q) \text{ for all Borel sets } Q.$$

An argument exactly parallel to that for P_X in the one-dimensional case shows that P_{XY} is a probability measure on the Borel sets, since it obeys the three axiomatic properties for a probability measure. We call P_{XY} the *probability measure induced by* $W = (X, Y)$. The resulting distribution of the probability mass on the plane is called the *joint probability distribution induced by* (X, Y). In the next section, we introduce joint distribution functions and joint density functions to describe joint distributions.

9.2. The Joint Distribution Function and Marginal Distributions

Definition 9.2.1. *The* joint distribution function F_{XY} *for random vector* $W = (X, Y)$ *is given by*

$$F_{XY}(a, b) = P(X \le a, Y \le b) = P((X, Y) \in Q_{ab}) = P_{XY}(Q_{ab})$$

where

$$Q_{ab} = \{(t, u) : t \le a, u \le b\}.$$

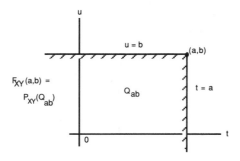

FIGURE 9.2. The region Q_{ab} for $F_{XY}(a,b)$.

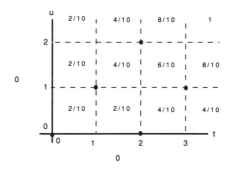

FIGURE 9.3. Distribution function for discrete distribution.

The region Q_{ab} (shown in Figure 9.2) is the set of points on the plane that are on or to the left of the vertical line $t = a$ and on or below the horizontal line $u = b$. These lines meet at the point (a, b). Thus, the value $F_{XY}(a, b)$ designates that part of the probability mass "covered" by an "infinite sheet" with a corner at point (a, b). This geometric picture makes it possible to evaluate the distribution function graphically in many simple cases.

Example 9.2.1: A Discrete Joint Distribution

The pair $\{X, Y\}$ produces a joint distribution that places mass $2/10$ at the five points $(0,0)$, $(1,1)$, $(2,0)$, $(2,2)$, and $(3,1)$ (see Figure 9.3). The joint distribution function F_{XY} is constant over the rectangular regions bounded by the grid through the points in the range of the random vector (X, Y). Values of F_{XY} are found geometrically by identifying the probability mass covered by the "infinite sheet." These values are shown on the various regions on Figure 9.3.

□

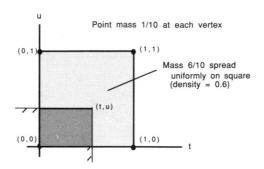

FIGURE 9.4. Determination of the joint distribution function for the mixed distribution of 9.2.2.

Example 9.2.2: A Mixed Distribution

The pair $\{X, Y\}$ produces a mixed distribution as follows (see Figure 9.4):

Point masses of $1/10$ at points $(0,0)$, $(1,0)$, $(1,1)$, and $(0,1)$.
Mass of $6/10$ is spread uniformly over the square have these points as vertices.

The joint distribution function F_{XY} is zero in the second, third, and fourth quadrants.

- If the point (t, u) is in the square, the infinite sheet with corner at (t, u) covers the point mass at $(0,0)$ and point mass 0.6 times the area of the portion of the square included. Thus, in this region (including left and lower boundaries)
$$F_{XY}(t, u) = (1 + 6tu)/10.$$

- If the point (t, u) is above the square (including its upper boundary), with $0 \le t < 1$, the infinite sheet covers two point masses plus the portion of the mass in the square to the left of the vertical line through point (t, u). Hence
$$F_{XY}(t, u) = (2 + 6t)/10.$$

- If the point (t, u) is to the right of the square (including its right boundary), with $0 \le u < 1$, the infinite sheet covers two of the point masses and the portion of the mass in the square below the line through (t, u), to give
$$F_{XY}(t, u) = (2 + 6u)/10.$$

- If (t, u) is both above and to the right of the square, i.e., both $1 \leq t$ and $1 \leq u$, all of the probability mass is covered, so that the joint distribution function has value one in this region.

——   □

As in the case of the distribution function for a real random variable, the joint probability distribution function determines uniquely the joint distribution for all Borel sets. This is true for any dimension n. We accept this fact from a measure-theoretic treatment of the probability measure.

MARGINAL DISTRIBUTIONS

The joint distribution function F_{XY} determines the probabilities associated with the pair $\{X, Y\}$, considered jointly as coordinates of a random vector. It also gives information about each random variable, considered separately, as the following argument shows.

$$F_X(t) = P(X \leq t) = P(X \leq t, Y < \infty) = F_{XY}(t, \infty)$$

and

$$F_Y(u) = P(Y \leq u) = P(X < \infty, Y \leq u) = F_{XY}(\infty, u).$$

The event that X does not exceed t is the event that X is no greater than t and Y takes any finite value. Similarly for Y no greater than u. Consider the probability mass covered by the infinite sheet with corner at (t, u). If we hold t fixed and let u increase without limit (see Figure 9.5), the sheet moves upward and ultimately covers all of the probability mass in the half plane $Q_{t\infty}$, consisting of all points on or to the left of the vertical line through $(t, 0)$. Then $F_X(t) = P_{XY}(Q_{t\infty})$. Similarly, if we hold u fixed and let t increase without limit, we cover the probability mass in the half plane $Q_{\infty u}$ consisting of all points on or below the horizontal line through $(0, u)$, and $F_Y(u) = P_{XY}(Q_{\infty u})$.

We carry the geometric interpretation one step further. The joint mapping by $W = (X, Y)$ maps point ω into point (t, u) on the plane. The first coordinate is the value of X and the second coordinate is the value of Y. Now we can think of the *coordinate mappings* in which X maps into the first coordinate axis and Y maps into the second coordinate axis. If, for each real t, we collapse all the probability mass in the half plane $Q_{t\infty}$ into the interval $(-\infty, t]$ on the first coordinate axis, we have on that axis the distribution induced by X. Likewise, if for each real u we collapse all the probability mass in the half plane $Q_{\infty u}$ into the interval $(-\infty, u]$ on the second coordinate axis, we have on that axis the probability distribution for Y. These distributions for the coordinate random variables on the coordinate axes are called the *marginal distributions*. The joint distribution determines the marginal distributions. The converse is *not* true, as the

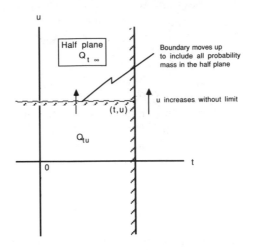

$$F_X(t) = \text{probability in half plane} = F_{XY}(b, \infty)$$

FIGURE 9.5. Construction for obtaining the marginal distribution for X.

card example in Example 9.1.1 shows clearly. In each case, both X and Y are distributed uniformly on the integers 0 through 9. However, the joint distributions are quite different in the various cases.

Example 9.2.3: Marginal Distributions for Example 9.2.1

If we collapse the probability mass into the first coordinate axis, we obtain point masses 0.2, 0.2, 0.4, 0.2 at $t = 0, 1, 2, 3$, respectively. This is the marginal distribution for X (see Figure 9.6). On the other hand, if we collapse the probability mass into the second coordinate axis, we have point masses 0.4, 0.4, 0.2 at $u = 0, 1, 2$, respectively. This is the marginal distribution for Y.

—   □

Example 9.2.4: Marginal Distributions for Example 9.2.2

In this case, it is probably easier to visualize $F_X(t)$ in terms of the mass included in the half plane on and to the left of the vertical line through $(t, 0)$.

- For $t < 0$, no mass is covered, so $F_X(t) = 0$.

- At $t = 0$, the two point masses on the vertical axis are "picked up," so the value of the distribution function jump to $F_X(0) = 0.2$.

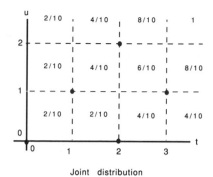

Joint distribution

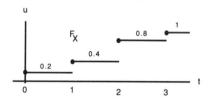

FIGURE 9.6. Marginal distribution for X in Examples 9.2.1 and 9.2.3.

- For $0 \leq t < 1$, a portion of the probability mass in the square is included. Geometrically (see Figure 9.7), we obtain

$$F_X(t) = 0.2 + 0.6t.$$

- For $t \geq 1$, all the probability mass is covered, so $F_X(t) = 1$.

By symmetry, it is apparent that the marginal distribution for Y is the same as that for X.

—   □

9.3. Joint Density Functions

The notion of absolute continuity of a random quantity extends to random vectors of any finite dimension n. The situation is quite analogous to the one-dimensional case. If the joint probability distribution assigns zero

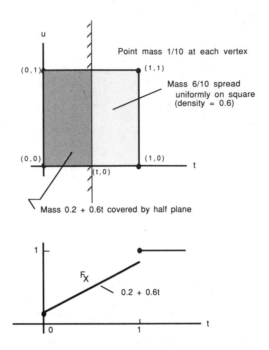

FIGURE 9.7. Marginal distribution for X in Examples 9.2.2 and 9.2.4.

probability mass to all Borel sets with zero size (as determined by Lebesgue measure), then there is a density function for the joint distribution.

In the two dimensional case, zero Lebesgue measure means zero area. Sets of zero area include discrete point sets (finite or countably many points), segments of lines or curves, or countable unions of such sets. If the joint probability distribution assigns zero probability mass to each Borel set with zero area, then there is a joint density function f_{XY} with the properties

(f1) $f_{XY} \geq 0$
(f2) $\iint f_{XY} = 1$
(f3) $P_{XY}(Q) = \int_Q \int f_{XY}$ for all Borel sets Q.

As in the one-dimensional case, the integrals are Lebesgue integrals; but these coincide with Riemann integrals for any ordinary functions. We have

$$F_{XY}(t, u) = \int_{Q_{tu}} \int f_{XY} = \int_{-\infty}^{t} \int_{-\infty}^{u} f_{XY}(r, s)\, ds\, dr.$$

At a point of continuity for f_{XY}, F_{XY} is twice differentiable, and

$$f_{XY}(t, u) = \frac{\partial^2 F_{XY}(t, u)}{\partial t \partial u} \qquad \text{[in either order].}$$

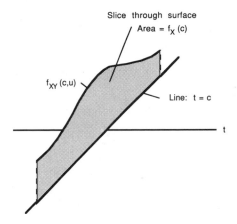

FIGURE 9.8. Graphical interpretation of integral expression for f_X.

Now we may use the integrals and the relation between the joint and marginal distribution functions to get

$$F_X(t) = \int_{-\infty}^{t} \int_{-\infty}^{\infty} f_{XY}(r, s) \, ds \, dr$$

and

$$F_Y(u) = \int_{-\infty}^{u} \int_{-\infty}^{\infty} f_{XY}(r, s) \, dr \, ds \qquad \text{(note the order)}.$$

Interpreting the integrals as Riemann integrals, we use the fundamental theorem of calculus and a change of the dummy variable to obtain

$$f_X(t) = F_X'(t) = \int_{-\infty}^{\infty} f_{XY}(t, u) \, du \qquad \text{[Integrate out } u\text{]}.$$

Likewise,

$$f_Y(u) = F_Y'(u) = \int_{-\infty}^{\infty} f_{XY}(t, u) \, dt \qquad \text{[Integrate out } t\text{]}.$$

We may give these quantities a graphical interpretation.

- f_{XY} defines a surface above the plane $\Re^2$ that encloses unit volume

- $P_{XY}(Q)$ is the volume in the space over region Q in the plane and under the f_{XY} surface.

- To visualize $f_X(c)$, slice through the volume with a vertical plane containing the line $t = c$. Then $f_X(c)$ is the area of the "slice" (see Figure 9.8).

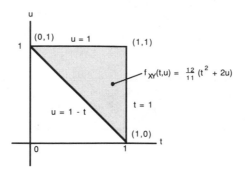

FIGURE 9.9. The joint density for Example 9.3.2.

Example 9.3.1:

Suppose the joint distribution function for the pair $\{X, Y\}$ is

$$F_{XY}(t, u) = (1 - e^{-\lambda t})u \qquad \text{for } 0 \le t < \infty, \ 0 \le u \le 1.$$

Then the joint density is defined in the same region and is determined by differentiating first with respect to u then with respect to t, to give

$$f_{XY}(t, u) = \lambda e^{-\lambda t} \qquad \text{for } 0 \le t < \infty, \ 0 \le u \le 1.$$

The marginal densities are obtained by integration to be

$$f_X(t) = \int_0^1 \lambda e^{-\lambda t}\, du = \lambda e^{-\lambda t} \quad \text{for } 0 \le t < \infty$$

and

$$f_Y(u) = \int_0^\infty \lambda e^{-\lambda t}\, dt = 1 \quad \text{for } 0 \le u \le 1.$$

Thus X is exponential (λ) and Y is uniform on $(0, 1)$.

—   □

Example 9.3.2:

The pair $\{X, Y\}$ has joint density function

$$f_{XY}(t, u) = \frac{12}{11}(t^2 + 2u)$$

on the triangle bounded by $t = 1$, $u = 1$, $u = 1 - t$. The vertices of the triangle are $(1, 0)$, $(1, 1)$, $(0, 1)$ (see Figure 9.9). As usual, we take f_{XY}

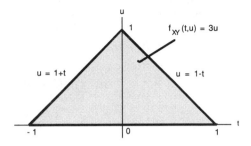

FIGURE 9.10. The joint density for Example 9.3.3.

to be zero outside designated region. Now, for each fixed t on the unit interval, u goes from $1 - t$ to 1, so that

$$f_X(t) = \int f_{XY}(u, u)\, du = \frac{12}{11} \int_{1-t}^{1} (t^2 + 2u)\, du = \frac{12}{11}(t^3 - t^2 + 2t)$$

for $0 \le t \le 1$. Similarly,

$$f_Y(u) = \frac{12}{11} \int_{1-u}^{1} (t^2 + 2u)\, dt = \frac{4}{11}(u^3 + 3u^2 + 3u) \qquad (0 \le u \le 1)$$

— ☐

Often the density function must be defined by different expressions for different values of the argument.

Example 9.3.3:

The pair $\{X, Y\}$ has joint density function $f_{XY}(t, u) = 3u$ on the triangle bounded by the lines $u = 0$, $u = 1 + t$, and $u = 1 - t$ (see Figure 9.10). Determine the marginal densities f_X and f_Y.

Solution

For f_X, we must consider two cases:

- For $-1 \le t \le 0$, $f_X(t) = \int f_{XY}(t, u)\, du = 3 \int_0^{1+t} u\, du = \frac{3}{2}(1+t)^2$.
- For $0 < t \le 1$, $f_X(t) = \int f_{XY}(t, u)\, du = 3 \int_0^{1-t} u\, du = \frac{3}{2}(1-t)^2$.

These may be combined in the two following ways into a single expression:

$$f_X(t) = I_{[-1,0]}(t)\frac{3}{2}(1+t)^2 + I_{(0,1]}(t)\frac{3}{2}(1-t)^2 = \frac{3}{2}(1-|t|)^2, \quad -1 \le t \le 1.$$

The first form, using indicator functions to pick out the correct expression for any given value of t, is useful in a variety of situations. The second form exhibits the fact that f_X is an even function. A single expression suffices for f_Y, which is obtained as follows:

$$f_Y(u) = \int f_{XY}(t, u)\, dt = 3u \int_{u-1}^{1-u} dt = 6u(1-u), \qquad 0 \le u \le 1.$$

—   □

Example 9.3.4: The Joint Normal Distribution

One of the most important joint distributions in applied probability is the *joint normal distribution*. The marginal distributions are each normal, and the joint character is of a particular form. We consider the two-dimensional case. The joint density function is

$$f_{XY}(t, u) = \frac{1}{2\pi\sigma_X\sigma_Y(1-\rho^2)^{1/2}} e^{-Q(t,u)/2}$$

where

$$Q(t, u) = \frac{1}{1-\rho^2}\left(\left(\frac{t-\mu_X}{\sigma_X}\right)^2 - 2\rho\left(\frac{t-\mu_X}{\sigma_X}\right)\left(\frac{u-\mu_Y}{\sigma_Y}\right) + \left(\frac{u-\mu_Y}{\sigma_Y}\right)^2\right).$$

We may put this in a useful form as follows. Setting $r = (t-\mu_X)/\sigma_X$ and $s = (u-\mu_Y)/\sigma_Y$, we have

$$Q(t, u) = q(r, s) = \frac{r^2 - 2\rho rs + s^2}{1-\rho^2}.$$

By "completing the square" in the numerator, we may write this in the form

$$q(r, s) = r^2 + \frac{(s - \rho r)^2}{1-\rho^2} = \left(\frac{t-\mu_X}{\sigma_X}\right)^2 + \frac{(u - \alpha(t))^2}{\sigma_Y^2(1-\rho^2)}$$

where

$$\alpha(t) = \frac{\rho\sigma_Y}{\sigma_X}(t - \mu_X) + \mu_Y.$$

Thus $f_{XY}(t, u) = h(t)g(t, u)$ where

$$h(t) = f_X(t) = \frac{1}{\sigma_X\sqrt{2\pi}} e^{-\frac{1}{2}\left(\frac{t-\mu_X}{\sigma_X}\right)^2}$$

and

$$g(t, u) = \frac{1}{\sigma_Y(1-\rho^2)^{1/2}\sqrt{2\pi}} e^{-v^2/2} \quad \text{with} \quad v = \frac{u - \alpha(t)}{\sigma_Y(1-\rho^2)^{1/2}}.$$

Inspection shows $h = f_X$ is the density for a random variable $N(\mu_X, \sigma_X^2)$. Also, for each fixed t, the function $g(t, \cdot)$ is the density for a random variable normal with mean $\alpha(t)$ and variance $\sigma_Y^2(1 - \rho^2)$. To obtain the marginal for X, we evaluate the integral

$$\int f_{XY}(t, u)\, du = h(t) \int g(t, u)\, du = h(t) = f_X(t)$$

since the integral of the density function $g(t, \cdot)$ is one for each real t. Thus, X is $N(\mu_X, \sigma_X^2)$. Because of symmetry with respect to r and s, a similar treatment shows the random variable Y is $N(\mu_Y, \sigma_Y^2)$.

— $\qquad\qquad\qquad\qquad\qquad\qquad\qquad\qquad\qquad\qquad\qquad\square$

One should not suppose that every pair of normally distributed random variables has the joint normal distribution. It is true, as we show in Chapter 10, that any independent pair of normally distributed random variables has the joint normal distribution. But consider the following

Example 9.3.5: A Pair of Normal Random Variables that is not Joint Normal

The function

$$h(t, u) = \frac{1}{2\pi} \exp\left(-\frac{t^2}{2} - \frac{u^2}{2}\right)$$

is a joint normal density function for a pair of random variables, each $N(0, 1)$, with $\rho = 0$. Now define the joint density function for the pair $\{X, Y\}$ by $f_{XY}(t, u) = 2h(t, u)$ in the first and third quadrants, and zero elsewhere. Then both X and Y, considered singly, are $N(0, 1)$, but they do *not* have the joint normal distribution. If they were joint normal they could take any pair of values; but since there is no probability mass in the second and fourth quadrants they cannot take any pair of values in those quadrants.

— $\qquad\qquad\qquad\qquad\qquad\qquad\qquad\qquad\qquad\qquad\qquad\square$

PROBLEMS

9-1. Two cards are selected without replacement from a standard deck. Random variable X is the number of kings in the hand and Y is the number of diamonds in the hand. Determine the joint and marginal distributions for (X, Y).

9-2. Two positions are open for campus jobs. Two sophomores, three juniors, and three seniors apply. They seem equally qualified, so it is decided to select two at random. Let X, Y, and Z be the respective numbers of sophomores, juniors, and seniors selected.

(a) Determine the marginal distributions for X, Y, and Z.

(b) Determine the joint distribution for (X, Y).

9-3. For each of the following discrete joint distributions:

(a) Determine F_{XY} and indicate its values on appropriate regions of the plane.

(b) Determine the marginal distribution functions F_X and F_Y.

(c) Determine $P(-1 < X \le 1, Y > 1)$ and $P(X > 2, Y \le 1)$.

Discrete joint distributions for problem 9-3

(1)

$t =$	1	2	3	4	5	6
$u = 5$	0	0	0	0.1	0	0.1
$u = 4$	0.1	0	0	0	0.1	0
$u = 3$	0	0.1	0.1	0	0.1	0
$u = 2$	0	0	0	0.1	0.1	0
$u = 1$	0.1	0	0	0	0	0

$P(X = t, Y = u)$

(2)

$t =$	0	1	4	9
$u = 2$	0.05	0.04	0.21	0.15
$u = 0$	0.05	0.01	0.09	0.10
$u = -1$	0.10	0.05	0.10	0.05

$P(X = t, Y = u)$

(3)

$t =$	-2	0	1	3
$u = 1$	0.15	0.05	0.10	0
$u = 0$	0.05	0.08	0.10	0.12
$u = -1$	0.10	0.02	0	0.23

$P(X = t, Y = u)$

9-4. For the following joint density functions listed:

(a) Determine the marginal density functions f_X and f_Y.

(b) Determine $P(X > 1/2, Y > 1/2)$, $P(0 \le X \le 1/2, Y > 1/2)$, and $P(-1 < X < 1/2, Y \le 1/2)$.

Joint density functions for problem 9-4 (zero outside indicated regions)

(1) $f_{XY}(t,u) = 1/\pi$ on circle with radius 1, center at $(0,0)$.

(2) $f_{XY}(t,u) = 1$ on triangle bounded by $t = 0$, $u = 0$, $u = 2(1-t)$.

(3) $f_{XY}(t,u) = 1/2$ on square with vertices at $(1,0)$, $(2,1)$, $(1,2)$, and $(0,1)$.

(4) $f_{XY}(t,u) = 4t(1-u)$ for $0 \le t \le 1, 0 \le u \le 1$.

(5) $f_{XY}(t,u) = \frac{1}{8}(t+u)$ for $0 \le t \le 2, 0 \le u \le 2$.

(6) $f_{XY}(t,u) = 2\alpha u e^{-\alpha t}$ for $0 \le t, 0 \le u \le 1$.

(7) $f_{XY}(t,u) = 8tu$ on triangle bounded by $t = 1$, $u = 0$, $u = t$.

(8) $f_{XY}(t,u) = \frac{24}{5}tu$ on triangle bounded by $t = 1$, $u = 1$, $u = 1-t$.

(9) $f_{XY}(t,u) = 12t^2u$ on the parallelogram with vertices $(-1,0)$, $(0,0)$, $(1,1)$, $(0,1)$.

(10) $f_{XY}(t,u) = \frac{6}{5}(t^2 + u)$ for $0 \le t \le 1, 0 \le u \le 1$.

9-5. For the following mixed joint distributions,

(a) Determine the joint and marginal densities for the absolutely continuous part.

(b) Determine $F_{XY}(0,0)$, $F_{XY}(0,1)$, and $F_{XY}(1/2,1/2)$.

(c) Determine $P(-1 < X \le 0, Y > 0)$, $P(X \le 1/2, Y \le 1/2)$, $P(X > 1/2, Y > 1/2)$, and $P(0 \le X \le 1/2, 1/2 \le Y \le 1)$.

Mixed joint distributions for problem 9-5

(1) Mass 0.1 at points $(-1,0)$, $(0,0)$, and $(1,0)$.
    Mass 0.7 spread uniformly over the square with vertices at $(-1,0)$, $(0,-1)$, $(1,0)$, and $(0,1)$.

(2) Mass 0.1 at points $(0,0)$, $(0,1)$, and $(1,0)$.
    Mass 0.7 spread uniformly over the triangle with vertices at those points.

10

Independence of Random Vectors

In Chapter 5, the concept of stochastic independence for classes of events is developed in terms of a simple product rule. With due attention to some subtleties, this concept is remarkably successful in capturing the intuitive notion of independence as lack of conditioning of one event by another. In this chapter we extend the concept of stochastic independence to families of random variables and random vectors. Since a random variable is a special case of a random vector, we formulate the concept of independence in terms of random vectors.

10.1. Independence of Random Vectors

In order to be able to make probability statements about random quantities, it is necessary to direct attention to the events they determine. We suppose two or more random vectors to be independent in the stochastic or probabilistic sense if the classes of events they determine form an independent family of classes. We begin by defining independence for a pair of random vectors and examining the implications for distribution functions and the joint mass distributions. Then we extend the idea of independence to an arbitrary family of random vectors. Further extensions are found in Section 11.3, where we consider functions of independent random variables.

We recall that for any random variable or random vector X, the set $X^{-1}(M) = \{\omega : X(\omega) \in M\}$ is an event (determined by X) for each Borel set M in the codomain of X. As in Section 7.1, we let $\sigma(X)$ be the class of all such events determined by X. By Theorem 7.1.4, $\sigma(X)$ is a sigma algebra of events.

Definition 10.1.1. *A pair $\{X, Y\}$ of random vectors is (stochastically) independent iff any of the following equivalent conditions hold.*

- *For each pair $\{M, N\}$ of Borel sets in the respective codomains, the corresponding pair $\{X^{-1}(M), Y^{-1}(N)\}$ of events determined by X, Y is independent.*

- *For each pair $\{M, N\}$ of Borel sets on the respective codomains, the following product rule holds*

$$P(X \in M, Y \in N) = P(X \in M)P(Y \in N).$$

- *The pair $\{\sigma(X), \sigma(Y)\}$ is an independent family of classes of events. That means that if $E \in \sigma(X)$ and $F \in \sigma(Y)$, then $\{E, F\}$ is an independent pair of events.*

Recall that $F_X(t) = P(X \le t) = P(X \in M_t)$, where M_t is the semi-closed interval $(-\infty, t]$, and similarly $F_Y(u) = P(Y \in N_u)$, where N_u is the interval $(-\infty, u]$. Hence, for any independent pair $\{X, Y\}$ of real random variables and any pair $\{t, u\}$ of real numbers, we must have

$$\begin{aligned} F_{XY}(t, u) &= P(X \in M_t, Y \in N_u) = P(X \in M_t)P(Y \in N_u) \\ &= F_X(t)F_Y(u). \end{aligned}$$

Measure theory shows that the structure of the classes of events determined by the random vectors is such that

$$P(X \in M_t, Y \in N_u) = P(X \in M_t)P(Y \in N_u) \qquad \text{for all } t, u$$

implies

$$P(X \in M, Y \in N) = P(X \in M)P(Y \in N) \qquad \text{for all Borel } M, N.$$

We therefore have the following fundamental result.

Theorem 10.1.2 (Product Rule for Distribution Functions) *A pair $\{X, Y\}$ of real random variables is independent iff*

$$F_{XY}(t, u) = F_X(t)F_Y(u) \qquad \text{for all real } t, u.$$

$\square$

Example 10.1.1:

Suppose the pair $\{X, Y\}$ has the joint distribution function

$$F_{XY}(t, u) = \begin{cases} (1 - e^{-\alpha t})u & \text{for } 0 \le t, \, 0 \le u \le 1 \\ (1 - e^{-\alpha t})1 & \text{for } 0 \le t, \, 1 < u. \end{cases}$$

Show that the pair $\{X, Y\}$ is independent.

Solution

We obtain the marginal distribution functions and show the product rule holds.

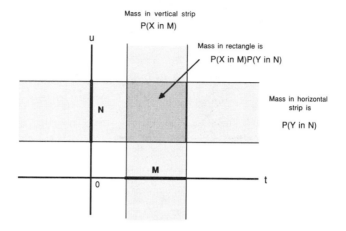

FIGURE 10.1. Joint distribution for independent pair of random variables.

- $F_X(t) = F_{XY}(t, \infty) = 1 - e^{-\alpha t}$ for $0 \leq t$.
- $F_Y(u) = F_{XY}(\infty, u) = u$ for $0 \leq u \leq 1$ and 1 for $1 < u$.

Obviously, the product rule holds.

—   □

INDEPENDENCE AND THE JOINT MASS DISTRIBUTION

One of the significant consequences of the product rule for distribution functions is that the joint distribution is determined by the marginal distributions. This places important restrictions on the nature of the joint distributions. Figure 10.1 illustrates the relationship for the case M and N are simple intervals on the respective axes. In the vertical strip whose projection on the horizontal axis is the interval M, the total probability mass is $P(X \in M)$. In the horizontal strip whose projection on the vertical axis is the interval N, the total probability mass is $P(Y \in N)$. In the rectangle $M \times N$ on the plane, which is the intersection of these two strips, the probability mass $P(X \in M, Y \in N)$ is the product of the marginal masses. This pattern extends to general rectangles obtained when M and N are not simple intervals.

A TEST FOR NONINDEPENDENCE

The product rule provides a simple test for failure of independence that is often quite useful. If any rectangle on the plane has zero probability,

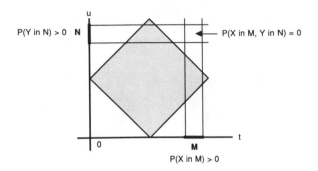

FIGURE 10.2. A test for nonindependence (see Example 10.1.2).

but both marginal probabilities are positive, then the pair $\{X, Y\}$ is *not* independent.

Example 10.1.2: A Test for Nonindependence

Consider the distribution illustrated in Figure 10.2. Suppose probability mass is distributed smoothly over the square shown shaded. Consider the small rectangle, outside the shaded region, with projections M and N on the axes. Clearly, $P(X \in M) > 0$ and $P(Y \in N) > 0$. But since the probability mass in the small rectangle is zero, the product rule cannot hold. That is, for this rectangle

$$P(X \in M, Y \in N) \neq P(X \in M)P(Y \in N)$$

so that the pair $\{X, Y\}$ cannot be independent.

—   □

Remark: While this test is not always applicable, it is quite useful for examining many joint distributions.

INDEPENDENT CLASSES

The notion of independence for an arbitrary class of events is defined by requiring that the product rule hold for every *finite* subclass of two or more members of the class. The concept of independence of random vectors is extended to arbitrary classes in a similar way.

Definition 10.1.3. *A class (finite or infinite)* $\{X_t : t \in T\}$ *of random vectors is (stochastically) independent iff any following equivalent conditions holds:*

- *For each class $\{M_t : t \in T\}$, where M_t is a Borel set in the codomain of the X_t, it is true that $\{X_t^{-1}(M_t) : t \in T\}$ is an independent class of events.*

- *For each finite subclass $\{X_{t_i} : 1 \le i \le n\}$ with $n \ge 2$ and each class $\{M_i : 1 \le i \le n\}$, where M_i is a Borel set in the codomain of the X_{t_i}, the following product rule holds:*

$$P(X_{i_1} \in M_1, X_{i_2} \in M_2, \ldots, X_{i_n} \in M_n) = \prod_{i=1}^{n} P(X_{t_i} \in M_i).$$

- $\{\sigma(X_t) : t \in T\}$ *is an independent family of classes of events.*

Theorem 10.1.2 extends readily to the general case, as follows.

Theorem 10.1.4. *A class $\{X_t : t \in T\}$ is independent iff for each finite subclass $\{X_{t_i} : 1 \le i \le n\}$ the product rule holds for the joint distribution function, i.e., the joint distribution function is the product of the marginal distribution functions.*

—   □

We often have independent classes of random variables, each member of which has the same marginal distribution. The following designation is useful and is increasingly common in the literature.

Definition 10.1.5. *A class $\{X_t : t \in T\}$ of random variables is said to be* iid, *an acronym for "independent, identically distributed," iff the class is independent and each member has the same distribution.*

Example 10.1.3: Random Samples

In Section 18.7, we characterize a random sample of size n as a class $\{X_i : 1 \le i \le n\}$ of n independent variables, each have the same distribution as the "population." This may be expressed by saying the class is iid, with the population distribution.

—   □

Example 10.1.4: Payoff on a Repetitive Game

If a game is played repeatedly and Y_i is the net gain (possibly negative) on the ith play, we often assume $\{Y_i : 1 \le i\}$ is iid. This means the results of successive plays are independent— the result on any play is not affected by, nor does it affect, any other play—and the distribution for each random variable Y_i is the same.

—   □

10.2. Simple Random Variables

We examine the independence condition for classes of simple random variables. We consider first the case of indicator functions.

Example 10.2.1: Indicator Functions

Consider the pair $\{I_A, I_B\}$ of indicator functions. Show that this pair of random variables is independent iff the pair of events $\{A, B\}$ is independent.

Solution

- Suppose $\{I_A, I_B\}$ is independent. Since A is the event $I_A = 1$ and B is the event $I_B = 1$, we must have

$$P(AB) = P(I_A = 1, I_B = 1) = P(I_A = 1)P(I_B = 1) = P(A)P(B)$$

  so that $\{A, B\}$ is independent.

- The class of events determined by the random variable I_A is $\{A, A^c, \emptyset, \Omega\}$. This means that $\{I_A \in M\}$ is one of the events A, A^c, $\emptyset$, or Ω. Similarly $\{I_B \in N\}$ is one of the sets B, B^c, $\emptyset$, or Ω. By the replacement Theorem 4.2.2, if $\{A, B\}$ is independent, then $P(I_A \in M, I_B \in N) = P(I_A \in M)P(I_B \in N)$ for any Borel sets M and N.

—   □

An extension of the argument used in this example shows the result holds for arbitrary classes of indicator functions.

Theorem 10.2.1   (Indicator Functions) *A class* $\{I_{A_t} : t \in T\}$ *of indicator functions is independent iff the class* $\{A_t : t \in T\}$ *of events is independent.*

—   □

THE JOINT DISTRIBUTION FOR INDEPENDENT SIMPLE RANDOM VARIABLES

The product rule for an independent pair of simple random variables takes the form

$$P(X = t_i, Y = u_j) = P(X = t_i)P(Y = u_j)$$

with corresponding extension to any independent family of simple random variables. This rule implies a simple structure for the joint probability distribution. We illustrate some of the essential facts in the following elementary example.

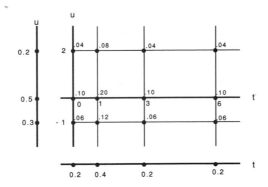

FIGURE 10.3. Joint and marginal probabilities for Example 10.2.2.

Example 10.2.2: Joint Distribution for an Independent Pair of Simple Random Variables

Suppose the pair $\{X, Y\}$ of simple random variables is independent, with the following distributions:

X has values $\{0, 1, 3, 6\}$ with probabilities $\{0.2, 0.4, 0.2, 0.2\}$, respectively.
Y has values $\{-1, 0, 2\}$ with probabilities $\{0.3, 0.5, 0.2\}$, respectively.

The marginal and joint distributions are shown in Figure 10.3. At each grid point (t_i, u_j), we have probability mass

$$P(X = t_i, Y = u_j) = P(X = t_i)P(Y = u_j).$$

Thus, at point $(0, 0)$ we have probability $0.2 \cdot 0.5 = 0.10$, and at point $(3, 2)$ we have probability $0.2 \cdot 0.2 = 0.04$. Values at other grid points are determined similarly.

—   □

Remarks

The following general properties are illustrated in this example.

1. Each point on the grid must have nonzero joint probability mass. If any grid point has zero joint probability mass, the pair cannot be independent.

2. Marginal distributions determine the joint distribution by the product rule. Since joint distributions always determine the marginals,

there is a unique, one-one relationship between the marginal distributions and the joint distributions in the independent case. Sometimes, partial information may be sufficient to complete the description of both marginal and joint distributions.

Example 10.2.3: Partial Information on a Joint Distribution

The following table contains partial information about the marginal and joint probabilities for a pair of simple random variables $\{X, Y\}$. Entries in the top row, above the double line, are values of X. Corresponding entries on the bottom row, below the double line, are values of the marginal distribution for X. Entries in the left column are values of Y. Entries in the last column, to the right of the double line, are values of the marginal distribution for Y. Entries in the body of the table are joint probabilities $P(X = k, Y = j)$. We wish to complete the entries in the table by completing the marginal and joint distributions.

$X =$	0	1	2	3	$P(Y = j)$
$Y = 2$	0.04				
$Y = 1$		0.20			
$Y = 0$				0.03	
$P(X = k)$		0.40	0.10		

If we can determine the complete marginal distributions, we can then complete the table by the product rule. Now, use of the product rule gives

$$P(Y = 0) = \frac{0.03}{0.10} = 0.30 \qquad P(Y = 1) = \frac{0.20}{0.40} = 0.50$$

$$P(X = 0) = \frac{0.04}{0.20} = 0.20.$$

Since the sum of the marginals, in each case, must be one, we have

$$P(Y = 2) = 1 - 0.30 - 0.50 = 0.20$$

and

$$P(X = 2) = 1 - 0.20 - 0.40 - 0.10 = 0.30.$$

With the complete marginals, we take products to determine the remaining joint probabilities.

$X =$	0	1	2	3	$P(Y = j)$
$Y = 2$	0.04	0.08	0.06	0.02	0.20
$Y = 1$	0.10	0.20	0.15	0.05	0.50
$Y = 0$	0.06	0.12	0.09	0.03	0.30
$P(X = k)$	0.20	0.40	0.30	0.10	

—   □

SOME CONDITIONS THAT ENSURE INDEPENDENCE OF SIMPLE RANDOM VARIABLES

We get the following general result for simple random variables

Theorem 10.2.2   (Simple Random Variables) *Suppose that* $X = \sum_{i=1}^{n} a_i I_{E_i}$ *and* $Y = \sum_{j=1}^{m} b_j I_{F_j}$. *Let* $\{M_p : 0 \le p \le 2^n - 1\}$ *be the class of minterms generated by* $\{E_i : 1 \le i \le n\}$ *and* $\{N_q : 0 \le q \le 2^m - 1\}$ *be the class of minterms generated by* $\{F_j : 1 \le j \le m\}$.
If each pair $\{M_p, N_q\}$ *is independent, then* $\{X, Y\}$ *is independent.*

Proof

Any event $X^{-1}(M)$ is a Boolean combination of the events in the class $\{E_i : 1 \le i \le n\}$ and any event $Y^{-1}(N)$ is a Boolean combination of events in the class $\{F_j : 1 \le j \le m\}$. By Theorem 4.2.5, $\{X^{-1}(M), Y^{-1}(N)\}$ is independent. Since M and N are arbitrary, this establishes the independence.

—   □

Remark: A parallel argument, using Theorem 4.2.7, extends this theorem to an arbitrary class. The only difficulty is that notation becomes cumbersome in the general case.

If the random variables are in canonical form, the following assertion can be made.

Corollary 10.2.3   (Simple Random Variables in Canonical Form) *Consider* $X = \sum_{i=1}^{n} t_i I_{A_i}$ *(canonical form) and* $Y = \sum_{j=1}^{n} u_j I_{B_j}$ *(canonical form).*
The pair $\{X, Y\}$ *is independent iff each pair* $\{A_i, B_j\}$ *is independent.*

Proof

• If $\{X, Y\}$ is independent, then for each permissible (i, j)

$$
\begin{aligned}
P(A_i B_j) &= P(X = t_i, Y = u_j) = P(X = t_i)P(Y = u_j) \\
&= P(A_i)P(B_j).
\end{aligned}
$$

• On the other hand, suppose each pair $\{A_i, B_j\}$ is independent. Since the A_i form a partition, the nonempty minterms generated by the class are just the events themselves. Thus, any pair $\{M_p, N_q\}$ is independent. By Theorem 10.2.2, the pair $\{X, Y\}$ is independent.

Remark: Again, the result may be extended to any class of simple random variables, each in canonical form.

Corollary 10.2.4. *Suppose* $X = \sum_{i=1}^{n} a_i I_{E_i}$ *and* $Y = \sum_{j=1}^{m} b_j I_{F_j}$.
If the class $\{E_i, F_j : 1 \le i \le n, 1 \le j \le m\}$ *is independent, then the pair* $\{X, Y\}$ *is independent.*

Proof

If M_p is any minterm generated by the E_i and N_q is any minterm generated by the F_j, then the pair $\{M_p, N_q\}$ is independent. Thus, Theorem 10.2.2 applies to ensure the independence of $\{X, Y\}$.

—   □

Remark: Once more we note the possibility of extension to a general class of simple random variables.

Example 10.2.4: Compound Bernoulli Trials

In Section 6.3 we consider two Bernoulli sequences operating independently.

- *First sequence.* E_i is the event of a success on the ith trial in the first sequence. As a Bernoulli trial, we suppose
  $$\{E_i : 1 \le i\} \text{ is independent and } P(E_i) = p_1.$$

- *Second sequence.* F_j is the event of a success on the jth trial in the second sequence. We suppose
  $$\{F_j : 1 \le j\} \text{ is independent and } p(F_j) = p_2.$$

- As *independent sequences*, we suppose $\{E_i, F_j : 1 \le i, 1 \le j\}$ is an independent class.

If X_n is the number of successes in n trials in the first sequence and Y_m is the number of successes in m trials in the second sequence, then Corollary 10.2.4 shows that $\{X_n, Y_m\}$ is independent.

—   □

Bernoulli Trials and the Binomial Distribution

In Section 6.3, we obtain the binomial distribution as the distribution for the random variable that counts the number of successes in n Bernoulli trials. We reexamine that formulation and modify it slightly, in order to extend it to a more general case. On each component trial in a Bernoulli sequence, the result is one (and only one) of two kinds. We let

E_i be the event the result on the ith trial is of the first kind
(a "success").
E_i^c be the event the result on the ith trial is of the second kind
(a "failure").

Then

$S_n = \sum_{i=1}^n I_{E_i}$ is the number of results of the first kind in the first n trials.

$F_n = \sum_{i=1}^n I_{E_i^c}$ is the number of results of the second kind in the first n trials.

We note that $S_n + F_n = n$. If $\{E_i : 1 \leq i\}$ is independent and $P(E_i) = p$, invariant with i, then by Theorem 6.3.2

$$S_n \text{ is binomial } (n, p) \qquad \text{and} \qquad F_n \text{ is binomial } (n, 1 - p).$$

MULTINOMIAL TRIALS AND THE MULTINOMIAL DISTRIBUTION

Suppose the result of each component trial in a sequence is one of m kinds. Let

E_{ki} be the event the result on the ith trial is of the kth kind, $1 \leq k \leq m$

Then $\mathcal{E}_i = \{E_{ki} : 1 \leq k \leq m\}$ is a partition for each $i \geq 1$. The random variable

$$N_{kn} = \sum_{i=1}^n I_{E_{ki}}$$

is the number of results of the kth kind in the first n trials. We note that

$$\sum_{k=1}^m N_{kn} = \sum_{k=1}^m \sum_{i=1}^n I_{E_{ki}} = \sum_{i=1}^n \sum_{k=1}^m I_{E_{ki}} = n,$$

$$\text{since } \sum_{k=1}^m I_{E_{ki}} = 1 \quad \text{for each } i \geq 1.$$

If we assume

1. $\{\mathcal{E}_i : 1 \leq i\}$ is an independent family, and

2. $P(E_{ki}) = p_k \ 1 \leq k \leq m$, invariant with i and $p_1 + p_2 + \cdots + p_m = 1$

then we have an extension of the notion of Bernoulli trials. We refer to such trials as multinomial trials and to a sequence of such component trials as a multinomial sequence.

Since $N_{kn} = \sum_{i=1}^n I_{E_{ki}}$, it is apparent that each N_{kn} is binomial (n, p_k), $1 \leq k \leq m$. The joint distribution for random vector $(N_{1n}, N_{2n}, \ldots, N_{mn})$ is limited to the set of m-tuples of nonnegative integers

$$Q = \{(n_1, n_2, \ldots, n_m) : n_1 + n_2 + \cdots + n_m = n\}.$$

Consider the joint distribution for $(N_{1n}, N_{2n}, \ldots, N_{mn})$

- If $n_1 + n_2 + \cdots + n_m \neq n$, then $P(N_{1n} = n_1, N_{2n} = n_2, \ldots, N_{mn} = n_m) = 0$.

- If $n_1 + n_2 + \cdots + n_m = n$, then the result is one of the mutually exclusive events of the form.

$A_1 A_2 \ldots A_n$, where n_1 of the A_i satisfy $A_i = E_{1i}$, n_2 of the A_i satisfy $A_i = E_{2i}$, $\ldots$, n_m of the A_i satisfy $A_i = E_{mi}$

Each of these events has probability $p_1^{n_1} p_2^{n_2} \ldots p_m^{n_m}$. Now there are

$$C(n; n_1, n_2, \ldots, n_m) = n!/(n_1! \, n_2! \ldots n_m!)$$

such mutually exclusively events.

Hence, for $n_1 + n_2 + \cdots + n_m = n$,

$$P(N_{1n} = n_1, N_{2n} = n_2, \ldots, N_{mn} = n_m) = n! \prod_{k=1}^{m} \frac{p_k^{n_k}}{n_k!}.$$

This joint distribution is known as the *multinomial distribution.*

The basic assumptions for multinomial trials may be expressed in terms of the kinds of results.

Theorem 10.2.5   (Equivalent Conditions for Multinomial Trials)
Suppose the result of each component trial in a sequence is one of m kinds. Let E_{ki} be the event the result on the ith trial is of the kth kind, $1 \leq k \leq m$ and $\mathcal{E}_i = \{E_{ki} : 1 \leq k \leq m\}$. For each $i \geq 1$, let

$$T_i = \sum_{k=1}^{m} k I_{E_{ki}} = \text{the kind (or type) of the } i\text{th result.}$$

Then the pair of conditions

1. $\{\mathcal{E}_i : 1 \leq i\}$ is an independent family

2. $P(E_{ki}) = p_k$, $1 \leq k \leq m$, invariant with i and $p_1 + p_2 + \cdots + p_m = 1$

is equivalent to the condition

$$\{T_i : 1 \leq i\} \qquad \text{is iid, with} \quad P(T_i = k) = p_k, \quad 1 \leq k \leq m.$$

In words, the results on successive trials in the sequence are independent, with the same distribution of kinds on each trial.

Proof

- Suppose (1) and (2) hold. Since $\mathcal{E}_i$ is a partition, $T_i = \sum_{k=1}^{n} k I_{E_{ki}}$ is in canonical form. Since the family $\{\mathcal{E}_i : 1 \leq i\}$ is independent, by the extension of Corollary 10.2.3, the class $\{T_i : 1 \leq i\}$ must be independent. Since $P(T_i = k) = P(E_{ki}) = p_k$ is invariant with i, the T_i must all have the same distribution.

- Suppose the conditions on $\{T_i : 1 \leq i\}$ hold. Then condition (2) obviously holds. By the extension of Corollary 10.2.3 to arbitrary families, we must have independence of $\{A_i : 1 \leq i\}$ where each A_i is one of the events in $\mathcal{E}_i$. Thus, condition (1) holds.

— $\qquad\qquad\qquad\qquad\qquad\qquad\qquad\qquad\qquad\qquad\qquad$ $\square$

Example 10.2.5:

A device can fail in one of three mutually exclusive ways. If $p_k =$ the probability of a failure of the kth kind, suppose

$$p_1 = 0.3, \qquad p_2 = 0.5, \qquad p_3 = 0.2.$$

Ten devices are received for service. The devices fail independently of one another. What is the probability of four failures of the first kind, three failures of the second kind, and three failures of the third kind?

Solution

$$P(N_1 = 4, N_2 = 3, N_3 = 3) = 10! \frac{0.3^4}{4!} \cdot \frac{0.5^3}{3!} \cdot \frac{0.2^3}{3!} \approx 0.034.$$

— $\qquad\qquad\qquad\qquad\qquad\qquad\qquad\qquad\qquad\qquad\qquad$ $\square$

Remark: We use the multinomial distribution in establishing the important Poisson decomposition Theorem 20.3.1.

10.3.   Joint Density Functions and Independence

The joint distribution function for a pair $\{X, Y\}$ of random variables obeys the product rule

$$F_{XY}(t, u) = F_X(t) F_Y(u) \qquad \text{for all } t, u$$

iff the pair is stochastically independent. If there is a joint density function f_{XY}, a similar product rule characterizes independence.

Theorem 10.3.1   (Product Rule for Density Functions) *Suppose the pair $\{X, Y\}$ of real random variables has joint density function f_{XY}. Then the pair is independent iff the densities obey the product rule*

$$f_{XY}(t, u) = f_X(t) f_Y(u) \qquad \text{for all } t, u.$$

Proof

- If the pair $\{X, Y\}$ is independent, the product rule must hold for the distribution function. Then

$$f_{XY}(t, u) = \frac{\partial^2}{\partial u \partial t} F_{XY}(t, u) = F_X'(t) F_Y'(u) = f_X(t) f_Y(u).$$

- If the product rule holds for the densities, then

$$
\begin{aligned}
F_{XY}(t, u) &= \int_{-\infty}^{t} \int_{-\infty}^{u} f_{XY}(r, s) \, ds \, dr = \int_{-\infty}^{t} \int_{-\infty}^{u} f_X(r) f_Y(s) \, ds \, dr \\
&= \int_{-\infty}^{t} f_X(r) \, dr \int_{-\infty}^{u} f_Y(s) \, ds = F_X(t) F_Y(u)
\end{aligned}
$$

Hence, the pair $\{X, Y\}$ is independent.

— $\square$

Remark: It should be apparent how this theorem generalizes to larger classes.

Example 10.3.1:

Suppose $f_{XY}(t, u) = \alpha \beta e^{-(\alpha t + \beta u)}$ for $0 \le t, \, 0 \le u$. The expression for f_{XY} may be written

$$\alpha \beta e^{-(\alpha t + \beta u)} = \alpha e^{-\alpha t} \beta e^{-\beta u}.$$

To get the marginal density for X, we integrate out the u

$$f_X(t) = \int f_{XY}(t, u) \, du = \alpha e^{-\alpha t} \int_0^{\infty} \beta e^{-\beta u} \, du = \alpha e^{-\alpha t}.$$

In a similar fashion, we find $f_Y(u) = \beta e^{-\beta u}$. Thus, X is exponential (α) and Y is exponential (β), and because of the product rule on the densities, the pair $\{X, Y\}$ is independent.

— $\square$

Example 10.3.2: Joint Normal with Parameter ρ Set to Zero

If we set the parameter ρ to zero, the joint density for the joint normal distribution is (see Example 9.3.4)

$$\frac{1}{2\pi\sigma_X\sigma_Y} \exp\left(-\frac{1}{2}\left(\frac{t-\mu_X}{\sigma_X}\right)^2 - \frac{1}{2}\left(\frac{u-\mu_Y}{\sigma_Y}\right)^2\right).$$

It is known that the marginal densities are for $X \sim N(\mu_X, \sigma_X^2)$ and $Y \sim N(\mu_Y, \sigma_X^2)$. Use of the law of exponents shows that $f_{XY}(t,u) = f_X(t)f_Y(u)$, so that the pair $\{X, Y\}$ is independent.

$\square$

PROBLEMS

10–1. Simple random variables X, Y form an independent pair. The range of X is $\{-1, 0, 1\}$ and that of Y is $\{1, 2\}$. Suppose

$$p_X(0) = 0.5, \qquad p_Y(2) = 0.6, \qquad \text{and} \qquad p_{XY}(1, 2) = 0.18.$$

Determine p_X, p_Y, and p_{XY} completely.

10–2. Two fair dice are thrown. The dice are distinguishable. Let

X be the total number of spots that turn up on a throw.
Y be the number of spots that turn up on the first die.

Show whether or not the pair $\{X, Y\}$ is independent.

10–3. For each of the following discrete joint distributions, determine whether or not the random variables form an independent pair:

(a)

$$P(X = t, Y = u)$$

$t =$	1	2	3	4	5	6
$u = 5$	0	0	0	0.1	0	0.1
$u = 4$	0.1	0	0	0	0.1	0
$u = 3$	0	0.1	0.1	0	0.1	0
$u = 2$	0	0	0	0.1	0.1	0
$u = 1$	0.1	0	0	0	0	0

(b)

$$P(X = t, Y = u)$$

$t =$	0	1	4	9
$u = 2$	0.05	0.04	0.21	0.15
$u = 0$	0.05	0.01	0.09	0.10
$u = -1$	0.10	0.05	0.10	0.05

(c)

$$P(X = t, Y = u)$$

$t =$	-2	0	1	3
$u = 1$	0.15	0.05	0.10	0
$u = 0$	0.05	0.08	0.10	0.12
$u = -1$	0.10	0.02	0	0.23

10–4. For each of the following joint density function, determine whether or not the random variables form an independent pair.

Joint density functions for problem 10-4 (zero outside indicated region):

(1) $f_{XY}(t, u) = 1/\pi$ on circle with radius 1, center at $(0, 0)$.
(2) $f_{XY}(t, u) = 1$ on triangle bounded by $t = 0$, $u = 0$, and $u = 2(1 - t)$.

(3) $f_{XY}(t,u) = 1/2$ on square with vertices at $(1,0)$, $(2,1)$, $(1,2)$, and $(0,1)$.

(4) $f_{XY}(t,u) = 4t(1-u) for 0 \le t \le 1, 0 \le u \le 1$.

(5) $f_{XY}(t,u) = \frac{1}{8}(t+u) for 0 \le t \le 2, 0 \le u \le 2$.

(6) $f_{XY}(t,u) = 2\alpha u e^{-\alpha t} for 0 \le t, 0 \le u \le 1$.

(7) $f_{XY}(t,u) = 8tu$ on triangle bounded by $t = 1$, $u = 0$, and $u = t$.

(8) $f_{XY}(t,u) = \frac{24}{5}tu$ on triangle bounded by $t = 1$, $u = 1$, and $u = 1 - t$.

(9) $f_{XY}(t,u) = 12t^2 u$ on the parallelogram with vertices $(-1,0)$, $(0,0)$, $(1,1)$, and $(0,1)$.

(10) $f_{XY}(t,u) = \frac{6}{5}(t^2 + u) for 0 \le t \le 1$, and $0 \le u \le 1$.

10–5. For each of the following mixed joint distributions, determine whether or not the random variables form an independent pair.

Mixed joint distributions for problem 10-5

(1) Mass 0.1 at points $(-1,0)$, $(0,0)$, $(1,0)$.
Mass 0.7 spread uniformly over the square with vertices at $(-1,0)$, $(0,-1)$, $(1,0)$, and $(0,1)$.

(2) Mass 0.1 at points $(0,0)$, $(0,1)$, $(1,0)$.
Mass 0.7 spread uniformly over the triangle with vertices at those points.

10–6. Orange Computer Corp. is planning to announce a new microcomputer in 180 days. Much of its sales success will depend upon the availability of good software. Two software companies, MicroWare and BusiCorp are working to bring out integrated business packages for release at the time of introduction of the computer. MicroWare has anticipated completion time, in days, exponential $(1/150)$. Busi-Corp has time to completion, in days, which is exponential $(1/120)$. What is the probability that at least one package will be ready on time?

10–7. The pair $\{X,Y\}$ is iid, uniform $(0,10)$. Determine $P(1 \le X \le 2, 5 \le Y \le 10)$.

10–8. A critical module in a Xerox copier has time to failure (in hours of machine run time) X exponential $(1/2000)$, so that the mean time to failure is 2000 hours. The copier is on continuously, except for repairs. The module is replaced routinely every 30 days (720 hours), unless failure occurs. If units fail independently, what is the probability the machine will operate for one year without breakdown due to failure of the critical module?

10–9. Five "identical" units are put into operation simultaneously. The service lifetimes in hours of the units may be considered iid, exponential (1/500). Thus, the expected lifetime of each unit is 500 hours. What is the probability that four or more are still operating after 458 hours?

Hint: View this as a Bernoulli trial. What is the probability of a success?

10–10. The location of ten points may be considered values of iid random variables with the symmetric triangular distribution on $[1, 3]$ (see Section 8.4). What is the probability all ten points lie within a distance $1/2$ of the point $t = 2$? What is the probability that exactly three lie in that interval?

11

Functions of Random Variables

A random variable X is a function defined on the basic space Ω. If the outcome of the experiment is ω, the corresponding number is $t = X(\omega)$. Frequently, if value t is observed, we are interested in a corresponding value $v = g(t)$ obtained by applying the rule for the function g to the observed value. If we make the assignment $v = g(X(\omega))$ for all ω, we have a new function $Z = g(X)$ defined on the basic space. Similarly, we may be interested in $Z = h(X, Y)$. If $X(\omega) = t$ and $Y(\omega) = u$, then $Z(\omega) = h(t, u)$. Elementary mapping properties show that if g is a Borel function, then $Z = g(X)$ is a random variable; similarly, if h is a Borel function of two variables, then $Z = h(X, Y)$ is a random variable. In this chapter, we address the basic problem: Given the distribution for X, or the joint distribution for (X, Y), how can we assign probabilities to events determined by the new random variable Z?

11.1. A Fundamental Approach and some Examples

We first use some typical examples to establish a simple approach to the basic problem. If X is a real-valued random variable and g is a Borel function, then $Z = g(X)$ is a new random variable on the basic space that has the value $g(t)$ whenever $X(\omega) = t$.

Problem: Given the distribution for X; determine $P(Z \in M)$. In particular, determine $F_Z(v) = P(Z \leq v)$.

Approach: Seek appropriate statements of the form: $Z(\omega) \in M$ iff $X(\omega) \in N$, so that $P(Z \in M) = P(X \in N)$.

Later, we obtain a general formal solution to this problem in terms of the mapping properties. But first we consider some important examples in which we work less formally (but no less precisely).

Example 11.1.1: Normal Distribution

Suppose $X \sim N(\mu, \sigma^2)$. Show that $Z = g(X) = (X - \mu)/\sigma \sim N(0, 1)$.

Solution

Since σ is positive, simple algebra shows $Z = (X - \mu)/\sigma \leq v$ iff $X \leq \sigma v + \mu$.

We may therefore assert $F_Z(v) = P(Z \leq v) = P(X \leq \sigma v + \mu) = F_X(\sigma v + \mu)$.

To obtain the density $f_Z(t)$, we evaluate the derivative

$$F_Z'(t) = \sigma f_X(\sigma t + \mu) = \frac{\sigma}{\sigma\sqrt{2\pi}} \exp\left(-\frac{1}{2}\left(\frac{\sigma t + \mu - \mu}{\sigma}\right)^2\right)$$

$$= \frac{1}{\sqrt{2\pi}} e^{-t^2/2} = \phi(t).$$

Thus, $Z \sim N(0,1)$. We may use this approach to obtain the pattern for finding F_X from the distribution function Φ for the standardized normal distribution.

$$F_X(t) = P(X \leq t) = P\left(\frac{X - \mu}{\sigma} \leq \frac{t - \mu}{\sigma}\right) = P\left(Z \leq \frac{t - \mu}{\sigma}\right)$$

$$= \Phi\left(\frac{t - \mu}{\sigma}\right).$$

This is the result obtained in Section 8.5 by change of variables techniques.

—   □

Example 11.1.2: Affine Function

Suppose $Z = aX + b$, with $a \neq 0$. Determine F_Z.

Solution

We must consider two cases.

1. $a > 0$: $Z(\omega) \leq v$ iff $X(\omega) \leq (v - b)/a$
   implies $F_Z(v) = F_X((v - b)/a)$.

2. $a < 0$: $Z(\omega) \leq v$ iff $X(\omega) \geq (v - b)/a$
   implies $F_Z(v) = 1 - F_X((v - b)/a) + P(X = (v - b)/a)$.

If X is absolutely continuous, so is Z. In this case, $P(X = (v - b)/a) = 0$. Differentiating F_Z to obtain the density yields two cases that can be combined by the use of the absolute value to obtain

$$f_Z(v) = \frac{1}{|a|} f_X\left(\frac{v - b}{a}\right).$$

—   □

These results have a simple geometric interpretation. Adding a constant to a random variable shifts all of its values, and hence shifts the induced probability distribution. Multiplication by a positive constant greater than one spreads the probability mass further because of the wider variation in the values. Multiplication by a positive constant less than one decreases the spread of the values and hence compresses the probability distribution. Multiplication by a negative constant reverses the orientation of the distribution about the origin and either spreads or compresses the mass. These facts are consistent with the effects of the resulting change of variable for the distribution and density functions. Subtracting a constant from the argument produces a horizontal shift of the graph without changing its shape, and multiplying the argument by a constant produces a change of horizontal scale (and a reversal of orientation if negative).

Example 11.1.3: Square Root Function

Suppose X is nonnegative. Consider $Z = X^{1/2}$. Determine the distribution for Z.

Solution

The function $g(t) = t^{1/2}$ is increasing on $[0, \infty)$. Now $X^{1/2} \leq v$ iff $X \leq v^2$. Thus for $v \geq 0$,

$$F_Z(v) = F_X(v^2) \quad \text{and for } X \text{ absolutely continuous} \quad f_Z(v) = 2v f_X(v^2).$$

—   □

Example 11.1.4: Exponential Function

Consider $Z = e^{-aX} \, a > 0$. Determine the distribution for Z.

Solution

We note that $Z > 0$. Use of the fact that $\ln e^x = x$ gives

$$Z = e^{-aX} \leq v \quad \text{iff} \quad -aX \leq \ln v \quad \text{iff} \quad X \geq -\frac{\ln v}{a}.$$

Thus

$$F_Z(v) = P\left(X \geq -\frac{1}{a}\ln v\right) = 1 - F_X\left(-\frac{1}{a}\ln v\right) + P\left(X = -\frac{1}{a}\ln v\right)$$

for $v > 0$. In the absolutely continuous case, the last term is zero. Differentiation by v gives

$$f_Z(v) = \frac{1}{av} f_X \left(-\frac{1}{a} \ln v \right) \qquad \text{for } v > 0.$$

—   □

Example 11.1.5: Weibull Distribution and the Exponential Distribution

In Section 8.4 we introduce the Weibull distribution as a generalization of the exponential. Random variable X is Weibull (α, λ, ν) iff

$$F_X(t) = 1 - e^{-\lambda(t-\nu)^\alpha} \qquad \alpha > 0, \ \lambda > 0, \ \nu \geq 0, \ t \geq \nu.$$

Suppose Y is exponential (1). Show that $X = \lambda^{-1/\alpha} Y^{1/\alpha} + \nu$ is Weibull (α, λ, ν).

Solution

Our fundamental relation is $X = \lambda^{-1/\alpha} Y^{1/\alpha} + \nu \leq t$ iff $Y \leq \lambda(t-\nu)^\alpha$ so that

$$F_X(t) = F_Y(\lambda(t-\nu)^\alpha) = 1 - e^{-\lambda(t-\nu)^\alpha} \qquad \text{for } t - \nu \geq 0.$$

—   □

In statistics, the chi-square distribution is employed in procedures known as *hypothesis testing*. This distribution is a special case of the gamma distribution. Random variable X has the chi-square distribution with n degrees of freedom iff it is gamma $(n/2, 1/2)$. In the next example, we take a first step towards showing why the chi-square distribution is important in hypothesis-testing procedures.

Example 11.1.6: Normal and Gamma Distributions

Show that if random variable $X \sim N(0, 1)$, then X^2 is gamma $(1/2, 1/2)$.

Solution

Y is gamma(α, λ) iff $f_Y(t) = e^{-\lambda t}(\lambda^\alpha t^{\alpha - 1}/\Gamma(\alpha))$, $t \geq 0$. Suppose $X \sim N(0, 1)$. Then

$$
\begin{aligned}
P(X^2 \leq v) &= P(-v^{1/2} \leq X \leq v^{1/2}) = 2P(0 \leq X \leq v^{1/2}) \\
&= \frac{2}{\sqrt{2\pi}} \int_0^{v^{1/2}} e^{-t^2/2} \, dt.
\end{aligned}
$$

Make the change of variable $u = t^2$ or $t = u^{1/2}$. Then, formally $dt = \frac{1}{2}u^{-1/2}\,du$ and the limit $t = v^{1/2}$ becomes $u = v$. We thus have

$$P(X^2 \leq v) = \frac{1}{\sqrt{2\pi}} \int_0^v e^{-u/2} u^{\frac{1}{2}-1}\,du.$$

We use the fact that $P(X^2 \leq v) \to 1$ as $v \to \infty$ and the fact (see the appendix on mathematical aids) that

$$\int_0^\infty e^{-u/2} u^{\frac{1}{2}-1}\,du = \frac{\Gamma(1/2)}{(1/2)^{1/2}},$$

to assert

$$\frac{1}{\sqrt{2\pi}} = \frac{(1/2)^{1/2}}{\Gamma(1/2)}$$

so that

$$P(X^2 \leq v) = \frac{(1/2)^{1/2}}{\Gamma(1/2)} \int_0^v u^{1/2-1} e^{-u/2}\,du \qquad \text{for } v \geq 0.$$

Thus, X^2 is gamma $(1/2, 1/2)$.

—— $\square$

To complete the development on the chi-square distribution, we need a result on the sum of independent gamma random variables developed in Section 18.2.

THE GENERAL PATTERN

To put the previous special cases in perspective and to provide a basis for considering functions of two or more variables, we use fundamental mapping ideas to make explicit the general pattern. We consider $Z = g(X)$, where g is a Borel function. All the "reasonably well-behaved functions" we consider are Borel functions.

We want to determine $P(Z \in M)$, where M is a Borel set. First, we establish that Z must be a random variable. In fact, we obtain the result for W a random vector and g a vectore-valued Borel function. The theorem then holds in the special case of a real valued function of a real random variable.

Theorem 11.1.1   (Borel Functions of Random Vectors) *Suppose W is a random vector and g is a Borel function whose domain includes the range of W. Then*

$$Z = g(W) \qquad \textit{is a random vector.}$$

Proof

We need to show that $Z^{-1}(M)$ is an event for any Borel set M in the codomain of g, and hence of Z.

- As before, let $N = g^{-1}(M)$ be the set of all t in the domain of g such that $g(t) \in M$. In this case,

$$Z(\omega) \in M \quad \text{iff} \quad g(W(\omega)) \in M \quad \text{iff} \quad W(\omega) \in N = g^{-1}(M).$$

This may be expressed equivalently by writing

$$Z^{-1}(M) = W^{-1}(N) \qquad \text{where} \quad N = g^{-1}(M).$$

- Since g is Borel, then $N = g^{-1}(M)$ must be a Borel set if M is Borel.

- By definition, $W^{-1}(N)$ is an event for any Borel set N. Hence, $Z^{-1}(M)$ is an event for any Borel set M.

— $\square$

For the real valued case, Z is a random variable with the property

$$P(Z \in M) = P(X \in N) = P(X \in g^{-1}(M)).$$

This shows that the task of determining $P(Z \in M)$ breaks down into the twofold task of determining

$$(1) \quad N = g^{-1}(M) \qquad \text{then} \qquad (2) \quad P(X \in N).$$

Example 11.1.7:

Suppose $Z = \sin X$, with X uniform on $[0, 2\pi]$. Determine

$$P(0 \le Z \le a) = P(Z \in [0, a]), \qquad \text{for } 0 < a \le 1.$$

Solution

We have $g(t) = \sin t$, and $M = [0, a]$. Reference to Figure 11.1 shows that $N = [0, \sin^{-1} a] \cup [\pi - \sin^{-1} a, \pi]$. Because X is uniformly distributed on $[0, 2\pi]$, we have

$$P(Z \in M) = P(X \in N) = \frac{2 \sin^{-1} a}{2\pi} = \frac{\sin^{-1} a}{\pi}.$$

— $\square$

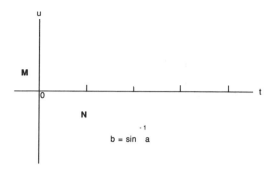

FIGURE 11.1. Mapping diagram for Example 11.1.7.

Theorem 11.1.1 shows that every event determined by $Z = g(W)$ is an event determined by W. Thus, the class $\sigma(Z)$ of events determined by Z is a subclass of the class $\sigma(W)$. There is a very important converse of this fact, which we use repeatedly in subsequent developments. Thus, we may assert

Theorem 11.1.2. *If Z and W are random vectors, then $\sigma(Z) \subset \sigma(W)$ iff there exists a Borel function g such that $Z = g(W)$.*
— $\square$

The "if" part is included in Theorem 11.1.1. We omit a proof of the converse, which requires some lengthy theoretical developments beyond the scope of this work.

11.2. Functions of More Than One Random Variable

We extend the strategy in the previous section to functions of more than one random variable. The essential theoretical patterns carry over from the one-variable case. In order to be able to exploit certain geometrical representations, we give principal attention to real-valued functions of two real-valued random variables. It should be clear how the strategy extends to higher dimensional cases.

Suppose the pair $\{X, Y\}$ is jointly distributed and let $W = (X, Y)$ be the random vector whose coordinates are the members of the pair. We consider $Z = g(X, Y)$, where g is any real-valued Borel function of two real variables. If the joint distribution for (X, Y) is known, we wish to determine $P(Z \in M)$ for various Borel sets M. In particular, we concentrate on the problem of determining the distribution function F_Z.

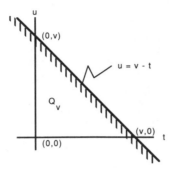

FIGURE 11.2. The region Q_V for $g(t,u) = t + u \leq v$.

According to the fundamental mapping ideas used in developing the notion of the joint distribution for (X,Y), each pair of values $\big(X(\omega), Y(\omega)\big)$ corresponds to a point on the plane of the distribution. Now $Z(\omega) = g\big(X(\omega), Y(\omega)\big) \leq v$ iff the pair of values $(t,u) = \big(X(\omega), Y(\omega)\big)$ is one of those pairs for which $g(t,u) \leq v$. These pairs are represented on the plane by the points in the region $Q_v = \{(t,u) : g(t,u) \leq v\}$.

We thus have $P(Z \leq v) = P\big((X,Y) \in Q_v\big)$.

This latter probability is the probability mapped into the region Q_v by *the joint mapping for* (X,Y). The problem reduces to two parts.

1. Find the region $Q_v = \{(t,u) : g(t,u) \leq v\}$.

2. Use the information provided by the joint distribution for (X,Y) to determine the probability mapped into the region Q_v.

Example 11.2.1: $Z = X + Y$

In this case $g(t,u) = t + u$, so that

$$Q_v = \{(t,u) : t + u \leq v\} = \{(t,u) : u \leq v - t\}.$$

The boundary is the line $u = v - t$, which is the line, slope -1, through the points $(0,v)$ and $(v,0)$. For each fixed v, the region Q_v consists of all points in the plane on or below this line (see Figure 11.2). Thus, $F_Z(v)$ is the probability mass in this region.

If there is a joint density function f_{XY}, then

$$F_Z(v) = \int_{Q_v} \int f_{XY} = \int_{-\infty}^{\infty} \int_{-\infty}^{v-t} f_{XY}(t,u) \, du \, dt.$$

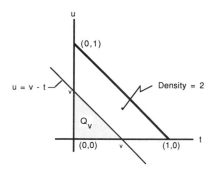

FIGURE 11.3. Density and region Q_V for Example 11.2.2.

The density function for Z is obtained by differentiation with respect to v to give

$$f_Z(v) = \int_{-\infty}^{\infty} f_{XY}(t, v - t)\, dt = \int_{-\infty}^{\infty} f_{XY}(v - u, u)\, du.$$

The second integral is obtained from the first by a standard change of variable.

—   □

Example 11.2.2:

Suppose $\{X, Y\}$ has joint distribution uniform on the triangular region with vertices at $(0,0)$, $(1,0)$, and $(0,1)$ (see Figure 11.3). Let $Z = X + Y$. Determine the distribution function F_Z.

Solution

Since the area is one half and the density is constant, $f_{XY}(t, u) = 2$ on the triangle, and zero elsewhere. For any v between 0 and 1, $F_Z(v) = 2\times$ the area in $Q_v = v^2$, $0 \le v \le 1$.

—   □

Example 11.2.3: $Z = XY$

To simplify writing, we consider only the case $X \ge 0$, $Y \ge 0$. Then $g(t, u) = tu$, $t \ge 0$, $u \ge 0$. We have for any fixed $v \ge 0$

$$Q_v = \{(t, u) : tu \le v\} = \{(t, u) : u \le v/t\}.$$

This is the region consisting of those points on or below the curve $u = v/t$

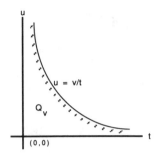

FIGURE 11.4. Region Q_V for $g(t, u) = tu \leq v$.

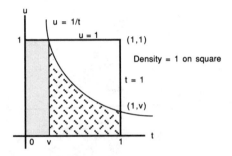

FIGURE 11.5. Density and regions for Example 11.2.4.

for $t > 0$ (see Figure 11.4).

—   □

Remark: If t can change sign, care must be taken to keep the direction of the inequality consistent with the sign of t. That is, if $t < 0$, then $u \geq v/t$.

Example 11.2.4:

Suppose $\{X, Y\}$ has uniform joint distribution over the unit square with vertices $(0, 0)$, $(1, 0)$, $(1, 1)$, and $(0, 1)$. Let $Z = XY$. Determine $F_Z(v)$ and $f_Z(v)$.

Solution

For any fixed v between 0 and 1, we have the geometry indicated in Figure 11.5. We need the probability in the rectangle with base $[0, v]$ and

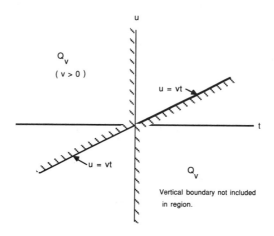

FIGURE 11.6. Region Q_V for $g(t, u) = v/t$, with $v > 0$.

side $[0, 1]$ plus the probability under the curve $u = v/t$ for $v \le t \le 1$. Since f_{XY} is constant at one over this region, we have

$$F_Z(v) = v \cdot 1 + \int_v^1 v/t\,dt = v - v \ln v, \quad 0 \le v \le 1.$$

Upon differentiating with respect to v, we obtain $f_Z(v) = \ln(1/v) = -\ln v$ for $0 < v \le 1$.

———   □

Example 11.2.5: $Z = Y/X$

In this case $Q_v = \{(t, u) : t > 0, u \le vt\} \bigcup \{(t, u) : t < 0, u \ge vt\}$.

This region is shown in Figure 11.6 for a positive value of v.

For the absolutely continuous case with joint density function f_{XY}, we have

$$F_Z(v) = \int_{-\infty}^0 \int_{vt}^\infty f_{XY}(t, u)\, du\, dt + \int_0^\infty \int_{-\infty}^{vt} f_{XY}(t, u)\, du\, dt.$$

Differentiating the integral with respect to v gives

$$f_Z(v) = -\int_{-\infty}^0 t f_{XY}(t, vt)\, dt + \int_0^\infty t f_{XY}(t, vt)\, dt = \int_{-\infty}^\infty |t| f_{XY}(t, vt)\, dt.$$

———   □

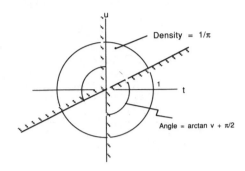

FIGURE 11.7. Density and region for Example 11.2.6.

Example 11.2.6:

Suppose f_{XY} is constant on the circle of unit radius, center at the origin. Since the area of the circle is π, the constant value of the density function is $1/\pi$ on the circle and zero elsewhere. Let $Z = Y/X$. Reference to Figure 11.7 shows

$$F_Z(v) = \frac{2\theta}{2\pi} \qquad \text{where } \theta = \frac{\pi}{2} + \arctan v$$

so that

$$F_Z(v) = \frac{1}{2} + \frac{1}{\pi} \arctan v \qquad \text{and} \qquad f_Z(v) = \frac{1}{\pi(1 + v^2)}.$$

□

Remark: This is a case of the *Cauchy distribution*, which has a density of the form

$$f_Z(v) = \frac{\alpha}{\pi\left(\alpha^2 + (v - \mu)^2\right)}.$$

A more important case of the Cauchy distribution arises as follows.

Example 11.2.7:

Suppose the pair $\{X, Y\}$ is independent, with $X \sim N(0, \sigma_X^2)$ and $Y \sim N(0, \sigma_Y^2)$. Put $Z = Y/X$. Determine the density f_Z.

Solution

We use the result of Example 11.2.5 to assert

$$f_Z(v) = \int_{-\infty}^{\infty} |t| f_{XY}(t, vt)\, dt$$

$$= \frac{1}{2\pi\sigma_X\sigma_Y} \int_{-\infty}^{\infty} |t| \exp\left(-\frac{1}{2}\left(\frac{t^2}{\sigma_X^2} + \frac{v^2 t^2}{\sigma_Y^2}\right)\right) dt.$$

Since the integrand is even in t, we may take twice the integral from 0 to ∞. We may write the exponent in a more compact form to get

$$f_Z(v) = \frac{1}{\pi\sigma_X\sigma_Y} \int_0^{\infty} t e^{-a^2 t^2/2} \, dt \quad \text{where } a^2 = \frac{1}{\sigma_X^2} + \frac{v^2}{\sigma_Y^2} = \frac{\sigma_Y^2 + \sigma_X^2 v^2}{\sigma_X^2\sigma_Y^2}.$$

Using the definite integral $\int_0^{\infty} u e^{-u^2} \, du = 1/2$ and making an appropriate change of variable, we find

$$f_Z(v) = \frac{1}{\pi\sigma_X\sigma_Y a^2} = \frac{\sigma_X\sigma_Y}{\pi(\sigma_Y^2 + \sigma_X^2 v^2)} = \frac{\alpha}{\pi(\alpha^2 + v^2)},$$

where $\alpha = \sigma_Y/\sigma_X$. Once again, we have a Cauchy distribution.

— 												□

Example 11.2.8: $Z = |X - Y|$

In this case,

$$Q_v = \{(t, u) : |t - u| \le v\} = \{(t, u) : t - v \le u \le t + v\}.$$

For each fixed $v \ge 0$, this is the strip bounded by two lines having unit positive slope. The upper line goes through the point $(0, v)$ and the lower line goes through the point $(v, 0)$.

— 												□

Example 11.2.9:

Suppose $\{X, Y\}$ is an independent pair, with X exponential (α) and Y exponential (β). Then both random variables are nonnegative and $f_{XY}(t, u) = f_X(t) f_Y(u)$. The strip is restricted to the first quadrant, as shown in Figure 11.8. If we integrate first with respect to the second variable u, it is convenient to consider two regions. Thus

$$
\begin{aligned}
F_Z(v) &= \int_0^v \int_0^{t+v} f_X(t) f_Y(u) \, du \, dt + \int_v^{\infty} \int_{t-v}^{t+v} f_X(t) f_Y(u) \, du \, dt \\
&= \int_0^v f_X(t) F_Y(t+v) \, dt + \int_v^{\infty} f_X(t) \big(F_Y(t+v) - F_Y(t-v)\big) \, dt \\
&= \int_0^v \alpha e^{-\alpha t} \big(1 - e^{-\beta(t+v)}\big) \, dt + \int_v^{\infty} \alpha e^{-\alpha t} \big(e^{-\beta(t-v)} - e^{-\beta(t+v)}\big) \, dt \\
&= \int_0^v \alpha e^{-\alpha t} \, dt - e^{-\beta v} \int_0^v \alpha e^{-(\alpha+\beta)t} \, dt
\end{aligned}
$$

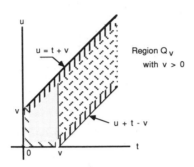

FIGURE 11.8. Density and subregions for Example 11.2.9.

$$+ \left(e^{\beta v} - e^{-\beta v}\right) \int_{v}^{\infty} \alpha e^{-(\alpha+\beta)t}\, dt$$

$$= \quad 1 - e^{-\alpha v} - e^{-\beta v}\frac{\alpha}{\alpha+\beta}\left(1 - e^{-(\alpha+\beta)v}\right)$$

$$+ \left(e^{\beta v} - e^{-\beta v}\right)\frac{\alpha}{\alpha+\beta}e^{-(\alpha+\beta)v}$$

$$= \quad 1 - \frac{\beta}{\alpha+\beta}e^{-\alpha v} - \frac{\alpha}{\alpha+\beta}e^{-\beta v}$$

for $v \geq 0$. In the special case that $\alpha = \beta$, we have

$$F_Z(v) = 1 - e^{-\alpha v} = F_X(v) = F_Y(v)$$

so that Z is also exponential with the common parameter value.

— $\qquad\qquad\qquad\qquad\qquad\qquad\qquad\qquad\qquad\qquad\quad$ $\square$

Example 11.2.10:

Lamps in an enclosed stairwell are installed in pairs, to reduce the likelihood that the stairway will be dark because of lamp failure. Suppose the lamps fail independently and the time (in days) to failure for each is exponential ($\lambda = 1/90$). A new pair is installed at the same time. What is the probability that they will not fail within ten days of each other?

Solution

Let X and Y be the respective times to failure. We assume $\{X,Y\}$ is iid, exponential (1/90). The problem is to determine $P(|X - Y| > 10)$. Now according to the result of Example 11.2.9, $Z = |X - Y|$ is also exponential (1/90). Thus

$$P(|X - Y| > 10) = 1 - F_Z(10) = e^{-10/90} \approx 0.89.$$

$\square$

The next problem can be solved without specific reference to the graphical representation, but the solution tacitly uses the graphical significance of the joint distribution function.

Example 11.2.11: Extreme Values

Suppose $V = \min\{X, Y\}$ and $W = \max\{X, Y\}$. This means that if we observe values $X(\omega)$ and $Y(\omega)$, we assign the larger of these as the value $W(\omega)$ and the smaller of these as the value $V(\omega)$. Determine the distribution functions F_V and F_W.

Solution

The defining conditions for V and W show that

$$W \le t \quad \text{iff both} \quad X \le t, \ Y \le t$$

and

$$V \le t \quad \text{iff either} \quad X \le t \quad \text{or} \quad Y \le t \quad \text{(or both)}.$$

Thus

$$F_W(t) = P(X \le t, Y \le t) = F_{XY}(t, t)$$

and

$$\begin{aligned} F_V(t) &= P(X \le t) + P(Y \le t) - P(X \le t, Y \le t) \\ &= F_X(t) + F_Y(t) - F_W(t). \end{aligned}$$

$\square$

Example 11.2.12:

For the stairwell lamps of Example 11.2.10, compare the probability that either lamp, considered individually, will fail within 90 days with the probability that at least one lamp will fail within 90 days. That is, compare $P(X \le 90) = P(Y \le 90)$ with $P(V \le 90)$.

Solution

For the assumed distributions and independence

$$P(X \le 90) = P(Y \le 90) = 1 - e^{-1} \approx 0.632$$

and

$$P(V \leq 90) = 2(1 - e^{-1}) - (1 - e^{-1})^2 \approx 0.865.$$

—   □

FUNCTIONS WITH COMPOSITE DEFINITIONS

Frequently we deal with a function that is defined differently for different values of the argument. We introduce a notational scheme using indicator functions we find useful in a variety of circumstances. In particular, this scheme is useful for expectations and conditional expectations considered in subsequent chapters.

Example 11.2.13:

Consider again the absolute value of the difference function, the minimum function, and the maximum function. These may be written

$$g(t,u) = |t - u| = \begin{cases} t - u & \text{for } t \geq u \\ u - t & \text{for } t < u \end{cases}$$

$$g(t,u) = \max\{t,u\} = \begin{cases} t & \text{for } t \geq u \\ u & \text{for } t < u \end{cases}$$

$$g(t,u) = \min\{t,u\} = \begin{cases} u & \text{for } t \geq u \\ t & \text{for } t < u \end{cases}.$$

If we let $Q_1 = \{(t,u) : t \geq u\}$ and $Q_2 = \{(t,u) : t < u\}$, then we may write

$$|t - u| = I_{Q_1}(t,u)(t - u) + I_{Q_2}(t,u)(u - t)$$
$$\min\{t,u\} = I_{Q_1}(t,u)u + I_{Q_2}(t,u)t$$
$$\max\{t,u\} = I_{Q_1}(t,u)t + I_{Q_2}(t,u)u.$$

For these functions, the region Q_1 is the region in the plane on or below the line $u = t$, and Q_2 is the region in the plane above that line (see Figure 11.9). These are disjoint regions and every possible pair of numbers for which the function is defined (i.e., the domain) is included in exactly one of these regions. We may write the corresponding functions of the random variables in this form. For example,

$$|X - Y| = I_{Q_1}(X,Y)(X - Y) + I_{Q_2}(X,Y)(Y - X).$$

In Q_1, $g(t,u) = t - u$ and the condition $g(t,u) \leq v$ is the condition $t - u \leq v$ or $u \geq t - v$. This is the strip on or above the line $u = t - v$ and on or below the line $u = t$.

In Q_2, $g(t,u) = u - t$ and the condition $g(t,u) \leq v$ is the condition $u - t \leq v$ or $u \leq t + v$. This is the strip above the line $u = t$ and on or below the line $u = t + v$.

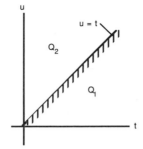

FIGURE 11.9. Regions for Example 11.2.13.

These two strips combine to give the region Q_v in Example 11.2.9 (Figure 11.8).

— ☐

The general case for functions of two variables may be written formally

$$Z = g(X, Y) = \sum_k I_{Q_k}(X, Y) g_k(X, Y)$$

where the Q_k form a disjoint class of regions on the plane whose union includes the domain of g. For $(t, u) \in Q_k$, the rule defining function g is the rule for g_k. To determine $Q_v = \{(t, u) : g(t, u) \leq v\}$, we consider the part of Q_v in each of the Q_k, where the relevant condition is $g_k(t, u) \leq v$. Thus,

$$Q_v = \bigcup_k Q_k Q_v^k \qquad \text{where} \quad Q_v^k = \{(t, u) : g_k(t, u) \leq v\}$$

so that

$$F_Z(v) = \sum_k P\big((X, Y) \in Q_k Q_v^k\big).$$

The following contrived example illustrates the procedure.

Example 11.2.14:

The pair $\{X, Y\}$ of random variables has joint distribution uniform over the unit square with vertices at $(0, 0)$, $(1, 0)$, $(1, 1)$, and $(0, 1)$. Consider the random variable

$$Z = \begin{cases} X & \text{for } X + Y \geq 1 \\ X^2 + Y^2 & \text{for } X + Y < 1 \end{cases} = I_{Q_1}(X, Y)X + I_{Q_2}(X, Y)(X^2 + Y^2).$$

Determine $P(Z \leq 1/4)$.

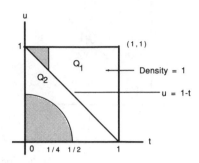

FIGURE 11.10. Regions for Example 11.2.14.

Solution

The region $Q_1 = \{(t, u) : t + u \geq 1\} = \{(t, u) : u \geq 1 - t\}$ is the region in the plane on or above the line $u = 1 - t$ (see Figure 11.10), and Q_2 is the complementary region. Let Q be the region in which $g(t, u) \leq 1/4$.

On Q_1, $g(t, u) \leq 1/4$ iff $t \leq 1/4$, so the part of Q in Q_1 is the triangular region to the left of the vertical line $t = 1/4$ and on or above the line $u = 1 - t$.

On Q_2, $g(t, u) \leq 1/4$ iff $t^2 + u^2 \leq 1/4$. The part of Q common to Q_2 is the region in the quarter circle, radius $1/2$, center at the origin.

Since the joint density is one, the probability in the region Q is the same as its area. Thus, geometrically,

$$P\left(Z \leq \frac{1}{4}\right) = \frac{1}{32} + \frac{\pi}{16} \approx 0.228.$$

—   □

This mode of representation is sometimes useful for functions of a single random variable.

Example 11.2.15: Simple Function Approximation to a Random Variable

Consider the simple function approximation to a random variable X as described in Section 7.3. We suppose, for simplicity, that X is bounded, so that its values lie in an interval $[a, b]$. Let $\{M_i : 1 \leq i \leq n\}$ be a partition of $[a, b]$, with

$$M_1 = [a, t_1), \qquad M_i = [t_{i-1}, t_i) 2 \leq i \leq n - 1 \qquad M_n = [t_{n-1}, b].$$

Then the approximating simple random variable X_s satisfies

$$X_s(\omega) = u_i \qquad \text{for } X(\omega) \in M_i$$

hence

$$X_s = \sum_{i=1}^{n} I_{M_i}(X)u_i = g(X)$$

where $g = \sum_{i=1}^{n} u_i I_{M_i}$. Since each interval M_i is a Borel set, g must be a Borel function.

— $\square$

FUNCTION OF A COMPOSITE FUNCTION

Frequently we need to consider a function of a random vector Z that has a composite definition. In this case, we have

$$h(Z) = h\big(g(X,Y)\big) = \sum_k I_{Q_k}(X,Y)h\big(g_k(X,Y)\big).$$

Example 11.2.16:

If $Z = I_Q(X,Y)XY + I_{Q^c}(X,Y)(X^3 + Y)$, then $Z^2 = I_Q(X,Y)X^2Y^2 + I_{Q^c}(X,Y)(X^3 + Y)^2$.

— $\square$

ORDER STATISTICS

In sampling statistics (see Section 18.7), we deal with an iid class $\{X_i : 1 \le i \le n\}$ of random variables, where n is a prescribed positive integer known as the sample size. An observation of this class gives an n-tuple of numbers $(t_1, t_2, \ldots, t_n)$. As an extension of the extreme values in the case of two variables (Example 11.2.11), it is often useful to define a random variable Y_1 whose value for any ω is the smallest of the $X_i(\omega)$; a second random variable Y_2 whose value at ω is the next smallest of the $X_i(\omega)$, and so on through Y_n whose value at ω is the largest of the $X_i(\omega)$. We would like to be able to obtain the distributions for these new random variables in terms of the common distribution for the X_i. We formulate the problem as follows.

Example 11.2.17: Order Statistics

Suppose $\{X_i : 1 \leq i \leq n\}$ is iid, with common distribution function F. Let

$$Y_1 = \text{smallest of} \quad X_1, X_2, \ldots, X_n$$
$$Y_2 = \text{next larger of} \quad X_1, X_2, \ldots, X_n$$
$$\vdots$$
$$Y_n = \text{largest of} \quad X_1, X_2, \ldots, X_n.$$

Then Y_k is called the kth *order statistic* for the class $\{X_i : 1 \leq i \leq n\}$. We wish to determine the distribution functions

$$F_k(t) = P(Y_k \leq t) 1 \leq k \leq n.$$

Now, $Y_k \leq t$ iff k or more of the X_i have values no greater than t. We may view the process as a Bernoulli sequence of n trials. There is a success on the ith trial iff $X_i \leq t$. The probability p of a success is $p = P(X \leq t) = F(t)$. Hence

$$F_k(t) = P(Y_k \leq t) = P(k \text{ or more of the } X_i \text{ lie in } (-\infty, t])$$
$$= \sum_{j=k}^{n} C(n,j) F^j(t) \big(1 - F(t)\big)^{n-j}.$$

— □

Remark: Once the common distribution function F for the X_i is known, then the F_k are calculated in a straightforward, though perhaps tedious, manner.

The following special case is important in characterizing the Poisson process (see Section 21.1).

Example 11.2.18: Order Statistics for Uniformly Distributed Random Variables

Suppose $\{U_i : 1 \leq i \leq n\}$ is iid, uniform on $(0, T]$. Determine the distribution functions for the order statistics.

Solution

The common distribution function for the U_i is given by

$$F(t) = \frac{t}{T} \quad 0 \leq t \leq T.$$

According to the result in Example 11.2.16, the kth order statistic Y_k has the distribution function

$$F_k(t) = P(Y_k \le t) = \sum_{j=k}^{n} C(n, j) \left(\frac{t}{T}\right)^j \left(\frac{T-t}{T}\right)^{n-j} \qquad \text{for } 0 < t < T.$$

—   □

VECTOR-VALUED FUNCTIONS

In the preceding examples, the functions g, h, etc., have been real-valued. This restriction is not necessary. We could have, for example

$$g(t, u) = \big(g_1(t, u), g_2(t, u)\big) \qquad \text{with } g_1,\ g_2 \text{ real-valued.}$$

The functions g_1 and g_2 are the first and second *coordinate functions* for g. Theoretical treatments of measure theory show that g is Borel iff both g_1 and g_2 are Borel. Such results extend to various dimensionalities of both domain and range. For example, g could be a three-dimensional function of a single variable

$$g(t) = \big(g_1(t), g_2(t), g_3(t)\big).$$

Or it could be a two-dimensional function of three variables

$$g(t, u, v) = \big(g_1(t, u, v), g_2(t, u, v)\big).$$

As a result of Theorem 11.1.1 and the fact noted in Section 9.1 that $W = (X_1, X_2, \ldots, X_n)$ is a random vector iff each X_i is a random variable, we know that such functions of random vectors are also random vectors.

Example 11.2.19: Some Typical Vector-Valued Functions of Random Variables

$$g(X, Y, Z) = (X, Z) \qquad g(X, Y) = (X, Y^2, X + Y)$$
$$g(X, Y) = (X, Y/X).$$

According to the preceding discussion, each function g is Borel, hence each function defines a random vector.

—   □

MAPPINGS AND MEASURABILITY

The discussion on the nature of the mappings and inverse images at the end of Section 11.1 leading to Theorems 11.1.1 and 11.1.2 apply to random vectors as well as random variables. Let $W = (X_1, X_2, \ldots, X_n)$, where the X_i are random vectors, and consider $Z = g(W)$, where g is a (real-valued or vector-valued) Borel function whose domain includes the set of all possible values of W. Suppose we want to determine $P(Z \in M)$, where M is a Borel set. Let N be the set of all multidimensional points t such that $g(t)$ is a point in M. Then

$$Z(\omega) \in M \quad \text{iff} \quad g\big(W(\omega)\big) \in M \quad \text{iff} \quad W(\omega) \in N = g^{-1}(M)$$

so that

$$P(Z \in M) = P(W \in N) = P\big(W \in g^{-1}(M)\big).$$

The problem of determining $P(Z \in M)$ is the two-stage process of determining $N = g^{-1}(M)$ and then calculating $P(W \in N)$.

11.3. Functions of Independent Random Variables

Suppose $Z = g(X)$, where g is a Borel function. By Theorem 11.1.1, any event determined by Z is an event determined by X. As a consequence, we have the following important proposition.

Theorem 11.3.1   (Functions of Independent Random Vectors)
Suppose $\{X_t : t \in T\}$ is an independent class of random vectors. For each $t \in T$, let $Z_t = g_t(X_t)$, where g_t is a Borel function on the codomain of X_t. Then the class $\{Z_t : t \in T\}$ is independent.

—
$\square$

Example 11.3.1: Simple Function Approximations to Independent Random Variables

Suppose $\{X, Y\}$ is an independent pair of real-valued random variables. Let $\{X_s, Y_s\}$ be a pair of simple approximating functions for X and Y of the type described in Section 7.3. Then this pair is independent.

Proof

By the results of Example 11.2.15

$$X_s = \sum_k I_{M_k}(X)t_k = g(X) \qquad \text{and} \qquad Y_s = \sum_j I_{N_j}(Y)u_j = h(Y).$$

The independence of $\{X_s, Y_s\}$ follows from Theorem 11.3.1.

—
$\square$

The following theorem relates independence of random vectors to independence of the coordinate random variables.

Theorem 11.3.2   (Independence of Coordinates of Random Vectors) *Consider an arbitrary class $\{X_t : t \in T\}$ of random vectors. For* *each* $t \in T$, *let* X_t^* *be a random vector whose coordinates form a nonempty subset of the coordinates for* X_t. *The choice of coordinates for any* X_t^* *is made without regard for the choices in the other cases.*

(a) *If* $\{X_t : t \in T\}$ *is independent, so is* $\{X_t^* : t \in T\}$.

(b) *If* $\{X_t^* : t \in T\}$ *is independent for every selection of finite subsets of the coordinates for each* X_t, *then* $\{X_t : t \in T\}$ *is independent.*

Idea of the proof

A projection function, which selects a subset of coordinates of a vector, is continuous and hence Borel. Each X_t^* is therefore a Borel function of X_t. Part (a) follows from Theorem 11.3.1.

Part (b) rests on the fact that independence on suitable subclasses of sigma algebras guarantees independence of the entire class. We assume this part without proof.

— $\qquad\qquad\qquad\qquad\qquad\qquad\qquad\qquad\qquad\qquad$ $\square$

Example 11.3.2:

Suppose $X = (X_1, X_2, X_3)$ and $Y = (Y_1, Y_2, Y_3, Y_4)$ with $\{X, Y\}$ independent. Let $V = X_1^3 + 3\sin X_2$ and $W = \max\{Y_1, Y_4\}$. Show that $\{V, W\}$ is independent.

Proof

Set $X^* = (X_1, X_2)$ and $Y^* = (Y_1, Y_4)$. By Theorem 11.3.2, the pair $\{X^*, Y^*\}$ is independent.

$V = g(X^*)$ where g is continuous, hence Borel. Similarly

$$W = \max\{Y_1, Y_4\} = \frac{1}{2}(Y_1 + Y_4 + |Y_1 - Y_4|) = h(Y^*).$$

Since h is continuous, hence Borel, Theorem 11.3.1 ensures the pair $\{V, W\}$ is independent.

— $\qquad\qquad\qquad\qquad\qquad\qquad\qquad\qquad\qquad\qquad$ $\square$

The next theorem is extremely useful in many situations. Its proof requires measure theoretic results similar to those needed for the proof of part (b) of Theorem 11.3.2. We assume the proposition without proof.

Theorem 11.3.3. *Suppose* $W = (X_1, X_2, \ldots, X_n)$ *and* $Z = (Y_1, Y_2, \ldots, Y_m)$ *are random vectors with the indicated coordinate random variables. If the class* $\{X_i, Y_j : 1 \leq i \leq n, 1 \leq j \leq m\}$ *consisting of all the coordinates of both random vectors is independent, then* $\{W, Z\}$ *is independent.*

— $\qquad\qquad\qquad\qquad\qquad\qquad\qquad\qquad\qquad\qquad$ $\square$

Remarks

1. The "coordinates" of W and Z may themselves be random vectors.

2. This theorem extends to an arbitrary class of random vectors for which the combined class of coordinates is independent.

Example 11.3.3:

Suppose $\{X, Y, Z\}$ is independent. Let $V = X^3$ and $W = 3Y + Z^2$. Show that $\{V, W\}$ is independent.

Solution

By Theorem 11.3.3, the pair $\{X, (Y, Z)\}$ is independent, where (Y, Z) is the random vector with coordinates Y, Z. Then $V = g(X)$ and $W = h(Y, Z)$. By Theorem 11.3.1, the pair $\{V, W\}$ must be independent.

—

$\square$

11.4. The Quantile Function

The quantile function is a generalized inverse of the distribution function for a random variable. It has a number of uses, both applied and theoretical. One important use is to transform observations of a random variable distributed uniformly on the unit interval into observations of a random variable with any prescribed distribution. This provides a powerful technique in simulation. On the theoretical side, we obtain a Poisson approximation theorem, which is used in Chapter 21 to establish a very simple set of axiomatic properties for the Poisson counting process. These properties show why the Poisson distribution appears so often in applications.

We first consider a special case of the quantile function, then show how to relax the restrictions. Let F be a probability distribution function such that:

> (1) F is continuous and strictly increasing on $[a, b]$
> (2) $F(a) = 0$ and $F(b) = 1$.

Let Q be the inverse function for the restriction of F to $[a, b]$. Then Q is defined on $[0, 1]$. Since F is strictly increasing, so is Q and

Q1) $Q(u) \le t$ iff $u \le F(t)$ $u \in [0, 1]$.

One of the most important consequences of property (Q1) is the following:

Q2) If U is a random variable that is distributed uniformly on $(0, 1)$, then $X = Q(U)$ has distribution function $F_X = F$.

To see this, note that $F_X(t) = P\big(Q(U) \le t\big) = P\big(U \le F(t)\big) = F(t)$.

Example 11.4.1: Simulated Sample with a Prescribed Distribution

A *random sample* of size n is an observation $(t_1, t_2, \ldots, t_n)$ of a random vector $(X_1, X_2, \ldots, X_n)$, where $\{X_i : 1 \le i \le n\}$ is iid with common distribution function F_X. Such a random sample is said to be from a *population* with distribution function F_X.

A sequence of n numbers from a random-number generator may be treated as an observation $(u_1, u_2, \ldots, u_n)$ of the random vector $(U_1, U_2, \ldots, U_n)$, where $\{U_i : 1 \le i \le n\}$ is iid, uniform on $(0, 1)$. Thus, we have a random sample from a population with distribution uniform on $(0, 1)$.

It follows from the preceding treatment, that if Q is the quantile function for F_X, the vector $(Q(u_1), Q(u_2), \ldots, Q(u_n))\}$ is a random sample from a population with distribution function F_X. Thus, a random sample from a uniformly distributed population (obtained with a random-number generator) can be transformed into a random sample from a population with prescribed distribution function F_X.

As an example, consider the exponential distribution.

If $F_X(t) = 1 - e^{-\alpha t} = u$, for $0 \le t$, then $t = -(1/\alpha) \ln(1-u) = Q(u)$ for $0 < u < 1$.

Simulation of the sample may be done easily with a programmable calculator that has a random-number generator.

— $\qquad\qquad\qquad\qquad\qquad\qquad\qquad\qquad\qquad\qquad$ □

Suppose we specify a probability u and wish to determine the point t such that the probability mass at or to the left of t is u. The desired value is $t = Q(u)$. Thus, the probability mass at or to the left of point $t = Q(0.5)$ is 0.5. For this reason, Q is called the *quantile function* for the distribution. We now remove the restrictions on F imposed earlier and generalize the inverse function, as follows.

Definition 11.4.1. *Suppose F is any probability distribution function. The quantile function Q associated with F is given by*

$$Q(u) = \inf\{t : u \le F(t)\} \qquad \text{for all } u \in (0, 1).$$

The property (Q1), remains valid, as the following argument shows.

If $F(t^*) \ge u^*$, then $t^* \ge \inf\{t : F(t) \ge u^*\} = Q(u^*)$.
If $F(t^*) < u^*$, then $t^* < \inf\{t : F(t) \ge u^*\} = Q(u^*)$.

As a consequence, property (Q2), remains valid in the general case.

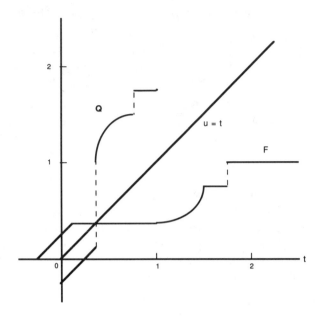

FIGURE 11.11. Construction of the graph of Q from the graph of F.

Remark: It is worth noting that property (Q2) establishes the fact that any distribution function F is the distribution function for a random variable, since $X = Q(U)$ is such a random variable.

A partial converse of (Q2) is also available. In the Appendix to this chapter, we show that if the distribution function F is continuous at $Q(u)$, then $F\big(Q(u)\big) = u$. Hence, we may assert

Q3) If random variable X has a continuous distribution function F, then $Z = F(X)$ is distributed uniformly on $(0, 1)$.

To see this, we note that $P(Z \leq t) = P\Big(F\big(Q(U)\big) \leq t\Big) = P(U \leq t)$.

In the Appendix to this chapter, we establish a number of other properties of the quantile function. For one thing, Q is left-continuous, whereas the distribution function F is right-continuous.

Construction of the graph of $u = Q(t)$ may be achieved by the simple device of reflecting the graph of $u = F(t)$ in the main diagonal $u = t$.

This is an extension of the standard procedure for generating the graph of an inverse function. The construction is illustrated in Figure 11.11. At a jump in the graph for F, the vertical section is represented by a line segment. The resulting continuous curve is then reflected in the main diagonal $u = t$. A horizontal section for the graph of F becomes a vertical section for the graph of Q, and a vertical section for the graph of F becomes a

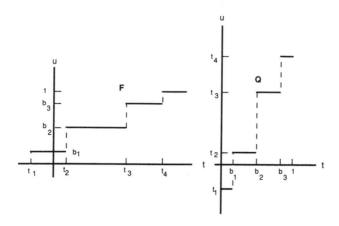

FIGURE 11.12. Distribution function F and quantile function Q for a simple random variable.

horizontal section for the graph of Q. Where the graph of F is increasing, the construction is that for obtaining the graph of the inverse function. We then make the resulting graph for Q left-continuous at jumps.

THE QUANTILE FUNCTION FOR A DISCRETE DISTRIBUTION

We use the graphical construction described earlier to exhibit the character of the quantile function for a discrete random variable. The distribution function F is a right-continuous step function with jump p_i at $t = t_i$, $1 \leq i \leq n$. The quantile function Q is a left-continuous step function, with domain $(0,1]$, having value t_i on the interval $(b_{i-1}, b_i]$, $1 \leq i \leq n$, where $b_0 = 0$ and $b_i = \sum_{j=1}^{i} p_j$. This is illustrated in Figure 11.12 for the case $n = 4$.

A THEORETICAL RESULT: A POISSON APPROXIMATION THEOREM

We first obtain a simple result, referred to as the *coupling lemma*.

Lemma 11.4.2   (Coupling lemma)

   (a) *For any random variables X and Y, $|P(X \in M) - P(Y \in M)| \leq P(X \neq Y)$ for all Borel sets M.*

(b) If $X = \sum_{i=1}^{n} X_i$ and $Y = \sum_{i=1}^{n} Y_i$, then

$$|P(X \in M) - P(Y \in M)| \le \sum_{i=1}^{n} P(X_i \ne Y_i)$$

for all Borel sets M.

Proof

1. If $P(X \in M) = P(Y \in M)$, then the inequality certainly holds.

2. If $P(X \in M) > P(Y \in M)$, the absolute value sign may be removed, to give

$$P(X \in M) - P(Y \in M) \le P(X \in M) - P(X \in M, Y \in M)$$
$$= P(X \in M, Y \in M^c) \le P(X \ne Y).$$

3. For the opposite inequality, interchange the role of X and Y.

4. $X \ne Y$ implies $X_i \ne Y_i$ for some i. That is, $\{X \ne Y\} \subset \bigcup_i \{X_i \ne Y_i\}$, so that

$$P(X \ne Y) \le P\left(\bigcup_i \{X_i \ne Y_i\}\right) \le \sum_i P(X_i \ne Y_i).$$

— $\square$

The next result is a lemma we extend to obtain the Poisson approximation theorem 11.4.4.

Lemma 11.4.3 (Approximation lemma) *If $X = I_E$ with $P(E) = p$ and Y is Poisson (p), then,*

$$|P(X \in M) - P(Y \in M)| \le p(1 - e^{-p}) < p^2 \qquad \text{for all Borel sets } M.$$

Proof

The strategy is to replace X and Y by $X^* = Q_X(U)$ and $Y^* = Q_Y(U)$, where Q_X, Q_Y are the quantile functions for X and Y, respectively, and U is uniform on $(0, 1)$. Since X^* and Y^* have the same distributions as X and Y, respectively,

$$|P(X \in M) - P(Y \in M)| = |P(X^* \in M) - P(Y^* \in M)|.$$

By the coupling lemma 11.4.2, this quantity is dominated by $P(X^* \ne Y^*)$. Because both X^* and Y^* are functions of the same uniformly distributed random variable U, $P(X^* \ne Y^*)$ is equal to the length of the subinterval

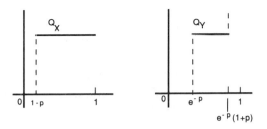

FIGURE 11.13. The quantile function for X and Y in Lemma 11.4.3.

of $(0, 1)$ on which $Q_X(u) \neq Q_Y(u)$. The graphs of the quantile functions are shown in Figure 11.13.

F_X has a jump of $1 - p$ at $t = 0$, and a jump of p at $t = 1$. Hence, Q_X has value 0 on $(0, 1 - p]$ and value 1 on $(1 - p, 1]$.

F_Y has a jump of e^{-p} at $t = 0$, a jump of pe^{-p} at $t = 1$, and jump of $\frac{1}{2}p^2e^{-p}$ at $t = 2$, with additional jumps at successive integer values. This means that Q_Y has value 0 on $(0, e^{-p}]$, value 1 on $(e^{-p}, (1 + p)e^{-p}]$, and values greater than one on $((1 + p)e^{-p}, 1]$.

We thus have $Q_X(u) \neq Q_Y(u)$ for $1 - p < u \leq e^{-p}$ and $(1 + p)e^{-p} < u \leq 1$. The probability that $X^* \neq Y^*$ is the probability that U has a value in one of these intervals. This probability is the sum of the lengths of these intervals. Thus

$$P(X^* \neq Y^*) = e^{-p} - (1 - p) + 1 - (1 + p)e^{-p} = p(1 - e^{-p}) < p^2.$$

—   □

We extend this theorem to the case $X = \sum_{i=1}^{n} I_{E_i} = \sum_{i=1}^{n} X_i$, with $\{E_i : 1 \leq i \leq n\}$ independent and Y a Poisson random variable with parameter $\lambda = \sum_{i=1}^{n} P(E_i)$. We use the fact, established easily with the methods of Chapter 18, that if $\{Y_i : 1 \leq i \leq n\}$ is independent, with each Y_i Poisson (μ_i), then $Y = \sum_{i=1}^{n} Y_i$ is Poisson (μ), with $\mu = \sum_{i=1}^{n} \mu_i$. This fact could have been developed with the convolution properties in Section 11.3, but the result is obtained much more elegantly and easily with the moment-generating functions or the generating functions studied in Chapter 18.

Theorem 11.4.4 (Poisson Approximation Theorem) *Suppose* $\{E_i : 1 \leq i \leq n\}$ *is independent with* $P(E_i) = p_i$, $1 \leq i \leq n$. *Set* $X = \sum_{i=1}^{n} I_{E_i}$, *and put* $\lambda = \sum_{i=1}^{n} p_i$. *Let* Y *be Poisson* (λ). *Then*

$$|P(X \in M) - P(Y \in M)| \leq \sum_{i=1}^{n} p_i(1 - e^{-p_i}) < \sum_{i=1}^{n} p_i^2.$$

Proof

As in lemma 11.4.3, we replace X and Y by suitable random variables having the same distributions and about which we can make the appropriate statements.

Let $\{U_i : 1 \leq i \leq n\}$ be iid, uniform on $(0,1)$. Let Q_i be the quantile function for I_{E_i}, and let R_i be the quantile function for the Poisson distribution with parameter (p_i). Put

$$X^* = \sum_{i=1}^n Q_i(U_i) = \sum_{i=1}^n X_i \quad \text{and} \quad Y^* = \sum_{i=1}^n R_i(U_i) = \sum_{i=1}^n Y_i.$$

Then X^* has the same distribution as X. Because of the independence of the U_i, the random variables $Y_i = R_i(U_i)$ are independent and Poisson. Hence, Y^* is Poisson (λ). By the coupling lemma 11.4.2

$$|P(X \in M) - P(Y \in M)| = |P(X^* \in M) - P(Y^* \in M)| \leq \sum_{i=1}^n P(X_i \neq Y_i).$$

By the approximation lemma 11.4.3, the last sum is dominated by

$$\sum_{i=1}^n p_i(1 - e^{-p_i}) < \sum_{i=1}^n p_i^2.$$

$\square$

The result of Theorem 11.4.4 may be put in the following form, which is frequently useful.

Theorem 11.4.5. *Suppose $\{E_i : 1 \leq i \leq n\}$ is independent with $P(E_i) = p_i$, $1 \leq i \leq n$. Set $X = \sum_{i=1}^n I_{E_i}$, and put $\lambda = \sum_{i=1}^n p_i$. Let Y be Poisson (μ). Then*

$$|P(X \in M) - P(Y \in M)| \leq \sum_{i=1}^n p_i(1 - e^{-p_i}) + 1 - e^{|\lambda - \mu|} < \sum_{i=1}^n p_i^2 + |\lambda - \mu|.$$

Proof

1. For $\lambda = \mu$, we have Theorem 11.4.4.

2. For $\lambda > \mu$, consider $\{W, Z\}$ independent, with W Poisson (μ) and Z Poisson $(\lambda - \mu)$. Then, $W + Z$ is Poisson (λ). Now

$$|P(X \in M) - P(Y \in M)| = |P(X \in M) - P(W \in M)|.$$

By the triangle inequality for absolute value

$$\begin{aligned} |P(X \in M) - P(W \in M)| \leq {} & |P(X \in M) - P(W + Z \in M)| \\ & + |P(W + Z \in M) - P(W \in M)|. \end{aligned}$$

By Theorem 11.4.4,

$$|P(X \in M) - P(W + Z \in M)| \le \sum_{i=1}^{n} p_i(1 - e^{-p_i}).$$

By the coupling lemma 11.4.2,

$$|P(W + Z \in M) - P(W \in M)| \le P(W + Z \ne W) \quad = \quad P(Z \ne 0)$$
$$= \quad 1 - e^{-(\lambda - \mu)}.$$

3. For $\mu > \lambda$, consider $\{W, Z\}$ independent, with W Poisson (λ) and Z Poisson $(\mu - \lambda)$. Then $W + Z$ is Poisson (μ) and

$$|P(X \in M) - P(Y \in M)| = |P(X \in M) - P(W + Z \in M)|.$$

The argument is completed as in 2, with the role of W and $W + Z$ interchanged. Also $\lambda - \mu$ is replaced by $\mu - \lambda$. The results in the two cases are combined by using $|\lambda - \mu|$.

—   □

Example 11.4.2: Poisson Approximation to the Binomial Distribution

As a very simple application of Theorem 11.4.4, we obtain the classical Poisson approximation to the binomial distribution. We consider a sequence of binomial distributions with increasing n. Let X_n have parameters (n, p_n), with $p_n = \mu/n$. Then, by Theorem 11.4.4

$$\left| P(X_n \in M) - e^{-\mu} \sum_{k \in M} \frac{\mu^k}{k!} \right| < \frac{\mu^2}{n}.$$

—   □

Remark: The convergence of the binomial distribution, under these conditions, to the Poisson distribution can be obtained by much more elementary means. Theorem 11.4.4 gives an error estimate, which shows the convergence is of the order of $1/n$. Moreover, it is not necessary to require that $p_n = \mu/n$. It is sufficient that $np_n^2 \to 0$. This is ensured if np_n is bounded.

In the appendix to Chapter 21, we use the approximation theorems to show that very simple axiomatic properties characterize the *Poisson process*, whose importance in applications would be hard to overemphasize.

11.5.   Coordinate Transformations

In this section we briefly treat a popular method of determining the joint density function for a vector-valued function of random variables. When it is applicable, the method has the advantage of providing a straightforward, though often tedious, formal procedure. Although the underlying structure has geometric significance, it is not necessary to examine the geometry in order to carry out the evaluation. For simplicity, we treat a special case. Generalization to higher dimensional cases is immediate.

Suppose (X, Y) has a joint density function f_{XY}. Let

$$(U, V) = h(X, Y) = \big(h_1(X, Y), h_2(X, Y)\big)$$

where is h is one-one, with appropriate differentiability properties. Then h has an inverse $g = h^{-1}$, so that

$$(X, Y) = h^{-1}(U, V) = g(U, V).$$

By the mapping properties for functions

$$(U, V) = h(X, Y) \in Q \qquad \text{iff} \qquad (X, Y) \in h^{-1}(Q)$$

so that

$$P\big((U, V) \in Q\big) = P\big((X, Y) \in h^{-1}(Q)\big).$$

If $g = h^{-1}$ satisfies

1. g has continuous first partial derivatives, and

2. $J_g(u, v) = \begin{vmatrix} \dfrac{\partial g_1(u, v)}{\partial u} & \dfrac{\partial g_1(u, v)}{\partial v} \\[2mm] \dfrac{\partial g_2(u, v)}{\partial u} & \dfrac{\partial g_2(u, v)}{\partial v} \end{vmatrix} \neq 0$

then a standard theorem on multiple integrals ensures

$$\int_{h^{-1}(Q)} \int f_{XY}(x, y)\, dx\, dy = \int_Q \int f_{XY}\big(g_1(u, v), g_2(u, v)\big) |J_g(u, v)|\, du\, dv.$$

Since this holds for all Borel sets Q in $\Re^2$, we must have

$$f_{UV}(u, v) = f_{XY}\big(g_1(u, v), g_2(u, v)\big) |J_g(u, v)|.$$

We illustrate this approach in some special cases, which we compare with results obtained in Section 11.2.

Example 11.5.1:

Set $U = XY = h_1(X, Y)$ and $V = Y = h_2(X, Y)$. Then

$$X = \frac{U}{V} = g_1(U, V) \qquad \text{and} \qquad Y = V = g_2(U, V).$$

According to the basic formula

$$J_g(u, v) = \begin{vmatrix} 1/v & -u/v^2 \\ 0 & 1 \end{vmatrix} = \frac{1}{v}.$$

We thus have $f_{UV}(u,v) = f_{XY}(u/v,v)|1/v|$, from which we obtain

$$f_U(u) = \int f_{UV}(u,v)\,dv = \int f_{XY}(u/v,v)\left|\frac{1}{v}\right|\,dv.$$

$\square$

Example 11.5.2: See Example 11.2.5

Set $U = Y/X = h_1(X,Y)$ and $V = X = h_2(X,Y)$. Then

$$X = V = g_1(U,V) \qquad \text{and} \qquad Y = UV = g_2(U,V).$$

Again we obtain the determinant

$$J_g(u,v) = \begin{vmatrix} 0 & v \\ 1 & u \end{vmatrix} = -v$$

so that $f_{UV}(u,v) = f_{XY}(v,uv)|v|$ and

$$f_U(u) = \int f_{UV}(u,v)\,dv = \int f_{XY}(v,uv)|v|\,dv.$$

$\square$

11a

Some Properties of the Quantile Function

We suppose throughout the following that F is a probability distribution function. We define the *quantile function* Q for the function F by the condition

$$Q(u) = \inf\{t : F(t) \geq u\} \qquad \text{for all } u \in (0, 1).$$

In Section 11.4, we show that

1. $Q(u) \leq t$ iff $u \leq F(t)$.

We establish several additional properties that exhibit the nature of the quantile function and serve as the basis of a procedure for determining its graph from that of the distribution function.

2. If F is continuous, strictly increasing on $[a, b]$, then Q is continuous, strictly increasing on $[F(a), F(b)]$ and $Q(u) = t$ iff $F(t) = u$.

   Proof

   (a) $\inf\{t : F(t) \geq u^*\} = t^*$ iff $F(t^*) = u^*$. This implies $Q(u) = t$ iff $F(t) = u$.

   (b) Thus, Q on $[F(a), F(b)]$ is the inverse of the restriction of F to $[a, b]$.

   (c) The continuity and increasing character of the inverse is well-known.

   —   □

3. If F has a jump at $t = a$, then $Q(u) = a$ for $u \in (F(a - 0), F(a)]$.

   Proof

   In this situation, $F(a) \geq u$, $F(a-0) < u$ for $u \in (F(a-0), F(a)]$, so that $\inf\{t : F(t) \geq u\} = a$.

   —   □

4. If $F(t) = F(a)$ for $t \in [a, b)$ with $F(a-h) < F(a)$ and $F(b+h) > F(a)$ for all $h > 0$, then $Q(F(a)) = a$ and $u > F(a)$ implies $Q(u) \geq b$.

Proof

> If $u < F(a)$, then $Q(u) < a$.
> If $u = F(a)$, then $Q(u) = a$.
> If $u > F(a)$, then $F(t) \geq u$ implies $t \geq b$, which implies $Q(u) \geq b$.

> —— $\square$

5. Q is left-continuous.

Proof

> The only discontinuities of Q occur at $u = F(a)$, when F has a constant interval beginning at $t = a$. By 4, the value of Q at the jump is the left-hand limit.

> —— $\square$

6. $F(Q(u)-0) \leq u \leq F(Q(u))$, so that if F is continuous at $Q(u)$, then $F(Q(u)) = u$.

Proof

> By definition, $F(Q(u)) \geq u$ and $F(Q(u) - h) < u$ for all $h > 0$.
> —— $\square$

Graph of the Quantile Function

The quantile function Q appears as a generalized inverse function for F. We may extend the well-known device of reflecting the graph of F in the main diagonal to get the graph of Q. We simply consider the graph of F at a jump to be a vertical line. A jump of F becomes a horizontal interval for Q and a horizontal interval for F becomes a jump of Q. The distribution function F is right-continuous; the quantile function Q is left-continuous. These facts are verified by properties 2 through 5.

Problems

11–1. Suppose X is Poisson (3).

Determine $P(4 < X < 9)$ and $P(X^2 + 4X \leq 21)$.

Caution. Be careful about endpoints of intervals.

11–2. Random variable X has density function $f_X(t) = \frac{1}{2}e^{-|t|}$ for all t. Determine the probability of each of the following events: $\{|X| + |X - 3| \leq 3\}$, $\{X^3 - X^2 - X - 2 \leq 0\}$, $\{e^{\sin \pi X} \geq 1\}$.

11–3. Suppose X is exponential (1). Determine the density function f_Y for
$$Y = \begin{cases} X & \text{for } X \leq 1 \\ 1/X & \text{for } X > 1 \end{cases}.$$

11–4. Random variable $X \sim N(0, 16)$. Determine $P(X^2 - 20X + 100 < 4)$.

11–5. Random variable $X \sim N(0, 4)$. Determine $P(X^2 < 2 - X)$.

11–6. The pair $\{X, Y\}$ has joint distribution

$$P(X = t, Y = u).$$

$t =$	-2	1	3	5
$u = 3$	0.15	0.05	0.10	0
$u = 1$	0.05	0.08	0.10	0.12
$u = -1$	0.10	0.02	0	0.23

(a) Determine $P(|X - Y| \leq 1)$.
(b) Determine $P(X^2 + Y \leq 2)$.
(c) Let $Z = \begin{cases} X & \text{for } X + Y \leq 1 \\ Y & \text{for } X + Y > 1 \end{cases} = I_{Q_1}(X,Y)X + I_{Q_2}(X,Y)Y$.
Determine $P(Z \leq 1)$.

11–7. The pair $\{X, Y\}$ has joint distribution

$$P(X = t, Y = u)$$

$t =$	0	1	3	4
$u = 3$	0.2	0.1	0.2	0.1
$u = 1$	0	0.1	0.1	0
$u = 0$	0.1	0	0	0.1

(a) Determine $P(|X - Y| > 1)$.
(b) Determine $P(4X - X^2 + Y \geq 5)$.
(c) Let $W = \max\{X, Y\}$, $Z = \begin{cases} X & \text{for } W \leq 2 \\ 2Y & \text{for } W > 2 \end{cases} = I_{Q_1}(X,Y)X + I_{Q_2}(X,Y)2Y$. Determine $P(Z \leq 1)$.

11-8. The pair $\{X, Y\}$ has joint density function $f_{XY}(t, u) = 8tu$ on the triangle bounded by the lines $u = 0$, $t = 1$, and $u = t$.

(a) Determine $P(|X - Y| \leq 0.5)$.

(b) Let $Z = \begin{cases} 4X^2 & \text{for } X + Y \leq 1 \\ 2Y & \text{for } X + Y > 1 \end{cases}$. Determine $P(Z \leq 1)$.

11-9. The pair $\{X, Y\}$ has joint distribution function $f_{XY}(t, u) = tu/36$, for $0 \leq t \leq 3$, $0 \leq u \leq 4$.

(a) Determine $P(|X - Y| \leq 1)$.

(b) Let $Z = \begin{cases} X & \text{for } X + Y \leq 3 \\ Y & \text{for } X + Y > 3 \end{cases}$. Determine $P(Z \leq 2)$.

11-10. The pair $\{X, Y\}$ has joint density function $f_{XY}(t, u) = 8tu$ on the triangle bounded by the lines $u = 0$, $t = 1$, and $u = t$.

(a) Determine $P(|X - Y| \leq 0.5)$.

(b) Let $Z = \begin{cases} 4X^2 & \text{for } X + Y \leq 1 \\ 2Y & \text{for } X + Y > 1 \end{cases}$. Determine $P(Z \leq 1)$.

11-11. The pair $\{X, Y\}$ has joint distribution uniform on the parallelogram with vertices $(1, 0)$, $(1, -2)$, $(-1, 0)$, and $(-1, 2)$. The sides of this parallelogram are segments of the lines $u = 1 - t$, $u = -1 - t$, $t = -1$, and $t = 1$. Determine $P(|X - Y| \leq 1)$ and $P(Y - X > 1)$.

11-12. Let $\{X, Y\}$ be an independent pair, each uniform on $(0, 1)$.

Let $Z = \max\{X, Y\}/\min\{X, Y\}$. Determine the distribution function F_Z for Z.

11-13. Suppose $\{X, Y\}$ is independent, with respective distribution functions F_X and F_Y. Let $V = \min\{X, Y\}$ and $W = \max\{X, Y\}$. Determine F_V and F_W in terms of F_X and F_Y.

11-14. The pair of random variables $\{X, Y\}$ is independent with joint density function

$$f_{XY}(t, u) = 1.5e^{-3t} \qquad \text{for } 0 \leq t, 0 \leq u \leq 2.$$

Let $W = \max\{X, Y\}$. Determine $P(0.5 < W \leq 1)$.

11-15. The number of automobile accidents in any given day on a certain stretch of freeway is a random quantity with Poisson (5) distribution. We suppose the numbers on four consecutive days are represented by the iid class $\{X_i : 1 \leq i \leq 4\}$. Let $V = \min\{X_i : 1 \leq i \leq 4\}$. Determine $P(2 \leq V \leq 5)$.

11–16. Suppose $\{X_i : 1 \le i \le 5\}$ is iid with symmetric triangular distribution on $(1, 3)$.

Let $W = \max\{X_i : 1 \le i \le 5\}$. Determine $P(1.5 \le W \le 2)$.

11–17. Suppose $\{X, Y\}$ is an independent pair with X exponential (α) and Y exponential (β).

Let $V = \min\{X, Y\}$ and $W = \max\{X, Y\}$. Show that the joint density function f_{VW} is given by

$$f_{VW}(t, u) = \begin{cases} \alpha\beta \left(e^{-(\alpha t + \beta u)} + e^{-(\alpha u + \beta t)} \right) & \text{if } 0 \le t \le u \\ 0 & \text{otherwise} \end{cases}.$$

11–18. A company is trying to get out a working prototype of a new personal computer for an upcoming trade show six months (25 weeks) away. It puts two design teams to work, trying independently to develop a unit. If X and Y are the times to completion for the two teams, respectively, suppose $\{X, Y\}$ is iid, exponential (0.1). What is the probability the company will have a satisfactory model ready for the show?

Hint: Consider $V = \min\{X, Y\}$.

11–19. In a quality control check, the operating life of n similar devices is measured. The life in hours of the ith unit is a random variable X_i. Suppose the units fail independently and have the same exponential distribution. That is, $\{X_i : 1 \le i \le n\}$ is iid, exponential (λ). Find the probability that the smallest observed value is greater than c/λ.

11–20. The time to failure in weeks for a certain device may be represented by a random variable uniform on $(9, 11)$. Two such devices are put into service at the same time. They fail independently. What is the probability that their failure times differ (in absolute value) by less than $1/2$ week?

11–21. Two students attempt independently to work a puzzle. They begin at the same time. Their working times, in minutes, are random variables X and Y, respectively. If both X and Y are distributed uniformly on $(10, 11)$, what is the probability they will finish within 15 seconds of each other?

11–22. The pair $\{X, Y\}$ has joint density function $f_{XY}(t, u) = t + u$, $0 \le t \le 1$, $0 \le u \le 1$. Let $W = \max\{X, Y\} = I_Q(X, Y)X + I_{Q^c}(X, Y)Y$. Determine the distribution function F_W.

11–23. The pair $\{X, Y\}$ has joint density $f_{XY}(t, u) = 3(t + u)$ on the triangle with vertices $(0, 0)$, $(1, 0)$, and $(0, 1)$.

(a) Determine the marginal density functions f_X and f_Y and show whether or not the pair $\{X, Y\}$ is independent.

(b) Determine $P(X + Y \leq 1/2)$.

11–24. The pair $\{X, Y\}$ has joint distribution uniform on the parallelogram that has vertices $(1, 0)$, $(1, -2)$, $(-1, 0)$, and $(-1, 2)$. Determine the values $P(|X - Y| \leq 1)$ and $P(Y - X > 1)$.

11–25. The pair $\{X, Y\}$ has joint density function $f_{XY}(t, u) = tu/2$ on the square with vertices $(1, 0)$, $(2, 1)$, $(1, 2)$, and $(0, 1)$. By symmetry, $f_X(t) = f_Y(t)$. Obtain an expression for the common density of the form $f_X(t) = I_{[0,1]}(t) f_1(t) + I_{(1,2]}(t) f_2(t)$. Use this to determine $P(1/2 \leq X \leq 5/4)$.

11–26. Suppose $f_{XY}(t, u) = (1 - u)e^{-t/2}$ for $0 \leq u \leq 1$, $0 \leq t$.

(a) Determine the marginal density functions f_X and f_Y and show whether or not the pair $\{X, Y\}$ is independent.

(b) Determine $P(X + Y \leq 2)$.

11–27. Suppose random variable U is distributed uniformly on $(0, 1)$. Determine a function g such that $X = g(U)$ has distribution function $F_X(t) = t^2$, $0 \leq t \leq 1$.

11–28. The following five numbers, obtained from a random-number generator, may be considered observations of five iid random variables, uniform on $(0, 1)$:

$$0.2789, \quad 0.3615, \quad 0.1074, \quad 0.2629, \quad 0.9252.$$

Obtain corresponding values for five iid random variables with common distribution function

$$F_X(t) = \sqrt{t - 1} \qquad 1 \leq t \leq 2.$$

11–29. Suppose U is a random variable distributed uniformly on $(0, 1)$. Determine a function g such that $X = g(U)$ has distribution function $F_X(t) = 1 - (1 - t)^2$ for $0 \leq t \leq 1$.

Part III

Mathematical Expectation

12

Mathematical Expectation

In this chapter, we introduce somewhat informally and intuitively the notion of mathematical expectation as a probability-weighted average. We present various mathematical forms and a useful interpretation of the concept, then obtain the mean values (expectations) for a number of standard distributions. In the process, we exhibit the integral character of mathematical expectation. In the next chapter, we elucidate that integral character by examining the relationship between mathematical expectation and the Lebesgue integral. The extensive literature on the Lebesgue integral provides important resources, for both applications and the further development of the probability model. For the current development, we operate quite informally.

12.1. The Concept

The notion of mathematical expectation is closely related to the idea of a weighted mean, used extensively in the handling of numerical data. Consider the arithmetic average x of the following ten numbers: 1, 2, 2, 2, 4, 5, 5, 8, 8, 8. We can write this in two equivalent ways, as follows:

$$x = \frac{1}{10}(1+2+2+2+4+5+5+8+8+8) = (0.1\cdot1+0.3\cdot2+0.1\cdot4+0.2\cdot5+0.3\cdot8).$$

Examination of the ten numbers to be added shows that five distinct values are included. One of the ten, or the fraction 1/10 of them, have the value 1, three of the ten, or the fraction 3/10 of them, have the value 2, 1/10 have the value 4, 2/10 have the value 5, and 3/10 have the value 8. The last form of the sum exhibits the MARK complete sentence

PATTERN

Multiply each possible value by the fraction of the numbers having that value and then sum these products.

Remark: The fractions are often referred to as the relative frequencies. A sum of this sort is known as a weighted average.

In general, suppose there are n numbers to be averaged. Of these, f_1 have value t_1, f_2 have value t_2, ..., f_m have value t_m. The f_i must add to

n. If we set $p_i = f_i/n$, then the fraction p_i is called the relative frequency of those numbers in the set that have the value t_i, $1 \le i \le m$. The average x of the n numbers may be written

$$x = \frac{1}{n} \sum_{k=1}^{n} x_k = \sum_{i=1}^{m} t_i p_i.$$

In probability theory, we have a similar averaging process, in which the relative frequencies of the various possible values are replaced by the probabilities that those values are observed on any trial.

> The mathematical expectation of a random variable X, designated $E[X]$, is the probability-weighted average of the values taken on by X.

This average takes on several forms that may seem quite different, but in fact are equivalent. We consider several cases.

A. SIMPLE RANDOM VARIABLE

For a simple random variable, the concept is exact. If the range of simple random variable X is $\{t_1, t_2, \ldots, t_n\}$ and $P(X = t_i) = p_i$, then we weight each possible value t_i by the probability $p_i = P(X = t_i)$ and add to obtain

$$E[X] = \sum_{i=1}^{n} t_i p_i.$$

Example 12.1.1:

- If X is the number of spots that turn up when a single die is thrown, then

$$E[X] = \frac{1}{6} \cdot 1 + \frac{1}{6} \cdot 2 + \frac{1}{6} \cdot 3 + \frac{1}{6} \cdot 4 + \frac{1}{6} \cdot 5 + \frac{1}{6} \cdot 6 = \frac{7}{2}.$$

- If X is the number of successes in n Bernoulli trials, then the probability of exactly k successes is $p_k = C(n, k)p^k(1 - p)^{n-k}$. In this case

$$E[X] = 0p_0 + 1p_1 + 2p_2 + \cdots + np_n = \sum_{k=1}^{n} kp_k = np.$$

Although the value np for the expectation is quite simple, sophisticated manipulation of the expression for the sum is required to obtain it. Properties of expectation developed in Section 14.2 make it possible to bypass this operation and obtain the expectation quite simply and directly.

—   □

B. Function of a Simple Random Variable

Suppose $Z = g(X)$, with range $\{v_1, v_2, \ldots, v_m\}$. Then according to the definition we have

$$E[Z] = \sum_{j=1}^{m} v_j P(Z = v_j).$$

Alternatively, this quantity may be expressed in terms of the distribution for X, as follows:

$$E[Z] = E[g(X)] = \sum_{i=1}^{n} g(t_i) P(X = t_i).$$

A direct demonstration of the equivalence requires only elementary facts about sums, but involves some notational complications. A simple demonstration of the equality of these two forms for E[Z] is included in the general development in Section 14.1.

C. Arbitrary Random Variables

For the general case, we may approximate random variable X as closely as desired by a simple random variable X_s. Following the treatment in Section 7.3, we consider the case of a bounded random variable X whose range is included in an interval $[a, b]$. We partition that interval into n subintervals by use of the subdivision points

$$a = b_0 < b_1 < \cdots < b_{n-1} < b_n = b.$$

For each i, $1 \leq i \leq n - 1$, we have the subintervals $M_i = [b_{i-1}, b_i)$. We set $M_n = [b_{n-1}, b_n]$. For each i, put $E_i = X^{-1}(M_i)$. Then $\{E_i : 1 \leq i \leq n\}$ is a partition. Consider any of the M_i and the corresponding inverse image E_i (see Figure 12.1 for a typical interval). Random variable X maps each point of E_i somewhere in the subinterval M_i. Pick an arbitrary point t_i in each subinterval M_i. Form a simple random variable X_s by assigning to each ω in E_i the value $X_s(\omega) = t_i$. Thus,

$$X_s = \sum_{i=1}^{n} t_i I_{E_i} = \sum_{i=1}^{n} t_i I_{M_i}(X).$$

Since X_s is simple, its expectation is given by

$$E[X_s] = \sum_{i=1}^{n} t_i P(X_s = t_i) = \sum_{i=1}^{n} t_i P(X \in M_i) = \sum_{i=1}^{n} t_i P_X(M_i).$$

Now $P_X(M_i) = F_X(b_i) - F_X(b_{i-1})$. Thus

$$E[X_s] = \sum_{i} t_i \big(F_X(b_i) - F_X(b_{i-1})\big) \approx \int t \, dF_X(t).$$

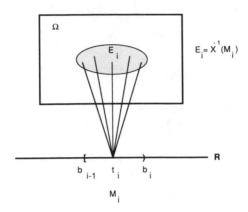

FIGURE 12.1. Mapping for approximation of simple random variable.

The approximation improves as the subdivisions become finer. We thus suppose $E[X]$ to be given by

$$E[X] = \int t\, dF_X(t) = \int t F_X(dt).$$

The careful treatment of Chapters 13 and 14 shows this to be the proper expression. If $Z = g(X)$, then its expectation may be expressed both in terms of F_Z and F_X, as follows:

$$E[Z] = \int v F_Z(dv) = \int g(t) F_X(dt).$$

The equality of the two forms is established in Section 14.1.

D. ABSOLUTELY CONTINUOUS RANDOM VARIABLES

In this case, we have a density function f_X, so that

$$P(X \in M_i) = P_X(M_i) \approx f_X(t_i)(b_i - b_{i-1}) = f_X(t_i)\Delta_i t.$$

This implies

$$E[X_s] \approx \sum_i t_i f_X(t_i)\Delta_i t \approx \int t f_X(t)\, dt.$$

As subdivisions become finer, both approximations improve. On the basis of this informal argument, we should have

$$E[X] = \int t f_X(t)\, dt.$$

Also, if $Z = g(X)$, then we should have

$$E[Z] = \int v f_Z(v)\, dv = \int g(t) f_X(t)\, dt.$$

E. $Z = h(X, Y)$, REAL-VALUED

In analogy to the case of a single random variable, we should have

$$E[Z] = \iint h(t, u)\, dF_{XY}(t, u) = \int v F_Z(dv).$$

In the absolutely continuous case, we have the joint density f_{XY}, so that

$$E[Z] = \iint h(t, u) f_{XY}(t, u)\, du\, dt.$$

These forms, arrived at somewhat intuitively, are validated in the more careful theoretical treatment sketched in the next two chapters. Before turning to that development, however, we obtain an important interpretation of the expectation, in terms of the probability distribution, and derive the mathematical expectations for a number of commonly used distributions.

12.2. The Mean Value of a Random Variable

We have seen that mathematical expectation is a probability-weighted average of the values of a random variable. Traditionally, this average has been called the mean or the mean value, of the random variable X. We therefore make the following definition.

Definition 12.2.1. *The* mean value μ_X, *or* $\mu[X]$, *of random variable* X *is the expectation* $E[X]$.

MECHANICAL INTERPRETATION

Before determining the expectation for several common distributions, we consider an important and helpful mechanical interpretation, in terms of the probability mass distribution induced by the random variable. In Example 14.2.6, we give an alternate interpretation in terms of mean-square estimation.

A. Discrete case

We suppose the random variable X has values $\{t_i : i \in J\}$, with $P(X = t_i) = p_i$. This produces a probability mass distribution, as shown in Fig-

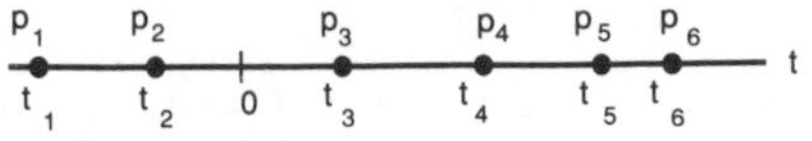

FIGURE 12.2. Probability distribution for a simple random variable.

ure 12.2, with point mass concentration in the amount of p_i at the point t_i. The expectation is

$$E[X] = \sum_i t_i p_i.$$

Mechanically, this is the moment of the probability mass distribution about the origin on the real line. From physical theory, this is known to be the same as the product of the total mass times the number that locates the center of mass. Since the mass is one, the mean value is the location of the center of mass. If the real line is viewed as a stiff, but weightless, rod with point masses p_i attached at each value t_i of X, then the mean value μ_X is the *point of balance*.

B. General case

We begin with the approximating simple random variable X_s. The moment of the mass distribution for X is approximately the moment for X_s, with the approximations improving as subdivisions become finer. For each approximating distribution, the expectation is the center of mass. The limiting position for these centers of mass is taken to be the center of mass for the general distribution. This is exactly the process by which physical center of mass is described and defined in the classical physics. Thus, for an arbitrary real-valued random variable X, the expectation or mean value $E[X]$ is the center of mass for its distribution.

MEAN VALUES FOR SOME DISCRETE RANDOM VARIABLES

We consider several of the commonly encountered discrete distributions described in Section 8.2. See the discussion in that section for some indication of the importance of these distributions.

1. *Uniform on* $\{t_1, t_2, \ldots, t_n\}$.

   Probability mass $1/n$ is placed at each t_i, $1 \leq i \leq n$. The center of mass is at the average of the t_i. Thus,

   $$\mu = E[X] = \frac{1}{n} \sum_{i=1}^{n} t_i \qquad \text{(arithmetic average)}.$$

2. *Binomial* (n, p). $(q = 1 - p)$

$$P(X = k) = p_k = C(n, k)p^k q^{n-k}, \qquad 0 \le k \le n$$

$$\mu = E[X] = \sum_{k=0}^{n} kC(n, k)p^k q^{n-k} = np.$$

The simple expression $\mu = np$ may be obtained by classical, but somewhat involved, manipulations of the sums. However, in Example 14.2.5, use of the fundamental linearity property of expectation makes the calculation almost trivial.

3. *Geometric* (p). $(q = 1 - p)$ $P(X = k) = p_k = pq^k, 0 \le k$

$$\mu = E[X] = \sum_{k=0}^{\infty} kpq^k = pq \sum_{k=1}^{\infty} kq^{k-1}.$$

Use of well-known results on the geometric series (see the appendix on mathematical aids) yields the result

$$\mu = \frac{pq}{(1 - q)^2} = \frac{pq}{p^2} = \frac{q}{p}.$$

4. *Negative binomial* (m, p).

See the discussion of this distribution in Section 8.2. We show later, with the aid of the moment-generating function, that

$$E[Y_m] = \frac{mq}{p} \qquad \text{and} \qquad E[X_m] = \frac{m}{p}.$$

5. *Poisson* (μ). $P(X = k) = p_k = e^{-\mu}\mu^k/k!$ $0 \le k$

$$\mu = E[X] = \sum_{k=0}^{\infty} ke^{-\mu}\frac{\mu^k}{k!} = \mu e^{-\mu} \sum_{k=1}^{\infty} \frac{\mu^{k-1}}{(k-1)!} = \mu e^{-\mu} \sum_{j=0}^{\infty} \frac{\mu^j}{j!}$$

$$= \mu e^{-\mu}e^{\mu} = \mu.$$

MEAN VALUES FOR SOME ABSOLUTELY CONTINUOUS RANDOM VARIABLES

1. *Uniform* $[a, b]$. $f_X(t) = 1/(b - a)$ for $a \le t \le b$.

The center of mass is at $(a + b)/2$. To calculate the mean value formally, we write

$$E[X] = \int tf_X(t)\, dt = \frac{1}{b - a} \int_a^b t\, dt = \frac{b^2 - a^2}{2(b - a)} = \frac{b + a}{2}.$$

2. *Symmetric triangular* on $[a, b]$.

   The graph of the density function, as the name implies, is an isosceles triangle with base on the interval $[a, b]$ (see Figure 8.1). In this case, the evaluation of the integral is straightforward, but somewhat cumbersome to write. There is no point in executing this computational exercise, unless one does not trust the center of mass idea. By symmetry, we have

   $$E[X] = \frac{(a+b)}{2} = c.$$

3. *Exponential* (λ). $f_X(t) = \lambda e^{-\lambda t}$, $0 \le t$.

   Using a well-known definite integral (see the appendix on mathematical aids), we obtain

   $$E[X] = \int t f_X(t)\, dt = \int_0^\infty \lambda t e^{-\lambda t}\, dt = 1/\lambda.$$

4. *Gamma* (α, λ). $f_X(t) = (1/\Gamma)(\alpha) t^{\alpha-1} \lambda^\alpha e^{-\lambda t}$, $0 \le t$.

   Again, we use one of the integrals provided in the appendix on mathematical aids to obtain

   $$E[X] = \int t f_X(t)\, dt = \frac{1}{\Gamma(\alpha)} \int_0^\infty t^\alpha \lambda^\alpha e^{-\lambda t}\, dt = \frac{\Gamma(\alpha+1)}{\lambda \Gamma(\alpha)} = \frac{\alpha}{\lambda}.$$

   The last equality comes from the fact that $\Gamma(\alpha+1) = \alpha \Gamma(\alpha)$.

5. *Beta* (r, s). $f_X(t) = \frac{\Gamma(r+s)}{\Gamma(r)\Gamma(s)}$ $t^{r-1}(1-t)^{s-1}$, $0 < t < 1$.

   We use the fact that $\int_0^1 u^{r-1}(1-u)^{s-1}\, du = \frac{\Gamma(r)\Gamma(s)}{\Gamma(r+s)}$, $r > 0$, $s > 0$.

   $$
   \begin{aligned}
   E[X] &= \int t f_X(t)\, dt = \frac{\Gamma(r+s)}{\Gamma(r)\Gamma(s)} \int_0^1 t^r (1-t)^{s-1}\, dt \\
   &= \frac{\Gamma(r+s)}{\Gamma(r)\Gamma(s)} \cdot \frac{\Gamma(r+1)\Gamma(s)}{\Gamma(r+s+1)} = \frac{r}{r+s}.
   \end{aligned}
   $$

   More generally,

   $$
   \begin{aligned}
   E[X^m] &= \frac{\Gamma(r+s)}{\Gamma(r)\Gamma(s)} \int_0^1 t^{r+m-1}(1-t)^{s-1}\, dt \\
   &= \frac{r(r+1)\ldots(r+m-1)}{(r+s)(r+s+1)\ldots(r+s+m-1)}.
   \end{aligned}
   $$

6. *Weibull* (α, λ, ν). $f_X(t) = \alpha \cdot \lambda \cdot (t-\nu)^{\alpha-1} e^{-\lambda(t-\nu)^\alpha}$, $t \ge \nu$.

   First consider Y exponential (λ) and determine

   $$E[Y^r] = \int_0^\infty t^r \lambda e^{-\lambda t}\, dt = \frac{\Gamma(r+1)}{\lambda^r}.$$

If Y is exponential (1), then $[Y/\lambda]^{1/\alpha} + \nu$ is Weibull (α, λ, ν). Hence,

$$E[X] = \frac{1}{\lambda^{1/\alpha}} E[Y^{1/\lambda}] + \nu = \frac{1}{\lambda^{1/\alpha}} \Gamma\left(\frac{1}{\alpha} + 1\right) + \nu.$$

Here we have used the fact that $E[aX + b] = aE[X] + b$, which follows from the integral nature of expectation. This is discussed more carefully in Chapters 13, 14.

7. *Normal* $\sim N(\mu, \sigma^2)$. $f_X(t) = (1/\sigma\sqrt{2\pi})e^{-\frac{1}{2}((t-\mu)/\sigma)^2}$.

The symmetry of the distribution about $t = \mu$ implies $E[X] = \mu$. This, of course may be verified by integrating

$$E[X] = \int_{-\infty}^{\infty} t f_X(t)\, dt = \int_{-\infty}^{\infty} (t - \mu) f_X(t)\, dt + \mu = \mu.$$

We have used the fact that $\int_{-\infty}^{\infty} f_X(t)\, dt = 1$. If we make the change of variable $x = t - \mu$ in the last integral, we see the integrand is an odd function, so the integral is zero, leaving only μ in the final result.

PROBLEMS

12–1. For the following discrete joint distributions, determine $E[X]$ and $E[Y]$.

Discrete joint distributions for problem 12-1

(a)

$$P(X = t, Y = u)$$

$t =$	1	2	3	4	5	6
$u = 5$	0	0	0	0.1	0	0.1
$u = 4$	0.1	0	0	0	0.1	0
$u = 3$	0	0.1	0.1	0	0.1	0
$u = 2$	0	0	0	0.1	0.1	0
$u = 1$	0.1	0	0	0	0	0

(b)

$$P(X = t, Y = u)$$

$t =$	0	1	4	9
$u = 2$	0.05	0.04	0.21	0.15
$u = 0$	0.05	0.01	0.09	0.10
$u = -1$	0.10	0.05	0.10	0.05

(c)

$$P(X = t, Y = u)$$

$t =$	-2	0	1	3
$u = 1$	0.15	0.05	0.10	0
$u = 0$	0.05	0.08	0.10	0.12
$u = -1$	0.10	0.02	0	0.23

12–2. For each of the following joint density f_{XY}, determine $E[X]$ and $E[Y]$.

Joint density functions for problem 12-2 (zero outside indicated region)

(a) $f_{XY}(t, u) = 1/\pi$ on circle with radius 1, center at $(0,0)$.

(b) $f_{XY}(t, u) = 1$ on triangle bounded by $t = 0$, $u = 0$, and $u = 2(1 - t)$.

(c) $f_{XY}(t, u) = 1/2$ on square with vertices at $(1,0)$, $(2,1)$, $(1,2)$, and $(0,1)$.

(d) $f_{XY}(t, u) = 4t(1 - u)$ for $0 \le t \le 1$, $0 \le u \le 1$.

(e) $f_{XY}(t, u) = \frac{1}{8}(t + u)$ for $0 \le t \le 2$, $0 \le u \le 2$.

(f) $f_{XY}(t, u) = 2\alpha u e^{-\alpha t}$ for $0 \le t$, $0 \le u \le 1$.

(g) $f_{XY}(t, u) = 8tu$ on triangle bounded by $t = 1$, $u = 0$, and $u = t$.

(h) $f_{XY}(t, u) = \frac{24}{5}tu$ on triangle bounded by $t = 1$, $u = 1$, and $u = 1 - t$.

(i) $f_{XY}(t, u) = 12t^2 u$ on the parallelogram with vertices $(-1,0)$, $(0,0)$, $(1,1)$, and $(0,1)$.

(j) $f_{XY}(t, u) = \frac{6}{5}(t^2 + u)$ for $0 \le t \le 1$, $0 \le u \le 1$.

12–3. For each of the following mixed joint distributions, determine $E[X]$ and $E[Y]$.

Mixed joint distributions for problem 12-3

   (a) Mass 0.1 at points $(-1, 0)$, $(0, 0)$, and $(1, 0)$.

       Mass 0.7 spread uniformly over the square with vertices at $(-1, 0)$, $(0, -1)$, $(1, 0)$, $(0, 1)$.

   (b) Mass 0.1 at points $(0, 0)$, $(0, 1)$, $(1, 0)$.

       Mass 0.7 spread uniformly over the triangle with vertices at those points.

13

Expectation and Integrals

In the previous chapter, the mathematical expectation of a real random variable X is characterized as the probability-weighted average of the possible values of X. In the case of a simple random variable, which has a finite set of possible values, the expectation is a sum. The idea of expectation as an average is extended by intuitive and heuristic arguments to the case of random variables with absolutely continuous distributions. The sum becomes an integral. For mixed distributions, we may use the Stieltjes integral, or a mixture of a sum (for the discrete part) and an integral (for the absolutely continuous part).

For many applications, such an informal treatment is quite effective and sufficient. However an extension of the concept to the expectation of a function of a random vector frequently leads to complications, in both the practical evaluation and the formal exposition of essential properties. A variety of possible forms for mathematical expectation may be developed, often as a result of intuitive arguments that are quite plausible, but lack the certainty and precision desired in a sophisticated mathematical model. Some of the most powerful features of the theory of mathematical expectation are difficult to formulate and even more difficult to establish by means of a naive, intuitive formulation.

In this chapter, we examine the notion of the Lebesgue integral of a measurable function with respect to a finite measure. If the measurable function is simple (i.e., has a finite range), the integral is a sum. If the measurable function is a simple random variable X and the finite measure is a probability measure P, the Lebesgue integral is exactly the sum we call the mathematical expectation of X. The extension of the integral to the general case involves a process of forming suitable approximations and taking limits. An examination of this procedure shows that it makes precise the ideas used in the informal extension of mathematical expectation from the simple case to the general case. We are thus led to define the mathematical expectation of a random variable X as the Lebesgue integral of X with respect to the probability measure P.

If we use the measure-theoretic character of probability, as developed in previous chapters, the concept of the Lebesgue integral is no more difficult than that of the ordinary Riemann integral used in the standard calculus. With a very modest investment of time and effort, we can sketch in considerable detail the essential features of the theory of the Lebesgue integral and can establish a number of powerful and useful properties of the integral, which then become properties of mathematical expectation. Such a treatment provides access to the extensive literature on the Lebesgue integral and mathematical probability. In fact, we borrow several significant theorems whose proofs require measure-theoretic preparation beyond the scope of this work, but that we can understand, state with precision, and use to considerable advantage.

In this chapter, we sketch the theory of the Lebesgue integral. In the next chapter, we specialize many of the general results to the case of mathematical expectation. Expectation takes expression in a variety of forms in various special cases. The formulation as a Lebesgue integral displays the essential unity of the concept, in spite of the diversity of forms, and it provides a wealth of mathematical resources for applications. A combination of general properties and an appropriate specific formulation often serves to make feasible an otherwise difficult and obscure solution.

It is not necessary to follow the treatment of this chapter in detail to use the resulting properties of mathematical expectation identified and employed in the following chapters. After reading the sketch of the development in Section 13.1 and the connection between Lebesgue integrals and mathematical expectation in Section 13.5, one may wish to proceed immediately to Chapter 14.

13.1.   A Sketch of the Development

We formulate the concept of Lebesgue integral in terms of a finite measure space. That is, we consider a finite measure μ, which assigns a nonnegative number $\mu(E)$ to each set E in a sigma algebra $\mathcal{F}$ of subsets of a basic space Ω. The measure μ has the properties

$$(1) \quad \mu(E) \geq 0.$$

$$(2) \quad \mu(\Omega) = C < \infty.$$

$$(3) \quad \text{if } E = \bigcup_{i=1}^{\infty} E_i \text{ then } \mu(E) = \sum_{i=1}^{\infty} \mu(E_i).$$

If $C = 1$, then μ is a probability measure, the subsets E are events, and F is the sigma algebra of events. The integrand X is a measurable function, which means that $X^{-1}(M)$ is in $\mathcal{F}$ for each Borel set M. A random variable is thus a measurable function. Because our primary interest is in integrals of random variables with respect to probability measures, we frequently refer to the integrands as random variables and the sets in the sigma algebra $\mathcal{F}$ as events. Two characteristics of measurable functions are exploited.

a. A measurable function X may be expressed as the difference $X^+ - X^-$ of nonnegative measurable functions.

b. A nonnegative measurable function is the limit of a nondecreasing sequence of simple measurable functions (see Section 7.3).

The development of the basic idea of the integral proceeds in several phases.

1. We begin with simple measurable functions, which play a role analogous to that played by the step functions used in introducing the ordinary Riemann integral. Suppose the possible values of X constitute the set $\{t_1, t_2, \ldots, t_n\}$. For each i, let A_i be the event $\{X = t_i\}$. Then the integral is the weighted sum

$$\int_\Omega X = \sum_{i=1}^n t_i \mu(A_i).$$

Six elementary properties are established. These are shown to be fundamental to the character of the integral.

2. For a nonnegative measurable function X, there is a nondecreasing sequence of simple measurable functions whose limit is X. For the formation of such a sequence, see Section 7.3. This implies that the integrals of the approximating simple functions form a nondecreasing sequence of real numbers. The nondecreasing sequence must have a limit (possibly infinite). The integral of X is defined to be the limit of the sequence of integrals. It is necessary to show this limit is unique, even though the sequence of approximating simple functions is not unique. This is accomplished with the aid of the six elementary properties of integrals of simple functions, referred to earlier. These properties survive this passage to the limit, with appropriate restrictions to account for the fact that the functions are nonnegative.

3. In the general case, X is expressed as the difference $X^+ - X^-$ of the nonnegative measurable functions X^+ and X^-. If either $\int X^+$ or $\int X^-$ is finite, the integral $\int X$ is defined to be the difference $\int X^+ - \int X^-$. Otherwise, $\int X$ is undefined. The six elementary properties identified for the simple function case survive these operations and hold in the general case.

An examination of the process of extension shows that the precise development in phases 2 and 3 corresponds to the informal extension of mathematical expectation from simple random variables to absolutely continuous random variables, as carried out in Chapter 12. We thus take the Lebesgue integral $\int X \, dP$ to be the appropriate vehicle for the concept of mathematical expectation of a random variable X.

The six elementary properties that play the key role in establishing the Lebesgue integral also are used to obtain a powerful set of properties and characteristics of that integral. When specialized to mathematical expectation, this extensive set of properties forms the basis for the central role of mathematical expectation in probability theory. A number of these properties are discussed in this chapter. Proofs of some are given as the properties are introduced. Others are provided in the Appendix to this chapter. In some cases, we borrow results whose proofs are beyond the scope of this work. We discuss them, perhaps in restricted form, in order to be able to use them intelligently and to be able to turn to the literature if more general forms are needed.

In the next three sections, we introduce the Lebesgue integral in three stages, as outlined earlier.

13.2.   Integrals of Simple Functions

We assume a fixed measure space $(\Omega, \mathcal{F}, \mu)$, where Ω is a basic space, $\mathcal{F}$ is a sigma algebra of subsets, which we call events, and μ is a finite measure, which assigns a nonnegative number $\mu(A)$ to each event A in $\mathcal{F}$. We say two measurable functions X and Y are equal almost everywhere, designated $X = Y$ a.e., iff the set of ω on which they fail to have the same value is a set whose measure is zero.

Recall that a simple measurable function can be expressed as a linear combination of indicator functions for events. In particular, we often find it useful to express the function in canonical form (see Section 7.3). If the range of X is $\{t_1, t_2, \ldots, t_n\}$ and A_i is the set (event) on which X has the value t_i, for $1 \leq i \leq n$, then

$$X = \sum_{i=1}^{n} t_i I_{A_i} \qquad \text{(canonical form)}.$$

The class of events determined by X consists of those sets that can be formed as the unions (disjoint) of various subclasses of the class $\{A_i : 1 \leq i \leq n\}$. This includes the empty set $\emptyset$, which may be viewed as the union of the empty subclass, which contains no sets.

The integral of a simple function X with respect to the finite measure μ is a weighted sum.

Definition 13.2.1. *If* $X = \sum_{i=1}^{n} t_i I_{A_i}$ *(canonical form), then* $\int X \, d\mu = \sum_{i=1}^{n} t_i \mu(A_i)$

Remarks

1. This is a weighted sum of the possible values of X. For a probability measure P and a simple random variable X, the weighted sum is exactly the mathematical expectation $E[X]$ defined in Section 12.1. Thus, for this case, we have

$$E[X] = \int X \, dP.$$

2. The symbol $\int X\,d\mu$ for the integral is often shortened to $\int X$ when the choice of measure μ is understood. In other situations, it may be desirable to use the expanded expression $\int X(\omega)\mu(d\omega)$.

3. It is customary to omit the designation of the domain of integration when it is the whole space Ω. Observe that $\int X$ is *not* an indefinite integral or antiderivative; it is an abbreviation of $\int_\Omega X\,d\mu$, used when the region of integration Ω is understood.

We concentrate on six elementary properties. These are properties of the weighted sums used in the definition. It turns out that these properties play a central role in the further development of the notion of the Lebesgue integral.

I0) If $X = Y$ a.e., then $\int X = \int Y$

I1) For any constant a and any event A, we have $\int aI_A = a\mu(A)$

I2) *Linearity* (with respect to integrand)

If $Z = aX + bY$, with a, b constants, then $\int Z = a\int X + b\int Y$

(This may be extended by induction to any linear combination.)

I3) *Positivity; monotonicity*

1. If $X \geq 0$ a.e., then $\int X \geq 0$, with equality iff $X = 0$ a.e.

2. If $X \geq Y$ a.e., then $\int X \geq \int Y$, with equality iff $X = Y$ a.e.

I4) *Fundamental lemma*

If X is a bounded, nonnegative measurable function and $\{X_n : 1 \leq n\}$ is an a.e. nonnegative, nondecreasing sequence of measurable functions with

$$\lim_n X_n(\omega) \geq X(\omega) \qquad \text{for} \quad \text{a.e.} \quad \omega,$$

then

$$\lim_n \int X_n \geq \int X.$$

Remark: Every nondecreasing sequence of real numbers is either bounded above and has a finite limit, or is unbounded. In the latter case, we say the limit is infinite. In this sense, every nondecreasing sequence of real numbers has a limit (possibly infinite).

I4a) *Monotone convergence* (MC)

If $0 \leq X_n \leq X_{n+1}$ a.e., for each $n \geq 1$ and $X_n \to X$ a.e., then $\int X_n \to \int X$.

IDEAS OF THE PROOFS OF THE ELEMENTARY PROPERTIES

- It should be clear that modifying the integrand X on a set of measure zero simply modifies one or more of the A_i, without changing any $\mu(A_i)$. Such a modification does not change the value of the defining sum. Thus, property (I0) is valid.

- If $X = aI_A$, then its canonical form is $X = aI_A + 0I_{A^c}$ so that property (I1) is just a special case of the defining relation.

- To verify (I2), express $Z = aX + bY$ in canonical form (using a joint partition). Some minor complications arise because Z may have the same value on several members of the joint partition. Handling this is simply a matter of careful use of notation. The property may be extended to any finite linear combination by a simple use of mathematical induction.

- Positivity (I3a) is a simple property of sums of real numbers. The almost-everywhere condition simply reflects the fact that changing the values of X on a set of ω whose measure is zero does not alter the value of the sum.

- Monotonicity (I3b) is a consequence of positivity and linearity, since

$$X \geq Y \quad \text{a.e.} \qquad \text{iff} \qquad X - Y \geq 0 \quad \text{a.e.}$$

and

$$\int X \geq \int Y \qquad \text{iff} \qquad \int X - \int Y = \int (X - Y) \geq 0.$$

Use of (I1) and (I2), extended by induction, gives

Theorem 13.2.2 (Integrals of Simple Measurable Functions)
If $X = \sum_{i=1}^{n} a_i I_{A_i}$, *(not necessarily in canonical form), then* $\int X \, d\mu = \sum_{i=1}^{n} a_i \mu(A_i)$.

— $\qquad\qquad\qquad\qquad\qquad\qquad\qquad\qquad\qquad\qquad\qquad\qquad$ □

Remark: This shows that the defining expression is valid whether or not X is in canonical form.

- The fundamental lemma (I4) plays an essential role in our development of the Lebesgue integral. Its proof is based on property (I1) and the following restricted versions of (I2) and (I3).

I2$^+$) If $X = aX + bY$, a, b positive constants, then $\int Z = a \int X + b \int Y$.

I3$^+$) If $X \geq Y \geq 0$ a.e., then $\int X \geq \int Y \geq 0$.

The argument to establish (I4), given in the appendix to this chapter, uses elementary ideas, but in a somewhat sophisticated manner. At this stage, all integrands are simple. Since properties (I1), (I2$^+$), and (I3$^+$) extend to general nonnegative integrands, the lemma generalizes automatically.

- Monotonicity and the fundamental lemma provide a very simple proof of monotone convergence (I4a), which we frequently denote (MC). A proof is given in the appendix.

13.3. Integrals of Nonnegative Functions

Suppose X is a nonnegative measurable function. As shown in Section 7.3, X is the limit of a nondecreasing sequence of nonnegative simple measurable functions. Thus, there is a sequence $\{X_n : 1 \leq n\}$ such that for every ω and every positive integer n, $0 \leq X_n(\omega) \leq X_{n+1}(\omega)$, with $X_n(\omega) \to X(\omega)$. Provisionally, we try the following.

Definition 13.3.1. *For $X \geq 0$, consider any nondecreasing sequence of nonnegative, simple, measurable functions that converges to X for every ω. The integral of X with respect to μ is given by*

$$\int X \, d\mu = \lim_n \int X_n \, d\mu \qquad \text{(may be infinite)}.$$

Remark: Before this can be accepted as a definition, two questions must be answered:

1. Since the sequence $\{X_n : 1 \leq n\}$ of approximating simple functions is not unique, is the limit of the sequence of integrals unique?

2. If X is a simple measurable function, does this new assignment agree with the value assigned by the original definition for integrals of simple measurable functions?

The fundamental lemma (I4) and monotone convergence (I4a) may be used to show quite easily that the answers to both questions are affirmative (see the appendix to this chapter).

Elementary properties of limits yield immediate extensions of (I0), (I2$^+$), and (I3$^+$). With these properties, the fundamental lemma extends immediately to nonnegative measurable functions. In turn, the monotone convergence theorem (I4a) follows immediately, provided the integrals are finite. In the appendix, a somewhat more sophisticated argument is used to show that (MC) holds even if X is unbounded.

13.4.   Integrable Functions

Extension of the integral to general measurable functions is now quite easy. We may represent a measurable function X as the difference between two nonnegative measurable functions, as follows. Set

$$X^+(\omega) = \max\{0, X(\omega)\} = X(\omega) \qquad \text{for } X(\omega) \geq 0 \quad \text{and zero elsewhere}$$
$$X^-(\omega) = \max\{0, -X(\omega)\} = -X(\omega) \quad \text{for } X(\omega) \leq 0 \quad \text{and zero elsewhere.}$$

Then, for any ω, $X^+(\omega) \geq 0$, $X^-(\omega) \geq 0$, and $X(\omega) = X^+(\omega) - X^-(\omega)$.

Elementary measure-theoretic arguments show that X^+ and X^- are measurable functions iff X is a measurable function.

Definition 13.4.1. *We say the integral of X exists and assign the value (possibly infinite)*

$$\int X = \int X^+ - \int X^-$$

provided at least one of the right-hand terms is finite. If both $\int X^+$ and $\int X^-$ are finite, then X is said to be integrable.

Remark: It is sometimes useful to say X is integrable-μ.

Theorem 13.4.2.   *a. Since $|X| = X^+ + X^-$, X is integrable iff $|X|$ is integrable.*

b. If $|X| \leq Y$ a.e. and Y is integrable, then X is integrable.

c. Since $(-X)^+ = X^-$ and $(-X)^- = X^+$, we have $\int(-X) = -\int X$.

$\square$

Property (I0) extends immediately. (I1) is a special case that needs no extension. Extension of (I2) requires a clever, though elementary, treatment of cases. The inequality portion of (I3a) is contained in (I3$^+$). Proof of the equality assertion uses a common type of argument that is instructive.

Proof of the Equality Provision in (I3a)

Suppose $X \geq 0$ a.e. Let $E_0 = \{X > 0\}$. Then $X = 0$ a.e. iff $\mu(E_0) = 0$.

Now $E_0 = \bigcup_{n=1}^{\infty} E_n$, where $E_n = \{X > 1/n\}$, since $X(\omega) > 0$ iff $X(\omega) > 1/n$ for some n.

Since $\mu(E_0) \leq \sum_{n=1}^{\infty} \mu(E_n)$, we have the desired result if we show that $\mu(E_n) = 0$ for each positive integer n.

Now $0 \leq \frac{1}{n} I_{E_n} \leq X$, for any positive n. Thus, by (I3$^+$) and (I1) we have

$$0 \leq \frac{1}{n}\mu(E_n) = \int \frac{1}{n} I_{E_n} \leq \int X = 0.$$

We conclude that each $\mu(E_n) = 0$, so that the theorem is established.

$\square$

Property (I3b) follows from (I3a) and linearity (I2), as in the case of simple functions. The fundamental lemma (I4) and monotone convergence (I4a) involve nonnegative functions, and have been established.

INTEGRALS OVER SUBSETS

The integrals studied so far have been over the entire basic space Ω. If we wish to integrate over some subset (event) E, we use the function $I_E X$, which agrees with X on the set E but has zero value elsewhere.

Definition 13.4.3. *The* integral of X over E *is given by*

$$\int_E X = \int I_E X.$$

Theorem 13.4.4.   *a. If X is integrable, so also is $I_E X$ for any event E.*

b. If $I_E X$ is integrable, so is $I_C X$ for any event C that is a subset of E.

— □

Remark: We may have a measurable function X such that $I_E X$ is integrable for some events E but not integrable for others. In this case, X is not integrable (over the whole space).

We note one very important property of integrals over subsets, which we extend later to obtain a property with far-reaching consequences.

Additivity: If X is integrable over $E = \bigcup_{i=1}^n E_i$ (disjoint union), then it is integrable over each E_i, and

$$\int_E X = \sum_{i=1}^n \int I_{E_i} X = \sum_{i=1}^n \int_{E_i} X.$$

Proof

Since $I_E X = \sum_{i=1}^n I_{E_i} X$, this property is a simple consequence of linearity.

— □

Remarks

1. We have used linearity to establish additivity. However, in the appendix to this chapter, we use additivity to establish linearity. An independent proof of additivity is provided. These facts display the close relationship between linearity and additivity for integrals.

2. Many properties of integrals (over the whole space) may be applied to integrals over events. All of the properties already developed, except the special case (I1), apply to integrals over events. Since $I_E I_A = I_{AE}$, property (I1) becomes

$$\int_E a I_A = a\mu(AE).$$

13.5. Mathematical Expectation and the Lebesgue Integral

In Chapter 12, the notion of mathematical expectation of a real random variable X is characterized as the probability-weighted average of the values taken on by X. For simple random variables, the concept is expressed exactly as the probability-weighted sum of the values. Since the total weight is one, the weighted average is the weighted sum. For a simple random variable X, the mathematical expectation is exactly the Lebesgue integral with respect to probability measure P. A careful examination shows that the process of extending the Lebesgue integral to nonnegative measurable functions and then to the general case is analogous to and consistent with the informal extension of mathematical expectation from simple random variables to the general case. Tracing these analogies does not constitute logical proof that mathematical expectation must be a Lebesgue integral. It does seem reasonable, however, to make this a matter of definition. Justification of this definition rests on the resulting mathematical theory of mathematical expectation.

Definition 13.5.1.   *The mathematical expectation $E[X]$ of real-valued random variable X is given by*

$$E[X] = \int X \, dP \qquad \text{(Lebesgue integral)}$$

provided the latter exists.

Remarks

1. Since the probability measure P is usually understood in a given discussion, no indication of this measure is used in the notation for expectation. If more than one probability measure is under consideration, various notational devices, such as $E_P[X]$, may be used to indicate the pertinent probability measure.

2. If the expectation of X is finite, we say the random variable X is integrable, understanding that it is integrable-P.

3. It may happen that we are considering a random vector X. If g is a real-valued Borel function whose domain includes the range of X, then $Z = g(X)$ is a real-valued random variable and

$$E[Z] = E[g(X)] = \int g(X)\,dP \qquad \text{(provided the latter exists).}$$

4. From a theoretical point of view, we usually consider the general random variable X as the primary object of interest, and consider the simple random variable as an approximation. On the other hand, it appears that nature knows only discrete quantities. Thus, from an empirical point of view, we may want to think of the simple random variable as the primary object and consider the general random variable as a useful approximation "to the real thing." The integral of the general random variable is then a convenient mathematical construct for purposes of analysis.

In the remainder of this chapter, we consider several additional properties of the Lebesgue integral. When applied to the special case of mathematical expectation, they both demonstrate the appropriateness of this definition and provide properties and characteristics of mathematical expectation that are the basis of its central role in probability. In the next chapter, we formulate, collect, and use an extensive list of such properties.

13.6.   The Lebesgue-Stieltjes Integral and Transformation of Integrals

Properties of the Lebesgue integral provide a very powerful resource for use of the concept of expectation. However, the integral on the basic space and its properties are rather formal and abstract. In many specific instances, we need to formulate expectations in terms of ordinary sums or integrals on the real line or in higher dimensional spaces, in the manner of the informal treatment in Chapter 12. We consider the Lebesgue version of these integrals, discuss their relationship to the ordinary Riemann and Stieltjes integrals, then establish a transformation theorem that relates the abstract integrals on the basic space to the corresponding Lebesgue integrals on Euclidean space.

The discussions in Chapters 8 and 9 on the probability mass distribution induced by a random vector show that this distribution on the Borel sets is described by the induced measure μ_X (often written P_X) or, equivalently, by the probability distribution function F_X. The probability system is transferred to the class $\mathcal{B}^n$ of Borel sets on $\Re^n$. Now the measure space

$(\Re^n, \mathcal{B}^n, \mu_X)$ is a perfectly good measure space, and if g is a real-valued Borel function on $\mathbf{R}^n$, then

$$\int_M g \, d\mu_X$$

is well-defined. Such an integral is known as a Lebesgue-Stieltjes integral on $\Re^n$ (over the set M). Since the measure is characterized by the distribution function F_X, it is customary to write

$$\int_M g \, d\mu_X = \int_M g \, dF_X = \int_M g(t) F_X(dt).$$

The measures on $\mathcal{B}^n$ need not be probability measures. They may be more general measures. The Lebesgue measure is particularly important. If $n = 1$, then $\Re$ is the real line, and Lebesgue measure λ assigns to each interval its length. For $n = 2$, $\Re^2$ is the plane, and Lebesgue measure assigns to each ordinary rectangle (two-dimensional interval) its area. Similarly for higher dimensional Euclidean spaces.

In the one-dimensional case, it is customary to write

$$\int_M g(t)\lambda(dt) = \int_M g(t) \, dt.$$

If Lebesgue measure is taken over the entire real line, then λ is not a finite measure. It does have the property that it assigns finite value to any bounded Borel set M, so that for such sets the integral theory is that for finite measures. There are some technical refinements needed to complete the theory for unbounded sets (including the whole real line). Most of the results for finite measures carry over to this case.

The second form of the integral in the preceding expression uses the same notation as that used for the ordinary Riemann integral (except for the manner in which the region of integration is designated). Also, the notation for the Lebesgue-Stieltjes integral is quite similar to that for the ordinary Stieltjes integral (sometimes referred to as the Riemann-Stieltjes integral). This raises the question of the relationship between the Riemann integrals and the Lebesgue integrals.

An examination of the manner in which the Riemann integrals and the corresponding Lebesgue integrals are formed shows some striking similarities, in spite of a significant difference. Both begin with "step" functions. A simple measurable function is a step function in the sense that it is constant over sets of elements in the domain that have positive measure. Although these sets are not simple intervals, they are measurable sets to which "size" can be assigned by the measure μ. Lebesgue's manner of forming step functions has some advantages over the Riemann approach, in that it allows the integral to be formed in more general situations. A careful analysis of the manner of formation of the integrals shows that whenever the Riemann (or Riemann- Stieltjes) integral exists, so does the corresponding Lebesgue integral; and their values are the same. In practical situations, whenever specific integrals are to be evaluated, they are almost always determined as Riemann integrals. The advantage of the more abstract Lebesgue integral is that it allows the identification and establishment of significant patterns and properties, many of which would be difficult or practically impossible to develop with the Riemann integral. These theoretical properties are important in applications as well as in the evolution of the theory, as subsequent developments show.

We need to establish the connection between integrals on the abstract space and integrals on $\Re^n$. This is a matter of examining the mappings and transfer of mass induced by the random vectors. Let

$X : \Omega \to \Re^n$ be a random vector.

$g : \Re^n \to \Re$ be a real-valued Borel function.

Set $Z = g(X)$. Then

$Z : \Omega \to \Re$ is a real-valued random variable.

X induces measure μ_X on the Borel sets $\mathcal{B}^n : \mu_X(M) = \mu(X^{-1}(M))$.

Z induces measure μ_Z on the Borel sets $\mathcal{B} : \mu_Z(N) = \mu(Z^{-1}(N))$.

For this system, we have the following

Theorem 13.6.1 (Basic Transformation Theorem)

$$\int_\Omega g(X)\,d\mu = \int_{\Re^n} g\,d\mu_X = \int_\Re u\mu_Z(du) \quad \textit{(if one exists, so do the others)}.$$

—   □

Remark: For clarity, we have indicated the spaces over which the integrals are taken.

Proof

1. Suppose $g = I_M$, where M is a Borel set on $\Re^n$, the codomain of X. Then $g(X) = I_M(X) = I_A$, where $A = X^{-1}(M) = \{\omega : X(\omega) \in M\}$. In this case

$$\int_\Omega g(X)\,d\mu = \int_\Omega I_A\,d\mu = \mu(A) = \mu(X^{-1}(M)),$$

and

$$\int_{\Re^n} I_M\,d\mu_X = \mu_X(M) = \mu(X^{-1}(M)).$$

2. Suppose $g \geq 0$. Then there exists a nondecreasing sequence of simple, nonnegative Borel functions g_n increasing to g. This implies that $g_n(X)$ increases to $g(X)$. We may write

$$g_n = \sum_i t_{in} I_{M_{in}}$$

so that by part 1 and linearity

$$\int_\Omega g_n(X)\, d\mu \;=\; \sum_i t_{in} \mu(X \in M_{in}) = \sum_i t_{in} \mu_X(M_{in})$$

$$=\; \int_{\Re^n} g_n\, d\mu_X.$$

By monotone convergence (MC),

$$\int_\Omega g_n(X)\, d\mu \to \int_\Omega g(X)\, d\mu$$

and

$$\int_{\Re^n} g_n\, d\mu_X \to \int_{\Re^n} g\, d\mu_X.$$

Since the two sequences are termwise equal, the limits are equal. This establishes the first equality for nonnegative g.

2. In the general case, let $g = g^+ - g^-$, to establish the first equality.

3. The second equality may be obtained as a special case of the first, as follows. Let $n = 1$ and take h to be the identity function; that is, $h(u) = u$ and $h(Z) = Z$. Thus

$$\int_\Omega Z\, d\mu = \int_\Omega h(Z)\, d\mu = \int_\Re h(u)\, d\mu_Z = \int_\Re u\, d\mu_Z.$$

□

The following extension is frequently useful.

Theorem 13.6.2 (Transformation Theorem) *Suppose* $E = X^{-1}(M)$. *Then* $\int_E g(X)\, d\mu = \int_M g\, d\mu_X$

□

Proof

$$\int_E g(X)\, d\mu \;=\; \int_\Omega I_E g(X)\, d\mu = \int_\Omega I_M(X) g(X)\, d\mu = \int_{\Re^n} I_M g\, d\mu_X$$

$$=\; \int_M g\, d\mu_X.$$

□

We use this result in the next chapter to establish the unity of several forms of mathematical expectation that appear in particular cases.

13.7. Some Further Properties of Integrals

In order to realize the full potential of the integral (hence expectation), we must have available a number of theoretical properties. In the next chapter, we tabulate an important list of these, translated into the notation and terminology of mathematical expectation. Our approach is twofold. On the one hand, we prove, or sketch the proofs of, some of the propositions. These are selected to exhibit the essential structure of the mathematical system and to illustrate the types of arguments that are employed. Many of the details are found in the appendix to this chapter. On the other hand, some properties either require a deeper grounding in measure theory than we presume or would require time and effort that is probably not warranted in a treatment whose goal is preparation for applications. In such cases, the precise formulation of the concepts enables us to borrow intelligently from the extensive mathematical literature.

Properties (I5) through (I9) are largely of theoretical interest. We include these, as well as other theoretical results, because many modern applications require a quite sophisticated knowledge of the underlying theory and an ability to turn to the mathematical literature for the "tools" needed in applications.

We make use of the following uniqueness condition in the development of the concept of conditional expectation (see Chapter 19), which is an essential tool in modern theory of stochastic processes and in important aspects of decision theory.

I5) *Uniqueness.*

The symbol ⊤ is to be read as one of the symbols $\leq$, $=$, or $\geq$.

a. $\int_E g(X) \top \int_E h(X)$ for all E determined by X iff $g(X) \top h(X)$ a.e.

b. $\int_E g(X,Y) = \int_E h(X,Y)$ for all E of the form $E = AB$, with A determined by X, and B determined by Y iff $g(X,Y) = h(X,Y)$ a.e.

I6) *Fatou's lemma.*

If $X_n \geq 0$ a.e., for all $n \geq 1$, then $\int \liminf X_n \leq \liminf \int X_n$.

I7) *Lebesgue's dominated convergence theorem.*

If for almost every ω, $X_n(\omega) \to X(\omega)$ and $|X_n(\omega)| \leq Y(\omega)$, with Y integrable, then X is integrable, $\lim_n \int |X - X_n| = 0$, and $\lim_n \int X_n = \int X$.

Fatou's lemma (I6) is used mainly as a theoretical tool in obtaining other, more directly useful, results. The dominated convergence theorem (I7) provides one of the important conditions for which the "integral of the limit is the limit of the integrals." This is an essential condition in many problems in stochastic processes. Countable additivity and countable sums (I8) are so important and the proof so easy and instructive that we give it here.

I8) *Countable additivity and countable sums.*

a. If X is integrable over E and $E = \bigcup\limits_{i=1}^{\infty} E_i$ (disjoint union), then
$\int_E X = \sum_{i=1}^{\infty} \int_{E_i} X$.

b. If $\sum_{n=1}^{\infty} \int |X_n| < \infty$, then $\sum_{n=1}^{\infty} |X_n| < \infty$ a.e. and
$\int \sum_{n=1}^{\infty} X_n = \sum_{n=1}^{\infty} \int X_n$.

Proof

a. First, consider $X \geq 0$. Then, $I_E X = \sum_{i=1}^{\infty} I_{E_i} X$.

Put $Y_n = \sum_{i=1}^{n} I_{E_i} X$, for each $n \geq 1$. Then $0 \leq Y_n \leq Y_{n+1}$ and $Y_n \to I_E X$.

By linearity (I2), $\int Y_n = \sum_{i=1}^{n} \int I_{E_i} X = \sum_{i=1}^{n} \int_{E_i} X$.

By (MC), $\int Y_n \to \int I_E X = \int_E X$.

By definition of infinite sums, $\sum_{i=1}^{n} \int_{E_i} X \to \sum_{i=1}^{\infty} \int_{E_i} X$.

Since the limit is unique, the theorem holds for $X \geq 0$.

For the general case, consider $I_{E_i} X = I_{E_i} X^+ - I_{E_i} X^-$, etc.

b. First suppose $X_i \geq 0$ for each $i \geq 1$. Put $Y_n = \sum_{i=1}^{n} X_i \to \sum_{i=1}^{\infty} X_i = Y$.

By linearity and (MC), $\int Y_n = \sum_{i=1}^{n} \int X_i \to \int Y = \int \sum_{i=1}^{\infty} X_i$.

This implies $\sum_{i=1}^{\infty} X_i < \infty$ a.e.

For the general case, let $X_n = X_n^+ - X_n^-$. Then

$$Y_n^+ = \sum_{i=1}^{n} X_i^+, \qquad Y_n^- = \sum_{i=1}^{n} X_i^-, \qquad Y = \lim_n Y_n^+ - \lim_n Y_n^-.$$

This implies $\sum_{i=1}^{\infty} X_i^+ < \infty$ and $\sum_{i=1}^{\infty} X_i^- < \infty$, hence $\sum_{i=1}^{\infty} (X_i^+ + X_i^-) < \infty$

The theorem follows from the nonnegative case and linearity.

—   □

Introduction of the concept of integrability suggests that not all random variables may be integrable. It is frequently useful to know the following conditions on integrability.

I9) *Some integrability conditions.*

a. X is integrable iff both X^+ and X^- are integrable iff $|X|$ is integrable.

b. For finite measure, X is integrable iff $\int_{\{|X|>a\}} |X| \to 0$ as $a \to \infty$.

c. If X is integrable, then X is finite a.e.

d. If the integral of X exists and $\mu(E) = 0$, then $\int_E X = 0$.

These conditions show that integrability requires that X not be "too large on too big a set," where the size of a set is determined by the measure assigned to the set. For elementary proofs, see the appendix to this chapter. In the next chapter we give an example of a finite random variable that has infinite expectation.

Inequalities play a significant role in probability theory, as in every area of analysis. In most cases, an appropriate inequality for random variables is set up by an analytic argument (which may or may not involve probabilistic ideas); then monotonicity (I3) is used to obtain the corresponding inequality for integrals. To illustrate, we establish the mean-value theorem.

MEAN-VALUE THEOREM

If $a \le X \le b$ a.e., on set A, then $a\mu(A) \le \int_A X \, d\mu \le b\mu(A)$.

Proof

Under the hypothesis, we have

$$aI_A \le I_A X \le bI_A$$

since all members are zero for ω not in A, and the inequalities of the hypothesis hold for ω in A. Using positivity (I3), property (I1), and the definition of $\int_A$, we have

$$\int aI_A = a\mu(A) \le \int I_A X = \int_A X \le \int bI_A = b\mu(A).$$

□

Other inequalities are established by a similar strategy. A number of the more important inequalities are listed in the Table of Properties of Mathematical Expectation, in the next chapter. We note two that are used in Section 13.9.

HOELDER'S INEQUALITY

For $1 \leq p$, q, with $1/p + 1/q = 1$, and X, Y real or complex,

$$\int |XY| \leq \left(\int |X|^p \right)^{1/p} \left(\int |Y|^q \right)^{1/q}.$$

—   □

MINKOWSKI'S INEQUALITY

For $1 < p$ and X, Y real or complex

$$\left(\int |X + Y|^p \right)^{1/p} \leq \left(\int |X|^p \right)^{1/p} + \left(\int |Y|^p \right)^{1/p}.$$

—   □

13.8. The Radon-Nikodym Theorem and Fubini's Theorem

We consider the celebrated Radon-Nikodym theorem (RN), whose importance for probability theory would be difficult to overemphasize. First, we state a basic theorem. Then, we obtain a simple extension that allows us to put the concept of density function on a sound footing. In Chapter 14, we formulate a special case of RN for expectation, which provides the basis for the concept of conditional expectation in Chapter 19. Since the proof of the basic theorem requires some technical aspects of measure theory beyond the scope of this treatment, we simply borrow it without proof. We note, however, the importance of the countable additivity property (I8) as an essential feature of the mathematical structure.

Consider a nonnegative, integrable function X. For any event $E = X^{-1}(M)$ determined by X, let $\nu(E) = \int_E X \, d\mu$. Then by positivity $\nu(E) \geq 0$. By countable additivity (I8) for integrals, if

$$E = \bigcup_{i=1}^{\infty} E_i \quad \text{(disjoint union), then} \quad \nu(E) = \sum_{i=1}^{\infty} \int_{E_i} X \, d\mu = \sum_{i=1}^{\infty} u(E_i)$$

so that the set function ν is countably additive. If $\mu(E) = 0$, then by the integrability condition (I9d), $\nu(E) = \int_E X \, d\mu = 0$, which implies $\nu(\emptyset) = 0$.

Thus, the set function ν formed in this manner is a measure on the class of sets determined by X. The Radon-Nikodym theorem provides a converse.

Theorem 13.8.1 (Basic Radon-Nikodym Theorem (RN)) *Suppose ν is a finite measure on a sigma algebra $\mathcal{G}$ of events and μ is a measure on $\mathcal{G}$. If ν has the property that $\mu(E) = 0$ implies $\nu(E) = 0$, then there is a measurable function $X \geq 0$ a.e. $[\mu]$ such that every event determined by X is an event in $\mathcal{G}$ (that is, X is measurable-$\mathcal{G}$), and*

$$\nu(E) = \int_E X \, d\mu \qquad \text{for all } E \in \mathcal{G}.$$

As a $\mathcal{G}$-measurable function, X is unique, except possibly for a set of ω having μ-measure zero.

— $\qquad\qquad\qquad\qquad\qquad\qquad\qquad\qquad\qquad\qquad\qquad\qquad\qquad$ $\square$

A classical proof considers the class of all those nonnegative random variables X that are measurable-$\mathcal{G}$ and such that $\int_E X = d\mu \leq \nu(E)$ for all $E \in \mathcal{G}$. From this class a maximal random variable is constructed. This variable is shown to be the desired function. The argument requires some sophisticated techniques, although the ideas are simple.

We may put the basic result in a form that leads to an important extension. We employ what we call a standard argument since the general pattern is used so often and is based on the fundamental properties of integrals.

1. By property (I1) and the basic (RN), we have for $Y = I_E$

$$\int Y \, d\nu = \nu(E) = \int I_E X \, d\mu = \int Y X \, d\mu.$$

2. By linearity of the integral, if $Y = \sum_{i=1}^{n} t_i I_{E_i}$, then $\int Y \, d\nu = \int XY \, d\mu$.

3. If Y is a nonnegative random variable with events $Y^{-1}(M)$ in $\mathcal{G}$, then Y is the limit of a nondecreasing sequence of simple functions of the type considered in step 2. By monotone convergence (MC), the second equality in step 2 must hold in this case, as well.

4. For the general case, set $Y = Y^{+} - Y^{-}$ and apply linearity, to establish the following.

Theorem 13.8.2 (Extended Radon-Nikodym Theorem) *Under the hypothesis of the RN theorem, if Y is integrable with respect to ν, then*

$$\int_E Y \, d\nu = \int_E Y X \, d\mu \qquad \text{for all } E \in \mathcal{G}.$$

— $\qquad\qquad\qquad\qquad\qquad\qquad\qquad\qquad\qquad\qquad\qquad\qquad\qquad$ $\square$

Density Functions

The extension of (RN) in Theorem 13.8.2 establishes conditions for the existence of a density function f for a measure m on the Borel sets on $\Re^n$. Suppose λ is the Lebesgue measure on the Borel sets on $\Re^n$. Let m be a measure on these sets such that $\lambda(B) = 0$ implies $m(B) = 0$. When this relationship holds between measure m and Lebesgue measure λ, the measure m (or its distribution) is said to be *absolutely continuous* (with respect to λ). By the extension of (RN), there is a nonnegative Borel function $f : \Re^n \to \Re$, unique except possibly on a set of Lebesgue measure zero, such that

$$m(B) = \int_B dm = \int_B f \, d\lambda \qquad \text{for all Borel sets } B \text{ on } \Re^n$$

and

$$\int g \, dm = \int g f \, d\lambda.$$

Definition 13.8.3. *The function f just described is the* density function *for the measure m (or for the distribution determined by m).*

Again, the integrals are Lebesgue integrals. In practice, this causes no problem, since we know that in most cases of interest a Lebesgue integral may be replaced by an ordinary Riemann integral. For example, on the plane we write

$$\iint g(t, u) f(t, u) \, dt \, du \qquad \text{or} \qquad \iint_B f(t, u) \, dt \, du$$

and treat these as the ordinary double integrals developed in the multivariable calculus.

In the next chapter, we specialize these results to the probability model and expectations.

Definition 13.8.4 (Fubini's theorem) *Frequently we deal with an indexed class $\{X_t : t \in T\}$ of random variables. If the index set T has an infinity of members (for example, if T is an interval), we have a* random process. *This may be viewed as a function*

$$X : T \times \Omega \to \Re, \qquad \text{with values } X_t(\omega) = X(t, \omega).$$

Suppose ν is a measure on the Borel subsets of T and μ is a measure on the sigma algebra $\mathcal{F}$ of events (subsets of Ω). Under very general conditions, we have

$$\int_\Omega \left(\int_T X(t, \omega) \nu(dt) \right) \mu(d\omega) = \int_T \left(\int_\Omega X(t, \omega) \mu(d\omega) \right) \nu(dt).$$

This is a special case of Fubini's theorem. Suppose $\mu = P$, a probability measure on the sigma algebra $\mathcal{F}$ of events. Since

$$\int_\Omega X(t, \omega)\mu(d\omega) = E[X_t]$$

the equality becomes

$$E\left[\int_T X_t \nu(dt)\right] = \int_T E[X_t]\nu(dt).$$

This interchange of order of integration is important in many topics in the theory of random processes. We use Fubini's theorem in the next chapter to obtain an important form of mathematical expectation, which has several useful special cases.

13.9.   Integrals of Complex Random Variables and the Vector Space $\mathcal{L}^2$

Recall that if $z = x + iy$, with x, y real, then the complex conjugate is $z = x - iy$ and $|z|^2 = zz = x^2 + y^2$. The coordinate x is called the real part of z, denoted $\operatorname{Re} z$, and the coordinate y is called the imaginary part of z, denoted $\operatorname{Im} z$. The complex number z may be expressed $z = |z|e^{i\theta}$, where $e^{i\theta} = \cos\theta + i\sin\theta$. The real number $|z|$ is the modulus of z, and θ in radians is the argument of z. Note that θ is determined only to within an additive constant $2k\pi$ for some integer k. If z is real-valued (i.e., $y = 0$), then $z = z$ and $|z| = |x|$.

Integrals are extended to complex-valued random variables (usually abbreviated "complex random variables") as follows:

Definition 13.9.1. *If $Z = X + iY$, where $X = \operatorname{Re} Z$ and $Y = \operatorname{Im} Z$, then Z is* integrable *iff both X and Y are integrable (as real random variables), in which case*

$$\int Z = \int X + i \int Y.$$

Remark: Since $|Z| \leq |X| + |Y|$, $|X| \leq |Z|$, and $|Y| \leq |Z|$, it follows that Z is integrable iff $|Z|$ is integrable iff both X and Y are integrable.

Examination of the proofs in the real case shows that the following properties extend to the complex case.

I1) $\int aI_A = a\mu(A)$, a real or complex.

I2) *Linearity.* Real or complex constants.

I5) *Uniqueness.*

I7) *Dominated convergence.*

I8) *Countable additivity.*

The triangle inequality has the following simple proof in the complex case, which, of course, includes the real case.

I10) *Triangle inequality*

For integrable Z, $|\int Z| \le \int |Z|$.

Proof

Put $W = \int Z$, a complex number. Then there is a complex constant c with $|c| = 1$ such that

$$cW = |W|. \qquad \text{(If } W = |W|e^{i\theta}, \text{ then } c = e^{-i\theta}\text{).}$$

Put $T = \operatorname{Re} cZ \le |cZ| = |Z|$. Then

$$\left| \int Z \right| = |W| = c \int Z = \int cZ = \int T + i \qquad 0 \le \int |Z|.$$

$\square$

The Ho:lder and Minkowski inequalities hold for complex as well as real random variables and constants. We consider, below, the Schwarz inequality in the quasi-geometric setting of vector spaces.

THE VECTOR SPACE $\mathcal{L}^2$

Many aspects of estimation and prediction in statistics, control, economic time series, communication theory, and other applications of random processes can be given a useful geometric character by introducing the concept of a vector space with an inner product. Real or complex random variables are viewed as elements of the vector space. Notions of orthogonality, distance, and projection are introduced in such a way that $\left(\int |X - Y|^2 \right)^{1/2}$ is interpreted as the distance between elements X and Y in the vector space. The integrals are taken with respect to a finite measure μ. When $\mu = P$, a probability measure, the integrals are expectations. Now $E[|X - Y|^2]$ is the probability-weighted sum (hence average) of the squares of the differences between values $X(\omega)$ and $Y(\omega)$. If Y is an estimator for X, then the quantity $E[|X - Y|^2]$ is the mean-squared error of estimation. We formulate these notions with complex random variables as elements of the vector space and complex constants as scalars.

We direct attention to complex random variables that are square-integrable—i.e., $\int |X|^2$ is finite. Since we are working with integrals, we do not distinguish between two random variables that have the same value for almost every ω. It is easy to see that almost-everywhere equality is an equivalence relation.

Definition 13.9.2. *The class of all random variables X with $\int |X|^2$ finite and X equivalent to Y iff the two are equal almost everywhere is denoted* $\mathcal{L}^2 = \mathcal{L}^2(\Omega, \mathcal{F}, \mu)$.

Remark: We ordinarily use the shorter designation when the measure space is agreed upon. $\mathcal{L}^2$ is usually read "L two," rather than "L squared."

Now $\mathcal{L}^2$ is a vector space over the complex numbers, in the sense that if X, Y are in $\mathcal{L}^2$, so then is $aX + bY$, for any complex numbers a, b. That is, $\mathcal{L}^2$ is closed under complex linear combinations of its elements. To verify this, we note that

1. $\int |cX|^2 = |c|^2 \int |X|^2$, so that X is in $\mathcal{L}^2$ iff cX is in $\mathcal{L}^2$.

2. By Minkowski's inequality,

$$\left(\int |X + Y|^2 \right)^{1/2} \le \left(\int |X|^2 \right)^{1/2} + \left(\int |Y|^2 \right)^{1/2}$$

   so that X, Y in $\mathcal{L}^2$ implies $X + Y$ is in $\mathcal{L}^2$.

The pair of conditions is equivalent to the desired closure under linear combinations.

Remark: The complex case includes the real case. If the random variables and the constants are real-valued, then $\mathcal{L}^2$ is a vector space over the reals.

INNER PRODUCT

We now introduce the notion of an inner product on the vector space $\mathcal{L}^2$ and note several characteristics.

Definition 13.9.3. *If X and Y are in $\mathcal{L}^2$, their* inner product $\langle X, Y \rangle$ *is given by*

$$\langle X, Y \rangle = \int X\overline{Y}$$

where $\overline{Y}$ is the complex conjugate of Y.

The following properties, which justify the term "inner product," are easy consequences of properties of integrals.

IP1) $\langle aX + bY, Z \rangle = a\langle X, Z \rangle + b\langle Y, Z \rangle$.

IP2) $\langle X, Y \rangle = \overline{\langle Y, X \rangle}$.

IP3) $\langle X, X \rangle \ge 0$.

IP4) $\langle X, X \rangle = 0$ iff $X = 0$ a.e.

From these properties we obtain easily a companion to (IP1).

IP5) $\langle X, aY + bZ \rangle = \overline{a}\langle X, Y \rangle + \overline{b}\langle X, Z \rangle$.

Remarks

1. A function $\langle \cdot, \cdot \rangle$ satisfying (IP2) is said to be *Hermitian symmetric*.

2. A function $\langle \cdot, \cdot \rangle$ satisfying (IP3) and (IP4) is said to be *positive definite*.

3. A function $\langle \cdot, \cdot \rangle$ satisfying (IP1) and (IP5) is called *conjugate linear*.

If attention is restricted to real random variables and real constants, $\mathcal{L}^2$ is a vector space over the reals. In this case, the complex conjugate is dropped from the definition and various expressions involving the inner product.

We now obtain the *Schwarz inequality* in its "natural setting." We express it both in its inner product form and in the equivalent integral form. The latter may be specialized to expectation form, when the measure is a probability measure.

SCHWARZ INEQUALITY

a. If X and Y are in $\mathcal{L}^2$, then $|\langle X, Y \rangle|^2 \leq \langle X, X \rangle \langle Y, Y \rangle$, with equality iff X and Y are linearly dependent.

b. If $\int |X|^2$ and $\int |Y|^2$ are both finite, then $|\int XY|^2 \leq \int |X|^2 \int |Y|^2$.

Equality holds iff one or both of X, Y are zero a.e. or if there is a nonzero constant a such that $X = aY$ a.e.

Proof

We carry out a classical proof in inner product form.

1. By definition $\langle X, 0 \rangle = \langle 0, Y \rangle = 0$. By the property (IP4) for inner product, equality holds if either $X = 0$ a.e. or $Y = 0$ a.e., or both.

2. Suppose $\langle X, X \rangle \neq 0$ and $\langle Y, Y \rangle \neq 0$. For any complex constant a we have

   $0 \leq \langle X - aY, X - aY \rangle$, with equality iff $X - aY = 0$ a.e. In the latter case, $a \neq 0$.

   Expansion of the product, using the basic properties of an inner product, yields

   $$0 \leq \langle X, Y \rangle - a\langle X, Y \rangle - a\langle Y, X \rangle + |a|^2 \langle Y, Y \rangle.$$

3. We now select the special value $a = \langle X, Y \rangle / \langle Y, Y \rangle$. Then

   $$0 \leq \langle X, X \rangle - \frac{|\langle X, Y \rangle|^2}{\langle Y, Y \rangle} - \frac{|\langle X, Y \rangle|^2}{\langle Y, Y \rangle} + \frac{|\langle X, Y \rangle|^2}{\langle Y, Y \rangle}.$$

This inequality holds iff $0 \leq \langle X, X \rangle \langle Y, Y \rangle - |\langle X, Y \rangle|^2$. We have equality iff $X = aY$ a.e.

$\square$

With $\mathcal{L}^2$ established as a vector space over the complex numbers (or over the real numbers in the real case) and $\langle X, Y \rangle$ established as the inner product of X, Y for any pair of elements of the vector space, the notions of norm and metric may be introduced in a standard manner. With a metric (distance function), the notion of convergence of a sequence of elements may be formulated. This is convergence in $\mathcal{L}^2$ discussed in Section 17.4. The result quoted in Theorem 17.4.2d ensures that $\mathcal{L}^2$ is complete in terms of such convergence. Hence, $\mathcal{L}^2$ is a Hilbert space. The concepts of orthogonality and of projection on a subspace likewise may be introduced in standard fashion, to provide important tools and interpretations in problems of estimation and prediction. We examine some aspects of this topic in connection with the regression problem in Sections 16.3 and 19.6.

13a

Supplementary Theoretical Details

13a.1.   Integrals of Simple Functions

Since the integral of a simple function is a finite sum, the properties of these integrals are, for the most part, properties of finite sums. The principal exceptions are the limit theorems (I4) and (I4a).

The fundamental lemma (I4) plays an essential role in our development of the Lebesgue integral. It is based on fundamental property (I1) and the following restricted versions of properties (I2) and (I3).

I2$^+$) If $Z = aX + bY$, a, b positive constants, then $\int Z = a \int X + b \int Y$.

I3$^+$) If $X \geq Y \geq 0$ a.e., then $\int X \geq \int Y \geq 0$

The proof of the theorem is instructive, for it illustrates typical arguments and reveals essential mathematical structure. At this stage, the random variables are simple. Since properties (I1), (I2$^+$), and (I3$^+$) extend to general nonnegative random variables, the lemma generalizes automatically. Here, and elsewhere, when we consider a nondecreasing sequence of real numbers, we consider the unbounded case as having infinite limit, so that a limit, in this extended sense, always exists.

I4) *Fundamental lemma.*

If X is a bounded, nonnegative measurable function and $\{X_n : 1 \leq n\}$ is an a.e. nonnegative, nondecreasing sequence of measurable functions with
$$\lim_n X_n(\omega) \geq X(\omega) \qquad \text{for} \quad \text{a.e.} \quad \omega,$$
then
$$\lim_n \int X_n \geq \int X.$$

Proof of the Fundamental Lemma

Since we may modify the integrands on sets of measure zero without affecting the values of their integrals, we may suppose the conditions of the hypotheses hold for all ω. Also, since X is nonnegative and bounded, there is a number b such that $0 \leq X(\omega) \leq b$ for all ω.

Consider any $\epsilon > 0$. For any $n \geq 1$, let $A_n = \{\omega : X_n(\omega) \geq X(\omega) - \epsilon\}$ and set $B_n = A_n^c$.

Since the X_n form a nondecreasing sequence, we must have $A_n \subset A_{n+1}$ for all n. In fact, the A_n increase to Ω so that the $\mu(A_n)$ increase to $\mu(\Omega)$. This requires that the measures $\mu(B_n)$ of the complements decrease to zero.

For any ω in A_n, we have $X(\omega) \leq X_n(\omega) + \epsilon$. Also, for any ω, $X(\omega) \leq b$. We may thus assert

$$X \leq (X_n + \epsilon)I_{A_n} + bI_{B_n} \leq X_n + \epsilon + bI_{B_n} \qquad \text{for all } n.$$

By (I1), (I2$^+$), and (I3$^+$), we have $\int X \leq \int X_n + \epsilon\mu(\Omega) + b\mu(B_n)$. Passing to the limit, we obtain

$$\int X \leq \lim \int X_n + \epsilon\mu(\Omega) + 0.$$

Since this inequality holds for any $\epsilon > 0$, we conclude that $\int X \leq \lim \int X_n$.
___
$\square$

As an easy consequence of this lemma, we obtain the monotone convergence theorem (I4), which establishes one of the most important cases in which the integral of the limit is the limit of the integrals. By monotonicity and elementary properties of limits of sequences of real numbers, $\lim \int X_n \leq \int X$. By (I4), the opposite inequality holds. Hence, we must have equality.

13a.2.   Integrals of Nonnegative Functions

The integral is extended to nonnegative measurable integrands. A nonnegative measurable function may be represented as the limit of a nondecreasing sequence of nonnegative simple measurable functions. The integral of nonnegative X is defined as the limit of the integrals of the approximating simple functions. This approach raises two questions.

1. Since the sequence of approximating simple functions is not unique, is the limit of the sequence of integrals unique?

2. If X is a simple measurable function, does this new assignment agree with the value assigned by the original definition for the integral of a simple measurable function?

The answer to each of these questions is affirmative and each is an easy consequence of the fundamental lemma (I4) and (MC).

Proof of the Uniqueness and Consistency of the Definition

1. Suppose $0 \le X_n(\omega) \le X_{n+1}(\omega)$ and $0 \le Y_n(\omega) \le Y_{n+1}(\omega)$, for any n, ω, with

$$\lim_n X_n(\omega) = \lim_n Y_n(\omega).$$

Then for any m, $\lim_n X_n(\omega) \ge Y_m(\omega)$. By the fundamental lemma, $\lim_n \int X_n \ge \int Y_m$.

Since this holds for any m, we have $\lim_n \int X_n \ge \lim_m \int Y_m$.

Because of the symmetry of the roles of the X and Y sequences, we may interchange them to obtain the opposite inequality, so that equality must hold.

2. If the limit function X in the definition is simple, then monotone convergence (I4a) for simple measurable functions ensures $\lim_n \int X_n = \int X$ [first definition]. Since by the second definition $\int X = \lim_n \int X_n$, both definitions assign the same value.

—   □

The basic properties extend, with some restrictions on (I2) and (I3), to maintain the condition of nonnegative integrands. Thus (I2) extends as (I2$^+$), which stipulates the coefficients must be nonnegative. Suppose $\{X_n : 1 \le n\}$ and $\{Y_n : 1 \le n\}$ are approximating sequences for X and Y, respectively. Then $\{aX_n + bY_n : 1 \le n\}$, with a, b nonnegative, is an approximating sequence for $aX + bY$. Since for each positive integer n,

$$\int (aX_n + bY_n) = a \int X_n + b \int Y_n$$

the linearity holds in the limit.

The restricted form of (I3) may be written: If $X \ge Y \ge 0$ a.e., then $\int X \ge \int Y \ge 0$.

To establish this, we argue as follows. We may suppose the approximating sequences are nonnegative, so that $\int X_n \ge 0$ and $\int Y_m \ge 0$ for any positive n, m.

Now $\lim_n X_n = X \ge Y \ge Y_m$, for any m. Hence, by the fundamental lemma

$$\int X = \lim_n \int X_n \ge \int Y_m \qquad \text{for all } m.$$

By fundamental properties of limits, we also have

$$0 \le \lim_m \int Y_m = \int Y \le \int X.$$

—   □

Given these properties, the fundamental lemma extends automatically if all $\int X_n$ are finite. If $\int X_n$ is infinite, for some n, then $\int X_m$ is infinite for each $m \geq n$, and the theorem holds.

The monotone convergence theorem (I4a) extends, as in the case of simple measurable functions, if X is bounded. In the unbounded case, a somewhat more sophisticated argument is needed. Because of the importance of this case, we give a proof that works for unbounded X.

Proof of Monotone Convergence

Since $0 \leq X_n \leq X_{n+1} \leq X$ a.e., we have by monotonicity (I3$^+$), $\int X_n \leq \int X_{n+1} \leq \int X = I$.
  Hence, $L = \lim_n \int X_n \leq I$.

1. Suppose I is finite. By definition, there is a nondecreasing sequence $\{Y_n : 1 \leq n\}$ of nonnegative simple measurable functions that converge to X, and such that $\int Y_n$ increases to I.

   Given any $\epsilon > 0$, there is an m such that $\int Y_m \geq I - \epsilon$. By the fundamental lemma (I4), with Y_m in place of X,

   $$L = \lim_n \int X_n \geq \int Y_m \geq I - \epsilon.$$

   Since this holds for any $\epsilon > 0$, we conclude $L \geq I$. We thus are assured that $L = I$.

2. For $I = \infty$, a similar argument shows L is greater than any preassigned M.

$$\text{---} \hspace{8cm} \square$$

13a.3.   Integrable Functions

We consider, next, the extension of the basic properties to the case of general integrable functions. Property (I1) is a special case, and needs no extension. To extend linearity (I2), we use (I2$^+$), additivity, and the fact that $\int(-X) = -\int X$. It is noted in Section 13.4 that additivity is a simple consequence of linearity. But since we want to use additivity to establish linearity, we must move somewhat more deliberately.

Additivity: If X is integrable over $E = \bigcup_{i=1}^{n} E_i$ (disjoint union), then it is integrable over each E_i, and

$$\int_E X = \sum_{i=1}^{n} \int_{E_i} X.$$

Proof

We suppose $E = A \bigcup B$ and use the fact that $I_E X = I_A X + I_B X$. We also note that

$$(I_E X)^+ = I_E X^+ = (I_A X)^+ + (I_B X)^+$$

with a similar decomposition of $(I_E X)^-$.

1. For X nonnegative, additivity is a special case of linearity.

2. From the decomposition noted earlier,

$$\int I_E X \;=\; \int I_E X^+ - \int I_E X^-$$

$$=\; \int (I_A X^+ + I_B X^+) - \int (I_A X^- + I_B X^-).$$

We may apply additivity for the nonnegative case to the first and second integrals in the last expression, rearrange, and use $I_A X^+ = (I_A X)^+$, etc., to get

$$\int I_E X \;=\; \int (I_A X)^+ - \int (I_A X)^- + \int (I_B X)^+ - \int (I_B X)^-$$

$$=\; \int I_A X + \int I_B X.$$

3. The general case follows from an inductive argument.

———   □

We now have the tools to establish linearity.

Proof of (I2a)

The proof requires some tricky selection of cases. First we show that $\int aX = a \int X$, where a is any constant. Then we show that $\int (X + Y) = \int X + \int Y$. These combine to give the desired linearity property (which may be extended by mathematical induction to any finite linear combination).

1. For $a \geq 0$, $(aX)^+ = aX^+$ and $(aX)^- = aX^-$, so that by (I2$^+$),

$$\int aX = \int aX^+ - \int aX^- = a\left(\int X^+ - \int X^- \right) = a \int X.$$

For $a \leq 0$, we have $-a \geq 0$, so that $\int aX = -\int (-a)X = -(-a)\int X = a \int X$.

2. To show $\int (X + Y) = \int X + \int Y$, we partition the basic space Ω in a special way and show that for each set E in the partition

we have $\int_E (X + Y) = \int_E X + \int_E X$. We then use additivity to obtain the desired result over the whole space. First, we partition the basic space as follows:

$$A = \{XY \geq 0\} \qquad B = \{X > 0, Y < 0\}$$

$$C = \{X < 0, Y > 0\}.$$

On A, either both X and Y have the same sign or at least one vanishes. On sets B and C, the random variables X and Y have opposite signs. We further partition B and C as follows:

$$B_1 = B \cap \{X + Y \geq 0\} \qquad B_2 = B \cap \{X + Y < 0\}$$

and

$$C_1 = C \cap \{X + Y \geq 0\} \qquad C_2 = C \cap \{X + Y < 0\}.$$

Consider the set A. Since X and Y have the same sign,

$$(I_A(X + Y))^+ = I_A X^+ + I_A Y^+$$
$$(I_A(X + Y))^- = I_A X^- + I_A Y^-.$$

Hence

$$\int I_A(X + Y) = \int I_A X^+ - \int I_A X^- + \int I_A Y^+ - \int I_A Y^-$$
$$= \int I_A X + \int I_A Y.$$

From the definition of $\int_A$, it is apparent the theorem holds on event A.

Next consider the set B_1. We have

$$I_{B_1} X = I_{B_1}(X + Y) + I_{B_1}(-Y).$$

Since the terms on the right-hand side are nonnegative, we may use $(I2^+)$ to assert

$$\int I_{B_1} X = \int I_{B_1}(X + Y) + \int I_{B_1}(-Y).$$

Using the fact that $\int(-Y) = -\int Y$ and rearranging algebraically, we have

$$\int I_{B_1}(X + Y) = \int I_{B_1} X + \int I_{B_1} Y.$$

On set B_2, we have $I_{B_2}(-Y) = I_{B_2}(-(X+Y)) + I_{B_2}X$. Since each term is nonnegative, we may use (I2$^+$), as in the previous case, then rearrange algebraically, to obtain the desired formula.

Parallel arguments on sets C_1 and C_2, with the roles of X and Y interchanged, gives the desired results on those sets.

By additivity, we conclude that the formula holds on the entire basic space.

——   □

I5) *Uniqueness.*

The symbol ⊤ is to be read throughout as one of the symbols $\leq$, $=$, or $\geq$.

(a) $\int_E g(X) \top \int_E h(X)$ for all E determined by X iff $g(X) \top h(X)$ a.e.

(b) $\int_E g(X,Y) = \int_E h(X,Y)$ for all E of the form $E = AB$, with A determined by X and B determined by Y, iff $g(X,Y) = h(X,Y)$ a.e.

Proof of (I5a)

Since $g(X) \top h(X)$ iff $g(X) - h(X) \top 0$, it is sufficient to show that $\int_E X \top 0$ for all E determined by X iff $X \top 0$ a.e.

The "if" part is included in (I0) and (I3), since $X \top 0$ implies $I_E X \top 0$, etc. We establish the "only if" part as follows.

(1) Suppose $\int_E X \geq 0$ for all E determined by X. Consider $E^- = \{X < 0\}$. Then

$$E^- = \bigcup_{n=1}^{\infty} E_n, \qquad \text{where } E_n = \{X \leq -1/n\}$$

since $X(\omega) < 0$ iff $X(\omega) < -\frac{1}{n}$ for some n. If we show that each $\mu(E_n) = 0$ for each n, then $\mu(E^-) = 0$ and therefore $X \geq 0$ a.e. Now,

$$I_{E_n} X \leq -\frac{1}{n} I_{E_n} \qquad \text{for each } n.$$

By hypothesis and monotonicity (I3),

$$0 \leq \int_{E_n} X \leq \int \left(-\frac{1}{n} I_{E_n} \right) = -\frac{1}{n} \mu(E_n).$$

We conclude that $\mu(E_n) = 0$ for each n.

(2) For ⊤ interpreted as $\leq$, use case 1 with X replace by $-X$.

(3) For ⊤ interpreted as $=$, combine cases 1 and 2.

——   □

COMMENT ON THE PROOF of (I5b)

If the proposition could be established for all Borel sets E determined by the random vector (X, Y), then we could simply invoke (I2a). The difficulty is that the class of all $E = AB$ is much smaller than the class of all Borel sets determined by the random vector (X, Y). Some sophisticated measure-theoretic arguments are required to make the extension. We simply borrow the result.

Monotone convergence (I4a) plays a critical role in establishing the next three properties. Because of their instructional value, we provide proofs.

I6) *Fatou's lemma.*

   If $X_n \geq 0$ a.e., for all $n \geq 1$, then $\int \liminf X_n \leq \liminf \int X_n$.

   Proof

   We recall that if $\{a_n : 1 \leq n\}$ is a sequence of real numbers, then $\liminf a_n = \sup_n(\inf_{k \geq n} a_k)$, and $\limsup a_n = \inf_n(\sup_{k \geq n} a_k)$. For any sequence, $\liminf a_n \leq \limsup a_n$, with equality iff the sequence converges, in which case the common value is the limit of the sequence. By $\liminf X_n$ we mean the function obtained by assigning to each ω the number $\liminf X_n(\omega)$. Measure-theoretic arguments ensure that this function is measurable. Let $X(\omega) = \liminf X_n(\omega) = \sup Y_n(\omega)$, where $Y_n(\omega) = \inf_{k \geq n} X_k(\omega) \leq X_n(\omega)$. Now for almost every ω, $0 \leq Y_n(\omega)$ and the latter sequence increases to $X(\omega)$. By (MC),

   $$\lim_n \int Y_n = \liminf \int Y_n = \int X.$$

   Since by monotonicity $\int Y_n \leq \int X_n$, we have

   $$\int X = \liminf \int Y_n \leq \liminf \int X_n.$$

   —   □

We use Fatou's lemma to obtain an important case in which the integral of the limit is the limit of the integrals. In the proof, we use the triangle inequality, which is proved in Section 13.9.

I7) *Lebesgue's dominated convergence theorem.*

   If for almost every ω, $X_n(\omega) \to X(\omega)$ and $|X_n(\omega)| \leq Y(\omega)$, with Y integrable, then X is integrable, $\lim_n \int |X - X_n| = 0$, and $\lim_n \int X_n = \int X$.

Proof

$|X_n| \leq Y$ a.e., all n, implies $|X| \leq Y$, which implies X and X_n are integrable, any n. We note that $|X - X_n| \leq 2Y$ and $\liminf(2Y - |X - X_n|) = 2Y$. We then apply Fatou's lemma to the nonnegative measurable functions $2Y - |X - X_n|$, to get

$$\int 2Y \leq \liminf \int (2Y - |X - X_n|)$$
$$= \int 2Y + \liminf \left(-\int |X - X_n| \right)$$
$$= \int 2Y - \limsup \int |X - X_n|.$$

From this we have

$$0 \leq \liminf \int |X - X_n| \leq \limsup \int |X - X_n| \leq 0$$

so that $\lim_n \int |X - X_n| = 0$.

By the triangle inequality (I10) and linearity,

$$\left| \int X - \int X_n \right| = \left| \int (X - X_n) \right| \leq \int |X - X_n|.$$

The first limit in the conclusion, just established, implies the second.

—   □

I9) *Some integrability conditions.*

(a) X is integrable iff both X^+ and X^- are integrable, iff $|X|$ is integrable.

(b) For finite measure, X is integrable iff $\int I_{\{|X|>a\}}|X| \to 0$ as $a \to \infty$.

(c) If X is integrable, then X is finite a.e.

(d) If $\int X$ exists and $\mu(E) = 0$, then $\int_E X = 0$.

Proof

(a) See Theorem 13.4.2.

(b) Let $Y_a = I_{\{|X|\leq a\}}|X|$ and $I_a = \int I_{\{|X|>a\}}|X|$. Then Y_a increases to $|X|$ as $a \to \infty$ and I_a is nonincreasing with a.

- Suppose X is integrable. Then $\int |X| = \int Y_a + I_a$.

  By monotone convergence, $\int Y_a \to \int |X|$, so $I_a \to 0$ as $a \to \infty$.

- Suppose $I_a \to 0$ as $a \to \infty$. Then there is an a_0 with $I_{a_0} < 1$.

  This implies $\int |X| = \int Y_{a_0} + I_{a_0} < 1 + a_0 \mu(\Omega) < \infty$.

(c) Let $E = \{|X| = \infty\}$ .If X is integrable, so is $I_E|X|$. On E, $|X| > k$, for all $k > 0$, so that $0 \leq kI_E < I_E|X|$. Hence, $0 \leq k\mu(E) < \int I_E|X| = M$, so that $0 \leq \mu(E) < M/k$, for all $k > 0$. We conclude that $\mu(E) = 0$.

(d) There exists $\{X_n : 1 \leq n\}$, simple, with $\lim_n \int_E X_n = \int_E X$.

  If $\mu(E) = 0$, then each $\int_E X_n = 0$. We conclude that $\int_E X = 0$.

—

$\square$

PROBLEMS

13–1. Show that $E[I_M(X)I_N(Y)] = P(X \in M, Y \in N)$.

13–2. Extend linearity (I2) to any finite linear combination of integrable random variables.

13–3. Use linearity (I2) and positivity (I3a) to establish monotonicity (I3b).

13–4. Prove Theorem 13.4.2.

13–5. Prove Theorem 13.4.4.

13–6. If X is real-valued, use monotonicity and the fact that both $X \le |X|$ and $-X \le |X|$ to establish the triangle inequality (I10).

13–7. In the proof of the basic transformation theorem 13.6.1, verify that $\int g_n(X)\, d\mu = \sum_i t_{in}\mu(X \in M_{in})$.

13–8. Establish (I5), (I7), and (I8) for complex integrals.

13–9. Establish properties (IP1) through (IP4) for inner products.

13–10. Use (IP1) through (IP4) for inner products to establish (IP5).

14

Properties of Expectation

In Chapter 12, we examine informally the concept of mathematical expectation, interpret it as the mean value or as the center of the probability mass distribution, and calculate the expectations for several common distributions. In Chapter 13, we examine the Lebesgue integral and express the mathematical expectation $E[X]$ as the Lebesgue integral $\int X\,dP$. In this chapter, we establish the resulting theory of mathematical expectation and demonstrate some of the ways that theory may be applied.

14.1. Some Basic Forms of Mathematical Expectation

The informal treatment of Chapter 12 leads to a variety of forms in various special cases. For simple random variables, expectation is a sum. For the general case, a Stieltjes integral is needed. If a probability density exists, expectation is an integral weighted by the density function. If $Z = g(X)$, then two different expressions for expectation of the same quantity are produced. We use the transformation theorem of Section 13.6, the Radon-Nikodym theorem of Section 13.8, and the equality of the Riemann and the Lebesgue integrals discussed in Section 13.6 to show the equivalence of the various forms encountered. The underlying abstract integral on the probability space provides the unifying concept and at the same time produces the variety of forms needed for special cases.

We suppose $Z = g(X)$. Then, by the basic transformation theorem 13.6.1,

$$E[g(X)] = \int g(t)F_X(dt) = \int u F_Z(du) = E[Z].$$

The Stieltjes integrals are Lebesgue-Stieltjes integrals. However, in practical cases, they may be evaluated as Riemann-Stieltjes integrals. In the proof of the basic transformation theorem, it is seen that the integrals of simple functions are sums. These sums may be expressed as follows:

$$E[g(X)] = \sum_i g(t_i)P(X = t_i) = \sum_j u_j P(Z = u_j) = E[Z].$$

If X is vector-valued, then the t_i are vector quantities. These sums are also obtained directly from extended linearity (I2) and the special case (I1).

Because of property (I8) on countable sums, the result can be extended to discrete random variables with a countably infinite set of values.

The Radon-Nikodym theorem gives a precise condition for the existence of a probability density function f_X for a distribution. This condition is *absolute continuity* of the probability measure induced by X, relative to Lebesgue measure. This means that if the Lebesgue measure $\lambda(M)$ of Borel set M is zero, then $P_X(M) = 0$. An absolutely continuous distribution assigns zero probability mass to sets of zero Lebesgue measure. On the line, Lebesgue measure is length; on the plane, Lebesgue measure is area; in three-dimensional space, Lebesgue measure is volume. Analogous statements hold for higher dimensional space. Practically, this means that on the line there are no point mass concentrations; on the plane there is no probability mass concentration on any point, arc, or line segment. Similar restrictions hold for higher dimensional spaces. If the distribution is absolutely continuous, there is a probability density function f_X such that

$$E[g(X)] = \int g(t) f_X(t)\, dt.$$

As a Borel function, the density function f_X is unique, except possibly on a set of Lebesgue measure zero; these possible exceptions have no effect on the value of the integral. Although the integral is a Lebesgue integral, in practice it is evaluated as an ordinary Riemann integral. For the real-variable case, if the derivative $f_X = F'_X$ of the distribution function is integrable, then f_X is the density. For two dimensions, the density is

$$f_{XY}(t, u) = \frac{\partial^2}{\partial t \partial u} F_{XY}(t, u)$$

provided f_{XY} is integrable.

If $Z = g(X)$ and X are both absolutely continuous, then

$$E[g(X)] = \int g(t) f_X(t)\, dt = \int u f_Z(u)\, du = E[Z].$$

Relationships between F_X and F_Z, or f_X and f_Z, are studied in Chapter 11 (see, also, the comments in Section 14.2).

Example 14.1.1:

Random variable X is exponential (0.3). Thus, it has density function $f_X(t) = u(t)0.3e^{-0.3t}$. The function g is defined by

$$g(t) = \begin{cases} t^2 & \text{for } 0 \le t \le 4 \\ 16 & \text{for } t > 4 \end{cases} = I_{[0,4]}(t)t^2 + 16 I_{(4,\infty)}(t).$$

Determine $E[g(X)]$.

Solution

Since X has a density function,

$$E[g(X)] = \int g(t) f_X(t) \, dt$$

$$= \int I_{[0,4]}(t) t^2 0.3 e^{-0.3t} \, dt + \int 16 I_{(4,\infty)}(t) 0.3 e^{-0.3t} \, dt.$$

The indicator functions serve to determine the limits of integration in each case, since they have zero value outside the designated intervals. The first integral goes from 0 to 4; the second is from 4 to infinity. The second integral is sixteen times the probability X is greater than 4. Thus,

$$E[g(X)] = \int_0^4 0.3 t^2 e^{-0.3t} \, dt + 16 P(X > 4) \approx 13.75.$$

Evaluation of the integral is straightforward, with the aid of the formula given in the appendix on mathematical aids. Also, $P(X > 4) = 1 - F_X(4) = e^{-1.2}$.

□

Remark: The use of the indicator functions in the expression for $g(t)$ and in determining the limits of integration is a very useful device that we employ frequently.

Example 14.1.2:

Suppose (X, Y) produces a uniform joint distribution over the triangular region that has vertices $(0,0)$, $(1,0)$, and $(1,1)$. Then $f_{XY}(t,u) = 2$ over the region (and is zero elsewhere).

Determine $E[XY]$.

Solution

In this case, $g(t, u) = tu$. The expectation is given by

$$E[XY] = \int tu f_{XY}(t, u) \, dt \, du.$$

To set up the double integral, it is necessary to give careful attention to the limits of integration. Suppose we integrate first with respect to the second argument u, then with respect to the first argument t. Reference to Figure 14.1 shows that for each fixed t, $0 \le t \le 1$, u varies from 0 to t. Thus, the integral is

$$E[XY] = \int_0^1 \int_0^t 2tu \, du \, dt.$$

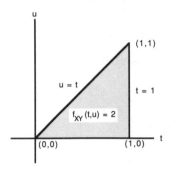

FIGURE 14.1. Joint distribution for Example 14.1.2.

Straightforward evaluation as an iterated integral shows $E[XY] = 1/4$.

—   □

14.2. A Table of Properties

The properties of mathematical expectation listed in the accompanying table constitute a powerful and convenient resource for the use of mathematical expectation. For the most part, we simply restate properties of the Lebesgue integral in the notation for mathematical expectation. When the measure is a probability measure, it is customary to replace the expression "almost everywhere" by *almost surely*, abbreviated a.s. Thus, for random variables we usually write $X = Y$ a.s. as an alternative to $X = Y$ a.e. Where applicable, a number such as (I3) or (I8) assigned to a property of the Lebesgue integral in Chapter 13 is simply changed to (E3) or (E8), respectively. Throughout the remainder of this book, we refer to these properties by the numbers in the table.

Property (E0) is absorbed into (E5), and no longer needs separate listing. We note a special case of (E1), which is used frequently:

Example 14.2.1: Expectation of a Constant

If X is almost surely a constant c, then $E[X] = c$.

Proof

Let $A = \{\omega : X(\omega) = c\}$. Then $X(\omega) = cI_A(\omega)$, with $P(A) = 1$. By property (E1),

$$E[X] = E[cI_A] = cP(A) = c.$$

— □

Example 14.2.2: Probability as Expectation

Probability may be expressed entirely in terms of expectation.

Proof

By (E1) and positivity (E3), we have $P(A) = E[I_A] \geq 0$.

As a special case of (E1), we have $P(\Omega) = E[I_\Omega] = 1$.

By the countable sums property (E8), if $A = \bigcup_i A_i$, then $P(A) = E[I_A] = \sum_i E[I_{A_i}] = \sum_i P(A_i)$.

Thus, the three defining properties for a probability measure are satisfied.

— □

Remark: There are treatments of probability that characterize mathematical expectation with axiomatic properties that are essentially (E0) through (E4a), then define probability as $E[I_A]$. Although such a program is quite feasible, this approach has not been popular.

Property (E1a) uses a notational scheme that we use repeatedly. Recall that

If $A = \{X \in M\} = X^{-1}(M)$, then $I_A = I_M(X)$.

Hence, by (E1) with $a = 1$, $E[I_M(X)] = E[I_A] = P(A) = P(X \in M)$. Now X can be a random vector of any dimension. It is sometimes useful to display this fact explicitly. Thus, if $W = (X, Y)$ is the random vector with real coordinate random variables X and Y, and Q is any Borel set on the plane, then

$$E[I_Q(W)] = E[I_Q(X, Y)] = P\big((X, Y) \in Q\big).$$

If we let $Q = M \times N$, the Cartesian product of M and N, then

$$A = \{(X, Y) \in M \times N\} = \{X \in M\} \cap \{Y \in N\}$$

and $I_A = I_M(X)I_N(Y)$, since both the left-hand and right-hand members have value one iff both X has a value in M and Y has a value in N. Thus,

$$P(X \in M, Y \in N) = E[I_M(X)I_N(Y)].$$

It should be clear that this result extends to higher dimensions.

PROPERTIES OF MATHEMATICAL EXPECTATION

$$E[g(X)] = \int g(X)\,dP.$$

We suppose, without repeated assertion, that the random variables and Borel functions of random variables or random vectors are integrable. Use of an expression such as $I_M(X)$ involves the tacit assumption that M is a Borel set on the codomain of X.

E1) $E[aI_A] = aP(A)$, any constant a, any event A.

E1a) $E[I_M(X)] = P(X \in M)$ and $E[I_M(X)I_N(Y)] = P(X \in M, Y \in N)$ for any Borel sets M, N.

   (Extends to any finite product of such indicator functions of random vectors.)

E2) *Linearity.* For any constants a, b, $E[aX + bY] = aE[X] + bE[Y]$.

   (Extends to any finite linear combination.)

E3) *Positivity; monotonicity.*

   1. $X \geq 0$ a.s. implies $E[X] \geq 0$, with equality iff $X = 0$ a.s.

   2. $X \geq Y$ a.s. implies $E[X] \geq E[Y]$, with equality iff $X = Y$ a.s.

E4) *Fundamental lemma.* If $X \geq 0$ is bounded, and $\{X_n : 1 \leq n\}$ is a.s. nonnegative, nondecreasing, with $\lim_n X_n(\omega) \geq X(\omega)$ for a.e. ω, then $\lim_n E[X_n] \geq E[X]$.

E4a) *Monotone convergence.* If for all $n, 0 \leq X_n \leq X_{n+1}$ a.s. and $X_n \to X$ a.s., then $E[X_n] \to E[X]$.

   (The theorem also holds if $E[X] = \infty$.)

E5) *Uniqueness.* ($\top$ is to be read as one of the symbols $\leq$, $=$, or $\geq$).

   1. $E[I_M(X)g(X)] \top E[I_M(X)h(X)]$ for all M iff $g(X) \top h(X)$ a.s.

   2. $E[I_M(X)I_N(Z)g(X,Z)] = E[I_M(X)I_N(Z)h(X,Z)]$ for all M, N iff $g(X,Z) = h(X,Z)$ a.s.

E6) *Fatou's lemma.* If $X_n \geq 0$ a.s., for all n, then $E[\liminf X_n] \leq \liminf E[X_n]$.

E7) *Dominated convergence.* If real or complex $X_n \to X$ a.s., $|X_n| \leq Y$ a.s. for all n, and Y is integrable, then $\lim_n E[X_n] = E[X]$.

E8) *Countable additivity and countable sums.*

1. If X is integrable over E and $E = \bigcup\limits_{i=1}^{\infty} E_i$ (disjoint union), then $E[I_E X] = \sum_{i=1}^{\infty} E[I_{E_i} X]$.

2. If $\sum_{n=1}^{\infty} E[|X_n|] < \infty$, then $\sum_{n=1}^{\infty} |X_n| < \infty$ a.s., and $E[\sum_{n=1}^{\infty} X_n] = \sum_{n=1}^{\infty} E[X_n]$.

E9) *Some integrability conditions*

1. X is integrable iff both X^+ and X^- are integrable iff $|X|$ is integrable.

2. X is integrable iff $E[I_{\{|X|>a\}}|X|] \to 0$ as $a \to \infty$.

3. If X is integrable, then X is a.s. finite.

4. If $E[X]$ exists and $P(A) = 0$, then $E[I_A X] = 0$.

E10) *Triangle inequality.* For integrable X, real or complex, $|E[X]| \le E[|X|]$.

E11) *Mean-value theorem.* If $a \le X \le b$ a.s. on A, then $aP(A) \le E[I_A X] \le bP(A)$.

E12) Suppose g is nonnegative, Borel, defined on the range of X. Consider $A = \{\omega : g(X(\omega)) \ge a\}$. Then $E[g(X)] \ge aP(A)$.

E13) *Markov's inequality.* If $g \ge 0$ and nondecreasing for $t \ge 0$ and $a \ge 0$, then
$$g(a)P(|X| \ge a) \le E[g(|X|)].$$

E14) *Jensen's inequality.* If g is convex on an interval I that contains the range of real random variable X, then $g(E[X]) \le F[g(X)]$.

E15) *Schwarz' inequality.* For X, Y real or complex,
$$|E[XY]|^2 \le E[|X|^2]E[|Y|^2],$$
with equality iff there is a constant c such that $X = cY$ a.s.

E16) *Hoelder's inequality.* For $1 \le p, q$, with $1/p + 1/q = 1$, and X, Y real or complex,
$$E[|XY|] \le E[|X|^p]^{1/p} E[|Y|^q]^{1/q}.$$

E17) *Minkowski's inequality.* For $1 < p$ and X, Y real or complex,
$$E[|X + Y|^p]^{1/p} \le E[|X|^p]^{1/p} + E[|Y|^p]^{1/p}.$$

E18) *Independence and expectation.* The following conditions are equivalent.

    1. The pair $\{X, Y\}$ is independent.

    2. $E[I_M(X)I_N(Y)] = E[I_M(X)]E[I_N(Y)]$ for all Borel M, N.

    3. $E[g(X)h(Y)] = E[g(X)]E[h(Y)]$ for all Borel g, h such that $g(X)$, $h(Y)$ are integrable.

E19) . *Special case of the Radon-Nikodym theorem*

If $g(Y)$ is integrable and X is a random vector, then there exists a real-valued Borel function $e(\cdot)$, defined on the range of X, unique a.s. $[P_X]$, such that $E[I_M(X)g(Y)] = E[I_M(X)e(X)]$ for all Borel sets M on the codomain of X.

E20) *Some special forms of expectation*

    1. Suppose F is nondecreasing, right-continuous on $[0, \infty)$, with $F(0^-) = 0$. Let $F^*(t) = F(t - 0)$. Consider $X \geq 0$ with $E[F(X)] < \infty$. Then 1) $E[F(X)] = \int_0^\infty P(X \geq t)F(dt)$ and 2) $E[F^*(X)] = \int_0^\infty P(X > t)F(dt)$.

    2. If X is integrable, then $E[X] = \int_{-\infty}^\infty \big(u(t) - F_X(t)\big)\, dt$.

    3. If X, Y are integrable, then $E[X - Y] = \int_{-\infty}^\infty \big(F_Y(t) - F_X(t)\big)\, dt$.

    4. If $X \geq 0$ is integrable, then

$$\sum_{n=0}^\infty P(X \geq n+1) \leq E[X] \leq \sum_{n=0}^\infty P(X \geq n) \leq N \sum_{k=0}^\infty P(X \geq kN),$$

    for all $N \geq 1$.

    5. If integrable $X \geq 0$ is integer-valued, then

$$E[X] = \sum_{n=1}^\infty P(X \geq n) = \sum_{n=0}^\infty P(X > n)$$

$$E[X^2] = \sum_{n=1}^\infty (2n - 1)P(X \geq n) = \sum_{n=0}^\infty (2n + 1)P(X > n).$$

    6. If Q is the quantile function for F_X, then

$$E[g(X)] = \int_0^1 g\big(Q(u)\big)\, du.$$

In the discussion of odds and betting odds in Section 2.3, we note that these ideas are related to the notion of a fair game. We can now clarify that relationship.

Example 14.2.3: Fair Games and Betting Odds

A player plays a game in which he gains an amount a if event A occurs and loses an amount b if event A^c occurs. This is often stated by saying the "opponent" gives odds a to b on outcome A. The net amount the player "gains" is expressed by the random variable

$$W = aI_A - bI_{A^c}.$$

If $E[W]$ is positive, then the expected winning on n plays is $nE[W]$, which becomes large if n is large. In this case, the player is very likely to win a large amount if he plays often enough. Similarly, if $E[W]$ is negative, the expected loss on a large number of plays is great. In either case, the game is not considered fair. However, the game or bet is considered fair iff $E[W] = 0$. The condition

$$E[W] = aP(A) - bP(A^c) = 0 \quad \text{implies} \quad O(A) = \frac{P(A)}{P(A^c)} = \frac{b}{a}.$$

Thus the betting odds are the reciprocal of the probability odds for a fair game.

— $\square$

We may use property (E1a) to obtain an alternate formulation of the problem of determining the distribution for a function of random vectors.

Example 14.2.4: Distribution for a Function of Random Variables

1. If $Z = g(X)$, then $F_Z(v) = P(Z \le v) = P(X \in Q_v) = E[I_{Q_v}(X)]$, where $Q_v = \{t : g(t) \le v\}$.

2. If $Z = g(X, Y)$, then $F_Z(v) = P(Z \le v) = P((X, Y) \in Q_v) = E[I_{Q_v}(X, Y)]$, where $Q_v = \{(t, u) : g(t, u) \le v\}$.

Proof

The first case follows from (E1a) and the fact that $\{Z \le v\} = \{X \in Q_v\}$. The second case uses the fact that $\{Z \le v\} = \{(X, Y) \in Q_v\}$.

— $\square$

Remark: It should be apparent that this pattern extends to random vectors of any finite dimension.

We may use property (E1) and linearity (E2), extended, to give a very simple determination of the expectation for the binomial distribution.

Example 14.2.5: Binomial Distribution

Historically, the binomial distribution arose from counting the successes in a Bernoulli sequence of n trials. If X is the number of successes in n trials, then X is binomial (n, p) and

$$X = \sum_{i=1}^{n} I_{E_i} \qquad \text{where } E_i \text{ is the event of a success on the } i\text{th trial.}$$

Since $P(E_i) = p$, we have $E[I_{E_i}] = p$, for each i. Thus, by linearity (E2), extended,

$$E[X] = \sum_{i=1}^{n} E[I_{E_i}] = np.$$

□

The next example establishes two simple patterns that we use in the theory of variance in the next chapter. The second pattern also provides an important interpretation of the mean value.

Example 14.2.6:

(a) $E[X + a] = E[X] + a$.

(b) $E[(X - a)^2]$ is a minimum iff $a = E[X]$, in which case $E\big[(X - E[X])^2\big] = E[X^2] - E^2[X]$.

Proof

The first statement follows from linearity and Example 14.2.1. For the second, we expand $(X - a)^2$ and apply linearity (E2) and Example 14.2.1 to get

$$E[X^2 - 2aX + a^2] = E[X^2] - 2aE[X] + a^2.$$

This is a quadratic in a, which has a minimum at $a = E[X]$. Substitution of this value for a gives the stated minimum value $E[X^2] - E^2[X]$.

□

INTERPRETATION

If we approximate the random variable X by a constant a, then for any observed value $X(\omega)$ the error of approximation is $X(\omega) - a$. The value of a that gives the smallest mean square of the error $E[(X - a)^2]$ is $a = E[X]$. In Sections 19.2 and 19.5 we extend this idea to give a fundamental interpretation to conditional expectation.

The use of several properties is illustrated in the next example.

Example 14.2.7: Proof of Jensen's Inequality

If g is a convex Borel function on an interval I that includes the range of a real random variable X, then $g(E[X]) \leq E[g(X)]$.

Proof

g is convex iff for each t_0 in the interval I there is a number $\lambda(t_0)$ such that

$$g(t) \geq g(t_0) + \lambda(t_0)(t - t_0).$$

If I includes the range of X, then, by the mean-value theorem (E11), I must include the constant $E[X]$. We chose the constant $t_0 = E[X]$ and designate the constant $\lambda(E[X])$ by a. We thus have

$$g(X) \geq g(E[X]) + a(X - E[X]).$$

Taking expectations of both sides, using linearity (E2) and monotonicity (E3), and recalling that the expectation of a constant is that same constant (Example 14.2.1), we get

$$E[g(X)] \geq g(E[X]) + a(E[X] - E[X]) = g(E[X]).$$

□

Remark: The function $\lambda(\cdot)$ is nondecreasing. This fact is used in establishing Jensen's inequality for conditional expectation in the appendix to Chapter 19.

Example 14.2.8:

A manufacturing process for ball bearings produces essentially spherical balls, with slight variations of radius from one ball to another. If the radius is a random variable R, uniformly distributed on the interval $[R_0 - a, R_0 + a]$, then the average radius is R_0 and the volume for that radius is $V_0 = \frac{4}{3}\pi R_0^3$. Let $V = \frac{4}{3}\pi R^3$ be the random variable whose value is the volume of the ball produced. Determine $E[V]$ and compare it with V_0.

Solution

$$V = \frac{4}{3}\pi R^3$$

so that

$$E[V] = \frac{4}{3}\pi E[R^3] \quad \text{and} \quad E[R^3] = \int t^3 f_R(t)\,dt = \frac{1}{2a}\int_{R_0-a}^{R_0+a} t^3\,dt.$$

Evaluation and some algebraic manipulations show that $E[R^3] = R_0^3 + a^2 R_0$.

This is a simple example of the fact that $E[g(R)] \neq g(E[R])$. Since $g(t) = t^3$ is convex for $t \geq 0$, we have an illustration of Jensen's inequality, since $E[R^3] > E^3[R] = R_0^3$. This, in turn, implies that the average volume is greater than the volume corresponding to the average radius.

$\square$

—

FUNCTIONS WITH COMPOSITE DEFINITIONS

Often we have a function of a random vector for which the function rule is expressed differently for various parts of the domain. We illustrate how the use of linearity and appropriate indicator functions can systematize the approach to such functions. Some of the following examples are simple enough that they could be formulated and solved without this elaborate machinery. But they demonstrate precise formal procedures that may be used in more complex situations for which intuition may be an inadequate guide.

Example 14.2.9:

Suppose X is absolutely continuous, with density f_X. Let

$$Y = \min\{X, c\} = \begin{cases} X & \text{for } X \leq c \\ c & \text{for } X > c \end{cases} = I_{(-\infty,c]}(X)X + cI_{(c,\infty)}(X).$$

Then

$$E[Y] = E[I_{(-\infty,c]}(X)X] + cE[I_{(c,\infty)}(X)].$$

The second expectation in the right-hand member is $P(X > c)$. The first can be written as an integral with the density function. Noting that $I_{(-\infty,c]}(t) = 1$ for $t \leq c$, we have

$$E[Y] = \int_{-\infty}^{c} t f_X(t)\, dt + cP(X > c).$$

For the case X is exponential (α), so that $f_X(t) = u(t)\alpha e^{-\alpha t}$, we have

$$\begin{aligned} E[Y] &= \int_0^c \alpha t e^{-\alpha t}\, dt + ce^{-\alpha c} = \frac{1}{\alpha}\left(1 - (1 + \alpha c)e^{-\alpha c}\right) + ce^{-\alpha c} \\ &= \frac{1}{\alpha}(1 - e^{-\alpha c}). \end{aligned}$$

—

$\square$

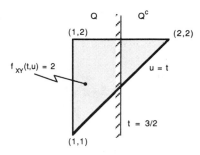

FIGURE 14.2. The joint distribution for Example 14.2.10.

Example 14.2.10:

Suppose $f_{XY}(t, u) = 2$ on the triangular region $1 \leq t \leq u \leq 2$ (zero elsewhere). See Figure 14.2.

$$Z = \begin{cases} X & \text{for } X \leq 3/2 \\ 2XY & \text{for } X > 3/2 \end{cases} = I_Q(X, Y)X + I_{Q^c}(X, Y)2XY.$$

Calculate $E[X]$, $E[Y]$, and $E[Z]$.

Solution

$$E[X] = 2 \int_1^2 \int_t^2 t \, du \, dt = 2 \int_1^2 t(2 - t) \, dt = \frac{4}{3}$$

and

$$E[Y] = 2 \int_1^2 \int_t^2 u \, du \, dt = \int_1^2 (4 - t^2) \, dt = \frac{5}{3}.$$

Since membership in Q depends only on the first coordinate, $I_Q(X, Y) = I_M(X)$, where $M = (-\infty, 3/2]$. We also have

$$f_X(t) = 2 \int_t^2 du = 2(2 - t) \qquad \text{for } 1 \leq t \leq 2.$$

Now,

$$E[Z] = E[I_M(X)X] + 2E[I_{M^c}(X)XY].$$

The first expectation on the right-hand side is of a function of X alone; the second is the expectation of a function of (X, Y). Thus,

$$E[Z] = \int_M t f_X(t) \, dt + 2 \iint I_{M^c}(t) t u f_{XY}(t, u) \, du \, dt$$

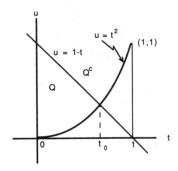

FIGURE 14.3. The joint distribution for Example 14.2.11.

$$
\begin{aligned}
&= \int_{1}^{3/2} 2t(2-t)\,dt + 4\int_{3/2}^{2}\int_{t}^{2} tu\,du\,dt \\
&= \frac{235}{96} \approx 2.45.
\end{aligned}
$$

—   □

Example 14.2.11:

Suppose $f_{XY}(t,u) = 3$ for $0 \le u \le t^2 \le 1$. See Figure 14.3.

$$
Z = \begin{cases} X & \text{for } X+Y \le 1 \\ 1 & \text{for } X+Y > 1 \end{cases} = I_Q(X,Y)X + I_{Q^c}(X,Y).
$$

where

$$
Q = \{(t,u) : u+t \le 1\} = \{(t,u) : u \le 1-t\}.
$$

Determine $E[X]$, $E[Y]$, and $E[Z]$.

Solution

$$
\begin{aligned}
E[X] &= \iint t f_{XY}(t,u)\,du\,dt = 3\int_0^1 t \int_0^{t^2} du\,dt = 3\int_0^1 t^3\,dt = \frac{3}{4} \\
E[Y] &= \iint u f_{XY}(t,u)\,du\,dt = 3\int_0^1 \int_0^{t^2} u\,du\,dt = 3\int_0^1 \frac{t^4}{2}\,dt = \frac{3}{10} \\
E[Z] &= E[I_Q(X,Y)X] + E[I_{Q^c}(X,Y)].
\end{aligned}
$$

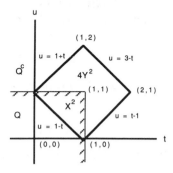

FIGURE 14.4. The joint distribution for Example 14.2.11.

Examination of Figure 14.3 shows that it is desirable to express the first expectation as the sum of two integrals: one for t from 0 to t_0 and the second for t from t_0 to 1. The second expectation is an integral for t from t_0 to 1. The figure shows that the point t_0 satisfies $t_0^2 = 1 - t_0$, so that $t_0 = (\sqrt{5} - 1)/2 \approx 0.6180$. Thus,

$$
\begin{aligned}
E[Z] &= 3 \int_0^{t_0} t \int_0^{t^2} du\, dt + 3 \int_{t_0}^1 t \int_0^{1-t} du\, dt + 3 \int_{t_0}^1 \int_{1-t}^{t^2} du\, dt \\
&= \frac{3}{4}(5t_0 - 2) \approx 0.8176.
\end{aligned}
$$

— $\square$

Example 14.2.12:

Suppose $f_{XY}(t, u) = 1/2$ on the square with vertices at $(1, 0)$, $(2, 1)$, $(1, 2)$, and $(0, 1)$ (see Figure 14.4). Let $H = \max\{X, Y\}$ and put

$$
W = \begin{cases} X & \text{for } N \le 1 \\ 2Y & \text{for } H > 1 \end{cases} = I_{Q_1}(X, Y)X + I_{Q_2}(X, Y)2Y.
$$

Determine $E[W^2]$.

Solution

$$
E[W^2] = E[I_{Q_1}(X, Y)X^2] + E[I_{Q_2}(X, Y)4Y^2]
$$

where

$$
Q_1 = \{(t, u) : t \le 1, u \le 1\} \qquad Q_2 = Q_1^c.
$$

The regions Q_1 and Q_2 are shown on Figure 14.4. From the geometry, we can determine the limits of integration to obtain

$$E[W^2] = \frac{1}{2}\int_0^1 t^2 \int_{1-t}^1 du\, dt + \frac{4}{2}\int_0^1 \int_1^{1+t} u^2\, du\, dt + \frac{4}{2}\int_1^2 \int_{t-1}^{3-t} u^2\, du\, dt.$$

Evaluation of the integrals gives the result $E[W^2] = 103/24$.

—   □

These examples are somewhat artificial and are intended to illustrate technique based on properties of expectation. The next example models a simple decision problem.

Example 14.2.13: Optimal Stocking of Merchandise

A merchant is preparing for the holiday sales period. He plans to stock m units of a particular item, at a cost of c per unit. Demand is a random quantity D with Poisson (μ) distribution. If units remain in stock at the end of the period, they may be returned for r per unit. If demand exceeds m, the extra units can be special-ordered at a cost of s each. If the units are sold at a price of p each, what is the expected gain?

Solution

If $X = g(D)$ is the gain from the sales, we want to determine $E[X]$. Now,

For $D \le m$, $X = D(p - c) - (m - D)(c - r) = D(p - r) + m(r - c)$
For $D > m$, $X = m(p - c) + (D - m)(p - s) = D(p - s) + m(s - c)$

Put $M^c = [0, m]$ and recall that $I_{M^c} = 1 - I_M$. Then

$$X = \big(1 - I_M(D)\big)\big(D(p - r) + m(r - c)\big) + I_M(D)\big(D(p - s) + m(s - c)\big).$$

Simple algebra shows that

$$X = D(p - r) - I_M(D)D(s - r) + I_M(D)m(s - r) - m(c - r).$$

By linearity and (E1), we have

$$E[X] = (p - r)E[D] - (s - r)E[I_M(D)D] + m(s - r)E[I_M(D)] - m(c - r).$$

For the Poisson (μ) distribution, $E[D] = \mu$. By (E1), $E[I_M(D)] = P(D > m) = P(D \ge m + 1)$.

A result on exponential series in the appendix on mathematical aids shows that

$$e^{-\mu} \sum_{k=m+1}^\infty k\frac{\mu^k}{k!} = \mu e^{-\mu} \sum_{k=m}^\infty \frac{\mu^k}{k!}$$

from which we obtain

$$E[I_M(D)D] = \mu P(D \geq m).$$

Thus, we have

$$E[X] = \mu(p-r) - \mu(s-r)P(D \geq m) + m(s-r)P(D \geq m+1) - m(c-r).$$

Suppose $\mu = 50$, $c = 30$, $p = 50$, $r = 20$, and $s = 40$. Then

$$E[X] = 1500 - 1000P(D \geq m) + 20mP(D \geq m+1) - 10m.$$

From a chart for the Poisson distribution we find the following data.

$$
\begin{array}{llll}
m = 50 & P(D \geq 50) = 0.5 & P(D \geq 51) = 0.49 \\
m = 60 & P(D \geq 60) = 0.09 & P(D \geq 61) = 0.07 \\
m = 70 & P(D \geq 70) = 0.004 & P(D \geq 71) = 0.003
\end{array}
$$

Substitution gives values $E[X] = 549$, 894, and 800 for $m = 50$, 60, and 70, respectively. It appears that an initial stock of about $m = 60$ should be optimal. If a better approximation to the optimum m is needed, simply calculate a few more points.

—   □

14.3. Independence and Expectation

A class of random vectors is stochastically independent iff the classes of events determined by the random variables form an independent family. A pair $\{X, Y\}$ of random vectors is independent iff

$$P(X \in M, Y \in N) = P(X \in M)P(Y \in N)$$

for each pair $\{M, N\}$ of Borel sets M and N on the respective codomains of X and Y. We may formulate equivalent conditions for independence in terms of expectations.

Theorem 14.3.1. *A pair $\{X, Y\}$ of random vectors is independent iff*

(a) $E[I_M(X)I_N(Y)] = E[I_M(X)]E[I_N(Y)]$ *for all Borel sets M, N iff*

(b) $E[g(X)h(Y)] = E[g(X)]E[h(Y)]$ *for all Borel functions g and h such that the expectations exist.*

Proof

1. Property (E1a) shows that (a) is equivalent to the defining product rule.

2. It is apparent that (b) implies (a) as a special case.

3. The argument that (a) implies (b) uses elementary properties of expectation in what we call a "standard argument."

   (a) If (a) holds, then condition (b) holds for $g = I_M$ and $h = I_N$.

   (b) By linearity (E2), condition (b) must hold for simple functions g, h.

   (c) For $g \geq 0$, $h \geq 0$, consider a sequence of simple functions g_n increasing to g, and a similar sequence h_n increasing to h.

   Condition (b) holds for each pair $\{g_n, h_n\}$.

By monotone convergence (MC),

$$E[g_n(X)] \to E[g(X)], \qquad E[h_n(Y)] \to E[h(Y)],$$

and

$$E[g_n(X)h_n(Y)] \to E[g(X)h(Y)].$$

The pairwise equality ensures that (b) holds in the limit.
   For the general case, let $g = g^+ - g^-$, $h = h^+ - h^-$, and use linearity.

—   □

Remarks

1. One must not suppose that $E[g(X)h(Y)] = E[g(X)]E[h(Y)]$ for one given function ensures independence. It is not difficult to construct counterexamples. The product rule must hold for all g, h such that the expressions have meaning.

2. The preceding argument, and hence the results, extends readily to any class $\{X_i : i \in J\}$ of random variables.

   Just as the product rule for events simplifies calculations for probabilities of events, the product rule for expectations often simplifies determination of expectations of functions of independent random vectors.

Example 14.3.1:

Suppose $\{X, Y\}$ is an independent pair, with X exponential (α) and Y exponential (β).

Determine $E[X^2 + Ye^{-X}]$.

Solution

$$E[X^2 + Ye^{-X}] = E[X^2] + E[Y]E[e^{-X}].$$

Use of integrals from the appendix on mathematical aids shows that

$$E[X^2] = \int_0^\infty t^2 \alpha e^{-\alpha t}\, dt = \frac{2}{\alpha^2} \quad \text{and} \quad E[Y] = \int_0^\infty t\beta e^{-\beta t}\, dt = \frac{1}{\beta}.$$

Also,

$$E[e^{-X}] = \int_0^\infty \alpha e^{-t} e^{-\alpha t}\, dt = \alpha \int_0^\infty e^{-(\alpha+1)t}\, dt = \frac{\alpha}{\alpha+1}.$$

Thus, we have

$$E[X^2 + Ye^{-X}] = \frac{2}{\alpha^2} + \frac{\alpha}{\beta(\alpha+1)}.$$

— □

The next example is illustrative of a problem that arises in certain theoretical developments.

Example 14.3.2:

Suppose $\{A, X, Y\}$ is an independent class, with A and X integrable and Y uniformly distributed on $[0, 2\pi]$. Show that

$$E[A\cos(X+Y)] = 0.$$

Proof

Since $\cos(u+v) = \cos u \cos v - \sin u \sin v$, we have by the product rule and linearity

$$E[A\cos(X+Y)] = E[A]E[\cos X]E[\cos Y] - E[A]E[\sin X]E[\sin Y].$$

Now,

$$E[\cos Y] = \frac{1}{2\pi}\int_0^{2\pi} \cos t\, dt = 0 \quad \text{and} \quad E[\sin Y] = \frac{1}{2\pi}\int_0^{2\pi} \sin t\, dt = 0.$$

The proposition follows immediately.

— □

14.4. Some Alternate Forms of Expectation

We obtain several special forms of mathematical expectation listed in the table under (E20). The first result (E20a) is obtained with the use of Fubini's theorem. Forms (E20b) through (E20e) are consequences of the first result. The form (E20f) is based on the quantile function, introduced in Section 11.4.

Suppose F is nondecreasing, right-continuous on $[0, \infty)$. Set

$$F^*(t) = F(t-0) = \lim_{h \to 0+} F(t-h).$$

Then F^* is nondecreasing, left-continuous on $[0, \infty)$, and $F(t) = F^*(t)$ at every point of continuity of F. If M_0 is the set of points of discontinuity of F and $P_X(M_0) = P(X \in M_0) = 0$, then $F(X) = F^*(X)$ a.s and $E[F(X)] = E[F^*(X)]$.

Theorem 14.4.1 (Property (E20a)) *Consider F, nondecreasing, right-continuous on $[0, \infty)$, and set $F^*(t) = F(t-0)$.*
  Suppose $F(0^-) = F^*(0) = 0$. Then

(a) $E[F(X)] = \int_0^\infty P(X \geq t)F(dt)$ (b) $E[F^*(X)] = \int_0^\infty P(X > t)F(dt)$

Proof

F determines a Lebesgue-Stieltjes measure ν, which assigns zero mass to the interval $(-\infty, 0)$. Now

$$\nu(-\infty, t] = \nu[0, t] = F(t) \quad \text{and} \quad \nu(-\infty, t) = \nu[0, t) = F(t-0) = F^*(t).$$

1. Consider the function $Y(t, \omega) = u_+(t)I_{[t,\infty)}(X(\omega))$.
    For fixed $t \geq 0$, $\int Y(t, \omega)\, dP = E[I_{[t,\infty)}(X)] = P(X \geq t)$, so that

    $$\int_0^\infty E[Y(t, \omega)]\nu(dt) = \int_0^\infty P(X \geq t)F(dt).$$

    For fixed ω, $Y(t, \omega) = 1$ for $0 \leq t \leq X(\omega)$ (and is zero elsewhere), so that

    $$\int_0^\infty Y(t, \omega)\nu(dt) = \int_0^{X(\omega)} \nu(dt) = F(X(\omega)).$$

    Hence,

    $$E\left[\int_0^\infty Y(t, \omega)\nu(dt)\right] = E[F(X)].$$

    By Fubini's theorem,

    $$\begin{aligned} E[F(X)] &= E\left[\int_0^\infty Y(t, \omega)\nu(dt)\right] = \int_0^\infty E[Y(t, \omega)]\nu(dt) \\ &= \int_0^\infty P(X \geq t)F(dt). \end{aligned}$$

2. Consider $Z(t, \omega) = I_{(t, \infty)}(X(\omega))$ and argue similarly.

□

Example 14.4.1:

Consider $X \geq 0$ and set $Y = \min\{X, c\}$. Show that $E[Y] = \int_0^c (1 - F_X(t))\, dt$

Proof

We have $Y = F(X)$ where

$$F(t) = \begin{cases} t & \text{for } 0 \leq t \leq c \\ c & \text{for } t > c. \end{cases}$$

Since $F'(t) = 1$ for $t \leq c$, and is zero elsewhere, Theorem 14.4.1 ensures

$$\begin{aligned} E[Y] &= \int_0^\infty P(X > t) F(dt) = \int_0^\infty (1 - F_X(t)) F'(t)\, dt \\ &= \int_0^c (1 - F_X(t))\, dt. \end{aligned}$$

For X exponential (α), $1 - F_X(t) = e^{-\alpha t}$, so that $E[Y] = \int_0^c e^{-\alpha t}\, dt = \frac{1}{\alpha}(1 - e^{-\alpha c})$.

□

Remark: For the exponential case, this solution is obtained in Example 14.2.9, by a different approach.

Example 14.4.2: (see the discussion of the Weibull distribution, Section 12.2)

Suppose X is exponential (α). Calculate $E[X^r]$ for any $r > 0$.

Solution

$X^r = F(X)$, where $F(t) = t^r$. For $t \geq 0$, F satisfies the hypothesis of property (E20a). Since $P(X > t) = e^{-\alpha t}$, we have

$$E[X^r] = \int_0^\infty e^{-\alpha t} F(dt) = \int_0^\infty e^{-\alpha t} r t^{r-1}\, dt = \frac{r\Gamma(r)}{\alpha^r} = \frac{\Gamma(r+1)}{\alpha^r}.$$

□

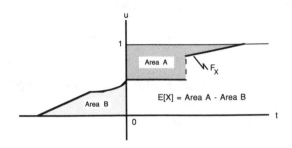

FIGURE 14.5. Graphical interpretation of Theorem 14.4.2.

In the special case $F(t) = t$, we may extend Theorem 14.4.1 as follows.

Theorem 14.4.2 (**Property (E20b)**) *If X is integrable, then $E[X] = \int_{-\infty}^{\infty} \big(u(t) - F_X(t)\big)\, dt.$*

Proof

1. For $X \geq 0$, we use Theorem 14.4.1, with $F(t) = t$ to get

$$E[X] = \int_0^{\infty} \big(1 - F_X(t)\big)\, dt = \int_{-\infty}^{\infty} \big(u(t) - F_X(t)\big)\, dt.$$

2. For the general case, $E[X] = E[X^+] - E[X^-]$, with

$$
\begin{aligned}
E[X^-] &= \int_0^{\infty} P(X^- \geq t)\, dt = \int_0^{\infty} P(-X \geq t)\, dt \\
&= \int_{-\infty}^0 P(X \leq t)\, dt = \int_{-\infty}^0 F_X(t)\, dt.
\end{aligned}
$$

Thus, since $u(t) = 0$ for $t < 0$, we have

$$-E[X^-] = \int_{-\infty}^0 \big(-F_X(t)\big)\, dt = \int_{-\infty}^0 \big(u(t) - F_X(t)\big)\, dt.$$

Using additivity of the integral, we obtain the desired result.

—   □

Remark: This result can be given a useful graphical interpretation. In Figure 14.5,

$$\text{Area } A = \int_0^{\infty} \big(1 - F_X(t)\big)\, dt \quad \text{and} \quad \text{Area } B = \int_{-\infty}^0 F_X(t)\, dt.$$

By Theorem 14.4.2, the difference is $E[X]$.

Example 14.4.3:

Suppose $F_X(t) = \big(1+\sin((\pi t)/2T)\big)/2$ for $-T \le t \le T$. Then $F_X(t) = 0$ for $t < -T$ and $F_X(t) = 1$ for $t > T$. Determine $E[X]$.

Solution

$$u(t) - F_X(t) = \begin{cases} -1/2 - \sin(\pi t/2T) & \text{for } -T \le t \le 0 \\ 1/2 - \sin(\pi t/2t) & \text{for } 0 \le t \le T \\ 0 & \text{elsewhere} \end{cases} .$$

Thus, $h(t) = u(t) - F_X(t)$ is an odd function (i.e., $h(-t) = -h(t)$). From this it follows that

$$E[X] = \int_{-T}^{T} h(t)\, dt = 0.$$

$\square$

Example 14.4.4: Reliability and Expectation

The reliability function R for nonnegative random variable X is given by

$$R(t) = 1 - F_X(t) \qquad \text{for } t \ge 0.$$

We therefore have as an alternate form of the conclusion for Theorem 14.4.2 in the nonnegative case

$$E[X] = \int_0^\infty R(t)\, dt.$$

$\square$

Corollary 14.4.3 (Property (E20c)) If X and Y are integrable, then $E[X - Y] = \int_{-\infty}^\infty \big(F_Y(t) - F_X(t)\big)\, dt.$

Proof

$$\big(u(t) - F_X(t)\big) - \big(u(t) - F_Y(t)\big) = F_Y(t) - F_X(t).$$

$\square$

Remark: This result includes Theorem 14.4.2, since if $Y = 0$ a.s., then $F_Y(t) = u_+(t)$.

Theorem 14.4.4 (Property (E20d)) If $X \ge 0$ is integrable, then

$$\sum_{n=0}^\infty P(X \ge n+1) \le E[X] \le \sum_{n=0}^\infty P(X \ge n) \le N \sum_{k=0}^\infty P(X \ge kN)$$

for any $N \ge 1$.

Proof

Let $E_n = [n, n+1)$ for any $n \geq 0$, and set $F(t) = t$ in Theorem 14.4.1.
Then,

$$E[X] = \int_0^\infty P(X \geq t)\, dt = \sum_{n=0}^\infty \int_{E_n} P(X \geq t)\, dt.$$

The first two inequalities follow from the fact that $P(X \geq t)$ is decreasing
with t and the interval E_n has unit length, so that by the mean-value
theorem

$$P(X \geq n+1) \leq \int_{E_n} P(X \geq t)\, dt \leq P(X \geq n).$$

The third inequality follows similarly from the fact that

$$\int_{kN}^{(k+1)N} P(X \geq t)\, dt \leq N \int_{E_{kN}} P(X \geq t)\, dt \leq N P(X \geq kN).$$

$\square$

Remark: Property (E20d) is primarily used in theoretical arguments. The
special case of integer-valued random variables is more frequently used.

Theorem 14.4.5 (Property (E20e)) *If integrable $X \geq 0$ is integer-valued, then*

(a) $E[X] = \displaystyle\sum_{n=1}^\infty P(X \geq n) = \sum_{n=0}^\infty P(X > n)$.

(b) $E[X^2] = \displaystyle\sum_{n=1}^\infty (2n-1)P(X \geq n) = \sum_{n=0}^\infty (2n+1)P(X > n)$.

Proof

(a) In the proof of Theorem 14.4.4, $P(X > t) = P(X > n)$ on E_n.

(b) Let $F(t) = t^2$ for $t \geq 0$. Then

$$
\begin{aligned}
E[X^2] &= \int_0^\infty P(X > t)F(dt) = 2\int_0^\infty tP(X > t)\, dt \\
&= 2\sum_{n=0}^\infty P(X > n) \int_n^{n+1} t\, dt.
\end{aligned}
$$

The theorem follows from the fact that $\int_n^{n+1} t\, dt = (2n+1)/2$.

$\square$

Example 14.4.5:

Calculate $E[X]$ in the case X has the geometric (p) distribution.

Solution

For the geometric (p) distribution, $P(X = k) = pq^k$ and $P(X \geq k) = q^k$. By (E20e)

$$E[X] = \sum_{n=1}^{\infty} P(X \geq k) = \sum_{k=1}^{\infty} q^k = \frac{q}{1-q} = \frac{q}{p}.$$

$\square$

Example 14.4.6: A Finite Random Variable with Infinite Expectation

A sequence of Bernoulli trials is performed repeatedly. Each sequence is performed until a success is observed, then the next sequence is started. Let X_0 be the waiting time for the first success on the initial sequence, and let X_k be the waiting time for the first success on the kth sequence.

Let N be the number of the first sequence whose waiting time exceeds that of the initial sequence. Determine $E[N]$.

Solution

The sequence $\{X_i : 0 \leq i\}$ is iid geometric (p). Hence, for each $i \geq 0$,

$$P(X_i = k) = pq^k, \quad P(X_i > k) = q^{k+1}, \quad \text{and} \quad P(X_i \leq k) = 1 - q^{k+1}.$$

Now

$$\{N = k\} = \bigcup_{i=0}^{\infty} \{X_0 = i\} \bigcap_{j=1}^{k-1} \{X_j \leq i\}\{X_k > i\}$$

so that

$$P(N = k) = p \sum_{i=0}^{\infty} q^i (1 - q^{i+1})^{k-1} q^{i+1}$$

and

$$P(N \geq k) = \sum_{j=k}^{\infty} \sum_{i=0}^{\infty} pq^i (1 - q^{i+1})^{j-1} q^{i+1}.$$

Since power series converge absolutely, we may interchange the order of summation, to obtain

$$P(N \geq k) \quad = \quad p \sum_{i=0}^{\infty} q^{2i+1} \sum_{j=k}^{\infty} (1 - q^{i+1})^{j-1}$$

$$= p \sum_{i=0}^{\infty} q^{2i+1}(1-q^{i+1})^{k-1} \sum_{j=0}^{\infty} (1-q^{i+1})^j.$$

As a geometric series,

$$\sum_{j=0}^{\infty} (1-q^{i+1})^j = \frac{1}{1-(1-q^{i+1})} = \frac{1}{q^{i+1}}$$

so that

$$P(N \geq k) = p \sum_{i=0}^{\infty} q^i (1-q^{i+1})^{k-1}.$$

By (E20e), for any $r > 0$,

$$E[N] = \sum_{k=1}^{\infty} P(N \geq k) \geq p \sum_{i=0}^{r} q^i \sum_{k=0}^{\infty} (1-q^{i+1})^k = p \sum_{i=0}^{r} \frac{1}{q} \to \infty$$

as $r \to \infty$. Thus, the expectation $E[N]$ cannot be finite.

— $\qquad\qquad\qquad\qquad\qquad\qquad\qquad\qquad\qquad\qquad$ □

Remark: Another application of (E20e) is found in the proof of Wald's equation in Theorem 20.1.1.

Property (E20f) is a consequence of properties of the quantile function, introduced in Section 11.4.

Theorem 14.4.6 (Property (E20f)) *If Q is the quantile function for the distribution function F_X, then*

$$E[g(X)] = \int_0^1 g\big(Q(u)\big)\, du.$$

Proof

The quantile function $Q(u) = \inf\{t : F_X(t) \geq u\}$ has the property that if $Y = Q(U)$, where U is uniformly distributed on $(0,1)$, then $F_Y(t) = F_X(t)$. Hence

$$
\begin{aligned}
E[g(X)] &= E[g(Y)] = E\big[g\big(Q(U)\big)\big] \\
&= \int g\big(Q(u)\big) f_U(u)\, du = \int_0^1 g\big(Q(u)\big)\, du.
\end{aligned}
$$

— $\qquad\qquad\qquad\qquad\qquad\qquad\qquad\qquad\qquad\qquad$ □

Remark: The geometric construction of the graph of Q from the graph of F_X, illustrated in Figure 11.11, yields an alternate derivation of the geometric interpretation of $E[X]$ given in Figure 14.5.

Example 14.4.7:

Suppose $F_X(t) = \sqrt{t}$ for $0 \le t \le 1$. Then $Q(u) = u^2$ and

$$E[X] = \int_0^1 u^2 \, du = \frac{1}{3}.$$

This result could, of course, be obtained in standard fashion, as follows.

$$f_X(t) = \frac{1}{2} t^{-1/2} \qquad \text{for } 0 \le t \le 1 \text{ and is zero elsewhere.}$$

Hence,

$$E[X] = \int t f_X(t) \, dt = \frac{1}{2} \int_0^1 t^{1/2} \, dt = \frac{1}{2} \cdot \frac{2}{3} = \frac{1}{3}.$$

$\square$

14.5. A Special Case of the Radon-Nikodym Theorem

In this section, we obtain a special case of the Radon-Nikodym theorem, which we use in Chapter 19 to establish the notion of conditional expectation, given a random vector.

Theorem 14.5.1 (Property (E19), a Special Case of the Radon-Nikodym Theorem) *If $g(Y)$ is integrable and X is a random vector, then there exists a real-valued Borel function $e(\cdot)$, defined on the range of X, unique a.s. $[P_X]$, such that*

$$E[I_M(X)g(Y)] = E[I_M(X)e(X)]$$

for all Borel sets M on the codomain of X.

IDEAS OF A PROOF

Suppose Y is a nonnegative, integrable random variable and X is any random vector. For each event $C = X^{-1}(M) = \{X \in M\}$, where M is a Borel set on the codomain of X, set

$$\nu(C) = \int_C Y \, dP = E[I_C Y] = E[I_M(X)Y].$$

As a result of the countable additivity of the Lebesgue integral, the set function $\nu(\cdot)$ is a measure on the class of events determined by X. By the integrability property (E9d), if $P(C) = 0$, then $\nu(C) = 0$, so that ν is absolutely continuous with respect to P. By the Radon-Nikodym theorem, there is a random variable Z, whose events are determined by Y, such that

$$\nu(C) = \int_C Z \, dP = E[I_C Z] = E[I_M(X)Z].$$

The random variable is unique, except possibly for a set of ω having zero probability. The events determined by Z are events determined by X; hence, by Theorem 11.1.2, there is a Borel function $e(\cdot)$, defined on the range of X, such that $Z = e(X)$. The a.s. uniqueness of Z implies that $e(X)$ is a.s. unique, and hence that $e(\cdot)$ is unique, except possibly on a set of P_X−measure zero. Thus,

$$E[I_M(X)Y] = E[I_M(X)e(X)]$$

for all Borel sets M on the codomain of X. We may remove the restriction that Y be nonnegative by considering $Y = Y^+ - Y^-$

We may then let Y be any random vector and g be any real-valued function on the codomain of Y such that $E[g(Y)]$ is finite. Upon replacing Y in the preceding argument by the real-valued random variable $g(Y)$, we have (E19).

— $\qquad\qquad\qquad\qquad\qquad\qquad\qquad\qquad\qquad\qquad\qquad\qquad$ $\square$

PROBLEMS

14-1. Show that $E\big[(X - E[X])(Y - E[Y])\big] = E[XY] - E[X]E[Y]$.

14-2. Use expectations and indicator functions to show that

$$
\begin{aligned}
P(A \cup B \cup C) \;=\;\; & P(A) + P(B) + P(C) \\
& - P(AB) - P(AC) - P(BC) \\
& + P(ABC).
\end{aligned}
$$

14-3. If the pair $\{X, Y\}$ is independent, A is an event determined by X (i.e., $\in \sigma(X)$), and B is an event determined by Y, show that $E[I_{AB}XY] = E[I_A X]E[I_B Y]$.

14-4. Show that for positive integers m, n, $E[|X|^n] \le 1 + E[|X|^{n+m}]$.

14-5. A sequence of n independent trials is performed. If E_i is the event of a success on the ith trial, then the "payoff" random variable for that trial is $X_i = aI_{E_i} + bI_{E_i^c}$. Let N be the number of successes in n trials and W be the net payoff. Determine $E[N]$ and $E[W]$. Express $E[W]$ in terms of $E[N]$.

14-6. A sports fan bets \$10 on each of three sporting events. Let E_i be the event he wins the ith bet. Suppose the class $\{E_1, E_2, E_3\}$ is independent and

$$
\begin{array}{lll}
P(E_1) = 0.2 & \text{and the payoff is} & \$100 \\
P(E_2) = 0.3 & \text{and the payoff is} & \$50 \\
P(E_3) = 0.5 & \text{and the payoff is} & \$20
\end{array}
$$

Let G be the random variable that expresses the net gain. Determine $P(G > 0)$ and $E[G]$.

14-7. A family is moving and has one month to sell its home. The market is such that the best offer may be considered a random variable X uniformly distributed between 100 and 150 (thousands of dollars). The family has a firm offer of 110, good any time during the month, so that if it is not offered a better price it will sell for 110. Let Y be the selling price. Determine $E[Y]$.

14-8. A device has time to failure X, in hours, exponentially distributed with $E[X] = 100$. The device will be used a maximum of 150 hours before replacement. It will be replaced immediately upon failure or at 150 hours, whichever comes first. Let Y be the time of replacement. Determine $E[Y]$.

14–9. The amount of a given item sold during the Christmas season is represented by a nonnegative random variable X. An amount M of the item is ordered at the beginning of the season. A fixed cost sM is associated with the ordering, handling, and storage. Revenue from sales is gX. If demand exceeds supply, a loss is experienced. This may be considered to be $c(X - M)$. The net gain on sales may thus be represented

$$G = \begin{cases} gX - sM & \text{for } 0 \le X \le M \\ (g - s)M - c(X - M) & \text{for } X > M \end{cases}.$$

(a) Express $E[G]$ in terms of $E[X]$, $P(X > M)$, g, s, c, and M.

(b) Obtain an expression for $E[G]$ in the case X is exponential (λ).

14–10. $\{X, Y\}$ has joint distribution uniform on the triangular region with vertices $(0, 0)$, $(2, 0)$, and $(0, 2)$.

Let $W = \max\{X, Y\}$.   Determine $E[X]$ and $E[W]$.

14–11. Let $V = \min\{X, Y\}$ and $W = \max\{X, Y\}$.

(a) Determine Q such that

$$E[V] = E[I_Q(X, Y)X] + E[I_{Q^c}(X, Y)Y]$$

and

$$E[W] = E[I_Q(X, Y)Y] + E[I_{Q^c}(X, Y)X].$$

(b) Determine Q_v such that $F_V(v) = E[I_{Q_v}(X, Y)]$.

(c) Determine M_v such that $F_W(v) = E[I_{M_v}(X, Y)]$.

14–12. The pair $\{X, Y\}$ has joint density $f_{XY}(t, u) = \frac{1}{28}(4t + 2u + 1)$ for $0 \le t \le 2$, $0 \le u \le 2$.

Let $W = \begin{cases} X & \text{for } X \le 1 \\ Y & \text{for } X > 1 \end{cases}$.

Determine F_W and $E[W]$.

14–13. Suppose $f_{XY}(t, u) = (1 - u)e^{-t/2}$ for $0 \le u \le 1$, $0 \le t$. Determine $E[X]$, $E[Y]$, $E[XY]$, $E[X - Y]$, and $E[X^2]$.

14–14. The pair $\{X, Y\}$ has joint distribution uniform on the parallelogram with vertices $(1, 0)$, $(1, 2)$, $(-1, 0)$, and $(-1, -2)$.

(a) Determine $E[X]$ and $E[Y]$. *Hint.* How are X and Y distributed?

(b) Let $W = \begin{cases} X & \text{for } X + Y \ge 1 \\ 2XY & \text{for } X + Y < 1 \end{cases}$.

Determine $E[W]$.

14–15. The pair $\{X, Y\}$ has joint density function $f_{XY}(t, u) = 8tu$ on the triangle bounded by the lines $u = 0$, $t = 1$, and $u = t$.

(a) Determine the marginal density functions f_X and f_Y and the mean values $E[X]$ and $E[Y]$.

(b) Determine $E[X^2 + 3XY]$.

(c) Let $Z = \begin{cases} P4X^2 & \text{for } X + Y \leq 1 \\ 2Y & \text{for } X + Y > 1 \end{cases}$.

Set up the integrals to determine $E[Z]$. Give careful attention to the limits of integration as well as the expressions for the integrands. It is not necessary to evaluate the integrals.

14–16. A regular rider of the city busses notices that the factors, such as weather, that tend to make him late (or early) similarly influence the time of arrival of the bus. Also, factors that delay the bus tend to produce more variability in the arrival time of the bus. Let T_0 be the scheduled time of arrival of the bus, X be the rider's arrival time, and Y be the arrival time of the bus. The rider assumes that the joint density function f_{XY} is given by

$$f_{XY}(t, u) = \frac{1}{12} \quad \text{for } T_0 - 2 \leq t \leq T_0 + 2, \ t - 1 \leq u \leq \frac{1}{2}(3t - T_0 + 4).$$

The event the rider arrives in time to catch the bus is $A = \{X \leq Y\}$. Determine $E[X]$ and $P(A) = E[I_Q(X, Y)]$.

14–17. A salesman deals principally with two items. Let X and Y be the respective numbers of the two items sold in a week. Sales record indicate the joint density f_{XY} can be approximated by $f_{XY}(t, u) = 3u/4000$ for (t, u) on the triangular region bounded by the lines $u = 0$, $u = t$, and $t = 20$. For total sales $X + Y \leq 20$, the salesman's commission is \$20 per item sold. For total sales greater than 20, he gets a bonus of \$5 on each item over 20. What is his expected weekly income?

That is, let $Z = \begin{cases} 20(X + Y) & \text{for } X + Y \leq 20 \\ 25(X + Y) - 100 & \text{for } X + Y > 20 \end{cases}$.

Determine $E[Z]$.

14–18. Suppose $f_{XY}(t, u) = 4t(1 - u)$ for $0 \leq t \leq 1$, $0 \leq u \leq 1$. Let $D = |X - Y|$. Determine $E[D]$.

14–19. Suppose $\{X_i : 1 \leq i \leq n\}$ is iid, with each X_i integrable. Let $S_n = \sum_{i=1}^{n} X_i$. Show that $E[X_i/S_n] = 1/n$, so that $E[(X_{i_1} + X_{i_2} + \cdots + X_{i_k})/S_n] = k/n$.

Hint. Show $E[X_k/S_n] = c$, invariant with k. Note that $X_k/S_n = g(X_k, S_{kn})$, where $S_{kn} = S_n - X_k$. Thus, each pair $\{X_k, S_{kn}\}$ has the same joint distribution.

14-20. The pair $\{X, Y\}$ of random variables has joint probability density $f_{XY}(t, u) = 1$ on the triangular region with vertices at $(0,0)$, $(1,0)$, and $(1,2)$, bounded by the lines $u = 0$, $t = 1$, and $u = 2t$.

Let $W = \begin{cases} X & \text{for } X + Y \le 1 \\ Y & \text{for } X + Y > 1 \end{cases}$.

Determine $E[W^2]$.

14-21. Suppose the joint density f_{XY} is constant over the square with vertices $(-1, 0)$, $(0, -1)$, $(1, 0)$, $(0, 1)$.

Let $Z = \begin{cases} X^2 Y & \text{for } X \le 0 \\ XY^2 & \text{for } X > 0 \end{cases}$.

Determine $E[Z]$.

14-22. Suppose $f_{XY}(t, u) = 1/2$ in the triangular region $0 \le t \le u \le 2$ (and zero elsewhere).

Let $Z = \begin{cases} X + Y & \text{for } Y \le 1 \\ X + 1 & \text{for } Y > 1 \end{cases}$.

Determine $E[Z]$.

14-23. Construct an example for which $E[g(X)h(Y)] = E[g(X)]E[h(Y)]$, but $\{X, Y\}$ is *not* independent.

15

Variance and Standard Deviation

In Chapter 12, we show that the mean value $\mu_X = E[X]$ of a random variable locates the center of mass of the induced probability distribution. As the center of mass, the mean value provides important information about the distribution, hence about the values likely to be taken on by the random variable. However, this location of the center of mass is not a sufficient characterization of the distribution for most purposes. Quite different probability distributions may share the same mean value. For example, the random variable X uniformly distributed on the interval $(-0.1, 0.1)$ and the random variable Y uniformly distributed on the interval $(-100, 100)$ both have zero mean value. But there is far less uncertainty about where to look for an observed value of X than there there is for Y. The difference is the spread of probability mass (hence possible values) about the mean. It is highly desirable to have an appropriate measure of this spread or variation.

15.1. Variance as a Measure of Spread

There are several possibilities for measures of the spread of the probability mass about the mean value (center of mass). One approach is to find an "average distance" from the mean value to the possible values of the random variable. Thus

$$E[|X - \mu_X|]$$

is a possible measure. If it were desired to weight large excursions more heavily than small ones in determining the spread, this could be modified by considering

$$E[|X - \mu_X|^p] \qquad p > 1.$$

It turns out that the choice of $p = 2$ is a highly satisfactory one. For one thing, it allows dropping the absolute value, which makes the mathematical handling more tractable. Also, with this choice of p, the mathematical entity has many useful and significant properties. We therefore define the following.

Definition 15.1.1. *The* variance *of random variable X, denoted* $\text{Var}[X]$ *or $\sigma^2[X]$ or σ_X^2, is*

$$\text{Var}[X] = E[(X - \mu_X)^2] \qquad \text{where} \quad \mu_X = E[X].$$

Remark: According to the result in Example 14.2.6, the choice of $a = \mu_X$ makes $E[(X - a)^2]$ a minimum.

SIMPLE RANDOM VARIABLE

The variance is the average of the squares of the distances from the mean value μ_X to the possible values of the random variable. This concept is quite straightforward and exact for a simple random variable. Suppose

$$X = \sum_{i=1}^{n} t_i I_{A_i} \quad \text{(canonical form) with } P(A_i) = P(X = t_i) = p_i, \ 1 \le i \le n.$$

Then the probability distribution consists of probability masses $p_1, p_2, \ldots, p_n$ located at points $t_1, t_2, \ldots, t_n$, respectively. The point μ_X is located at the center of mass for this distribution. Now

$$\text{Var}[X] = E[(X - \mu_X)^2] = (t_1 - \mu_X)^2 p_1 + (t_2 - \mu_X)^2 p_2 + \cdots + (t_n - \mu_X)^2 p_n.$$

In terms of the mass representation of probability, the variance may be interpreted physically as the *second moment* of the probability mass about the center of mass. This is known as the second central moment or the *moment of inertia* about the center of mass. Since the total mass is one, this is equal to the square of the radius of gyration.

As in the case of the mean value, this interpretation can be extended by intuitive arguments to more general distributions.

As a measure of the average spread of the probability distribution, we take the square root of the variance.

Definition 15.1.2. *The* standard deviation *of random variable X, denoted σ_X or $\sigma[X]$, is the square root of the variance.*

Example 15.1.1:

Consider the two distributions described in the introductory statement. X is uniform $(-0.1, 0.1)$ and Y is uniform $(-100, 100)$. From symmetry it is obvious that $\mu_X = \mu_Y = 0$, so that

$$\text{Var}[X] = E[X^2] \qquad \text{and} \qquad \text{Var}[Y] = E[Y^2].$$

Thus

$$\text{Var}[X] \quad = \quad \frac{1}{0.2} \int_{-0.1}^{0.1} t^2 \, dt = \frac{1}{300}$$

$$\text{Var}[Y] \quad = \quad \frac{1}{200} \int_{-100}^{100} t^2 \, dt = \frac{10,000}{3}.$$

The standard deviations are $\sigma_X = 1/10\sqrt{3}$ and $\sigma_Y = 100/\sqrt{3}$, indicating that the average spread for values of Y is 1000 times greater than the spread of values of X.

—   □

15.2. Some Properties

The variance for a random variable (or its distribution) may be calculated by direct use of the defining expression. That is, we may calculate the expectation of the indicated function of the random variable, using a form suitable for the type of distribution. However, before calculating variances for some common distributions, we find it expedient to derive a number of properties that may be exploited to simplify the process.

V1) $\mathrm{Var}[X] = E[X^2] - E^2[X] = E[X^2] - \mu_X^2$.

V2) $\mathrm{Var}[aX] = a^2\,\mathrm{Var}[X]$ for any constant a.

V3) $\mathrm{Var}[X + b] = \mathrm{Var}[X]$ for any constant b.

V4) $\mathrm{Var}\left[\sum_i a_i X_i\right] = \sum_i a_i^2\,\mathrm{Var}[X_i] + 2\sum_{i<j} a_i a_j c_{ij}$
where $c_{ij} = E[X_i X_j] - E[X_i]E[X_j]$.

In particular, for two variables X and Y, we have

$$\mathrm{Var}[X \pm Y] = \mathrm{Var}[X] + \mathrm{Var}[Y] \pm 2\big(E[XY] - E[X]E[Y]\big).$$

As we see in the next chapter, the constants c_{ij} provide a measure of interdependence of X_i and X_j. The case that the c_{ij} are all zero is important, since in that case the variances of the linear combination is much simpler.

Definition 15.2.1. *The pair* $\{X_i, X_j\}$ *is* uncorrelated *iff* $c_{ij} = 0$.

Remarks

1. If the pair $\{X_i, X_j\}$ is independent then it is uncorrelated. The converse is not true, since there are uncorrelated random variables that are not independent (see the remark after Example 16.2.1).

2. If the $a_i = \pm 1$ for all i and $\{X_i, X_j\}$ is uncorrelated for $i \neq j$, then the variance of the combination is the sum of the variances. In particular, if $\{X, Y\}\}$ is uncorrelated, then $\mathrm{Var}[X \pm Y] = \mathrm{Var}[X] + \mathrm{Var}[Y]$. Note that the variance for the sum is the same as the variance for the difference. This case occurs in many important applications.

PROOFS, DISCUSSIONS, AND INTERPRETATIONS

V1) is a direct consequence of part (ii) of Example 14.2.6. This is the form most often used for calculating variances, as the examples in the next section show.

V2) is a formal expression of the fact that multiplication of X by a real constant produces a change of scale on the probability mass distribution, with a corresponding change of the average square of the spread of this mass about its center. This intuitive argument may be substantiated precisely, with the aid of (V1) and linearity of expectation.

$$\text{Var}[aX] = E[(aX)^2] - E^2[aX] = a^2 E[X^2] - a^2 E^2[X] = a^2 \text{Var}[X].$$

V3) expresses the fact that adding a constant to a random variable shifts the distribution and the center of mass by the amount of that constant, without changing the spread of the mass about its center. Formally, we use linearity and (V1) to argue

$$\begin{aligned}
\text{Var}[X + b] &= E[(X + b)^2] - E^2[X + b] \\
&= E[X^2 + 2bX + b^2] - (E[X] + b)^2 \\
&= E[X^2] + 2bE[X] + b^2 - E^2[X] - 2bE[X] - b^2 \\
&= E[X^2] - E^2[X] = \text{Var}[X].
\end{aligned}$$

V4) is not particularly intuitive. It does show that for uncorrelated random variables, the variances for sums or differences add (see the second remark after the preceding definition).

To establish the property, consider $X = \sum_i a_i X_i$. Then

$$X^2 = \sum_i a_i^2 X_i^2 + \sum_{i \neq j} a_i a_j X_i X_j = \sum_i a_i^2 X_i^2 + 2 \sum_{i < j} a_i a_j X_i X_j.$$

Hence, by linearity

$$E[X^2] = \sum_i a_i^2 E[X_i^2] + 2 \sum_{i < j} a_i a_j E[X_i X_j].$$

Also, by a similar algebraic pattern

$$E^2[X] = \left(\sum_i a_i E[X_i] \right)^2 = \sum_i a_i^2 E^2[X_i] + 2 \sum_{i < j} a_i a_j E[X_i] E[X_j].$$

Using (V1) and taking the difference we obtain

$$\text{Var}[X] = \sum_i a_i^2 \left(E[X_i^2] - E^2[X_i] \right) + 2 \sum_{i < j} a_i a_j \left(E[X_i X_j] - E[X_i] E[X_j] \right).$$

Again applying (V1) to the first sum and the definition of c_{ij} to the second sum, we get (V4).

$\square$

These properties are used repeatedly in dealing with variances and standard deviations.

15.3.   Variances for Some Common Distributions

We use the properties to calculate the variances for some common distributions. In doing so, we both illustrate technique and obtain useful results.

DISCRETE DISTRIBUTIONS

1. *Indicator function* $X = I_E$ $P(X = 1) = P(E) = p$, $P(X = 0) = q = 1 - p$.

   $$E[X] = p \quad \text{and} \quad E[X^2] = p, \quad \text{since } X = X^2.$$

   Using (V1), we find

   $$\text{Var}[X] = p - p^2 = p(1 - p) = pq.$$

2. *Simple random variable* $X = \sum_{i=1}^{n} t_i I_{A_i}$ (canonical form)
   $P(X = t_i) = P(A_i) = p_i$.

   We apply (V4) to the linear combination of indicator functions and use the variance of the indicator function, as determined earlier. First we note that since $I_{A_i} I_{A_j} = 0$ for $i \neq j$, we have

   $$c_{ij} = E[I_{A_i} I_{A_j}] - E[I_{A_i}]E[I_{A_j}] = -p_i p_j \quad \text{for } i \neq j$$

   $$\text{Var}[X] = \sum_{i=1}^{n} t_i^2 \, \text{Var}[I_{A_i}] + 2 \sum_{i<j} t_i t_j c_{ij} = \sum_{i=1}^{n} t_i^2 p_i q_i - 2 \sum_{i<j} t_i t_j p_i p_j.$$

3. *Binomial* (n, p) $X = \sum_{i=1}^{n} I_{E_i}$ with $\{I_{E_i} : 1 \leq i \leq n\}$ iid
   $P(E_i) = p$.

   Since the indicator functions are independent, the variance for the sum is the sum of the variances. Hence

   $$\text{Var}[X] = n \, \text{Var}[I_{E_i}] = npq.$$

4. *Geometric* (p) $P(X = k) = pq^k$ for all $k \geq 0$ $E[X] = q/p$.

   $$E[X^2] = p \sum_{k=0}^{\infty} k^2 q^k = pq^2 \sum_{k=2}^{\infty} k(k - 1)q^{k-2} + pq \sum_{k=1}^{\infty} kq^{k-1}.$$

   Use of results on geometric series (see the appendix on mathematical aids) gives

   $$E[X^2] = \frac{2pq^2}{(1 - q)^3} + \frac{pq}{(1 - q)^2} = \frac{2q^2 + pq}{p^2}.$$

Hence,

$$\text{Var}[X] = E[X^2] - E^2[X] = \frac{2q^2 + pq}{p^2} - \frac{q^2}{p^2} = \frac{q}{p^2}.$$

If $Y - 1$ is geometric (p), then $\text{Var}[Y] = \text{Var}[X] = q/p^2$.

5. *Negative binomial* (m, p).

This distribution is best handled with the aid of the moment-generating function in Section 18.2. With the aid of that tool, it is shown that X has the same distribution as the sum of m independent random variables, each geometric (p). Thus

$$\text{Var}[X] = m \cdot \frac{q}{p^2}.$$

6. *Poisson* (μ) $P(X = k) = p_k = e^{-\mu} \mu^k / k!$ for all $k \geq 0$, $E[X] = \mu$.

We use a device similar to that for the geometric distribution to write

$$E[X^2] = E[X(X - 1) + X] = e^{-\mu} \sum_{k=2}^{\infty} k(k-1) \frac{\mu^k}{k!} + \mu.$$

Using a result listed in the appendix on mathematical aids under the exponential series, we have

$$E[X^2] = e^{-\mu} \mu^2 \sum_{k=0}^{\infty} \frac{\mu^k}{k!} + \mu = e^{-\mu} \mu^2 e^{\mu} + \mu = \mu^2 + \mu.$$

Thus,

$$\text{Var}[X] = E[X^2] - \mu^2 = \mu.$$

Absolutely Continuous Distributions

1. *Uniform* (a, b) $f_X(t) = 1/(b - a)$, $a < t < b$ (zero elsewhere)
$E[X] = (b + a)/2$.

$$E[X^2] = \int_a^b t^2 f_X(t)\, dt = \frac{1}{b - a} \int_a^b t^2\, dt = \frac{b^3 - a^3}{3(b - a)} = \frac{b^2 + ab + a^2}{3}.$$

Hence,

$$\text{Var}[X] = \frac{b^2 + ab + a^2}{3} - \frac{a^2 + 2ab + b^2}{4} = \frac{(b - a)^2}{12}.$$

2. *Symmetric triangular* (a, b).

Since shifting the distribution does not change the variance, we consider the symmetric triangular distribution $(-c, c)$ where $c = (b - a)/2$.

For this case, the density function f_X is an even function and $E[X] = 0$. Thus

$$\text{Var}[X] = E[X^2] = \int_{-c}^{c} t^2 f_X(t)\, dt = 2 \int_0^c t^2 f_X(t)\, dt.$$

Since $f_X(t) = 1/c - t/c^2$ for $0 \le t \le c$, we have

$$\text{Var}[X] = 2 \int_0^c \left(\frac{t^2}{c} - \frac{t^3}{c^2} \right) dt = \frac{c^2}{6} = \frac{(b-a)^2}{24}.$$

3. *Exponential* (λ) $f_X(t) = \lambda e^{-\lambda t}$ for $t \ge 0$, $E[X] = 1/\lambda$.

$$E[X^2] = \int_0^\infty \lambda t^2 e^{-\lambda t}\, dt = \frac{2}{\lambda^2} \quad \text{so that} \quad \text{Var}[X] = \frac{2}{\lambda^2} - \frac{1}{\lambda^2} = \frac{1}{\lambda^2}.$$

4. *Gamma* (α, λ) $f_X(t) = \lambda^\alpha t^{\alpha-1} e^{-\lambda t} / \Gamma(\alpha)$ for $t \ge 0$, $E[X] = \alpha/\lambda$.

$$E[X^2] = \frac{1}{\Gamma(\alpha)} \int_0^\infty \lambda^\alpha t^{\alpha+1} e^{-\alpha t}\, dt = \frac{\Gamma(\alpha+2)}{\lambda^2 \Gamma(\alpha)} = \frac{(\alpha+1)\alpha}{\lambda^2}.$$

From this it follows that

$$\text{Var}[X] = \frac{\alpha^2 + \alpha}{\lambda^2} - \frac{\alpha^2}{\lambda^2} = \frac{\alpha}{\lambda^2}.$$

5. *Beta* (r, s) $f_X(t) = \Gamma(r+s)/(\Gamma(r)\Gamma(s)) t^{r-1} (1-t)^{s-1}$ for $0 < t < 1$, $E[X] = r/(r+s)$.

Using the general result on $E[X^m]$ in Chapter 12, we have

$$E[X^2] = \frac{\Gamma(r+s)}{\Gamma(r)\Gamma(s)} \cdot \frac{\Gamma(r+2)\Gamma(s)}{\Gamma(r+s+2)} = \frac{r(r+1)}{(r+s)(r+s+1)}.$$

From this it is a matter of algebra to obtain

$$\text{Var}[X] = \frac{r(r+1)}{(r+s)(r+s+1)} - \frac{r^2}{(r+s)^2} = \frac{rs}{(r+s)^2 (r+s+1)}.$$

6. *Weibull* (α, λ, ν)

In Section 11.1 we show that if Y is exponential (1), then $1/\lambda^{1/\alpha} Y^{1/\alpha} + \nu$ is Weibull (α, λ, ν). Also, in Appendix A we show that if Y is exponential (λ), then $E[Y^r] = \Gamma(r+1)/\lambda^r$. Since the variance is indifferent to a shift (i.e., to the addition of a constant), we consider

$$X = \frac{1}{\lambda^{1/\alpha}} Y^{1/\alpha}$$

so that

$$E[X] = \frac{1}{\lambda^{1/\alpha}}\Gamma\left(\frac{1}{\alpha}+1\right) \quad \text{and} \quad E[X^2] = \frac{1}{\lambda^{2/\alpha}}\Gamma\left(\frac{2}{\alpha}+1\right).$$

From this it follows by (V1) that

$$\text{Var}[X] = \frac{1}{\lambda^{2/\alpha}}\left(\Gamma\left(\frac{2}{\alpha}+1\right) - \Gamma^2\left(\frac{1}{\alpha}+1\right)\right).$$

7. *Normal* (μ, σ) Designated $X \sim N(\mu, \sigma^2)$, $E[X] = \mu$.

For $Y \sim N(0,1)$, the density is $\phi(t) = (1/\sqrt{2\pi})e^{-t^2/2}$ and $E[Y] = 0$.
Thus, in this case

$$\text{Var}[Y] = E[Y^2] = \frac{1}{\sqrt{2\pi}}\int_{-\infty}^{\infty} t^2 e^{-t^2/2}\, dt = \frac{2}{\sqrt{2\pi}}\int_0^{\infty} t^2 e^{-t^2/2}\, dt = 1.$$

The value of the last integral is found in a table of integrals. $X \sim N(\mu, \sigma^2)$ iff $Y = X - \mu/\sigma \sim N(0,1)$. Since $X = \sigma Y + \mu$, we have by (V2) and (V3)

$$\text{Var}[X] = \sigma^2 \text{Var}[Y] = \sigma^2.$$

—   □

The variance of a function of a random variable may be calculated by the use of (V1) and standard techniques for expectation. As examples, we consider again the random variables in Examples 14.2.10 and 14.2.11.

Example 15.3.1: (See Example 14.2.10 and Figure 14.2)

Suppose $f_{XY}(t,u) = 2$ on the triangular region $1 \le t \le u \le 2$ (zero elsewhere). See Figure 14.2.

$$Z = \begin{cases} X & \text{for } X \le 3/2 \\ 2XY & \text{for } X > 3/2 \end{cases} = I_Q(X,Y)X + I_{Q^c}(X,Y)2XY.$$

From Example 14.2.10 we have $E[X] = 4/3$, $E[Y] = 5/3$, and $E[Z] = 235/96$.

Determine $\text{Var}[X]$, $\text{Var}[Y]$, and $\text{Var}[Z]$.

Solution

$$E[X^2] = 2\int_1^2 \int_t^2 t^2\, du\, dt = 2\int_1^2 t^2(2-t)\, dt = \frac{11}{6} \approx 1.833$$

so that

$$\text{Var}[X] = \frac{11}{6} - \left(\frac{4}{3}\right)^2 = \frac{1}{18}$$

$$E[Y^2] = 2 \int_1^2 \int_t^2 u^2 \, du \, dt = \frac{2}{3} \int_1^2 (8 - t^3) \, dt = \frac{17}{6},$$

which implies that

$$\text{Var}[Y] = \frac{17}{6} - \frac{25}{9} = \frac{1}{18}.$$

Now $Z^2 = I_M(X)X^2 + 4I_{M^c}(X)X^2Y^2$, so that the integral for $E[Z^2]$ is obtained by suitably modifying the integral for $E[Z]$. Thus

$$E[Z^2] = \int_1^{3/2} 2t^2(2 - t) \, dt + 8 \int_{3/2}^2 \int_t^2 t^2 u^2 \, du \, dt = \frac{3065}{288}$$

so that

$$\text{Var}[Z] = \frac{3065}{288} - \frac{235^2}{96^2} \approx 4.65.$$

—   □

Example 15.3.2: (See Example 14.2.11 and Figure 14.3)

Suppose $f_{XY}(t, u) = 3$ for $0 \le u \le t^2 \le 1$. See Figure 14.3.

$$Z = \begin{cases} X & \text{for } X + Y \le 1 \\ 1 & \text{for } X + Y > 1 \end{cases} = I_Q(X, Y)X + I_{Q^c}(X, Y).$$

where

$$Q = \{(t, u) : u + t \le 1\} = \{(t, u) : u \le 1 - t\}.$$

From Example 14.2.11 we have

$$E[X] = \frac{3}{4} \qquad E[Y] = \frac{3}{10} \qquad \text{and} \qquad E[Z] = \frac{3}{4}(5t_0 - 2)$$

with $t_0 = (\sqrt{5} - 1)/2$ since $t_0^2 = 1 - t_0$. Determine $\text{Var}[X]$, $\text{Var}[Y]$, and $\text{Var}[Z]$.

Solution

We use the calculating formula (V1) and results from Example 14.2.11.

$$E[X^2] = 3 \int_0^1 t^2 \int_0^{t^2} du \, dt = 3 \int_0^1 t^4 \, dt = \frac{3}{5}.$$

Thus

$$\text{Var}[X] = \frac{3}{5} - \left(\frac{3}{4}\right)^2 = \frac{3}{80} = 0.0375.$$

Similarly,

$$E[Y^2] = 3 \int_0^1 \int_0^{t^2} u^2 \, du \, dt = 3 \int_0^1 \frac{t^6}{3} \, dt = \frac{1}{7}$$

so that

$$\text{Var}[Y] = \frac{1}{7} - \left(\frac{3}{10}\right)^2 = \frac{37}{700} \approx 0.0529.$$

To evaluate $E[Z^2]$ we use the integral expression for $E[Z]$ with t replaced by t^2 and the fact that $t_0^2 = 1 - t_0$. Thus

$$
\begin{aligned}
E[Z^2] &= 3\int_0^{t_0} t^2 \int_0^{t^2} du\, dt + 3\int_{t_0}^1 t^2 \int_0^{1-t} du\, dt + 3\int_{t_0}^1 \int_{1-t}^{t^2} du\, dt \\
&= \frac{100t_0 + 16}{80}.
\end{aligned}
$$

Now

$$E^2[Z] = \frac{9}{16}(25t_0^2 - 20t_0 + 4) = \frac{-2025t_0 + 1305}{80}$$

so that

$$\text{Var}[Z] = E[Z^2] - E^2[Z] = \frac{2125t_0 - 1289}{80} \approx 0.304.$$

The standard deviation $\sigma_Z = \sqrt{0.304} \approx 0.551.$

—   □

15.4. Standardized Variables and the Chebyshev Inequality

Experience has shown that the standard deviation σ for a random variable is a natural measure of the spread of the probability mass, hence the spread of the possible values of the variable. We first encounter this idea in the relation of the normal distribution to the standardized normal distribution. Suppose X is $N(\mu, \sigma^2)$. We show in Section 8.5, and again in Section 11.1, that

$$Y = \frac{X - \mu}{\sigma} \sim N(0, 1).$$

Each time X varies from its mean value μ by $k\sigma$ units, the variable Y varies from its mean value 0 by k units. The essential properties of the normal distribution are contained in the distribution for the standardized variable. As we know, this is what makes possible the use of tables of the distribution for the standardized variable to serve for any normally distributed random variable. The central limit theorem, discussed in Section 18.6, shows the extreme importance of the normal distribution in many practical situations. Certainly the normal distribution plays a significant role in large sample statistics (see the brief discussion in Section 18.7).

A second indication of the significance of the standard deviation is provided by the widely used Chebyshev inequality.

Theorem 15.4.1 (Chebyshev Inequality) *If X is a real random variable with mean value $E[X] = \mu$ and variance $\mathrm{Var}[X] = \sigma^2$ (i.e., standard deviation σ), then*

$$P(|X - \mu| \geq a\sigma) = P\left(\frac{|X - \mu|}{\sigma} \geq a\right) \leq \frac{1}{a^2} \qquad \textit{for } a > 0$$

or, equivalently, with $b = a\sigma$,

$$P(|X - \mu| \geq b) \leq \frac{\sigma^2}{b^2} \qquad \textit{for } b > 0.$$

Proof

We note that since $b > 0$,

$$|X - \mu| \geq b \qquad \text{iff} \qquad (X - \mu)^2 \geq b^2.$$

Let

$$g(t) = (t - \mu)^2 \qquad \text{so that} \quad E[g(X)] = \sigma^2.$$

Put

$$A = \{\omega : g(X(\omega)) \geq b^2\} = \{(X - \mu)^2 > b^2\} = \{|X - \mu| \geq b\}.$$

By considering first $\omega \in A^c$ then $\omega \in A$, we see that for all ω, $b^2 I_A(\omega) \leq g(X(\omega))$. By monotonicity (E3b) and the basic property (E1), we have

$$E[b^2 I_A] = b^2 P(A) = b^2 P(|X - \mu| \geq b) \leq E[g(X)] = \sigma^2.$$

Division by b^2 gives the second inequality in the conclusion. Substitution of $a\sigma = b$ in this inequality gives the first inequality.

$$\square$$

Remark: This result further suggests that the proper unit of measure of spread of probability mass about the mean (the center of mass) is the standard deviation σ. The Chebyshev inequality is useful in certain theoretical investigations, since it requires nothing of the distribution for X except that it have a mean value and a variance (i.e., finite first and second moments). As an example of its theoretical usefulness, see the discussion of the sample average in Section 18.7.

The generality and ease of calculation of the Chebyshev estimate of the probability $P(|X - \mu| \geq b)$ should not lead to the conclusion that it is always the best estimate. If there is specific knowledge of the distribution of X, it may be possible to calculate the probability more precisely. And this more exact result may be much smaller than the Chebyshev estimate, which is essentially independent of the particular distribution.

Example 15.4.1: Chebyshev Estimate and Exact Value for a Normal Distribution

Suppose X is $N(20, 16)$. Obtain the Chebyshev estimate of

$$P(|X - 20| \geq 6)$$

and compare with the value determined from the actual distribution.

Solution

By Chebyshev's inequality

$$P(|X - 20| \geq 6) = P(|X - 20| \geq 1.5\sigma) \leq \frac{1}{1.5^2} = \frac{4}{9} \approx 0.444.$$

If we make use of the fact that X is normally distributed, then

$$P(|X - 20| \geq 6) = P\left(\frac{|X - 20|}{4} \geq 1.5\right) = 2(1 - \Phi(1.5)) \approx 0.134$$

[from table]. As anticipated, the value calculated from the known distribution is smaller than the Chebyshev estimate, which is based only on knowledge of μ and σ. Since in most applications we would desire to know how small the probability is, the Chebyshev estimate is not particularly good. But if the distribution is not known, then the Chebyshev estimate is better than none and is often sufficient for the purposes at hand.

—   □

CENTERED AND STANDARDIZED RANDOM VARIABLES

The variance depends upon the spread of the probability mass about the mean value. The appropriate unit for measuring spread seems to be the standard deviation. These facts suggest the usefulness of the following random variables related to random variable X with mean value μ and standard deviation σ.

Definition 15.4.2. *A random variable X' is said to be* centered *iff its mean value $E[X] = 0$.*

A random variable X^* is said to be standardized *iff it is centered and its variance $\text{Var}[X^*] = 1$.*

Any random variable with finite mean and variance may be centered and standardized by the following procedures.

Theorem 15.4.3 (Centering and Standardizing a Random Variable) *Suppose random variable X has mean value μ and variance σ^2. Then*

$X' = X - \mu$ is centered.

$X^* = \dfrac{X'}{\sigma} = \dfrac{X - \mu}{\sigma}$ is standardized.

Proof

To show X' is centered, we show its mean value is zero, as follows

$$E[X'] = E[X - \mu] = E[X] - \mu = \mu - \mu = 0.$$

We note that by (V3), $\mathrm{Var}[X'] = \mathrm{Var}[X - \mu] = \mathrm{Var}[X] = \sigma^2$. From this and (V2) it follows that

$$\mathrm{Var}[X^*] = \mathrm{Var}\left[\frac{1}{\sigma}X'\right] = \frac{1}{\sigma^2}\mathrm{Var}[X'] = \frac{\sigma^2}{\sigma^2} = 1.$$

It is also readily apparent that

$$E[X^*] = \frac{1}{\sigma}E[X'] = 0.$$

—   □

INTERPRETATION

Centering simply takes the center of mass as the origin for the line on which the probability is distributed. Standardization involves both centering and using the standard deviation as the unit of length. We make use of the centered and standardized versions of random variables in the next chapter in examining the concept of covariance and correlation.

PROBLEMS

15-1. For the following discrete joint distributions, determine Var[X] and Var[Y].

Discrete joint distributions for problem 15-1

(a)

$$P(X = t, Y = u)$$

$t =$	1	2	3	4	5	6
$u = 5$	0	0	0	0.1	0	0.1
$u = 4$	0.1	0	0	0	0.1	0
$u = 3$	0	0.1	0.1	0	0.1	0
$u = 2$	0	0	0	0.1	0.1	0
$u = 1$	0.1	0	0	0	0	0

$$E[X] = \frac{36}{10} \qquad E[Y] = \frac{32}{10}$$

(b)

$$P(X = t, Y = u)$$

$t =$	0	1	4	9
$u = 2$	0.05	0.04	0.21	0.15
$u = 0$	0.05	0.01	0.09	0.10
$u = -1$	0.10	0.05	0.10	0.05

$$E[X] = \frac{44}{10} \qquad E[Y] = \frac{6}{10}$$

(c)

$$P(X = t, Y = u)$$

$t =$	-2	0	1	3
$u = 1$	0.15	0.05	0.10	0
$u = 0$	0.05	0.08	0.10	0.12
$u = -1$	0.10	0.02	0	0.23

$$E[X] = \frac{65}{100} \qquad E[Y] = \frac{-5}{100}.$$

15-2. For the following joint densities, determine Var[X] and Var[Y].

Joint density functions for problem 15-2 (zero outside indicated regions)

(1) $f_{XY}(t, u) = 1/\pi$ on circle with radius 1, center at $(0,0)$

$$\int_0^1 t^2(1 - t^2)^{1/2} = \frac{\pi}{16}.$$

(2) $f_{XY}(t, u) = 1$ on triangle bounded by $t = 0$, $u = 0$, $u = 2(1 - t)$

$$E[X] = \frac{1}{3} E[Y] = \frac{2}{3}.$$

(3) $f_{XY}(t, u) = 1/2$ on square with vertices at $(1,0)$, $(2,1)$, $(1,2)$, and $(0,1)$.

(4) $f_{XY}(t, u) = 4t(1 - u)$ for $0 \le t \le 1$, $0 \le u \le 1$

$$E[X] = \frac{2}{3} E[Y] = \frac{1}{3}.$$

(5) $f_{XY}(t, u) = (t + u)/8$ for $0 \le t \le 2$, $0 \le u \le 2$

$$E[X] = E[Y] = \frac{7}{6}.$$

(6) $f_{XY}(t, u) = 2\alpha u e^{-\alpha t}$ for $0 \le t$, $0 \le u \le 1$

$$E[X] = \frac{1}{\alpha} E[Y] = \frac{2}{3}.$$

(7) $f_{XY}(t, u) = 8tu$ on triangle bounded by $t = 1$, $u = 0$, $u = t$

$$E[X] = \frac{4}{5} E[Y] = \frac{8}{15}.$$

(8) $f_{XY}(t, u) = (24/5)tu$ on triangle bounded by $t = 1$, $u = 1$, and $u = 1 - t$

$$E[X] = E[Y] = \frac{18}{25}.$$

(9) $f_{XY}(t, u) = 12t^2 u$ on the parallelogram with vertices $(-1, 0)$, $(0, 0)$, $(1, 1)$, and $(0, 1)$

$$E[X] = \frac{2}{5} E[Y] = \frac{11}{15}.$$

(10) $f_{XY}(t, u) = \frac{6}{5}(t^2 + u)$ for $0 \le t \le 1$, $0 \le u \le 1$

$$E[X] = E[Y] = \frac{3}{5}.$$

15-3. For the following mixed distributions, determine Var$[X]$ and Var$[Y]$.
Mixed joint distributions for problem 15-3

(1) Mass 0.1 at points $(-1, 0)$, $(0, 0)$, and $(1, 0)$.
Mass 0.7 spread uniformly over the square with vertices at $(-1, 0)$, $(0, -1)$, $(1, 0)$, and $(0, 1)$.

(2) Mass 0.1 at points $(0, 0)$, $(0, 1)$, and $(1, 0)$.
Mass 0.7 spread uniformly over the triangle with vertices at those points.

$$E[X] = E[Y] = \frac{1}{3}.$$

15–4. For the Beta (r, s) distribution

(a) Determine $E[X^k]$, k a positive integer.

(b) Use the result of part (a) to obtain $E[X]$ and $\mathrm{Var}[X]$.

15–5. Random variable X has distribution as follows: 1. Probability masses $1/10$ at $t = -1, 0, 1, 2$. 2. Probability mass $6/10$ spread uniformly on $[-1, 2]$. Determine $E[X])$ and $\mathrm{Var}[X]$.

15–6. Random variables X, Y have a joint distribution. Suppose

$$E[X] = 3, \ E[X^2] = 11, \ E[Y] = 10, \ E[Y^2] = 101, \ E[XY] = 30.$$

Determine $\mathrm{Var}[10X - 2Y]$.

15–7. Random variables X, Y have a joint distribution. Suppose

$$E[X] = 2, \quad E[X^2] = 5, \quad E[Y] = 1, \quad E[Y^2] = 2, \quad E[XY] = 1.$$

Determine $\mathrm{Var}[2X + 3Y]$.

15–8. The pair $\{X, Y\}$ is independent, with $E[X] = 2$, $E[Y] = 1$, $\mathrm{Var}[X] = 6$, and $\mathrm{Var}[Y] = 4$.

If $Z = X^2 + 2XY^2 - 3Y + 4$, determine $E[Z]$.

15–9. Suppose $\{X_1, X_2, X_3\}$ is iid with common mean value $\mu = 7$ and variance $\sigma^2 = 3$.

Let $A = \frac{1}{3}(X_1 + X_2 + X_3)$. Determine $E[A]$ and $\mathrm{Var}[A]$.

15–10. The pair $\{X, Y\}$ has joint density function $f_{XY}(t, u) = 8tu$ on the triangle bounded by the lines $u = 0$, $t = 1$, and $u = t$. Determine the marginal density functions f_X and f_Y, the mean values $E[X]$ and $E[Y]$, and the variances $\mathrm{Var}[X]$ and $\mathrm{Var}[Y]$.

16

Covariance, Correlation, and Linear Regression

The mean value $\mu = E[X]$ and the variance $\text{Var}[X] = E[(X - \mu)^2]$ provide important characterizations of the distribution for a single real random variable X. In this chapter, we seek a similar characterization of the joint distribution for a pair (X, Y). We can, of course, determine the mean and variance for each of the marginal distributions. This is important, but shows nothing of the probabilistic ties between X and Y. In view of the success of introducing the mean value and variance as expectations of appropriate functions of the single random variable, it is natural to look to expectations of appropriate functions of the pair. We examine one such function to obtain the covariance and a standardized form of that function to obtain the correlation coefficient.

16.1. Covariance and Correlation

We have seen that if $\{X, Y\}$ is an independent pair, then

$$E[XY] - E[X]E[Y] = 0 \qquad \text{(the product rule for expectation)}.$$

It is a simple exercise in the use of properties of expectation to show

$$E[XY] - E[X]E[Y] = E[(X - \mu_X)(Y - \mu_Y)]$$

where $\mu_X = E[X]$ and $\mu_Y = E[Y]$. From the treatment of variance in the previous chapter, we have as a special case of (V4)

$$\text{Var}[X \pm Y] = \text{Var}[X] + \text{Var}[Y] \pm 2\big(E[XY] - E[X]E[Y]\big).$$

This suggests that the quantity $E[(X - \mu_X)(Y - \mu_Y)]$ might be useful as a measure of stochastic dependence between X and Y. We examine that possibility.

We find it expedient to work with the centered and standardized forms of the random variables X and Y (see Theorem 15.4.3), since this simplifies considerably the handling of algebraic details.

Centered random variables: $X' = X - \mu_X$, $Y' = Y - \mu_Y$.

Standardized random variables: $X^* = \dfrac{X - \mu_X}{\sigma_X}$, $Y^* = \dfrac{Y - \mu_Y}{\sigma_Y}$.

We now define two related quantities whose role in characterizing the interdependence of X and Y we want to examine.

Definition 16.1.1. *The* covariance *of the pair* $\{X, Y\}$ *is*

$$\text{Cov}[X, Y] = E[(X - \mu_X)(Y - \mu_Y)] = E[X'Y'].$$

The correlation coefficient *of the pair* $\{X, Y\}$ *is*

$$\rho = \rho[X, Y] = \frac{E[(X - \mu_X)(Y - \mu_Y)]}{\sigma_X \sigma_Y} = \frac{\text{Cov}[X, Y]}{\sigma_X \sigma_Y} = E[X^* Y^*].$$

Remarks

1. The covariance is sometimes referred to as the *product moment*.

2. $\text{Cov}[X, X] = \text{Var}[X]$.

3. The more exact term *coefficient of linear correlation* is often used for ρ. The reason for that becomes apparent in the examination of the significance of the concept (see the discussion at the end of Section 16.3).

As a first step in examining ρ, we use Schwarz's inequality (E15) for expectation to assert

$$\rho^2 = E^2[X^* Y^*] \le E[(X^*)^2] E[(Y^*)^2] = 1$$

with equality iff there is a constant c such that $X^* = cY^*$. In the case of equality, we have

$$1 = c^2 E^2[(Y^*)^2] = c^2 \qquad \text{so that} \quad c = \pm 1.$$

Hence we have equality iff

$$\rho = E[cY^* Y^*] = cE[(Y^*)^2] = c = \pm 1.$$

We wish to relate the value of ρ to the nature of the the joint distribution for (X, Y). We consider first the joint distribution for (X^*, Y^*); then we map the results to the joint distribution for (X, Y) by the mapping

$$\begin{aligned} t &= \sigma_X r + \mu_X \\ u &= \sigma_Y s + \mu_Y. \end{aligned}$$

This affine mapping has the property of taking straight lines into straight lines.

Joint distribution for (X^*, Y^*) : $(r, s) = (X^*, Y^*)(\omega)$
First we note that (see Figure 16.1)

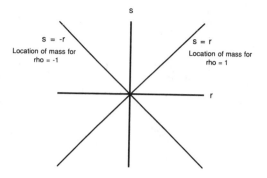

FIGURE 16.1. The $\rho = 1$ and $\rho = -1$ lines for $\{X^*, Y^*\}$.

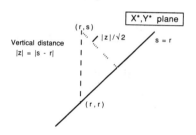

FIGURE 16.2. Geometric significance of $\frac{1}{2}E[(X^* - Y^*)^2]$.

$\rho = 1$ iff all probability mass lies on the line $s = r$.
$\rho = -1$ iff all probability mass lies on the line $s = -r$.

Now suppose $-1 < \rho < 1$. This means that some of the probability mass must be located at points not on either the line $s = r$ or the line $s = -r$. We seek an indication of the spread of probability mass about these lines. Consider the random variable

$$Z = Y^* - X^*.$$

This random variable has the value $z = s - r$ when $(X^*, Y^*) = (r, s)$. Figure 16.2 indicates the geometric significance. The quantity $|z| = |s - r|$ indicates the vertical distance from the point (r, s) to the point (r, r) on the line $s = r$ (the $\rho = 1$ line). The perpendicular distance from the point (r, s) to the line is $|z|/\sqrt{2}$. This means that

$$E\left[\left(Z/\sqrt{2}\right)^2\right] = \frac{1}{2}E[(Y^* - X^*)^2]$$

is the probability-weighted average of the squares of the distances of the image points (r, s) to the line $s = r$. In mechanical terms, this is the second moment of the mass distribution about the line $s = r$.

An exactly parallel argument shows that $\frac{1}{2}E[(Y^* + X^*)^2]$ is the second moment of the joint probability mass distribution about the line $s = -r$.

We now express these second moments or variances in terms of the correlation coefficient ρ.

$$
\begin{aligned}
\frac{1}{2}E[(Y^* \pm X^*)^2] &= \frac{1}{2}\left(E[(X^*)^2] + E[(Y^*)^2] \pm 2E[X^*Y^*]\right). \\
&= \frac{1}{2}(1 + 1 \pm 2\rho) = 1 \pm \rho.
\end{aligned}
$$

We thus have

$1 - \rho$ is the variance about $s = r$ on the (X^*, Y^*) plane.
$1 + \rho$ is the variance about $s = -r$ on the (X^*, Y^*) plane.

The condition $\rho = 0$ corresponds to equality of these variances. Thus, $\rho = 0$ indicates a quasi-symmetry about these two lines. In keeping with the preceding result, $\rho = 1$ means all probability mass is on the line $s = r$, since the variance about that line is zero. A similar statement holds for $\rho = -1$ and the line $s = -r$.

The affine transformation to the (t, u) plane and the joint distribution for (X, Y) shows that:

The $\rho = 1$ line becomes $u = \dfrac{\sigma_Y}{\sigma_X}(t - \mu_X) + \mu_Y$.

The $\rho = -1$ line becomes $u = -\dfrac{\sigma_Y}{\sigma_X}(t - \mu_X) + \mu_Y$.

Both lines go through the point (μ_X, μ_Y). The respective slopes for $\rho = \pm 1$ are $\pm\sigma_Y/\sigma_X$. Elementary analysis shows that in the plane for the joint distribution for (X, Y), the quantities $1 - \rho$ and $1 + \rho$ are proportional to the variance of the joint probability mass about the lines for $\rho = 1$ and $\rho = -1$, respectively.

16.2. Some Examples

The following examples are intended to illustrate both techniques of calculation and the use of geometric interpretation as an aid, and sometimes a shortcut, to obtaining numerical results.

Example 16.2.1:

Suppose (X, Y) has joint uniform distribution on the unit circle, center at (a, b), see Figure 16.3. It is apparent from symmetries that $\mu_X = a$, $\mu_Y = b$, and $\sigma_X = \sigma_Y$. The $\rho = 1$ line has slope $\sigma_Y/\sigma_X = 1$ and the

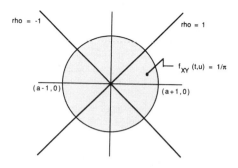

FIGURE 16.3. Distribution for Example 16.2.1.

$\rho = -1$ line has slope -1. The variance about these two lines is the same, so that we must have $\rho = 0$. The random variables are uncorrelated, since $\text{Cov}[X, Y] = \sigma_X \sigma_Y \rho = 0$. These result may be obtained by direct calculations, which are straightforward but somewhat tedious.

——

$\square$

Remark: This joint distribution provides a simple example of an uncorrelated pair that is not independent. The symmetry provides the lack of correlation. The failure of the product rule on rectangles with zero joint probability but nonzero marginals precludes independence.

Example 16.2.2: Joint Distributions with Uniform Marginals

This standard example compares several joint distributions, each of which has both X and Y distributed uniformly on (-1.1). Refer to Figure 16.4 for the geometry of the joint distributions.

Case (a) Probability mass is distributed uniformly on the line segment $u = t$ for $-1 \le t \le 1$.

Case (b) Probability mass is distributed uniformly on the line segment $u = -t$ for $-1 \le t \le 1$.

Case (c) Probability mass is distributed uniformly on the square with vertices at $(-1, -1)$, $(1, -1)$, $(1, 1)$, and $(-1, 1)$.

Case (d) Probability mass is distributed uniformly over those portions of the square in Case (c) that lie in the first and third quadrants.

Determine the correlation coefficient ρ in each case.

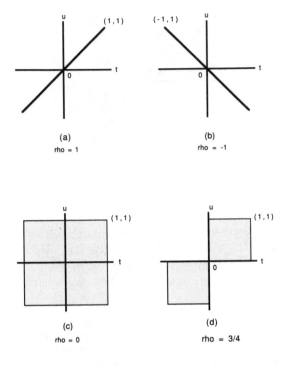

FIGURE 16.4. Distributions for Example 16.2.2.

Solution

In each case, the marginal distributions are uniform on $(-1,1)$ for both X and Y. Both have mean zero and the same variance, so the lines for $\rho = \pm 1$ are $u = \pm t$, respectively. From the result for the uniform distribution in Section 15.3, we have $\sigma_X^2 = \sigma_Y^2 = 1/3$.

Case (a) Since all probability lies on the $\rho = 1$ line, we have $\rho = 1$.

Case (b) Since all probability lies on the $\rho = -1$ line, we have $\rho = -1$.

Case (c) The joint uniform distribution over the square ensures the independence of $\{X, Y\}$. This implies Cov$[X, Y] = 0$, so that $\rho = 0$.

Case (d) The variance about the $\rho = 1$ line is apparently less than that about the $\rho = -1$ line, so that we expect $0 < \rho < 1$. To determine the exact value, we need to evaluate the appropriate integral. Since the mean values are both zero and $\sigma_X \sigma_Y = 1/3$, we have

$$\rho = \frac{E[XY]}{1/3} = 3E[XY] = 3 \iint tu f_{XY}(t,u) \, du \, dt.$$

Since $(-t)(-u) = tu$, for each point (t,u) in the first quadrant there is a corresponding point $(-t,-u)$ in the third quadrant that gives the same value to the integrand. Thus, we may take twice the integral over the square in the first quadrant.

Since the total area is 2, the constant value of f_{XY} is $1/2$, so that

$$\rho = 6 \int_0^1 \int_0^1 tu\frac{1}{2} \, du \, dt = \frac{3}{4}.$$

This result is consistent with our geometric interpretation and gives the precise number.

—   □

For somewhat less intuitive examples, we again consider several joint distributions used in previous examples.

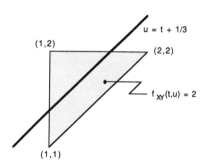

FIGURE 16.5. Joint distribution for Example 16.2.3.

Example 16.2.3: (See Example 14.2.10, Figure 14.2, Example 15.3.1, and Figure 16.5)

Suppose $f_{XY}(t,u) = 2$ on the triangular region $1 \le t \le u \le 2$ (zero elsewhere). From Examples 14.2.10 and 15.3.1,

$$E[X] = \frac{4}{3}, \quad E[Y] = \frac{5}{3}, \quad \text{Var}[X] = \text{Var}[Y] = \frac{1}{18}.$$

Determine $\text{Cov}[X,Y]$ and $\rho = \rho[X,Y]$.

Solution

In addition to the values for means and variances from the earlier examples, we need $E[XY]$.

$$E[XY] = \iint tu f_{XY}(t,u)\, du\, dt = 2 \int_1^2 t \int_t^2 u\, du\, dt = \frac{9}{4}.$$

Using this result, we get

$$\text{Cov}[X,Y] = E[XY] - E[X]E[Y] = \frac{9}{4} - \frac{20}{9} = \frac{1}{36}.$$

and

$$\rho = \frac{\text{Cov}[X,Y]}{\sigma_X \sigma_Y} = \frac{1}{36} \cdot \frac{18}{1} = \frac{1}{2}.$$

The $\rho = 1$ line is $u = (t - 4/3) + 5/3 = t + 1/3$.

— $\square$

Example 16.2.4: (See Example 14.2.11, Figure 14.3, and Example 15.3.2)

Suppose $f_{XY}(t,u) = 3$ for $1 \le u \le t^2 \le 1$. From Examples 14.2.11 and 15.3.2 we have

$$E[X] = \frac{3}{4}, \qquad E[Y] = \frac{3}{10}, \qquad \text{Var}[X] = \frac{3}{80}, \qquad \text{Var}[Y] = \frac{37}{700}.$$

Determine $\text{Cov}[X,Y]$ and $\rho = \rho[X,Y]$.

Solution

We need to calculate

$$E[XY] = \iint tu f_{XY}(t,u)\, dt\, du = 3 \int_0^1 t \int_0^{t^2} u\, du\, dt = \frac{3}{2} \int_0^1 t^5\, dt = \frac{1}{4}.$$

Then

$$\text{Cov}[X,Y] = E[XY] - E[X]E[Y] = \frac{1}{4} - \frac{3}{4} \cdot \frac{3}{10} = \frac{1}{40}.$$

From this we calculate

$$\rho = \frac{\text{Cov}[X,Y]}{\sigma_X \sigma_Y} = \frac{1}{40} \left(\frac{80}{3} \cdot \frac{700}{37} \right)^{1/2} \approx 0.562.$$

— $\square$

In Section 9.3, we introduce the joint normal distribution for a pair of random variables. This distribution has five parameters, μ_X, μ_Y, σ_X, σ_Y, and ρ. The marginal distribution for X is $N(\mu_X, \sigma_X^2)$ and the marginal for Y is $N(\mu_Y, \sigma_Y^2)$. Thus, four of the parameters are accounted for. We now show that the fourth parameter ρ is indeed the correlation coefficient, as the notation suggests.

Example 16.2.5: The Joint Normal Distribution

The pair (X, Y) has the joint normal distribution iff the joint density is given by

$$f_{XY}(t, u) = \frac{1}{2\pi\sigma_X\sigma_Y(1 - \rho^2)^{1/2}} e^{-Q(t,u)/2}$$

where

$$Q(t, u) = \frac{1}{1 - \rho^2}\left(\left(\frac{t - \mu_X}{\sigma_X}\right)^2 - 2\rho\left(\frac{t - \mu_X}{\sigma_X}\right)\left(\frac{u - \mu_Y}{\sigma_Y}\right) + \left(\frac{u - \mu_Y}{\sigma_Y}\right)^2\right).$$

Determine the correlation coefficient $\rho = \rho[X, Y]$.

Solution

In Section 9.3, we put this into the following form. Setting $r = (t - \mu_X)/\sigma_X$ and $s = (u - \mu_Y)/\sigma_Y$, we have

$$Q(t, u) = q(r, s) = \frac{r^2 - 2\rho r s + s^2}{1 - \rho^2}.$$

Upon completing the square, we may write this in the form

$$q(r, s) = r^2 + \frac{(s - \rho r)^2}{1 - \rho^2} = \left(\frac{t - \mu_X}{\sigma_X}\right)^2 + \frac{(u - \alpha(t))^2}{\sigma_Y^2(1 - \rho^2)}$$

where

$$\alpha(t) = \frac{\rho\sigma_Y}{\sigma_X}(t - \mu_X) + \mu_Y.$$

Thus $f_{XY}(t, u) = h(t)g(t, u)$ where

$$h(t) = f_X(t) = \frac{1}{\sigma_X\sqrt{2\pi}}\exp\left(-\frac{1}{2}\left(\frac{t - \mu_X}{\sigma_X}\right)^2\right)$$

and

$$g(t, u) = \frac{1}{\sigma_Y(1 - \rho^2)^{1/2}\sqrt{2\pi}}\exp(-v^2/2)$$

with

$$v = \frac{u - \alpha(t)}{\sigma_Y(1 - \rho^2)^{1/2}}.$$

Inspection shows $h = f_X$ is the density for a random variable $N(\mu_X, \sigma_X^2)$. Also, for each fixed t, the expression for $g(t, u)$ is the expression for the density of a random variable normal with mean $\alpha(t)$ and variance $\sigma_Y^2(1 - \rho^2)$. From this it follows that f_X is the marginal density for X. Because of symmetry with respect to r and s, a similar treatment shows the random variable Y is $N(\mu_Y, \sigma_Y^2)$. To determine the correlation coefficient ρ, we consider the standardized variables

$$X^* = \frac{X - \mu_X}{\sigma_X} \quad \text{and} \quad Y^* = \frac{Y - \mu_Y}{\sigma_Y}.$$

The correlation coefficient is given by

$$E[X^*Y^*] = \iint \left(\frac{t - \mu_X}{\sigma_X}\right)\left(\frac{t - \mu_Y}{\sigma_Y}\right) f_{XY}(t, u)\, du\, dt.$$

Use of the special form of f_{XY}, yields

$$E[X^*Y^*] = \frac{1}{\sigma_X \sigma_Y} \int (t - \mu_X) f_X(t) \int (u - \mu_Y) g(t, u)\, du\, dt.$$

Now we may write $u - \mu_Y = u - \alpha(t) + \alpha(t) - \mu_Y$, so that

$$\int (u - \mu_Y) g(t, u)\, du = 0 + \alpha(t) - \mu_Y = \rho \frac{\sigma_Y}{\sigma_X}(t - \mu_X).$$

Hence,

$$E[X^*Y^*] = \frac{1}{\sigma_X \sigma_Y} \cdot \frac{\rho \sigma_Y}{\sigma_X} \int (t - \mu_X)^2 f_X(t)\, dt = \frac{\rho}{\sigma_X^2} \cdot \sigma_X^2 = \rho.$$

Thus, the correlation coefficient is the parameter called ρ. This, of course, is why that parameter is so designated.

— $\square$

16.3.   Linear Regression

Suppose random variables X and Y have a known (or assumed) joint probability distribution. If a value of X is observed, we want to estimate the corresponding value of Y. To do this, we seek to determine a decision rule that is a function r such that

If the value $X(\omega)$ is observed, the best estimate of $Y(\omega)$ is $r\big(X(\omega)\big)$.

In order to formulate the problem precisely, we must agree on what is meant by "best." That is, we must agree on a criterion of goodness that determines which of two possible decision rules is better. It is clear that we want the difference between the actual value $Y(\omega)$ and the estimated value $r(X(\omega))$ to be small. This difference is the error of estimation. In a probabilistic situation, characterized by uncertainty, we cannot determine a rule that forces all individual estimates to be small. However, we may hope to make the error small in some average sense. Thus, we seek to minimize the average, in the probabilistic sense, of some function of the error.

THE REGRESSION PROBLEM

The most commonly employed measure of failure to predict exactly is the mean-squared error $E\big[(Y - r(X))^2\big]$. This is the probability-weighted average of the squares of the errors. Positive and negative errors are treated alike; larger errors carry a greater penalty than do smaller ones. The popularity of this measure is based both on these intrinsic characteristics and the fact that its use leads to a feasible mathematical theory.

If we apply the mean-squared error criterion to two rules r_1 and r_2, we say r_1 is the better of the two iff

$$E\big[(Y - r_1(X))^2\big] < E\big[(Y - r_2(X))^2\big].$$

In this case we say $r_1(X)$ is the better *estimator* of Y, in the mean-squared sense. The general problem is to find a function r for which the mean-squared error is minimal. This function must have the property that $E[r^2(X)] < \infty$. The problem of finding the best mean-square estimator is called the *regression problem*. In Section 19.5, we show that this problem is solved completely by the use of conditional expectation. In this section, we consider an important special case.

LINEAR REGRESSION

With the aid of the theory of correlation, we show how to obtain the best affine decision rule of the form $r(t) = at + b$. That is, if $X(\omega)$ is observed, the estimate for $Y(\omega)$ is given by

$$Y(\omega) \approx r(X(\omega)) = aX(\omega) + b.$$

The problem reduces to finding the coefficients a and b such that $E[(Y - aX - b)^2]$ is minimized. Since the graph of r is a straight line with slope a and vertical-coordinate intercept b, it is customary to refer to the best linear rule $aX + b$. Generally, the very best "fit" requires a nonlinear rule. However, there are cases in which the linear rule is optimal among all rules based on the mean-squared criterion. The joint normal distribution is a notable example.

To determine the optimum values for a and b, we first perform some manipulations, based on experience, to express

$$Y - aX - b = (Y - \mu_Y) - a(X - \mu_X) - \beta, \qquad \text{where } \beta = b - \mu_Y + a\mu_X.$$

Upon squaring, we have

$$(Y - \mu_Y)^2 + a^2(X - \mu_X)^2 + \beta^2$$
$$- 2\beta(Y - \mu_Y) + 2a\beta(X - \mu_X) - 2a(X - \mu_X)(Y - \mu_Y).$$

Taking expectations and using the fact that $E[X - \mu_X] = E[Y - \mu_Y] = 0$, we obtain

$$E[(Y - aX - b)^2] = \sigma_Y^2 + a^2\sigma_X^2 + \beta^2 - 2a\,\mathrm{Cov}[X, Y].$$

We note that any nonzero value of β increases the expectation, so we set $\beta = 0$, leaving a quadratic in a. Setting to zero the derivative with respect to a, we find

$$2a\sigma_X^2 - 2\,\mathrm{Cov}[X, Y] = 0.$$

Since the second derivative is $2\sigma_X^2 = 2\,\mathrm{Var}[X] > 0$, the value $a = \mathrm{Cov}[X, Y]/\mathrm{Var}[X]$ yields a minimum. Then the condition $\beta = 0$ corresponds to $b = \mu_Y - a\mu_X$. Therefore, the best linear rule may be expressed

$$u = \frac{\mathrm{Cov}[X, Y]}{\mathrm{Var}[X]}(t - \mu_X) + \mu_Y = \rho\frac{\sigma_Y}{\sigma_X}(t - \mu_X) + \mu_Y = \alpha(t).$$

The affine function determines a line on the plane of the joint distribution, which is named, as follows.

Definition 16.3.1. *The regression line of Y on X is the line*

$$u = \alpha(t) = \rho\frac{\sigma_Y}{\sigma_X}(t - \mu_X) + \mu_Y = \frac{\mathrm{Cov}[X, Y]}{\mathrm{Var}[X]}(t - \mu_X) + \mu_Y.$$

Remark: The form using $\mathrm{Cov}[X, Y]$ is generally more useful for calculations, since it is not necessary to determine σ_Y or ρ.

Example 16.3.1: The Joint Normal Distribution

If (X, Y) has the joint normal distribution, we simply substitute directly the parameters, determined earlier, to get

$$u = \rho\frac{\sigma_Y}{\sigma_X}(t - \mu_X) + \mu_Y = \alpha(t).$$

$\square$

Example 16.3.2: (See Example 14.1.2 and Figure 14.1)

Suppose $f_{XY}(t,u) = 2$ over the triangular region with vertices $(0,0)$, $(1,0)$, and $(1,1)$. Then the marginal densities are

$$f_X(t) = 2 \int_0^t du = 2t, \quad 0 \le t \le 1$$

and

$$f_Y(u) = 2 \int_u^1 dt = 2(1-u), \quad 0 \le u \le 1.$$

From this it is easy to show that $E[X] = 2/3 E[Y] = 1/3$ and $\mathrm{Var}[X] = 1/18$. We also have

$$E[XY] = 2 \int_0^1 \int_0^t tu\, du\, dt = \frac{1}{4}.$$

From this we find

$$\mathrm{Cov}[X,Y] = \frac{1}{4} - \frac{2}{9} = \frac{1}{36} \quad \text{so that} \quad \frac{\mathrm{Cov}[X,Y]}{\mathrm{Var}[X]} = \frac{1}{2}.$$

Thus, the regression line is

$$u = \frac{1}{2}\left(t - \frac{2}{3}\right) + \frac{1}{3} = \frac{t}{2} \quad 0 \le t \le 1.$$

——   □

Example 16.3.3:

Suppose (X,Y) has the joint density function

$$f_{XY}(t,u) = \frac{1}{28}(4t + 2u + 1) \quad 0 \le t \le 2 0 \le u \le 2.$$

Then the marginal densities are readily shown to be

$$f_X(t) = \frac{1}{14}(4t+3),\ 0 \le t \le 2 \quad \text{and} \quad f_Y(u) = \frac{1}{14}(2u+5),\ 0 \le u \le 2.$$

From these we obtain $E[X] = 25/21$, $\mathrm{Var}[X] = 131/441$, $E[Y] = 23/21$, and $E[XY] = 9/7$. Hence, $\mathrm{Cov}[X,Y] = -8/441$, so that the regression line is

$$u = -\frac{8}{131}\left(t - \frac{25}{21}\right) + \frac{23}{21} = -\frac{8}{131}t + \frac{3213}{2751}.$$

——   □

Remark: For convenience, the examples have used absolutely continuous distributions with joint density functions. Similar examples may be worked out with discrete or mixed distributions.

COEFFICIENT OF LINEAR CORRELATION

In the second remark after the definition of $\text{Cov}[X, Y]$ and ρ, we note that the term coefficient of linear correlation is sometimes used. The discussion of linear regression shows that ρ is tied intimately to the best linear rule. It is easy to suppose that large $|\rho|$, i.e., $|\rho|$ near one, means the random variables are closely related, and that small $|\rho|$ means that there is little probabilistic tie. The latter statement is not generally true. The condition $\rho = 0$, i.e., the random variables are uncorrelated, does not imply independence. The uniform joint distribution over a circle (see the remark after Example 16.2.1) is a very simple example of an uncorrelated pair that is not independent. The following example shows that a pair may be functionally related and yet be uncorrelated in the sense the term is used in probability.

Example 16.3.4:

Suppose X is distributed uniformly on $(-1, 1)$. Set

$$Y = g(X) = \cos X.$$

In the usual sense of the term, one would think of X and Y as highly correlated. If the value of $X(\omega)$ is known, the value of $Y(\omega)$ is known exactly. Consider, however, the calculated value of the covariance. It is apparent that $E[X] = 0$. Then

$$\text{Cov}[X, Y] = E[XY] = E[Xg(X)] = \frac{1}{2} \int_{-1}^{1} t \cos t \, dt = 0.$$

The value of the integral is seen immediately from the fact that the integrand is an odd function, integrated from -1 to 1.

— $\square$

The correlation coefficient ρ is an indicator of the linear correlation of the joint distribution. It provides a measure of how well the joint distribution clusters about a line. Much of its usefulness arises from its application to problems in which the random variables have the joint normal distribution. In this case, $\rho = 0$ corresponds to independence, as we show in Section 10.3 (see also Example 16.2.5).

AFFINE TRANSFORMATIONS OF UNCORRELATED RANDOM VARIABLES

Any two real-valued random variables X and Y whose mean values and variances exist may be represented as an affine transformation of a pair of uncorrelated random variables. To see this, we begin with the standardized variables X^* and Y^* and set $U = X^* + Y^*$ and $V = X^* - Y^*$. It is an easy exercise to show that

$$E[U] = E[V] = 0 \qquad \text{and} \qquad \text{Cov}[U, V] = E[UV] = E[(X^*)^2 - (Y^*)^2] = 0$$

so that $\{U, V\}$ is an uncorrelated pair. We then have

$$X = \sigma_X X^* + \mu_X = \frac{\sigma_X}{2}(U + V) + \mu_X$$

$$Y = \sigma_Y Y^* + \mu_Y = \frac{\sigma_Y}{2}(U - V) + \mu_Y.$$

This is the asserted affine transformation.

Example 16.3.5: The Joint Normal Distribution

If $\{X, Y\}$ has the joint normal distribution, then it is known that $\{U, V\}$ also has the joint normal distribution (see comment at end of Section 18.5). This implies that $\{U, V\}$ is independent, with both variables standardized normal. Thus, any pair $\{X, Y\}$ with the joint normal distribution may be represented as an affine transformation of an independent pair of standardized, normal random variables.

COVARIANCE AND ORTHOGONALITY

In Section 13.9 we introduce the notion of an inner product of random variables. In the real-variable case, $\langle X, Y \rangle = E[XY]$. If the random variables are centered (i.e., have zero mean values), then

$$\text{Cov}[X, Y] = E[XY] = \langle X, Y \rangle.$$

The condition of zero inner product indicates orthogonality of the random variables as elements in the vector space L^2 (see Section 13.9). Thus uncorrelated, centered random variables are orthogonal in L^2. The random variables

$$U = X^* + Y^* \qquad \text{and} \qquad V = X^* - Y^*$$

thus constitute an orthogonal pair.

General orthogonalization procedures, such as the Gram-Schmidt procedure, make it possible to start with any set $\{X_i : 1 \leq i \leq n\}$ and produce an orthonormal set (i.e., orthogonal set, with each member having unit norm or length) such that the original set is obtained by an affine transformation of the orthonormal set. This is a standard topic in functional analysis or the theory of operators in linear spaces.

———   □

PROBLEMS

16–1. For the following discrete joint distributions

    (a) Determine $\text{Cov}[X, Y]$ and $\rho = \rho[X, Y]$.

    (b) Determine the $\rho = 1$ and the $\rho = -1$ lines.

    (c) Write a formula for the regression line of Y on X.

    (d) Let $Z = X + Y$. Determine $\text{Var}[Z]$.

Discrete joint distributions for problem 16-1

(1)

$$P(X = t, Y = u)$$

$t =$	1	2	3	4	5	6
$u = 5$	0	0	0	0.1	0	0.1
$u = 4$	0.1	0	0	0	0.1	0
$u = 3$	0	0.1	0.1	0	0.1	0
$u = 2$	0	0	0	0.1	0.1	0
$u = 1$	0.1	0	0	0	0	0

$$E[X] = \frac{36}{10} \qquad E[Y] = \frac{32}{10}$$

$$\text{Var}[X] = \frac{284}{100} \qquad \text{Var}[Y] = \frac{156}{100}.$$

(2)

$$P(X = t, Y = u)$$

$t =$	0	1	4	9
$u = 2$	0.05	0.04	0.21	0.15
$u = 0$	0.05	0.01	0.09	0.10
$u = -1$	0.10	0.05	0.10	0.05

$$E[X] = \frac{44}{10} \qquad E[Y] = \frac{6}{10}.$$

$$\text{Var}[X] = \frac{1144}{100} \qquad \text{Var}[Y] = \frac{174}{100}.$$

(3)

$$P(X = t, Y = u)$$

$t =$	-2	0	1	3
$u = 1$	0.15	0.05	0.10	0
$u = 0$	0.05	0.08	0.10	0.12
$u = -1$	0.10	0.02	0	0.23

$$E[X] = \frac{65}{100} \qquad E[Y] = \frac{-5}{100}$$

$$\text{Var}[X] = \frac{41,275}{10,000} \qquad \text{Var}[Y] = \frac{6,475}{10,000}.$$

16-2. For the following joint densities

(a) Determine $\text{Cov}[X,Y]$ and $\rho = \rho[X,Y]$.

(b) Determine the $\rho = 1$ and the $\rho = -1$ lines.

(c) Write a formula for the regression line of Y on X.

(d) Let $Z = X + Y$. Determine $\text{Var}[Z]$.

Joint density functions for problem 16-2 (zero outside indicated regions)

(1) $f_{XY}(t,u) = 1/\pi$ on circle with radius 1, center at $(0,0)$

$$\text{Var}[X] = \text{Var}[Y] = \frac{1}{4}.$$

(2) $f_{XY}(t,u) = 1$ on triangle bounded by $t = 0$, $u = 0$, $u = 2(1-t)$

$$E[X] = \frac{1}{3}, \quad E[Y] = \frac{2}{3}, \quad \text{Var}[X] = \frac{1}{18}, \quad \text{Var}[Y] = \frac{2}{9}.$$

(3) $f_{XY}(t,u) = 1/2$ on square with vertices at $(1,0)$, $(2,1)$, $(1,2)$, $(0,1)$

$$\text{Var}[X] = \text{Var}[Y] = \frac{1}{6}.$$

(4) $f_{XY}(t,u) = 4t(1-u)$ for $0 \le t \le 1$, $0 \le u \le 1$

$$E[X] = \frac{2}{3}, \quad E[Y] = \frac{1}{3}, \quad \text{Var}[X] = \text{Var}[Y] = \frac{1}{18}.$$

(5) $f_{XY}(t,u) = \frac{1}{8}(t+u)$ for $0 \le t \le 2$, $0 \le u \le 2$

$$E[X] = E[Y] = \frac{7}{6}, \quad \text{Var}[X] = \text{Var}[Y] = \frac{11}{36}.$$

(6) $f_{XY}(t,u) = 2\alpha u e^{-\alpha t}$ for $0 \le t$, $0 \le u \le 1$

$$E[X] = \frac{1}{\alpha}, \quad E[Y] = \frac{2}{3}, \quad \text{Var}[X] = \frac{1}{\alpha^2}, \quad \text{Var}[Y] = \frac{1}{18}.$$

(7) $f_{XY}(t,u) = 8tu$ on triangle bounded by $t = 1$, $u = 0$, $u = t$

$$E[X] = \frac{4}{5}, \quad E[Y] = \frac{8}{15}, \quad \text{Var}[X] = \frac{2}{75}, \quad \text{Var}[Y] = \frac{11}{225}.$$

(8) $f_{XY}(t,u) = \frac{24}{5}tu$ on triangle bounded by $t = 1$, $u = 1$, $u = 1-t$

$$E[X] = E[Y] = \frac{18}{25}, \quad \text{Var}[X] = \text{Var}[Y] = \frac{26}{625}.$$

(9) $f_{XY}(t,u) = 12t^2u$ on the parallelogram with vertices $(-1,0)$, $(0,0)$, $(1,1)$, $(0,1)$

$$E[X] = \frac{2}{5}, \quad E[Y] = \frac{11}{15}, \quad \text{Var}[X] = \frac{6}{26}, \quad \text{Var}[Y] = \frac{14}{225}.$$

(10) $f_{XY}(t,u) = \frac{6}{5}(t^2 + u)$ for $0 \le t \le 1, 0 \le u \le 1$

$$E[X] = E[Y] = \frac{3}{5}, \quad \text{Var}[X] = \frac{2}{25}, \quad \text{Var}[Y] = \frac{11}{150}.$$

16–3. For the following mixed distributions

(a) Determine $\text{Cov}[X,Y]$ and $\rho = \rho[X,Y]$.
(b) Determine the $\rho = 1$ and the $\rho = -1$ lines.
(c) Write a formula for the regression line of Y on X.
(d) Let $W = X - Y$. Determine $\text{Var}[W]$.

Mixed joint distributions for problem 16-3

(1) Mass 0.1 at points $(-1,0)$, $(0,0)$, $(1,0)$.
Mass 0.7 spread uniformly over the square with vertices at $(-1,0)$, $(0,-1)$, $(1,0)$, and $(0,1)$.

$$\text{Var}[X] = \frac{19}{60}, \quad \text{Var}[Y] = \frac{7}{60}.$$

(2) Mass 0.1 at points $(0,0)$, $(0,1)$, $(1,0)$.
Mass 0.7 spread uniformly over the triangle with vertices at those points.

$$E[X] = E[Y] = \frac{1}{3}, \quad \text{Var}[X] = \text{Var}[Y] = \frac{19}{180}.$$

16–4. A receptacle contains N objects; N_1 are of one kind; $N_2 = N - N_1$ are of a second kind. A sample of size n is taken without replacement. Let X be the number of the first kind and Y be the number of the second kind selected. Determine $\rho[X,Y]$.

Hint: Note that $X + Y = n$.

16–5. Suppose $Y = aX+b$, where a and b are constants. Determine $\rho[X,Y]$.

16–6. The pair $\{X,Y\}$ has joint distribution with parameters

$$E[X] = 3, \quad E[Y] = -10, \quad E[X^2] = 13,$$
$$E[Y^2] = 125, \quad \text{and} \quad E[XY] = -34.4.$$

Determine the regression line of Y on X.

16-7. Suppose $\text{Var}[X] = 36$, $\text{Var}[X + Y] = 62$, and $\text{Var}[X - y] = 42$. Determine $\rho[X, Y]$.

16-8. Suppose $E[X] = 1$, $E[Y] = -1$, $\text{Var}[X] = 4$, $\text{Var}[X + Y] = 7$, and $\text{Var}[X - Y] = 19$.

Determine the regression line of Y on X.

16-9. For density (4) in problem 16-2, let $W = 3X + 4Y - 2$. Determine $\text{Cov}[W, X]$.

16-10. Suppose $\text{Var}[X] = 3$, $\text{Var}[Y] = 7$, and $\text{Cov}[X, Y] = 13$. Put $W = 3X + 5Y$ and $Z = -2X + Y$. Determine $\text{Cov}[W, Z]$.

16-11. Suppose $\{X_i : 1 \leq i \leq n\}$ is a class of random variables with $\text{Var}[X_i] = \sigma_i^2$. Set

$$W = \sum_{i=1}^{n} a_i X_i \qquad \text{and } Z = \sum_{i=1}^{n} b_i X_i.$$

Obtain a formula for $\text{Cov}[W, Z]$. Show that if each $a_i = b_i$, so that $W = Z$, the formula reduces to that for $\text{Var}[W]$.

16-12. Suppose $\{X, Y\}$ has the joint normal distribution. Let $U = aX + bY + c$ and $V = dX + eY + f$ where a, b, c, d, e, f are constants. Assume that $\{U, V\}$ have a joint normal distribution.

(a) Determine $\text{Cov}[U, V]$.

(b) Suppose $X \sim N(0, 4)$ and $Y \sim N(1, 9)$. Determine values of the constants to make $\{U, V\}$ independent.

16-13. Suppose $\{X, Y, Z\}$ is independent. Put $U = X + Y$ and $V = Y + Z$.

(a) Determine $\text{Cov}[U, V]$.

(b) Is there a condition on the joint distribution so that $\{U, V\}$ is independent?

16-14. A professor attempts to show that the number of hours a student spends preparing for the final examination in his course and the grade earned on that examination are related. Let X be the random variable whose value is the number of hours of preparation and Y be the random variable whose value is the grade on the exam. To make possible use of numbers, the grades A, B, C, D, and F are assigned numbers 1, 2, 3, 4, and 5, respectively. From data taken over the past several years, the professor calculates the following joint probabilities $p_{XY}(t, u)$.

$t =$	1	2	3	4	5
$u = 1$	0.00	0.01	0.06	0.02	0.01
2	0.02	0.09	0.00	0.06	0.03
3	0.02	0.15	0.03	0.16	0.04
4	0.02	0.07	0.01	0.08	0.02
5	0.04	0.01	0.01	0.00	0.04

Calculate the correlation coefficient $\rho = \rho[X, Y]$. Do you find the result surprising? If so, why?

16–15. Suppose the class $\{X_i : 1 \leq i \leq n\}$ is pairwise uncorrelated and all members have the same variance σ^2.

Let $Y = \sum_{i=1}^{n} a_i X_i$ and $Z = \sum_{i=1}^{n} b_i X_i$. Show that $\mathrm{Cov}[Y, Z] = \sigma^2 \sum_{i=1}^{n} a_i b_i$.

16–16. Suppose X is uniform $(0, 1)$. Let $Y = X^2$. Determine the correlation coefficient $\rho = \rho[X, Y]$. Why is the value not one?

16–17. Suppose the class $\{X, Y, Z\}$ is independent. Let $U = X + Y$ and $V = Y + Z$. Determine $\mathrm{Cov}[U, V]$.

(a) Are there conditions under which $\{U, V\}$ could be independent? If so, describe them.

(b) Suppose $\{X, Y, Z\}$ is pairwise uncorrelated, but not necessarily independent. Is $\mathrm{Cov}[U, V]$ changed?

17

Convergence in Probability Theory

Several notions of convergence play essential roles in probability theory. The theory of such convergence is extensive and sophisticated, although much of it could be developed on the basis of the background provided by this text. We simply examine the underlying concepts and state, largely without proof, some of the principal properties of and relationships between the more important modes of convergence. The goal of this treatment is to provide a minimal background for reading the literature in which these concepts are used. Detailed developments are readily available in many standard works on mathematical probability for those who need a more thorough grounding.

While we are primarily interested in convergence of sequences of random variables, we first examine the notion of the limit of a sequence of events. Then, for sequences of random variables, we consider four types of convergence: almost sure convergence, convergence in probability, convergence in the mean, and convergence of probability distributions. The first of these is pointwise convergence; the others involve quite different modes of convergence.

17.1. Sequences of Events

Consider a sequence of events $\{A_n : 1 \leq n\}$, which we represent mathematically as a sequence of subsets of the basic space Ω. The following concepts are useful.

Definition 17.1.1. *The* limit superior $A^* = \limsup A_n$ *of the sequence is the set of all elements that are in an infinity of the subsets.*

Remark: If the outcome of the experiment ω is in A^*, then it is in an infinity of the sets A_n. This is often expressed in terms of occurrence of the events by saying the A_n occur "infinitely often." This terminology arises naturally in the case the A_n are component events associated with a sequence of component trials (see Section 6.1). For this reason, it is sometimes useful to write A_n i.o. instead of $\limsup A_n$. We will show, how to express $\limsup A_n$ as a Boolean combination of the A_n.

Definition 17.1.2. *The* limit inferior $A_* = \liminf A_n$ *of the sequence is the set of all elements that are in all but a finite number of the sets A_n.*

Remark: It is clear that if ω is in all but a finite number of the A_n it must be in an infinity of them, so that $A_* = \liminf A_n \subset \limsup A_n = A^*$. However, ω could be in an infinity of the sets (for example, in all of the odd numbered members) and not be in all but a finite number of them. So the inclusion does not necessarily go in the opposite direction. If it does, then the two sets are equal. This is a common and important case, which we identify as follows.

Definition 17.1.3. *If* $\liminf A_n = \limsup A_n$, *we say the sequence has a* limit. *The common set is designated* $A = \lim A_n$.

The limit inferior and the limit superior may be expressed in terms of the members of the sequence.

Theorem 17.1.4.

$$A_* = \liminf A_n = \bigcup_{n=1}^{\infty} \bigcap_{k=n}^{\infty} A_k \qquad and \qquad A^* = \limsup A_n = \bigcap_{n=1}^{\infty} \bigcup_{k=n}^{\infty} A_k.$$

Proof

The proof amounts to reading the symbols. Element ω is in all but a finite number of the A_n iff for some n it is in all A_k for $k \geq n$ iff for some n it is in $\bigcap_{k=n}^{\infty} A_k$ iff it is in $\bigcup_{n=1}^{\infty} \bigcap_{k=n}^{\infty} A_k$.

A similar reading establishes the expression for the limit superior.

—   □

Theorem 17.1.5 (Limits of Monotone Sequences) *Every monotone sequence of sets has a limit.*

1. *If $A_n \subset A_{n+1}$ for all n, then* $\lim A_n = \bigcup_{n=1}^{\infty} A_n = \bigcup_{n=k}^{\infty} A_n$ *for any $k \geq 1$.*

2. *If $A_n \supset A_{n+1}$ for all n, then* $\lim A_n = \bigcap_{n=1}^{\infty} A_n = \bigcap_{n=k}^{\infty} A_n$ *for any $k \geq 1$.*

Proof

1. $A_k \subset A_{k+1}$ all k implies $\bigcup_{k=1}^{\infty} A_k = \bigcup_{k=n}^{\infty} A_k$ for all n and $\bigcap_{k=n}^{\infty} A_k = A_n$ for all n. Hence

$$\limsup A_n = \bigcap_{n=1}^{\infty} \bigcup_{k=n}^{\infty} A_k = \bigcup_{k=n}^{\infty} A_k = \bigcup_{n=1}^{\infty} A_n = \liminf A_n.$$

2. The case $A_k \supset A_{k+1}$ all k implies $\bigcup_{k=n}^{\infty} A_k = A_n$ for any n and $\bigcap_{k=1}^{\infty} A_k = \bigcap_{k=n}^{\infty} A_k$ for each n, so that $\limsup A_n = \liminf A_n = \bigcap_{k=n}^{\infty} A_k$ for any n.

—   □

Remark: Property (P10) for probability may be expressed: For monotone sequences of events, $P(\lim A_n) = \lim_n P(A_n)$. By monotonicity, if $\{A^n : 1 \leq n\}$ is nondecreasing (nonincreasing), then so is $\{P(A_n) : 1 \leq n\}$. The following extension is often important.

Theorem 17.1.6. *If $\{A_n : 1 \leq n\}$ is any sequence of events and P is a probability measure, then*

$$P(\liminf A_n) \leq \liminf P(A_n) \leq \limsup P(A_n) \leq P(\limsup A_n).$$

In particular, if $\lim A_n$ exists, then $P(\lim A_n) = \lim_n P(A_n)$.

Proof

Put $B_n = \bigcap_{k=n}^{\infty} A_k$ and $C_n = \bigcup_{k=n}^{\infty} A_k$. Then $B_n \subset B_{n+1}$ and $C_n \supset C_{n+1}$ for all n. Also, $A_* = \lim B_n$ and $A^* = \lim C_n$. Hence, by (P10), $P(A_*) = \lim_n P(B_n) = \sup_n P(B_n)$ and $P(A^*) = \lim_n P(C_n) = \inf_n P(C_n)$. Now,

$$P(B_n) \leq P(A_k), \quad \text{for all } k \geq n, \qquad \text{so} \qquad P(B_n) \leq \inf_{k \geq n} P(A_k),$$

which implies

$$P(A_*) \leq \sup_n \inf_{k \geq n} P(A_k) = \liminf P(A_n).$$

A similar argument shows

$$P(A^*) \geq \inf_n \sup_{k \geq n} P(A_k) = \limsup P(A_n) \geq \liminf P(A_n).$$

The special case follows from the fact that the existence of a limit for a sequence of events is tantamount to the condition $A_* = A^*$ and the condition for the existence of the limit of a sequence of numbers is the equality of the limit superior and the limit inferior.

— □

We use this result to obtain an important extension of the product rule for independent events.

Corollary 17.1.7 (Product Rule for an Independent Sequence) *If $\{A_n : 1 \leq n\}$ is independent, then $P\left(\bigcap_{n=1}^{\infty} A_n\right) = \prod_{n=1}^{\infty} P(A_n)$.*

Proof

Let $E_n = \bigcap_{k=1}^{n} A_k$ and $E = \bigcap_{k=1}^{\infty} A_k$. Then $E_n \supset E_{n+1}$ and $\lim E_n = E$ so that $P(E_n) \to P(E)$.

Now

$$P(E_n) = \prod_{k=1}^{n} P(A_k) \to \prod_{k=1}^{\infty} P(A_k) = P(E).$$

— □

One of the celebrated results on probabilities for sequences of sets is the following.

Theorem 17.1.8 (The Borel-Cantelli Lemma)

a. If $\sum_{n=1}^{\infty} P(A_n) < \infty$, then $P(A^*) = P(A \; i.o.) = 0$.

b. If $\{A_n : 1 \leq n\}$ is independent and $\sum_{n=1}^{\infty} P(A_n) = \infty$, then $P(A^*) = 1$

Proof

1. For any n

$$0 \leq P(A^*) \leq P\left(\bigcup_{k=n}^{\infty} A_k\right) \leq \sum_{k=n}^{\infty} P(A_k) \qquad \text{by subadditivity.}$$

Since $\sum_{k=1}^{\infty} P(A_k)$ converges, the tail sums $\sum_{k=n}^{\infty} P(A_k) \to 0$ as $n \to \infty$. We conclude $P(A^*) = 0$.

2. We use complements to write $(A^*)^c = \bigcup_{n=1}^{\infty} F_n$, where $F_n = \bigcap_{k=n}^{\infty} A_k^c$.

Then by subadditivity and other basic properties $0 \leq 1 - P(A^*) \leq \sum_{n=1}^{\infty} P(F_n)$.

By the product rule for independent sequences,
$P(F_n) = \prod_{k=n}^{\infty} P(A_k^c) = \prod_{k=n}^{\infty} (1 - P(A_k))$.

To obtain a bound, we use the fact that $1 - a \leq e^{-a}$ for $0 \leq a$. Hence, for any $N > n$,

$$\prod_{k=n}^{N} (1 - P(A_k)) \leq \exp\left(-\sum_{k=n}^{N} P(A_k)\right).$$

For each n, the sum $\sum_{k=n}^{N} P(A_k)$ must grow without bound as N increases, so the right-hand expression must go to zero. As a consequence, each $P(F_n) = 0$, so that $1 - P(A^*) = 0$.

—   □

Remark: For the independent case, we have $P(A^*) = 0$ or 1, according as $\sum_{n=1}^{\infty} P(A_n)$ converges or diverges. This is an important example of a zero-one law in probability.

17.2. Almost Sure Convergence

Consider a sequence $\{X_n : 1 \leq n\}$ of real random variables. For each outcome ω, $\{X_n(\omega) : 1 \leq n\}$ is a sequence of real numbers. Such a sequence may converge for some ω and diverge for others. We will show that the set D of ω for which the sequence $\{X_n(\omega) : 1 \leq n\}$ fails to converge, the divergence set, and its complement C, the convergence set, are both events. If we define the function

$$X(\omega) = \begin{cases} \lim X_n(\omega) & \text{for } \omega \text{ in } C \\ 0 & \text{for } \omega \text{ in } D \end{cases}$$

some standard measure-theoretic analysis shows X must be a random variable.

Conceptually the simplest, and usually the most desirable, mode of convergence is "almost sure" convergence, or "convergence with probability one."

Definition 17.2.1. *A sequence* $\{X_n : 1 \leq n\}$ *of random variables is said to converge almost surely, or to converge with probability one, iff the probability of the divergence set is zero.*

Example 17.2.1: An Almost Surely Convergent Sequence

We take as our basic space Ω the unit interval $[0, 1]$ on the real line. The class of events is the class of Borel sets on the interval and the probability measure P is Lebesgue measure, which assigns to each subinterval of $[0, 1]$ its length. For each $n \geq 1$, define X_n on the basic space by

$$X_n = \begin{cases} I_{[0,1/n]} & \text{for } n \text{ odd} \\ I_{(0,1/n]} & \text{for } n \text{ even} \end{cases} .$$

Then, $X_n(\omega) \to 0$ as $n \to \infty$ for every ω in the interval except for $\omega = 0$. The values of $X_n(0)$ alternate between 0 and 1, so that there is no limit. The divergence set D is the single-element set $\{0\}$, which has probability zero. Thus, $X_n \to 0$ a.s..

— □

To analyze the almost sure convergence condition, we study the divergence set D. The sequence $\{X_n(\omega) : 1 \leq n\}$ fails to converge to $X(\omega)$ iff

There is some k such that for any n there is at least one $i \geq n$ such that
$$|X(\omega) - X_i(\omega)| \geq \frac{1}{k}.$$

Let $E_i(k) = \{\omega : |X(\omega) - X_i(\omega)| \geq 1/k\}$. Then $A_{kn} = \bigcup_{i=n}^{\infty} E_i(k)$ is the set of those ω such that $|X(\omega) - X_i(\omega)| \geq 1/k$ for all $i \geq n$. Now

$$D_k = \bigcap_{n=1}^{\infty} A_{kn} = \bigcap_{n=1}^{\infty} \bigcup_{i=n}^{\infty} E_i(k) = \limsup_i E_i(k)$$

is the set of those ω for which there is at least one $i \geq n$ such that the inequality holds. The sequence fails to converge iff ω belongs to at least one of the D_k. Hence

$$D = \bigcup_{k=1}^{\infty} D_k.$$

If X is a random variable, then each $E_i(k)$ must be an event; therefore D and $C = D^c$ are events.

The following theorems provide important conditions for almost sure convergence.

Theorem 17.2.2. (a) $X_n \to X$ a.s.   iff   (b) $\lim_n P(A_{kn}) = 0$ for every $k \geq 1$.

Proof
(a) implies (b). $0 = P(D) \geq P(D_k) = \lim_n P(A_{kn}) \geq 0$.
  (b) implies (a). Each $P(D_k) = \lim_n P(A_{kn}) = 0$. Hence, $P(\bigcup_{k=1}^{\infty} D_k) \leq \sum_{k=1}^{\infty} P(D_k) = 0$.
  $\square$

Theorem 17.2.3. If $\sum_{i=1}^{\infty} P(E_i(k))$ converges for each $k \geq 1$, then $X_n \to X$ a.s.

Proof
By the Borel-Cantelli lemma, $P(D_k) = P(\limsup_i E_i(k)) = 0$ for each k.
  $\square$

THE STRONG LAW OF LARGE NUMBERS

A variety of related theorems are referred to by this name. With the use of an inequality known as *Kolmogorov's inequality*, an analytical lemma known as *Kronecker's lemma*, and ideas similar to the preceding ones, the following general result has been established.

Theorem 17.2.4   (Kolmogorov's Strong Law of Large Numbers)
  Let $\{X_n : 1 \leq n\}$ be an independent class of random variables, each with finite mean and variance. Let $\{a_n : 1 \leq n\}$ be an increasing sequence of positive numbers with $a_n \to \infty$. Let $S_n = \sum_{k=1}^{n} X_k$.
  If $\sum_{n=1}^{\infty} \mathrm{Var}[X_n]/a_n^2 < \infty$ then $(S_n - E[S_n])/a_n \to 0$ a.s.
  In particular, if $E[X_n] = \mu$ and $\mathrm{Var}[X_n] \leq \sigma^2$ for all n, then $S_n/n \to \mu$ a.s.
  $\square$

The special case is the usual form of the strong law. Convergence in the strong law is almost sure convergence. If independent random variables have a common mean value, then for almost every ω the average of a large number of the $X_n(\omega)$ will lie close to the mean value and the approximation improves as the number of terms increases. This notion was operative at an intuitive level long before the probability model was formulated precisely and the mathematical theory developed carefully. Early forms of the theorem were, in fact, versions of what is now known as the weak law of large numbers. The weak law involves a second type of convergence, which we examine in the next section.

ALMOST UNIFORM CONVERGENCE

Let ϵ be any positive number. Suppose $X_n \to X$. This means that for each ω there is an integer N such that $|X(\omega) - X_n(\omega)| < \epsilon$ for all $n \geq N$. Now, in general, the value of N depends on ω (for a given ϵ). If for each ϵ there is an N for which the inequality holds for all ω, then we say the convergence is *uniform*. It frequently occurs that we may have uniform convergence for all ω in some subset A of the basic space. In this case, we say the sequence converges uniformly over A. The following condition is important in the study of sequences of random variables.

Definition 17.2.5. *A sequence $\{X_n : 1 \leq n\}$ of random variables converges almost uniformly to a random variable X, denoted $X_n \to X$ a. unif., iff for each $\epsilon > 0$, there is an event A with $P(A) < \epsilon$, such that the sequence converges uniformly over the set A^C.*

The following relationship has been established.

Theorem 17.2.6 (Almost Sure Convergence and Almost Uniform Convergence) *For a probability measure, a sequence converges almost surely iff it converges almost uniformly.*
—   □

17.3.   Convergence in Probability

Frequently, we cannot say whether a sequence $\{X_n : 1 \leq n\}$ of random variables converges in a pointwise sense, i.e., converges for any given ω. But we may be able to assert that if we are far enough out in the sequence (i.e., n is large enough), then with high probability the value selected will lie close to some limiting value. This leads to the notion of convergence in probability. This is a new kind of convergence. To understand its characteristics and to relate it to pointwise convergence, we need to extend the analysis of the convergence and divergence sets.

We recall that a sequence of real numbers converges iff it is a fundamental sequence, or a Cauchy sequence. This means that the terms of the sequence must lie increasingly close together as n increases.

Definition 17.3.1. *A sequence* $\{a_n : 1 \le n\}$ *of real numbers is said to be* fundamental, *or* Cauchy, *iff for every integer* $k \ge 1$, *there is an integer* n *such that for all* $i \ge n$, $|a_i - a_n| < 1/k$.

It is a an essential property of real numbers that every fundamental sequence must have a finite limit. An extension of the argument in Section 17.2 shows that for a sequence of real random variables the divergence set may be expressed

$$D = \bigcup_{k=1}^{\infty} \bigcap_{n=1}^{\infty} \bigcup_{i=1}^{\infty} G_{in}(k) = \bigcup_{k=1}^{\infty} \bigcap_{n=1}^{\infty} \bigcup_{i=1}^{\infty} E_i(k)$$

where

$$
\begin{aligned}
E_i(k) &= \{\omega : |X(\omega) - X_i(\omega)| \ge 1/k\} \\
G_{in}(k) &= \{\omega : |X_i(\omega) - X_n(\omega)| \ge 1/k\}.
\end{aligned}
$$

In terms of this analysis, we introduce the concept of convergence in probability.

Definition 17.3.2. *A sequence* $\{X_n : 1 \le n\}$ *of real random variables*

a.   *Is* fundamental in probability *iff* $\lim_{m,n} P(G_{mn}(k)) = 0$ *for all* $k \ge 1$

b.   *Converges in probability* to random variable X *iff* $\lim_i P(E_i(k)) = 0$ *for all* $k \ge 1$.

One question immediately comes to mind. How is convergence in probability different from almost sure convergence? We attempt to form a visual representation, then produce a simple counterexample, and finally look at the mathematical conditions for the two types of convergence.

A VISUAL REPRESENTATION

In the informal discussion in Section 1.1, the basic set Ω of possible outcomes of an experiment or trial is represented by a jar of balls—each ball representing one possible outcome. In other words, an element ω of the basic set is represented by a ball. Performance of the experiment is represented by the selection of one of the balls in the jar. In the case of sequences of random variables, we may employ that analogy as follows. Instead of a ball, we represent each possible outcome ω by a "tape" of infinite length. For each ω we have a sequence of numbers $\{X_n(\omega) : 1 \le n\}$. We write those numbers sequentially on the tape corresponding to the given ω. Making an observation of the sequence of random variables is represented by selecting a tape. A probability measure assigns probabilities to various subsets of the tapes.

Almost Sure Convergence

In this case, for almost every tape, the sequence on the tape converges. The exceptional tapes make up a set whose total probability is zero. This means that it is extremely unlikely, although perhaps not impossible, that one of these tapes will be selected on any random trial.

Convergence in Probability

This condition does not guarantee convergence in the usual (pointwise) sense on any tape. It simply says that if the nth number on the tape selected is to be observed, for large n the probability is high that number on the tape selected will lie close to the "limit value" $X(\omega)$. It does not guarantee anything about the terms $X_i(\omega)$ for $i > n$.

Example 17.3.1: A Sequence that Converges in Probability but not Almost Surely

As in Example 17.2.1, we take as our basic space Ω the unit interval $[0, 1]$ on the real line. The class of events is the class of Borel sets on the interval and the probability measure P is Lebesgue measure, which assigns to each subinterval of $[0, 1]$ its length. We define a sequence of random variables, each of which is a step function on the unit interval. These are formed in groups, as follows.

1. The first group consists of two functions:

$$X_1 = I_{[0,1/2)}, \qquad X_2 = I_{[1/2,1]}.$$

2. The second group consists of four functions:

$$X_3 = I_{[0,1/4)}, \quad X_4 = I_{[1/4,1/2)}, \quad X_5 = I_{[1/2,3/4)}, \quad X_6 = I_{[3/4,1]}.$$

The nth group consists of 2^n functions obtained as follows:

1. Partition the unit interval into 2^n equal subintervals, with all but the last open on the right. Each ω in the unit interval belongs to exactly one of these subintervals.

2. Let the rth function in the nth subgroup be the indicator function for the rth subinterval.

Let $X(\omega) = 0$ for all ω in the unit interval. Then $E_k(k) = \{\omega : |X_i(\omega)| \geq 1/k\}$. For any X_i in the nth group,

$$P\big(E_i(k)\big) = P(|X_i| \geq 1/k) = 2^{-n}.$$

As n increases, this probability approaches zero. By definition, $X_n \xrightarrow{P} 0$.

But consider any ω in the unit interval. By construction, this number will be in exactly one subinterval in each group, so that exactly one $X_i(\omega)$ in each group will have value one. All others in the group will have value zero. Consider the sequence of values $\{X_n(\omega) : 1 \leq n\}$. This will consist of a string of zeros and ones. The ones will be increasingly widely separated, but beyond any point there will always be additional ones. Thus, the sequence cannot converge in the ordinary sense for any ω. We have convergence in probability, but no pointwise convergence, so certainly not almost sure convergence.

In terms of the tape representation, each tape will have a string of zeros and ones, with the ones increasingly separated as the numbers on the tape are read sequentially. On no tape does the sequence converge. Yet for large enough n, the probability can be made arbitrarily small that the nth number on a tape selected at random has value one.

—   ☐

The example shows that convergence in probability does not imply almost sure convergence. However, we can easily establish the following.

Theorem 17.3.3. *If a sequence $\{X_n : 1 \leq n\}$ converges almost surely to X, then it converges in probability to X.*

Proof

Since $A_{nk} = \bigcup_{i=n}^{\infty} E_i(k)$, we have $P(A_{nk}) \geq P(E_n(k))$ for all n, $k \geq 1$. Now

$$X_n \to X \quad \text{a.s.} \quad \text{implies} \quad P(A_{nk}) \to 0 \quad \text{for all } k$$

$$\text{implies} \quad P(E_n(k)) \to 0 \quad \text{for all } k \quad \text{implies} \quad X_n \xrightarrow{P} X.$$

—   ☐

Since convergence in probability is a new kind of convergence, several questions need to be settled to establish the potential usefulness of the concept. We gather a number of important propositions that answer these questions. Proofs may be found in many standard works on mathematical probability.

Theorem 17.3.4   (Properties of Convergence in Probability)

a. *Uniqueness. If $X_n \xrightarrow{P} X$ and $X_n \xrightarrow{P} Y$, then $X = Y$ a.s.*

b. $X_n \to X$ *a.s. implies $X_n \xrightarrow{P} X$.*

c. $X_n \xrightarrow{P} X$ *implies the sequence is fundamental in probability.*

d. *If the sequence is fundamental in probability, then there is a subsequence that converges almost surely to a random variable X.*

e. *If the sequence is fundamental in probability, then there is a random variable X such that $X_n \xrightarrow{P} X$, and this random variable is almost surely equal to the random variable in part d.*

—   □

WEAK LAW OF LARGE NUMBERS

The so-called weak laws involve convergence in probability. Historically, these were the first to be proved, since they can be established by simpler arguments. A major tool in most proofs is the Chebyshev inequality (see Section 15.4). In fact, it seems that for a long period, the weak laws were interpreted as if they ensured strong convergence. This seems to have been true of Bernoulli's form of the law of large numbers for Bernoulli sequences (ca. 1713), which involves weak convergence. It was not until the first decade of the twentieth century (ca. 1909) that Borel established the strong form, which drew attention to the difference. As Example 17.3.1 shows, there are cases in which there is weak convergence but not strong convergence.

17.4.   Convergence in the Mean

In this section, we sketch the theory of a type of convergence intimately tied to the theory of integrals or expectation. To formulate the concept precisely, we adopt some standard terminology and notation. We suppose throughout that we have a given probability measure defined on the events in a basic space. That is, we assume a probability space $(\Omega, \mathcal{F}, P)$. For any positive number $p \geq 1$, we consider the class $\mathcal{L}^p$ of random variables such that $E[|X|^p]$ is finite. That is, X^p is integrable. The symbol $\mathcal{L}^p$ is usually read "L p".

Definition 17.4.1. *Consider a sequence $\{X_n : 1 \leq n\}$ of random variables in $\mathcal{L}^p$*

1. *The sequence is* fundamental in the mean of order p *iff for each $\epsilon > 0$, there is an integer N such that for all $m, n \geq N$, $E[|X_n - X_m|^p] < \epsilon$.*

2. *The sequence* converges in the mean of order p *iff there is a random variable X such that $E[|X - X_n|^p] \to 0$ as $n \to \infty$. We denote this convergence by $X_n \xrightarrow{L^p} X$ or by $\lim_n X_n = X$ $[\mathcal{L}^p]$.*

Remark: Usually we are interested in $p = 1$, in which case we speak of convergence in the mean, or $p = 2$, in which case we speak of mean-square convergence.

INTERPRETATION

To determine whether $X_n \xrightarrow{L^p} X$, we study the probability-weighted average value of the random variables $|X - X_n|^p$. For a given ω, the difference $X(\omega) - X_n(\omega)$ is the error of approximation of $X(\omega)$ by $X_n(\omega)$ and $|X(\omega) - X_n(\omega)|^p$ is the pth power of the magnitude of that error. Taking the expectation determines the probability-weighted average of these values. If these averages go to zero as n increases, we have mean convergence of order p.

Example 17.4.1: Mean Convergence and Average Difference

The averaging process just described may be visualized in terms of random variables defined on the probability space used in Example 17.2.1 and Example 17.3.1. As in those examples, we take as our basic space Ω the unit interval $[0, 1]$ on the real line. The class of events is the class of Borel sets on the interval and the probability measure P is Lebesgue measure, which assigns to each subinterval of $[0, 1]$ its length. Suppose the X_n and X are continuous, hence Borel, on the unit interval. Then $E[|X - X_n|] = \int_0^1 |X(t) - X_n(t)|\, dt$ is the area of the region between the two curves. If we take p greater than one, then large differences are penalized more than smaller ones.

—   □

Since $\mathcal{L}^p$ convergence is quite different from pointwise convergence, a number of properties must be established. We gather several of the more important in the following.

Theorem 17.4.2   (Some Properties of Mean Convergence)

a. *If* $X \xrightarrow{L^p} X$, *then* X *is in* $\mathcal{L}^p$.

b. *Uniqueness: If* $X_n \xrightarrow{L^p} X$ *and* $X_n \xrightarrow{L^p} Y$, *then* $X = Y$ *a.s.*

c. *If* $1 \le p < q$, *then* X *in* $\mathcal{L}^q$ *implies* X *in* $\mathcal{L}^p$ *and* $X_n \xrightarrow{L^q} X$ *implies* $X_n \xrightarrow{L^p} X$.

d. *The sequence is fundamental in the mean of order* p *iff there is an* X *in* $\mathcal{L}^p$ *for which* $X_n \xrightarrow{L^p} X$.

e. $X_n \xrightarrow{L^p} X$ *and* $Y_n \xrightarrow{L^p} Y$ *implies* $aX_n + bY_n \xrightarrow{L^p} aX + bY$.

—   □

Uniform Integrability

Membership in $\mathcal{L}^p$ is a question of integrability. In dealing with arbitrary classes of random variables, the property of uniform integrability is frequently crucial. According to the property (E9b) for expectation

$$X \text{ is integrable iff } E[I_{\{|X|>a\}}|X|] \to 0 \quad \text{as } a \to \infty.$$

We use this characterization of integrability of a single random variable to define the notion of uniform integrability of a class.

Definition 17.4.3. *An arbitrary class $\{X_t : t \in T\}$ of random variables is uniformly integrable (abbreviated u.i.) with respect to a probability measure P iff*

$$\sup_t E[I_{\{|X_t|>a\}}|X_t|] \to 0 \quad \text{as } a \to \infty.$$

Remark: Not only is uniform integrability important for certain questions of mean convergence, but it also plays a key role in many other areas– for example in theory of submartingales.

We next list several propositions that show some of the ties between the various types of convergence. While this list is important, it is far from complete.

Theorem 17.4.4 (Some Relationships between Convergence Types)
Consider a sequence $\{X_n : 1 \leq n\}$ of random variables in $\mathcal{L}^p, p \geq 1$.

 a. $X_n \overset{L^p}{\to} X$ *iff the sequence is u.i. and $X_n \overset{P}{\to} X$ in $\mathcal{L}^p$.*

 b. $X_n \overset{L^1}{\to} X$ *iff $E[I_A X_n] \to E[I_A X]$ uniformly in $A \in \mathcal{F}$.*

 c. *If $X_n \to X$ a.s., X in $\mathcal{L}^1$, and the sequence is u.i., then $X_n \overset{L^1}{\to} X$.*

—   □

For a summary of relationships between the principal types of convergence, see the end of the next section.

17.5.   Convergence in Distribution

The final type of convergence we consider is known as convergence in distribution of sequences of random variables, or as weak convergence of the sequences of probability distribution functions for those random variables. We begin by considering a sequence $\{F_n : 1 \leq n\}$ of probability distribution functions. We suppose, as usual, $F_n(-\infty) = 0$ and $F_n(\infty) = 1$, for every n.

Definition 17.5.1. *The sequence $\{F_n : 1 \leq n\}$ of probability distribution function is said to converge weakly to F, denoted $F_n \overset{w}{\to} F$, iff $F_n(t) \to F(t)$ at every point of continuity for F. If $F(-\infty) = 0$ and $F(\infty) = 1$, the sequence is said to converge completely, denoted $F_n \overset{c}{\to} F$.*

If $\{X_n : 1 \leq n\}$ is a sequence of random variables, with F_n the distribution function for X_n, and X is a random variable with distribution function F, then the sequence of random variables is said to *converge in distribution* to X, denoted $X_n \overset{d}{\to} X$, iff $F_n \overset{c}{\to} F$.

Theorem 17.5.2. *Convergence in probability implies convergence in distribution.*

Proof

Let t be a point of continuity of F. Then,

$$
\begin{aligned}
P(X \leq t - \epsilon) &= P(X \leq t - \epsilon, X_k \leq t) + P(X \leq t - \epsilon, X_k > t) \\
&\leq P(X_k \leq t) + P(|X_k - X| > \epsilon).
\end{aligned}
$$

Similarly

$$
\begin{aligned}
P(X_k \leq t) &= P(X_k \leq t, X > t + \epsilon) + P(X_k \leq t, X \leq t + \epsilon) \\
&\leq P(|X - X_k| > \epsilon) + P(X \leq t + \epsilon).
\end{aligned}
$$

Now

$$
P(X_k \leq t) = F_k(t), \qquad P(X \leq t - \epsilon) = F(t) - O_1(\epsilon),
$$

and

$$
P(X_t \leq t + \epsilon) = F(t) + O_2(\epsilon).
$$

The continuity of F at t ensures that the positive functions O_1 and O_2 decrease to zero as ϵ decreases to zero. The convergence in probability of the X_n ensures that for any $\epsilon > 0$, $P(|X - X_k| > \epsilon)$ can be made as small as desired by keeping k large enough. We combine the inequalities just listed to assert

$$
F(t) - O_1(\epsilon) - P(|X - X_k| > \epsilon) \leq F_k(t) \leq F(t) + O_2(\epsilon) + P(|X - X_k| > \epsilon).
$$

By selecting ϵ sufficiently small and taking k sufficiently large, we can guarantee $F_n(t)$ as close as we please to $F(t)$ for all $n \geq k$. $\qquad \square$

Among the important characterizations of convergence in distribution is the following.

Theorem 17.5.3. *The sequence $\{X_n : 1 \leq n\}$ of real random variables converges in distribution to random variable X iff $E[g(X_n)] \to E[g(X)]$ for all real-valued functions g that are bounded, continuous, and go to zero at $\pm\infty$.* $\qquad \square$

The notion of convergence in distribution of random variables, or weak convergence of distributions, is important in the theory of characteristic functions. Theorem 18.5.1 on limits of characteristic functions is a prime example. The central limit theorem, as stated in Section 18.6, is a theorem on convergence in distribution. The following recapitulation shows that convergence in distribution occurs whenever any of the other principal types considered herein occurs.

Theorem 17.2.6 shows almost sure convergence *iff* almost uniform convergence.

Theorem 17.3.3 shows almost sure convergence *implies* convergence in probability.

Theorem 17.4.4 shows mean convergence *iff* convergence in probability and uniform integrability.

Theorem 17.5.2 shows convergence in probability *implies* convergence in distribution.

PROBLEMS

17–1. Let $X_n = I_{E_n}$, where E_n is the event of a success on the nth trial in a Bernoulli sequence, with $P(E_n) = p$. Let $A_n = \sum_{i=1}^{n} X_i$, the fraction of successes in the first n trials. Use the strong law of large numbers to show $A_n \to p$ a.s. as $n \to \infty$. [This is Borel's theorem.]

17–2. Suppose $\{X_n : 1 \le n\}$ is iid, exponential (λ). Use the strong law of large numbers to show that

$$\lim_n A_n = \lim_n \frac{1}{n} \sum_{k=1}^{n} X_k = \frac{1}{\lambda}.$$

17–3. Suppose $X_n \le X_{n+1}$ for all $n \ge 1$ and $X_n \to^P X$. Show that $X_n \to X$ a.s.

17–4. Consider a discrete probability space. That is, suppose there is a countable number of possible outcomes and each event $\{\omega\}$ consisting of one element has positive probability. Show that convergence in probability implies almost sure convergence.

17–5. Show that if $X_n \to^P X$ and $X_n \to^P Y$, then $X = Y$ a.s.

17–6. Suppose $X_n \to^P X$ and $Y_n \to^P Y$, with $X = Y$ a.s. Show that $P(|X_n - Y_n| \ge \epsilon) \to 0$ as $n \to \infty$, for any $\epsilon > 0$.

17–7. Suppose $X_n \to^P X$ and $Y_n \to^P Y$. Show

    (a) $aX_n + bY_n \to^P aX + bY$.

    (b) $|X_n| \to^P |X|$.

    (c) $X_n Y_n \to^P XY$.

18

Transform Methods

In previous chapters we show that $E[X]$ and $E[(X - \mu_X)^2]$ give important information about the distribution for a real random variable X. In some statistical studies, the third central moment $E[(X - \mu_X)^3]$ is used to indicate a lack of symmetry of the distribution about the center of mass. These are expectations of functions of the random variable X. It is natural to ask whether there are other functions h such that $E[h(X)]$ provides useful information.

In this chapter, we consider three important cases in which the function h has a parameter, so that we have a one-parameter family of functions of X. One might suppose that for suitable choice of h the resulting family of expectations would contain much more information about the distribution than could the expectation of any single function of X or the expectations of a small set functions of X. In the subsequent development, we see that this is indeed true.

18.1. Expectations and Integral Transforms

We consider a Borel function $h(\cdot, \cdot)$ such that $E[h(s, X)]$ is finite for each s in a parameter set S. For each value of the parameter s, the expectation $E[h(s, X)]$ has a value that depends upon s. Thus $T(s) = E[h(s, X)]$ defines a function T over the parameter set. This is not just a mathematical curiosity. In each of the three cases considered later, the resulting function has properties that make it useful as an analytical tool for studying distributions. The expectations, as integrals, have the forms of certain standard *integral transforms*, which are used extensively as tools for analysis in many fields of engineering and physical science. The usefulness of these transforms is thus enhanced by the existence of a considerable literature devoted to their theory and applications.

The functions considered herein play an essential role in modern probability theory and its applications. We begin with formal definitions, examine some relationships between the functions introduced, then consider their relationships to well-known integral transforms. We then obtain some properties upon which their usefulness is based. To illustrate techniques and to obtain some useful results for later reference, we determine the transforms for some common distributions.

The significance and usefulness of these transforms become apparent only gradually. It is the nature of transform theory to be abstract and formal. Insight is a result of familiarity with certain correlations between patterns of "behavior" of the transforms and corresponding patterns in the structure of the probability distributions represented. This insight comes with perseverance and practice.

We introduce the moment-generating function, the characteristic function, and the generating function. The transforms are related to *distributions* rather than to random variables. However, it is notationally convenient to designate a distribution by the symbol for a random variable that has that distribution.

Moment generating function:
$$M_X(s) = E[e^{sX}] \quad s \text{ is a } complex \text{ parameter.}$$

Characteristic function:
$$\phi_X(u) = E[e^{iuX}] \quad u \text{ is a } real \text{ parameter.}$$

Generating function:
$$g_X(s) = E[s^X] \quad s \text{ is a } real \text{ parameter.}$$

Remarks about the generating function

- The generating function is also known by the following names: probability generating function, factorial moment-generating function, and z-transform (since z is commonly used to represent the parameter).

- Although defined for a more general class of random variables, the generating function is most useful when X is a nonnegative, integer-valued random variable.

It should be clear from the defining expressions that these transforms are related. For reference, we summarize as follows:

RELATIONSHIPS

1. Since $i^2 = -1$, so that $1/i = -i$, we have

$$\phi_X(u) = M_X(iu) \quad \text{and} \quad M_X(s) = \phi_X(-is) \quad (s \text{ real}).$$

2. Since $s^t = e^{t \ln s}$, we have

$$g_X(s) = M_X(\ln s) \quad \text{and} \quad M_X(s) = g_X(e^s) \quad (s \text{ real}).$$

RELATIONSHIPS TO INTEGRAL TRANSFORMS

We now examine the relationships of these concepts to certain widely used integral transforms, in order to facilitate reference to the extensive literature on those subjects, should the need arise. It is not necessary to be familiar with that literature in order to understand the developments in this and subsequent sections.

Many works on probability theory limit the parameter s in the moment-generating function M_X to real values. This is an unnecessary restriction. If we allow complex values, we may exploit the relationships to classical integral transforms.

In the general case:
$$M_X(-s) = \int e^{-st} F_X(dt) \quad \text{(Laplace-Stieltjes transform of } F_X)$$
$$\phi_X(-u) = \int e^{-iut} F_X(dt) \quad \text{(Fourier-Stieltjes transform of } F_X).$$
In the absolutely continuous case:
$$M_X(-s) = \int e^{-st} f_X(t)\, dt \quad \text{(bilateral Laplace transform of } f_X)$$
$$\phi_X(-u) = \int e^{-iut} f_X(t)\, dt \quad \text{(Fourier transform of } f_X).$$
In the discrete case:
$$M_X(s) = \sum_i p_i e^{st_i} \quad \text{(Dirichlet series).}$$
In the nonnegative, integer-valued case:
$$g_X(s) = \sum_i p_i s^i \quad \text{(power series or polynomial).}$$

Remarks

1. In each case, the transform is determined by the distribution. If two random variables have the same distribution, they have the same M_X, the same ϕ_X, and the same g_X.

2. Use of certain well-known analytical results shows an essentially one-one relationship between any one of these transforms and the associated distribution. Thus, if two distributions have the same M_X, then they are essentially the same. Remember, however, that quite different random variables may have the same distribution.

3. The moment-generating function and the characteristic function may be used to obtain the moments $E[X^n]$ for random variable X.

4. There are certain operational properties and limit theorems that make these functions useful. For one of the latter, see the treatment of the characteristic function in Section 18.5.

MOMENTS

1. One important characteristic of the moment-generating function (from which it gets its name) is the manner in which it may be used to determine the moments of the distribution. If $E[X^k]$ is finite, then

$$M^{(k)}(0) = E[X^k].$$

2. For the characteristic function, the formula is only slightly complicated by the presence of the imaginary unit i in the expression.

$$\phi^{(k)}(0) = i^k E[X^k].$$

3. The generating function g_X may also be used to obtain moments, although not as easily. For the first moment, we have

$$g_X'(1) = E[X].$$

It is more difficult to use the generating function to find higher moments. It is usually better to transform $g_X(s)$ into $M_X(s)$ by the substitution of e^s for s (see Example 18.3.3).

Derivation

We derive the rule for the moment-generating function as follows:

$$M^{(k)}(s) = \frac{d^k}{ds^k} E[e^{sX}] = E[X^k e^{sX}].$$

The differentiation "inside the integral" with respect to the parameter s is justified by the regularity of the exponential function and the simple structure of the distribution function F_X. Upon setting $s = 0$, we obtain the desired result.

A similar argument holds for the characteristic function.

—   □

OPERATIONAL PROPERTIES

Among the important properties of these transforms, there are two that we refer to as operational properties. These are

T1) If $Z = aX + b$, a, b constants, then

$$M_Z(s) = e^{sb} M_X(as), \quad \phi_Z(u) = e^{iub} \phi_X(au), \quad g_Z(s) = s^b g_X(s^a).$$

Proof

To see why this is so, consider the moment-generating function. We have
$$E[e^{s(aX+b)}] = e^{sb} E[e^{(as)X}]$$

from which the property follows. Similar arguments hold for the characteristic function and the generating function.

—   □

T2) If $\{X, Y\}$ is independent then

$$M_{X+Y} = M_X M_Y, \qquad \phi_{X+Y} = \phi_X \phi_Y, \qquad \text{and} \qquad g_{X+Y} = g_X g_Y.$$

Proof

For the moment-generating function, we have

$$E[e^{s(X+Y)}] = E[e^{sX} e^{sY}] = E[e^{sX}] E[e^{sY}].$$

Similar arguments hold for the other transforms.

— $\qquad\qquad\qquad\qquad\qquad\qquad\qquad\qquad\qquad\qquad\qquad\qquad\qquad\qquad$ □

A partial converse to (T2) is obtained as follows:

T3) If $M_{X+Y}(s) = M_X(s)M_Y(s)$, then $E[XY] = E[X]E[Y]$ (i.e., $\{X, Y\}$ is uncorrelated).

Proof

To prove this, we obtain two expressions for $E[(X + Y)^2]$, one by direct expansion and use of linearity, the other by taking the second derivative of the moment-generating function.

$$E[(X + Y)^2] = E[X^2] + E[Y^2] + 2E[XY]$$

$$
\begin{aligned}
M''_{X+Y}(s) &= \bigl(M_X(s)M_Y(s)\bigr)'' \\
&= M''_X(s)M_Y(s) + M_X(s)M''_Y(s) + 2M'_X(s)M'_Y(s).
\end{aligned}
$$

Setting $s = 0$ and recalling that $M_X(0) = M_Y(0) = 1$, we get

$$E[(X + Y)^2] = E[X^2] + E[Y^2] + 2E[X]E[Y].$$

Taking the difference of the two expressions for $E[(X + Y)^2]$ yields the desired result.

— $\qquad\qquad\qquad\qquad\qquad\qquad\qquad\qquad\qquad\qquad\qquad\qquad\qquad\qquad$ □

BASIS FOR CHOICE

Fundamental theory shows that each of these transforms characterizes completely the probability distribution. On what basis does one choose from among the alternatives? We summarize some of the primary considerations.

- Since $\phi_X(u)$ exists for all real u and the theory of Fourier integrals is highly developed, there are many mathematical advantages in using the characteristic function—for both theoretical and applied purposes.

- $M_X(s)$ exists for s in a vertical strip on the complex plane; this strip includes the origin. If the strip is not degenerate (i.e., reduced to the vertical axis), then M_X has a power series expansion about the origin. Expressions for M_X are usually "cleaner" than corresponding ones for the characteristic function ϕ_X, because of the suppression of the imaginary unit i.

  We ordinarily use the moment-generating function. The expression for the characteristic function ϕ_X can be obtained by substituting iu for s in the expression for the moment-generating function M_X.

- g_X is used almost exclusively for nonnegative, integer-valued random variables. If the random variable is simple, the generating function is a polynomial; otherwise, it is a power series. The coefficients for s^k in the power series are the probabilities $p_k = P(X = k)$.

18.2.   Transforms for Some Common Distributions

We calculate the moment-generating function for some standard distributions. If the random variable is nonnegative and integer-valued, we also calculate the generating function. In obtaining these results, we illustrate the use of some of the properties of these transforms.

DISCRETE DISTRIBUTIONS

1. *Indicator function*: $X = I_E$ $P(X = 1) = P(E) = p$, $P(X = 0) = q = 1 - p$

$$E[X] = p \qquad \text{Var}[X] = pq$$

$$M_X(s) = E[e^{sX}] = q + pe^s \qquad g_X(s) = E[s^X] = q + ps.$$

2. *Simple random variable*: $X = \sum_{i=1}^n t_i I_{A_i}$ (canonical form) $P(X = t_i) = P(A_i) = p_i$

$$E[X] = \sum_{i=1}^n t_i p_i \qquad \text{Var}[X] = \sum_{i=1}^n t_i^2 p_i q_i - 2 \sum_{i<j} t_i t_j p_i p_j$$

$$M_X(s) = \sum_{i=1}^n p_i e^{st_i}.$$

3. *Binomial (n, p)*: $X = \sum_{i=1}^n I_{E_i} = \sum_{i=1}^n X_i$ with $\{I_{E_i} : 1 \le i \le n\}$ iid $P(E_i) = p$

$$E[X] = np \qquad \text{Var}[X] = npq.$$

Use of the result for the indicator function and the fact that X is the sum of independent random variables gives

$$M_X(s) = \prod_{i=1}^{n} M_{X_i}(s) = (q + pe^s)^n.$$

A similar argument shows that

$$g_X(s) = (q + ps)^n.$$

We could also have used the fact that $g_X(s) = M_X(\ln s)$ and $e^{\ln s} = s$.

4. *Geometric(p):* $P(X = k) = pq^k$ for all $k \geq 0$

$$E[X] = q/p \qquad \text{Var}[X] = \frac{q}{p^2}.$$

The moment-generating function is given by an infinite series that may be put into the form of a geometric series, and hence may be expressed in closed form.

$$M_X(s) = p \sum_{k=0}^{\infty} e^{sk} q^k = p \sum_{k=0}^{\infty} (qe^s)^k = \frac{p}{1 - qe^s} \qquad |qe^s| < 1.$$

The same transformation as used in the binomial case shows that

$$g_X(s) = \frac{p}{1 - qs} \qquad |qs| < 1.$$

If $Y - 1$ is geometric, so that $P(Y = k) = pq^{k-1}$ for all $k \geq 1$, then

$$E[Y] = \frac{1}{p} \qquad \text{Var}[Y] = \frac{q}{p^2}$$

$$M_Y(s) = \frac{pe^s}{1 - qe^s} |qe^s| < 1 \qquad \text{and} \qquad g_Y(s) = \frac{ps}{1 - qs} |qs| < 1.$$

5. *Negative binomial (m, p):*

We show in Section 8.2 that if Y_m is the number of the trial in a Bernoulli sequence on which the mth success occurs and $X_m = Y_m - m$ is the number of failures before the mth success, then

$$P(X_m = k) = P(Y_m - m = k) = C(-m, k)(-q)^k p^m$$

so that

$$M_{X_m}(s) = p^m \sum_{k=0}^{\infty} C(-m, k)(-q)^k e^{sk} = p^m \sum_{k=0}^{\infty} C(-m, k)(-qe^s)^k.$$

Using the fact that

$$(1-x)^{-m} = \sum_{k=0}^{\infty} C(-m,k)(-x)^k \qquad m \geq 1|x| < 1$$

we have

$$M_{X_m}(s) = \frac{p^m}{(1-qe^s)^m} = \left(\frac{p}{1-qe^s}\right)^m \qquad |qe^s| < 1.$$

Remark: $X_m = Y_m - m$ has the same distribution as the sum of m independent random variables, each geometric (p). This suggests that the sequence is characterized by independent, successive waiting times to success. This allows us to obtain the expectation and variance for X_m and Y_m. Thus

$$E[X_m] = m \times \text{expectation of geometric distribution} = \frac{mq}{p}.$$

Hence,

$$E[Y_m] = E[X_m + m] = m(1 + q/p) = \frac{m}{p}$$

and

$$\text{Var}[Y_m] = \text{Var}[X_m] = \frac{mq}{p^2}.$$

This result may be verified by using the derivatives of M_{X_m} to obtain the needed moments.

6. *Poisson*(μ) $P(X = k) = e^{-\mu}\mu^k/k!$ for all integers $k \geq 0$. We have

$$E[X] = \mu \, \text{Var}[X] = \mu$$

$$M_X(s) \;=\; E[e^{sX}] = e^{-\mu}\sum_{k=0}^{\infty}\frac{\mu^k}{k!}e^{sk} = e^{-\mu}\sum_{k=0}^{\infty}\frac{(\mu e^s)^k}{k!} = e^{-\mu}e^{\mu e^s}$$

$$=\; e^{\mu(e^s - 1)}$$

for all s. Similarly,

$$g_X(s) = e^{\mu(s-1)} \qquad \text{for all } s.$$

ABSOLUTELY CONTINUOUS DISTRIBUTIONS

1. *Uniform* (a, b): $f_X(t) = 1/(b-a)$, $a < t < b$ (zero elsewhere)

$$E[X] = \frac{b+a}{2} \, \text{Var}[X] = \frac{(b-a)^2}{12}$$

$$M_X(s) = \frac{1}{b-a}\int_a^b e^{st}\, dt = \frac{e^{sb} - e^{sa}}{s(b-a)}.$$

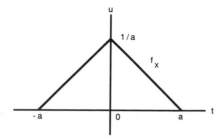

FIGURE 18.1. The density function for the symmetric triangular distribution.

2. *Symmetric triangular* $(-a, a)$:   See Figure 18.1

$$f_X(t) = \begin{cases} \dfrac{a+t}{a^2} & -a \le t < 0 \\ \dfrac{a-t}{a^2} & 0 \le t \le a \end{cases}$$

$$E[X] = 0 \qquad \mathrm{Var}[X] = \frac{a^2}{6}.$$

$$\begin{aligned} M_X(s) &= \frac{1}{a^2} \int_{-a}^{0} (a+t)e^{st}\, dt + \frac{1}{a^2} \int_{0}^{a} (a-t)e^{st}\, dt \\ &= \frac{e^{as} + e^{-as} - 2}{a^2 s^2} \\ &= \frac{e^{as} - 1}{as} \cdot \frac{1 - e^{-as}}{as} \end{aligned}$$

Remark: The last expression shows that X has the same distribution as the difference of two independent random variables X, Y, each uniform $[c, c+a]$ for any c. To see this, consider

$$M_{X_1}(s) = \frac{e^{(a+c)s} - e^{cs}}{as} \qquad \text{and} \qquad M_{-X_2}(s) = M_{X_2}(-s).$$

Then

$$\begin{aligned} M_{X_1 - X_2}(s) &= M_{X_1}(s) M_{-X_2}(s) = \frac{e^{cs}(e^{as} - 1)}{as} \cdot \frac{e^{-cs}(1 - e^{-as})}{as} \\ &= \frac{(e^{as} - 1)(1 - e^{-as})}{a^2 s^2} \end{aligned}$$

3. *Exponential* (λ): $f_X(t) = \lambda e^{-\lambda t}, \; t \geq 0$

$$E[X] = \frac{1}{\lambda} \qquad \mathrm{Var}[X] = \frac{1}{\lambda^2}$$

$$M_X(s) = \int_0^\infty e^{st} \lambda e^{-\lambda t} \, dt = \lambda \int_0^\infty e^{-(\lambda - s)t} \, dt = \frac{\lambda}{\lambda - s} \quad \mathrm{Re}\, s < \lambda.$$

4. *Gamma* (α, λ): $f_X(t) = \lambda^\alpha t^{\alpha - 1} e^{-\lambda t} / \Gamma(\alpha) \quad t \geq 0, \, \alpha > 0, \, \lambda > 0.$

$$E[X] = \frac{\alpha}{\lambda} \qquad \mathrm{Var}[X] = \frac{\alpha}{\lambda^2}$$

$$M_X(s) = \frac{\lambda^\alpha}{\Gamma(\alpha)} \int_0^\infty t^{\alpha - 1} e^{-(\lambda - s)t} \, dt = \left(\frac{\lambda}{\lambda - s} \right)^\alpha \quad \mathrm{Re}\, s < \lambda.$$

Remark: For $\alpha = n$, a positive integer, $M_X(s) = \left(\lambda / (\lambda - s) \right)^n$.

If $X = \sum_{i=1}^n Y_i$, with $\{Y_i : 1 \leq i \leq n\}$ independent, each distributed gamma (α_i, λ), then

$$M_X(s) = \prod_{i=1}^n M_{Y_i}(s) = \left(\frac{\lambda}{\lambda - s} \right)^\alpha, \qquad \text{where } \alpha = \sum_{i=1}^n \alpha_i.$$

Thus, X is gamma (α, λ). In particular, if X is the sum of n independent random variables, each exponential (λ), then X is gamma (n, λ).

5. *Normal* with parameters (μ, σ), often designated $X \sim N(\mu, \sigma^2)$: $E[X] = \mu$ and $\mathrm{Var}[X] = \sigma^2$.

For $X \sim N(0, 1)$, the density is $\phi(t) = (1/\sqrt{2\pi}) e^{-t^2/2}$. Thus $M_Z(s) = \frac{1}{\sqrt{2\pi}} \int_{-\infty}^\infty e^{-t^2/2} e^{st} \, dt$

Now $st - t^2/2 = s^2/2 - \frac{1}{2}(t - s)^2$, so that,

$$M_Z(s) = e^{s^2/2} \frac{1}{\sqrt{2\pi}} \int_{-\infty}^\infty e^{-\frac{1}{2}(t-s)^2} \, dt = e^{s^2/2}$$

since $\frac{1}{\sqrt{2\pi}} e^{-\frac{1}{2}(t-s)^2}$ is the probability density for $N(s, 1)$, hence integrates to 1. Now consider the general case that $X \sim N(\mu, \sigma^2)$. Then $X = \sigma Z + \mu$, where Z is standardized normal.

By the operational property (T1) of the moment-generating function

$$M_X(s) = e^{\mu s} M_Z(\sigma s) = \exp\left(\frac{\sigma^2 s^2}{2} + \mu s \right).$$

$\square$

Example 18.2.1: Affine Combinations of Independent Normal Random Variables

As an important application, we show that an affine combination of independent, normally distributed random variables is normally distributed.

Suppose $\{X, Y\}$ is independent, with $X \sim N(\mu_X, \sigma_X^2)$ and $Y \sim N(\mu_Y, \sigma_Y^2)$. Let $Z = aX + bY + c$, with a, b, c constants. Then

$$E[Z] = a\mu_X + b\mu_Y + c \qquad \text{and} \qquad \text{Var}[Z] = a^2\sigma_X^2 + b^2\sigma_Y^2.$$

Also,

$$
\begin{aligned}
M_Z(s) &= e^{cs} M_X(as) M_Y(bs) \\
&= e^{cs} \exp\left(\frac{a^2\sigma_X^2 s^2}{2} + a\mu_X s\right) \exp\left(\frac{b^2\sigma_Y^2 s^2}{2} + b\mu_Y s\right) \\
&= \exp\left(\frac{\sigma^2 s^2}{2} + \mu s\right),
\end{aligned}
$$

where

$$\sigma^2 = a^2\sigma_X^2 + b^2\sigma_Y^2 \qquad \text{and} \qquad \mu = a\mu_X + b\mu_Y + c.$$

Thus, $Z \sim N(\mu, \sigma^2)$. The fact that Z has a normal distribution is determined by the form of M_Z. It should be clear from the argument that the result extends to any affine combination of any finite, independent class of normally distributed random variables.

—   □

18.3.   Generating Functions for Nonnegative, Integer-Valued Random Variables

As pointed out in Section 18.1, the generating function g_X is used primarily for nonnegative, integer-valued random variables. If X in canonical form is

$$X = \sum_{k=0}^{\infty} k I_{A_k} \qquad \text{with} \quad A_k = \{X = k\} \quad \text{and} \quad p_k = P(A_k) = P(X = k)$$

for all integers $k \geq 0$, then g_X is the power series

$$g_X(s) = \sum_{k=0}^{\infty} p_k s^k.$$

Significant Facts about the Generating Function

We note several significant facts.

- The coefficient of s^k is $p_k = P(X = k) = g_X^{(k)}(0)/k!$.

- Since $\sum_{k=0}^{\infty} p_k = 1$, the power series always converges for $|s| \leq 1$. Usually, the radius of convergence is greater than one, and often it is infinite.

- If X has only a finite set of possible values, g_X is a polynomial.

- In many important cases (e.g., the binomial, geometric, negative binomial, Poisson (see Section 18.2)) the power series converges to a function that is expressible in closed form. Since the power series expansion exists and is unique, there is a one-one relationship between the distribution and the generating function.

- As a result of the unique relationship, when $g_X(s)$ is available in closed form, this form serves to identify the type of distribution.

Example 18.3.1: Identification of a Distribution by the Form of the Generating Function

X is *binomial* (n, p)	iff	$g_X(s) = (ps + q)^n$.
X is *geometric* (p)	iff	$g_X(s) = p/(1 - qs)$.
$Y - 1$ is *geometric* (p)	iff	$g_Y(s) = ps/(1 - qs)$.
X is *negative binomial* (m, p)	iff	$g_X(s) = (p/(1 - qs))^m$.
X is *Poisson* (λ)	iff	$g_X(s) = e^{\lambda(s-1)}$.

—   □

- If $g_X(s)$ is known in closed form, the coefficients of a power series expansion about the origin (which always exists for nonnegative, integer-valued random variables) yields the distribution.

Example 18.3.2: Distribution from the Power Series Expansion of the Generating Function

Suppose $g_X(s) = e^{\lambda(s-1)}$. Determine the corresponding probability distribution.

Solution

We could expand in a power series by using the derivative formulas for the coefficients. In this case, however, we may make use of the known power series for the exponential.

$$g_X(s) = e^{\lambda(s-1)} = e^{-\lambda}e^{\lambda s} = e^{-\lambda}\sum_{k=0}^{\infty}\frac{(\lambda s)^k}{k!} = e^{-\lambda}\sum_{k=0}^{\infty}\frac{\lambda^k}{k!}s^k.$$

Since p_k is the coefficient of s^k in the expansion, we have

$$p_k = P(X = k) = e^{-\lambda}\frac{\lambda^k}{k!} \qquad \text{for all } k \geq 0.$$

The distribution is, as we know, Poisson (λ). In Section 18.2 we show that the Poisson distribution must have the generating function under consideration. We could have used the generating function to define the Poisson distribution. Had we done so, the distribution could have been obtained, as shown earlier.

—   □

In Section 18.1, we show that the generating function shares two operational properties with the moment-generating function and the characteristic function. Also, we establish the simple relationship between the generating function and the moment-generating function.

$$M_X(s) = g_X(e^s) \qquad \text{and} \qquad g_X(s) = M_X(\ln s).$$

Only a change of variable is required to change one transform to the other. This is particularly useful if one wishes to determine moments. The generating function is not well suited for determining moments $E[X^k]$ other than the first moment $E[X]$.

Example 18.3.3: Generating Functions and Moments

Suppose
$$g_X(s) = 0.3 + 0.2s + 0.3s^2 + 0.2s^3.$$

Then X has values 0, 1, 2, 3 with probabilities 0.3, 0.2, 0.3, 0.2, respectively. The moment-generating function is obtained by replacing s with e^s. Thus
$$M_X(s) = 0.3 + 0.2e^s + 0.3e^{2s} + 0.2e^{3s}.$$

To obtain the moments, we first differentiate $M_X(s)$, then set $s = 0$.

$$M_X'(s) = 0.2e^s + 0.3 \cdot 2e^{2s} + 0.2 \cdot 3e^{3s}$$
$$M_X''(s) = 0.2e^s + 0.3 \cdot 2 \cdot 2e^{2s} + 0.2 \cdot 3 \cdot 3e^{3s}.$$

Upon setting $s = 0$, we obtain

$$
\begin{aligned}
E[X] &= M_X'(0) = 0.2 + 0.6 + 0.6 = 1.4 \\
E[X^2] &= M_X''(0) = 0.2 + 1.2 + 1.8 = 3.2.
\end{aligned}
$$

From this, we obtain the variance

$$
\text{Var}[X] = E[X^2] - E^2[X] = 3.2 - 1.4^2 = 1.24.
$$

— $\qquad$ $\square$

We may use the operational properties to obtain distributions for sums of independent random variables.

Example 18.3.4: Distributions for Sums of Independent Random Variables

Suppose X, Y is an independent pair. X has possible values 0, 1, 2, 5 with probabilities 0.2, 0.3, 0.3, 0.2, respectively, and Y has values 1, 2, 3 with probabilities 0.2, 0.4, 0.4, respectively. Determine the distribution for $Z = X + Y$.

Solution

The generating functions for X and Y are

$$
g_X(s) = \frac{1}{10}(2 + 3s + 3s^2 + 2s^5) \qquad \text{and} \qquad g_Y(s) = \frac{1}{10}(2s + 4s^2 + 4s^3).
$$

Then $g_Z(s) = g_X(s) g_Y(s)$. Straightforward multiplication of the two polynomials gives

$$
g_Z(s) = \frac{1}{100}(4s + 14s^2 + 26s^3 + 24s^4 + 12s^5 + 4s^6 + 8s^7 + 8s^8).
$$

As a check, we note that the sum of the coefficients inside the parentheses is 100, corresponding to the fact that the sum of the probabilities is one. The resulting distribution for Z is tabulated as follows.

Distribution for Z

$k =$	1	2	3	4	5	6	7	8
$p_k =$	0.04	0.14	0.26	0.24	0.12	0.04	0.08	0.08

The multiplication of the generating functions has automatically solved the combinatorial problem of determining which probability combinations go with each possible value.

— $\qquad$ $\square$

The generating function and the moment-generating function provide powerful analytical aids in a variety of applied problems. In Chapter 20, we find them particularly useful in the solution of an important class of problems.

18.4. Moment Generating Function and the Laplace Transform

In this section, we present a few facts about the Laplace transform that make it possible to obtain probability distribution functions or density functions from moment-generating functions by using readily available tables of Laplace transform pairs. We do not presume previous experience with the Laplace transform. We limit our examples to nonnegative random variables, so that the distribution function is zero for negative values of its argument. This is an important class of random variables and it allows direct use of standard tables. Readers familiar with the bilateral Laplace transform will be able to employ the various devices and theorems needed for the treatment of distributions for random variables that take on negative values.

Suppose F is a probability distribution function with $F(-\infty) = 0$. The bilateral Laplace transform for F is given by

$$\int_{-\infty}^{\infty} e^{-st} F(t)\, dt.$$

The Laplace-Stieltjes transform for F is

$$\int_{-\infty}^{\infty} e^{-st} F(dt).$$

Thus, if M is the moment-generating function for F, $M(-s)$ is the Laplace-Stieltjes transform for F.

The theory of Laplace-Stieltjes transforms shows that under conditions sufficiently general to include all practical distribution functions

$$M(-s) = \int_{-\infty}^{\infty} e^{-st} F(dt) = s \int_{-\infty}^{\infty} e^{-st} F(t)\, dt.$$

Hence

$$\frac{1}{s} M(-s) = \int_{-\infty}^{\infty} e^{-st} F(t)\, dt.$$

The right-hand expression is the bilateral Laplace transform of F. We may use tables of Laplace transforms to recover F when M is known. This is particularly straightforward when $F(t) = 0$ for $t < 0$, since in this case the bilateral Laplace transform reduces to the one-sided Laplace transform found in most tables. For this condition on F, the random variable is nonnegative. We consider a simple and somewhat artificial example to illustrate the procedure. For a more interesting application, see the treatment in Section 20.7 on composite demand in a random time period.

Example 18.4.1:

The moment-generating function for a nonnegative random variable X is

$$M_X(s) = \frac{10s^2 - 21s + 3}{(1 - 10s)(3 - 10s)}.$$

The Laplace transform of the distribution function is

$$\frac{1}{s}M_X(-s) = \frac{10s^2 + 21s + 3}{s(10s + 1)(10s + 3)} = \frac{0.1s^2 + 0.21s + 0.03}{s(s + 0.1)(s + 0.3)}.$$

We may put this in more usable form by use of the partial fraction expansion for a rational function. This is the algebraic technique used in the integration of rational functions.

$$\frac{1}{s}M_X(-s) = \frac{1}{s} - \frac{0.5}{s + 0.1} - \frac{0.4}{s + 0.3}.$$

From a table of Laplace transforms, we find that $1/s$ is the Laplace transform for the unit step, which has value one for $t \geq 0$. Also, $1/(s-a)$ is the Laplace transform for the function defined by e^{at} for $t \geq 0$. As an integral transform, the Laplace transform of a linear combination of functions is that linear combination of the transforms. Using in one case $a = -0.1$ and in the other case $a = -0.3$, we find that

$$F_X(t) = 1 - 0.5e^{-0.1t} - 0.4e^{-0.3t} \qquad \text{for } t \geq 0.$$

—   □

If the distribution has a density function f, then

$$\int h(t)F(dt) = \int h(t)f(t)\, dt.$$

Thus we have

$$M(-s) = \int_{-\infty}^{\infty} e^{-st} f(t)\, dt$$

so that $M(-s)$ is the bilateral Laplace transform of the density function. For a nonnegative random variable, this reduces to the ordinary Laplace transform. As an example, we consider the moment-generating function for the gamma distribution. Of course, we know from Section 18.2 what the associated density is. But suppose we were not familiar with this moment-generating function and wanted to find the density function. We could proceed as follows.

Example 18.4.2:

Suppose the moment-generating function for nonnegative random variable X is given by

$$M_X(s) = \left(\frac{\lambda}{\lambda - s}\right)^\alpha \qquad \text{with} \quad \alpha > 0.$$

Now

$$M_X(-s) = \left(\frac{\lambda}{\lambda + s}\right)^\alpha.$$

From a table of Laplace transforms we find that for $\alpha \geq 0$

$$\frac{\Gamma(\alpha)}{(s - a)^\alpha} \qquad \text{is the Laplace transform of} \quad t^{\alpha-1} e^{at}.$$

If we put $a = -\lambda$ and do some simple algebraic manipulations, we find that

$$f_X(t) = \frac{\lambda^\alpha t^{\alpha-1} e^{-\lambda t}}{\Gamma(\alpha)} \qquad \text{for } t \geq 0.$$

This is, indeed, the density function for the gamma (α, λ) distribution.

— $\qquad\qquad\qquad\qquad\qquad\qquad\qquad\qquad\qquad\qquad\qquad\qquad$ $\square$

18.5. Characteristic Functions

In this section we summarize, usually without proof, some important mathematical facts about the characteristic function. For detailed development, one may consult any of a number of advanced texts on probability theory or mathematical statistics.

1. In integral form, $\phi_X(-u) = \int e^{-iut} F_X(dt)$.

   This is the Fourier-Stieltjes transform of F_X. Thus, the characteristic function is determined by the distribution function rather than the random variable. Two quite different random variables may have the same distribution and hence the same characteristic function.

   If the distribution has a density function f_X, then

   $$\phi_X(-u) = \int e^{-iut} f_X(t)\, dt,$$

   which is the Fourier transform of f_X. The Fourier and the Fourier-Stieltjes transforms have been studied extensively, and the pertinent literature provides a powerful body of theory. These transforms have been used widely in physics and engineering, so there are important resources for applications.

2. Since $|e^{iuX(\omega)}| = 1$ for any ω for which $X(\omega)$ is finite, $\phi_X(u)$ is defined for any real u for any probability distribution function F_X. This condition ensures a great deal of regularity for the characteristic function. For example, it may be used to show that ϕ_X is uniformly continuous on the entire real line.

3. There is a well-known inversion integral for the Fourier integral. This shows that every distribution function is determined by its characteristic function. The form of the characteristic function often serves to establish the form of the distribution function. We use this in establishing a special case of the central limit theorem in the next section.

4. Since for a complex random variable $E[\bar{Z}] = \overline{E[Z]}$, we must have $\phi_X(u) = \overline{\phi_X(-u)}$. The overbar indicates the complex conjugate. If the distribution of X is symmetric with respect to the origin, so that X and $-X$ have the same distribution, then

$$\phi_X(u) = \phi_{-X}(u) = \phi_X(-u) = \overline{\phi_X(u)},$$

which ensures that ϕ_X is real-valued and even.

5. We point out, in Section 18.1, two operational properties, (TO1) and (TO2), dealing with affine functions of a random variable and sums of independent random variables.

6. We may sharpen the results in Section 18.1 on moments. If $E[|X|^n] < \infty$, then

(a) $\displaystyle \phi_X(u) = \sum_{k=0}^{n} \frac{(iu)^k}{k!} E[X^k] + o(u^n)$ as $u \to 0$

(b) $\phi_X^{(k)}(0) = i^k E[X^k]$ for $k \le n$.

Theorem 18.5.1 (Fundamental Convergence Theorem) *Consider a sequence $F_n : 1 \le n$ of probability distribution functions, with ϕ_n the characteristic function for F_n for each n.*

a. *If F is a distribution function such that $F_n(t) \to F(t)$ at every point of continuity for F and ϕ is the characteristic function for F, then $\phi_n(u) \to \phi(u)$ for all u.*

b. *If $\phi_n(u) \to \phi(u)$ for all u and ϕ is continuous at 0, then ϕ is the characteristic function for a distribution function F such that $F_n(t) \to F(t)$ at each point of continuity of F.*

—   □

Remark: This may be stated in terms of weak convergence (see Section 17.5).

JOINT CHARACTERISTIC FUNCTION

For many problems, it is useful to have a characteristic function for a random vector with real coordinates. This is a joint characteristic function for the coordinate random variables.

Definition 18.5.2. *Suppose* $W = (X_1, X_2, \ldots, X_n)$ *is a random vector with values in* $\Re^n$. *The characteristic function* ϕ *for* W *is given by*

$$\phi(u_1, u_2, \ldots, u_n) = E\left[e^{i(u_1 X_1 + \cdots + u_n X_n)}\right].$$

This can be given a compact form by use of matrix and inner product notation of linear algebra. Suppose

$$\boldsymbol{u} = (u_1, u_2, \ldots, u_n) \in \Re^n.$$

We represent $\boldsymbol{u}$ by the column matrix $[u_1 u_2 \ldots u_n]^T$. The superscript T indicates the transpose of the row matrix, producing a column matrix.

Definition 18.5.3. *The* Euclidean inner product *of two vectors* $\boldsymbol{t}$ *and* $\boldsymbol{u}$ *in* $\boldsymbol{R}^n$ *is*

$$\langle \boldsymbol{t}, \boldsymbol{u} \rangle = \sum_{i=1}^{n} t_i u_i$$

the sum of the products of the respective coordinates.

In these terms, we have

$$\phi_W(\boldsymbol{u}) = E[e^{i\langle \boldsymbol{u}, W \rangle}] = E[e^{i\langle W, \boldsymbol{u} \rangle}].$$

Note that, as in the one-dimensional case, the characteristic function is determined by the joint probability distribution, so that two random vectors with the same joint distribution have the same characteristic function. It is known that the characteristic function for a multidimensional distribution determines that distribution, as in the one-dimensional case. Many properties of ϕ_W are similar to those for the real-variable case.

We list, for reference, three important properties:

(a) Consider the class $\{X_1, X_2, \ldots, X_n\}$ of real random variables and let $W = (X_1, X_2, \ldots, X_n)$. The class is independent iff $\phi_W(\boldsymbol{u}) = \prod_{i=1}^{n} \phi_{X_i}(u_i)$ for all $\boldsymbol{u}$ in $\boldsymbol{R}^n$.

(b) Suppose W is n-dimensional and m-dimensional V is given by $V = AW + B$, where A is an $(m \times n)$ matrix and B is an $(m \times 1)$ column matrix. That is, V is obtained from W by an affine transformation. We then have

$$\phi_V(\boldsymbol{v}) = E[e^{i\langle \boldsymbol{v}, AW + B \rangle}] = e^{i\langle \boldsymbol{v}, B \rangle} E[e^{i\langle A^T \boldsymbol{v}, W \rangle}] = e^{i\langle \boldsymbol{v}, B \rangle} \phi_W(A^T \boldsymbol{v}).$$

(c) If the characteristic function ϕ_W is p times differentiable, then moments of all integer orders $q \leq p$ are obtained from the partial derivative of ϕ_W evaluated at $\boldsymbol{u} = 0$. For example, if $p = 2$, then

$$E[X_r] = -i \left. \frac{\partial \phi}{\partial u_r} \right|_0 , \quad E[X_r^2] = - \left. \frac{\partial^2 \phi}{\partial u_r^2} \right|_0 , \quad E[X_r X_s] = \left. \frac{\partial^2 \phi}{\partial u_s \partial u_r} \right|_0 .$$

The similarity of these properties to the corresponding ones for the one-dimensional case is apparent.

JOINT NORMAL DISTRIBUTION

The joint characteristic function is particularly useful in studying the joint normal distribution. This distribution plays a central role in the theory of Gaussian random processes, which serve as models for many phenomena in communication and control theory. Some authors use the characteristic function to define the joint normal distribution, as follows.

Definition 18.5.4. $W = (X_1, X_2, \ldots, X_n)$ has the joint normal distribution iff its characteristic function ϕ is given by

$$\phi(\boldsymbol{u}) = \exp\left(i \langle \boldsymbol{m}, \boldsymbol{u} \rangle - \frac{1}{2} \langle \mathcal{K}\boldsymbol{u}, \boldsymbol{u} \rangle \right) = \exp\left(i \sum_j \mu_j u_j - \frac{1}{2} \sum_{j,k} u_j K_{jk} u_k \right)$$

where $\boldsymbol{m} = (\mu_1, \mu_2, \ldots, \mu_n)$ and $\mathcal{K} = [K_{jk}]$ is a positive semidefinite matrix.

Utilization of these formulas shows that $\mu_j = E[X_j]$, the mean value of the jth coordinate random variable, and $K_{jk} = \text{Cov}[X_j, X_k]$, the covariance of X_j, X_k. It is necessary, of course, to show that this definition is consistent with the alternative definition in terms of the joint density. Either of these equivalent formulations may be taken as definition, in which case the other becomes a theorem. The characteristic function provides a powerful analytical aid in the study of the remarkable properties of the joint normal distribution. For example, if $\{X_i : 1 \leq i \leq n\}$ has the joint normal distribution, so does any affine transformation of these variables. Many treatments of Gaussian random processes include extensive developments of such properties.

18.6.   The Central Limit Theorem

The central limit theorem (CLT) asserts that if random variable X is the sum of a large number of independent random variables, each with reasonable distributions, then X is approximately normally distributed. This celebrated theorem has been the object of prodigious research effort directed toward the discovery of the most general conditions under which it is valid. On the other hand, this theorem serves as the basis of an extraordinary amount of applied work. In the statistics of large samples, the sample average is a constant times the sum of the random variables in the sampling process (see Section 18.7). Thus, for large samples, the sample average is approximately normal—whether or not the population distribution is normal. In much of the theory of errors of measurement, the observed error is the sum of a large number of independent random quantities that contribute to the result. Similarly, in the theory of noise, the noise signal is the sum of a large number of random components, independently produced. In such situations, the assumption of a normal population distribution may be quite appropriate.

We consider a form of the CLT under hypotheses that are reasonable assumptions in many practical situations. The proof of this theorem, known as the Lindeberg-Lévy theorem, uses characteristic functions. The proof is not difficult; it illustrates the kind of argument used in more sophisticated proofs required for more general cases of the CLT.

Suppose Z is a standardized normal random variable. That is, $Z \sim N(0, 1)$, with distribution function Φ. Consider an independent sequence $\{X_n : 1 \leq n\}$ of random variables. Form the sequence $\{S_n : 1 \leq n\}$ of partial sums

$$S_n = \sum_{i=1}^{n} X_i \qquad \text{for all } n \geq 1.$$

Then, by properties of expectation and of variance,

$$E[S_n] = \sum_{i=1}^{n} E[X_i] \qquad \text{and} \qquad \text{Var}[S_n] = \sum_{i=1}^{n} \text{Var}[X_i] \qquad \text{for all } n \geq 1.$$

Let S_n^* be the standardized sum obtained by subtracting the mean and dividing by the standard deviation for S_n. Let F_n be the distribution function for S_n^*. The CLT asserts that under appropriate conditions $F_n(t) \to \Phi(t)$ as $n \to \infty$ for any real t.

We prove the simple case referred to earlier, using properties of characteristic functions and a simple lemma from the theory of complex variables (see the appendix on mathematical aids).

Theorem 18.6.1. *If $X_n : 1 \leq n$ is iid, with*

$$E[X_n] = \mu, \qquad \text{Var}[X_n] = \sigma^2, \qquad and \qquad S_n^* = \frac{S_n - n\mu}{\sigma\sqrt{n}}$$

then

$$F_n(t) \to \Phi(t) \qquad as\ n \to \infty\ for\ all\ t.$$

Proof

There is no loss of generality in assuming $\mu = 0$. Let ϕ be the common characteristic function for the X_n, and for each n let ϕ_n be the characteristic function for S_n^*. Thus

$$\phi(t) = E[e^{itX}] \qquad \text{and} \qquad \phi_n(t) = E[e^{itS_n^*}] = \phi^n(t/\sigma\sqrt{n}).$$

Since $E[X] = 0$ and $E[X^2] = \sigma^2$, we have

$$\phi(t) = 1 - \frac{\sigma^2 t^2}{2} + \beta(t) \qquad \text{with} \quad \beta(t) = o(t^2) \qquad \text{as } t \to 0.$$

This implies

$$\left| \phi(t/\sigma\sqrt{n}) - \left(1 - \frac{t^2}{2n}\right) \right| = |\beta(t/\sigma\sqrt{n})| = o\left(\frac{t^2}{\sigma^2 n}\right).$$

Hence

$$n\left| \phi(t/\sigma\sqrt{n}) - \left(1 - \frac{t^2}{2n}\right) \right| \to 0 \qquad \text{as } n \to \infty \qquad \text{for all } t.$$

Now by a simple lemma (see the appendix on mathematical aids),

$$\left| \phi^n(t/\sigma\sqrt{n}) - \left(1 - \frac{t^2}{2n}\right)^n \right| \leq n\left| \phi(t/\sigma\sqrt{n}) - \left(1 - \frac{t^2}{2n}\right) \right| \to 0 \quad \text{as } n \to \infty.$$

It is a well-known property of the exponential that

$$\left(1 - \frac{t^2}{2n}\right)^n \to e^{-t^2/2} \qquad \text{as } n \to \infty$$

so that

$$\phi^n(t/\sigma\sqrt{n}) \to e^{-t^2/2} \qquad \text{for all } t.$$

By the fundamental convergence theorem for characteristic functions, we must have $F_n(t) \to \Phi(t)$, since $e^{-t^2/2}$ is the characteristic function for Φ. □

Example 18.6.1: Random Walk

Suppose $\{Y_i : 1 \leq i\}$ is iid, with $E[Y_i] = 0$ and $\text{Var}[Y_i] = \sigma^2$. For each $n \geq 1$, let

$$X_n = \sum_{i=1}^{n} Y_i.$$

The sequence $\{X_n : 1 \leq n\}$ forms a *random walk*. This designation comes from the fact that such a sequence may be used to model the following behavioral system. At discrete time $t = i$, a particle moves a distance Y_i

along a line. The net distance moved after the nth displacement is X_n. Individual movements are independent, with the same distribution, and the average movement is zero.

Show that as time increases, the probability of being within distance c of the starting position goes to zero, no matter how large c is.

Solution

For large n, $X_n^* = X_n/\sigma\sqrt{n}$ is approximately $N(0,1)$. Thus,

$$P(|X_n| \leq c) = P\left(|X_n^*| \leq \frac{c}{\sigma\sqrt{n}}\right) \approx 2\Phi\left(\frac{c}{\sigma\sqrt{n}}\right) - 1.$$

As $n \to \infty$, $c/\sigma\sqrt{n} \to 0$ and $\Phi(c/\sigma\sqrt{n}) \to 0.5$

Hence, $P(|X_n| \leq c) \to 0$ as $n \to \infty$.

—   □

18.7. Random Samples and Statistics

In this section, we formulate the notion of a random sample, which underlies much of classical statistics. We then apply probability theory to the resulting model of the sampling process, in order to exhibit a few of the ideas of statistical analysis.

Underlying classical mathematical statistics is the notion of a population distribution. This is the idea that some quantity associated with the members of a population varies according to a given distribution.

EXAMPLES OF POPULATION DISTRIBUTIONS

1. The ages of members of a given human population at a specified time.

2. A critical dimension of items from a production line.

3. Response to a question on an opinion poll. Here the observed entity may not be inherently numerical. However, if the answer is to be in one of several categories (e.g., yes, no, no opinion), these categories can be given numbers and the result of an observation is a number. The item of interest in the population is the fraction of responses in each category.

This short list is merely suggestive. We could expand it indefinitely.

In order to obtain information about the population distribution, we select, in a "random" manner, a subset of the population and observe the values associated with these members. In doing so, we hope that the distribution of quantities among the selected members is representative, or typical, in the sense that it approximates the population distribution. Such an experiment is modeled in mathematical statistics by a random sampling process, which we formulate in terms of random variables and their distributions.

INDEPENDENT RANDOM SAMPLING PROCESS

The quantity of interest associated with the members of a population has the population distribution. We take a sample of size n, which means that we select n members of the population and observe the quantity associated with each. The selection is done in such a way that on any trial each member is equally likely to be selected. Also, the selection and observation is done in such a manner that the value of any one observation does not affect, nor is it affected by, the values of any others. The result of each observation is a number. Suppose the first number observed is t_1, the second t_2, and so on, with the last or nth one t_n. Now if a new sample were to be taken, most likely we would observe different values on the various component trials making up the sampling operation. This means that before the sampling experiment is performed, there is uncertainty about the values to be observed. We model this uncertainty by considering each t_i to be an observed value of a random variable X_i. We suppose the the selection and observation process to be such that we may make the following assumptions:

1. Each of the X_i has the population distribution.

2. The variables form an independent class.

This provides a *model* for either sampling with replacement from a small population, or sampling (with or without replacement) from a very large population (often referred to as an infinite population). In either case, the random choice of an object or element in the population to be examined does not appreciably affect the population distribution. We formalize the model as follows.

Definition 18.7.1. *A* random sampling process of size n *is an iid class* $\{X_i : 1 \leq i \leq n\}$ *of random variables, each of which has the* population distribution.

A random sample of size n *is an observation, or realization, $(t_1, t_2, \ldots, t_n)$ of the sampling process.*

The set of observed values can be characterized by various numerical parameters derived from the sample. For example, consider the sample average:

$$\frac{1}{n}\sum_{i=1}^{n} t_i.$$

This is an observed value of a random variable that is a function of the variables in the sampling process

$$A_n = g(X_1, X_2, \ldots, X_n) = \frac{1}{n}\sum_{i=1}^{n} X_i.$$

We call this random variable A_n the *sample average*. If the observed set of sample values provides a satisfactory approximation to the population distribution, then it seems reasonable to suppose that the observed sample average (which is an observed value of A_n) should be a useful approximation to the population mean. If the assumptions are valid, we may use probability theory to examine the question of how good an estimate such an observation is. Direct use of the iid character of the class $\{X_i : 1 \leq i \leq n\}$ shows that

$$E[A_n] = \mu \quad \text{and} \quad \text{Var}[A_n] = \frac{\sigma^2}{n}$$

where μ is the mean and σ^2 is the variance of the population distribution. These facts tell us that if sampling is repeated, the observed sample averages will vary about the population mean and that they will do so with a variability, or spread, that is reduced by the sample size n. Recall that the standard deviation is a measure of the spread of the distribution. The standard deviation for the sample-average random variable A_n is the standard deviation for the population distribution reduced by a factor $1/\sqrt{n}$. Thus, for large samples, the probability is high that the observed value of the sample average will lie close to the population mean.

One of the central questions of large sample statistics is how large the sample size n should be to ensure a high probability that the observed value of the sample average A_n will lie within a given distance from the population mean μ. If the population standard deviation σ can be estimated reasonably well, then we may use the central limit theorem to solve this problem.

Example 18.7.1: Sample Size

A sample is to be taken of a population distribution whose standard deviation $\sigma \approx 3$. How large a sample size n is required to obtain a probability of 0.95 that the observed value of the sample average A_n is within 0.2 of the population mean value μ?

Solution

We want to determine the smallest integer n such that

$$P(|A_n - \mu| \le 0.2) \ge 0.95.$$

For any reasonable population distribution and any moderate value of n, the central limit theorem ensures that A_n is approximately normal. Since A_n has mean value μ and standard deviation $3/\sqrt{n}$,

$$Z_n = \frac{A_n - \mu}{3/\sqrt{n}} \quad \text{is approximately} \quad N(0, 1).$$

Thus

$$P(|A_n - \mu| \le 0.2) = P\left(\frac{|A_n - \mu|}{3/\sqrt{n}} \le \frac{0.2\sqrt{n}}{3}\right) = P(|Z_n| \le \sqrt{n}/15).$$

From the symmetry of the standardized normal distribution,

$$P(|Z_n| \le \sqrt{n}/15) \approx 2\Phi(\sqrt{n}/15) - 1.$$

We thus want to determine the smallest integer n such that

$$2\Phi(\sqrt{n}/15) - 1 \ge 0.95 \quad \text{or} \quad \Phi(\sqrt{n}/15) \ge 0.975.$$

From a table of the standardized normal distribution, we find that $\Phi(1.96) = 0.9750$. Since Φ is increasing, we want

$$\frac{\sqrt{n}}{15} \ge 1.96 \quad \text{or} \quad n \ge (1.96 \cdot 15)^2 \approx 864.4.$$

Taking the nearest larger integer, we have a sample size $n = 865$.

—   □

As a function of the random variables in the sampling process, A_n is an example of a statistic.

Definition 18.7.2. *A statistic G is a function of the class $\{X_i : 1 \le i \le n\}$ such that no unknown population parameters are used in the rule for the function.*

Example 18.7.2: Statistics as Functions of the Sampling Process

The random variable

$$H = \frac{1}{n}\sum_{i=1}^{n}(X_i - \mu)^2 \quad \text{where} \quad \mu = E[X_i]$$

is not a statistic, since its evaluation involves the unknown population parameter μ. The following random variable is a statistic, since it does not use any parameter of the distribution.

$$V_n^* = \frac{1}{n} \sum_{i=1}^{n} (X_i - A_n)^2 = \frac{1}{n} \sum_{i=1}^{n} X_i^2 - A_n^2.$$

$\square$

We are interested in statistics that may be used as estimators for population parameters. As the preceding discussion shows, the statistic A_n is a good estimator for μ if n is sufficiently large. In Example 18.7.1, an estimate of the population standard deviation (the square root of the population variance) is needed. The result of Example 18.7.2 suggests a possible estimator.

Example 18.7.3: An Estimator for Population Variance

The statistic

$$V_n = \frac{1}{n-1} \sum_{i=1}^{n} (X_i - A_n)^2$$

is an estimator for the population variance σ^2, as the following development shows.

Consider first the statistic introduced in Example 18.7.2

$$V_n^* = \frac{1}{n} \sum_{i=1}^{n} (X_i - A_n)^2 = \frac{1}{n} \sum_{i=1}^{n} X_i^2 - A_n^2.$$

Use of properties of expectation and variance shows that

$$E[V_n^*] = \frac{1}{n} n(\sigma^2 + \mu^2) - \left(\frac{\sigma^2}{n} + \mu^2 \right) = \frac{n-1}{n} \sigma^2.$$

Use of linearity shows that

$$E[V_n] = E\left[\frac{n}{n-1} V_n^* \right] = \frac{n}{n-1} E[V_n^*] = \sigma^2.$$

Since V_n has mean value σ^2, it should be considered a possible estimator for the population variance. There remains the question: How good is this estimator? As a random variable, V_n has a variance. An evaluation similar to that for the mean, although much more complicated in detail, shows that

$$\text{Var}[V_n] = \frac{1}{n} \left(\mu_4 - \frac{n-3}{n-1} \sigma^4 \right) \qquad \text{where} \quad \mu_4 = E[(X - \mu)^4].$$

Again, for large n, $\text{Var}[V_n]$ is small, so that V_n is a good large-sample estimator for the population variance.

—   □

A statistic, as a function of the sample random variables, has a distribution derived from the population distribution. The probability distributions for various statistics used in statistical analysis are known as sampling distributions. A significant task of mathematical statistics is the development of sampling distributions for various classes of statistics. The study of these distributions involves sophisticated use of many of the basic techniques and ideas of probability theory that we consider in this text.

Because of the central limit theorem, it should be no surprise that populations with a normal, or Gaussian, distribution receive considerable attention. We state some of the important results, as illustrations.

For a population $N(\mu, \sigma^2)$:

- The sample average $A_n \sim N(\mu, \sigma^2/n)$.

  We show in Section 18.2 that an affine combination of independent, normally distributed random variables is normally distributed. The value of the mean and the variance were noted.

- The class $\{A_n, \sum_{i=1}^n (X_i - A_n)^2\}$ is independent.

  This result requires a detailed analysis based on properties of the joint normal distribution.

- The random variable $U = ((n-1)/\sigma^2)\,V_n$ has the chi-square distribution with $n-1$ degrees of freedom.

  The chi-square distribution with n degrees of freedom is, by definition, the same as the gamma $(n/2, 1/2)$ distribution. Now U can be put into a special form that shows it to be the sum of $n-1$ independent random variables, each of which is the square of a standardized normal random variable. In Example 11.1.6, we show that if $X \sim N(0,1)$, then X^2 has the gamma $(1/2, 1/2)$ distribution. The result in Section 18.2 on the sum of independent gamma random variables shows that U must be gamma $((n-1)/2, 1/2)$.

This brief excursion into statistics is intended merely to illustrate the manner in which fundamental probability theory provides a conceptual and theoretical foundation for classical mathematical statistics. In many statistical studies, much more complicated sampling procedures are used; these require extension of the theory of simple random samples illustrated earlier. In Section 22.2, we introduce some of the central ideas of the so-called Bayesian statistics. There we not only introduce the ideas, but provide some of the probabilistic structure underlying the subject. Reasonable mastery of the probability theory developed in this book should enable one to move quickly and efficiently into mathematical statistics and the development of statistical method.

PROBLEMS

18–1. Calculate directly $g_X(s)$ for the geometric(p) distribution (rather than transforming $M_X(s)$).

18–2. Calculate directly $g_X(s)$ for the Poisson(μ) distribution.

18–3. A projection bulb has life (in hours) represented by random variable X exponential $(1/50)$. The device will be replaced immediately upon failure or at 60 hours, whichever comes first. Let Y be the time of replacement. Determine $M_Y(s)$.

18–4. The pair $\{X, Y\}$ has joint density function $f_{XY}(t, u) = t + u$, $0 \leq t \leq 1$, $0 \leq u \leq 1$. This implies $f_X(t) = t + \frac{1}{2}$ for $0 \leq t \leq 1$, $E[X] = 7/12$, $E[X^2] = 5/12$.

Determine the moment-generating function $M_X(s)$.

18–5. Simple random variable X takes on values $-3, -2, 0, 1, 4$ with probabilities $0.1, 0.25, 0.3, 0.25, 0.1$, respectively.

    (a) Write an expression for the moment-generating function $M_X(s)$.

    (b) Show by direct calculation that $M_X'(0) = E[X]$ and $M_X''(0) = E[X^2]$.

18–6. Show that $\phi^{(k)}(0) = i^k E[X^k]$.

18–7. If X is a nonnegative, integer-valued random variable, express $g_X(s)$ as a power series.

    (a) Show that $g_X^{(k)}(1) = E[X(X-1)(X-2)\ldots(X-k+1)]$.

    (b) Use this fact to show that $\operatorname{Var}[X] = g_X''(1) + g_X'(1) - [g_X'(1)]^2$.

18–8. If $M_X(\cdot)$ is the moment-generating function for X, show that $\operatorname{Var}[X]$ is the second derivative of $e^{-s\mu} M_X(s)$ with respect to s, evaluated at $s = 0$. Use this to show that if X is $N(\mu, \sigma^2)$, then $\operatorname{Var}[X] = \sigma^2$.

18–9. Use derivatives of $M_{X_m}(s)$ to obtain the mean and variance for the negative binomial (m, p) distribution.

18–10. Use the moment-generating function to obtain the variance for the following distributions.

    (a) Exponential(λ) (b) Gamma(α, λ) (c) Normal $N(\mu, \sigma^2)$.

18–11. Random variable X has moment-generating function

$$M_X(s) = p^2/(1 - qe^s)^2.$$

Determine $E[X]$ and $\operatorname{Var}[X]$.

18–12. Random variable X has a double exponential (α, ν) distribution iff

$$f_X(t) = \frac{\alpha}{2} e^{-\alpha|t-\nu|} \qquad \text{for all } t \text{ and } \nu, \text{ and for all } \alpha > 0.$$

(a) Show that $Y = X - \nu$ has double exponential $(\alpha, 0)$ distribution.

(b) Use M_Y to obtain $E[Y]$ and $\text{Var}[Y]$. From these obtain $E[X]$ and $\text{Var}[X]$.

18–13. The pair $\{X, Y\}$ of random variables is independent. X is Poisson (3) and Y is geometric (0.3). Let $Z = 2X + 3Y$. Determine the generating function $g_Z(s)$.

18–14. Random variable X has moment-generating function

$$M_X(s) = \frac{1}{1 - 2s} e^{(9s^2/2 + 4s)}.$$

Determine $E[X]$ and $\text{Var}[X]$.

Hint: This may be done by using M_X to obtain the appropriate moments. Or it may be done by recognizing forms and using rules of combination. Recall that

For Y exponential (λ), $M_Y(s) = \dfrac{\lambda}{\lambda - s}$.

For $Z \sim N(\mu, \sigma^2)$, $M_Z(s) = e^{(\sigma^2 s^2/2 + \mu s)}$.

18–15. Suppose $M_X(s) = e^{3(e^s - 1)}/(1 - 10s)$. Determine $E[X]$ and $\text{Var}[X]$.

Hint: Recall that for Y exponential(λ),

$$M_Y(s) = \frac{\lambda}{\lambda - s} = \frac{1}{(1 - s/\lambda)}$$

and for Z Poisson(μ),

$$M_Z(s) = e^{\mu(e^s - 1)}.$$

18–16. The pair $\{X, Y\}$ of simple random variables is independent. X takes on values -2, 0, 1, 2 with probabilities 0.2, 0.3, 0.3, 0.2, respectively. Y takes on values 0, 1, 2 with probabilities 0.3, 0.4, 0.3, respectively.

(a) Determine the moment-generating functions M_X and M_Y.

(b) Use the moment-generating functions to determine the distribution for $Z = X + Y$.

18–17. Use the moment-generating function to show that variances add for the sum or difference of independent random variables.

18–18. $\{X, Y\}$ is iid, with common moment-generating function $M(s) = \lambda^3/(\lambda - s)^3$.

Let $Z = 3X - 2Y + 2$. Determine $M_Z(s)$.

18–19. $\{X, Y\}$ is iid, with common moment-generating function $M(s) = (0.7 + 0.3e^s)^{10}$.

Let $Z = 3X + 4Y - 3$. Determine $M_Z(s)$.

18–20. The pair $\{X, Y\}$ is independent. X is distributed uniformly on the integers $\{0, 1, 2, 3\}$ and Y is distributed on $\{0, 1, 2, 3, 4\}$ with probabilities 0.1, 0.2, 0.3, 0.2, 0.2, respectively.

(a) Determine the generating function g_X and g_Y.

(b) Determine g_{X+Y} and from this obtain the distribution for $X+Y$.

18–21. Suppose X is binomial $(3, 0.4)$ and Y is binomial $(4, 0.4)$. The pair $\{X, Y\}$ is independent. Let $Z = X + Y$. Use the generating function g_Z to determine $P(Z \geq 2)$.

Hint. What does the form of g_Z show about the distribution for Z?

18–22. Suppose $\{X, Y\}$ is independent. Let $Z = X + Y$. For some of the following distributions, under some parameter conditions, the distributions for the sum Z is of the same class as that for X and Y. In each case, show whether or not this is true.

(a) Binomial (n, p).
(b) Geometric (p).
(c) Poisson (μ).
(d) Uniform (a, b).
(e) Exponential (λ).
(f) Gamma (n, λ).

18–23. Use the moment-generating function for a distribution symmetric triangular on $(-a, a)$ to determine the moment-generating function for a distribution symmetric triangular on (A, B).

18–24. Use the result of problem 18-23 to show that the sum of two iid random variables, each uniform on (a, b), has the symmetric triangular distribution on $(2a, 2b)$.

18–25. Suppose $\{X, Y\}$ is an independent pair, with X Poisson (μ) and Y Poisson (λ). Then $X + Y$ is Poisson $(\mu + \lambda)$. Does $X - Y$ have a Poisson distribution? Why?

18–26. Suppose the pair $\{X, Y\}$ is independent, Y is nonnegative, and both X and $X + Y$ are Poisson. Show that Y must also be Poisson.

18–27. The pair $\{X, Y\}$ is iid $N(2, 4)$. That is, each random variable is normal with mean $\mu = 2$ and variance $\sigma^2 = 4$. Let $Z = 3X - 5Y$. Obtain $M_Z(s)$ and use this to show that Z is normally distributed.

18–28. Show that for large sample size n, the sample average A_n is approximately $N(\mu, \sigma^2/n)$ for any reasonable population distribution having mean μ and standard deviation σ.

18–29. A sample of size 30 is taken from a population $N(\mu, 2.25)$. It is desired to estimate the probability that the sample average will lie within 0.6 of the population mean.

 (a) Use the Chebyshev inequality to estimate the probability.

 (b) Use a table of normal distributions to estimate the probability.

18–30. A population has standard deviation of approximately three. It is desired to determine how large a sample is required to ensure that with probability 0.95 the sample average will not differ from the population mean by more than 0.5.

 (a) Use the Chebyshev inequality to estimate the needed sample size.

 (b) Use the central limit theorem and the normal approximation to estimate n.

18–31. A population has mean μ and standard deviation $\sigma \approx 2.4$. Use the central limit theorem to determine how large a sample must be taken in order that the sample average A_n will satisfy $P(|A_n - \mu| < 0.2) \geq 0.98$.

18–32. On the basis of the central limit theorem, how large should the sample size n be in order to ensure that with probability 0.95 the observed value of the sample average A_n will lie within 0.1σ of the population mean? That is, how large must n be to ensure $P(|A_n - \mu| \leq 0.1\sigma) \geq 0.95$.

18–33. A sample is taken from a population known to be uniform on an interval of length six. The location of the interval, hence its mean, is not known.

 (a) Use Chebyshev's inequality to estimate how large a sample is needed to ensure that with probability 0.95 the sample mean lies within 0.1 of the population mean.

 (b) Repeat (a), using the central limit theorem.

Part IV

Conditional Expectation

19

Conditional Expectation, Given a Random Vector

The concept of conditional expectation, given a random vector, is essential to many modern topics in random processes, decision theory, and statistics. The usual introductory treatment of conditional expectation is intuitive and straightforward, but severely limited in scope. More general treatments assume considerable familiarity with measure theory and abstract integration theory. We seek to bridge the gap, in order to make appropriate aspects of a general treatment more readily accessible. We build upon the theory of integration and expectation sketched in Chapters 13 and 14, while trying to maintain some intuitive feel for what is an essentially abstract concept.

We begin with conditioning by an event. Then, we use intuitive notions of a conditional distribution, given a value of the conditioning random vector. Operating from this point of view, we obtain the usual elementary formulations in two important special cases. We organize the treatment of these cases to exhibit a fundamental pattern, which provides the basis for both extending the concept and establishing significant properties. One of these properties provides the basis for a new interpretation of the conditional expectation as a mean-square estimator. This and other properties serve as powerful theoretical tools for a wide variety of applications.

19.1. Conditioning by an Event

If a conditioning event C occurs, we modify the original probabilities by introducing the conditional probability measure $P(\cdot|C)$. Thus, $P(A)$ is replaced by $P(A|C) = P(AC)/P(C)$. In making this change, we accomplish two things:

- We limit possible outcomes to those ω in event C.

- We "normalize" the probability mass in C to make it the new unit of mass.

It seems reasonable to make a corresponding modification of mathematical expectation, which we view as a probability-weighted average of the values taken on by a random variable. We may make the modification in one of two ways.

- We could replace the prior probability measure $P(\cdot)$ by the conditional probability measure $P(\cdot|C)$, then take the expectation (i.e., weighted average) with respect to this new probability mass assignment.

- We could continue to use the original probability measure $P(\cdot)$ and modify the averaging process in the following manner.

  - For a real random variable X, we consider the values $X(\omega)$ for only those ω in the event C. We do this by using the random variable $I_C X$, which has the value $X(\omega)$ for ω in C and has the value zero for any ω outside C. This means that the integral $E[I_C X]$ is the probability-weighted sum of the values taken on by X in the event C.
  - We divide the weighted sum by $P(C)$ to obtain the weighted average over C.

As shown in Theorem 19.1.2, these two approaches are equivalent. For reasons that become more apparent in subsequent developments, we take the second approach as the basis for definition. For one thing, we can do the "weighted summing" with the prior probability measure, no matter what the conditioning event C, then obtain the average by dividing by $P(C)$ for the particular conditioning event. In addition, this approach facilitates relating the present concept to the more general concept developed in the next two sections.

Definition 19.1.1. *If the event C has positive probability and has indicator function I_C, the conditional expectation of X, given C, is the quantity*

$$E[X|C] = \frac{E[I_C X]}{P(C)}.$$

Several properties may be established easily.

Theorem 19.1.2.

a. *$E[X|C]$ is expectation with respect to conditional probability measure $P(\cdot|C)$.*

b. *$E[I_A|C] = P(A|C)$.*

c. *If $C = \bigcup\limits_i C_i$, then $E[X|C]P(C) = \sum_i E[X|C_i]P(C_i)$*

Proof of proposition a.
Consider simple random variable $\sum_k t_k I_{A_k}$ in canonical form. Then

$$E[X|C] = \frac{E[I_C X]}{P(C)} = \frac{E[\sum_k t_k I_C I_{A_k}]}{P(C)} = \sum_k t_k \frac{E[I_{C A_k}]}{P(C)} = \sum_k t_k P(A_k|C).$$

The final sum is $E_C[X]$, where $E_C[\cdot]$ indicates expectation relative to the conditional probability measure $P(\cdot|C)$. We may use a "standard argument" to extend the result to any nonnegative random variable X, then to any integrable random variable X.

Propositions b and c are established easily from properties of mathematical expectation.

—   □

The following simple result provides a link between the present concept and the more general concept developed in the next two sections.

Theorem 19.1.3. *Let* $C = X^{-1}(M) = \{X \in M\}$, *for any Borel set* M *on the codomain of* X. *If* $P(C) > 0$, *then*

$$E[I_M(X)g(Y)] = E[g(Y)|X \in M]P(X \in M).$$

Proof

Recall that $I_M(X) = I_C$. By definition, $E[I_C g(Y)] = E[g(Y)|C]P(C)$. Hence, $E[I_M(X)g(Y)] = E[g(Y)|X \in M]P(X \in M)$.

—   □

Remark: It should be noted that both X and Y can be vector-valued. The function g must be real-valued and M may be any Borel set on the codomain of X. Also, the case $X = Y$ is quite permissible.

Example 19.1.1:

Suppose X is exponential (0.1), so that $F_X(t) = 1 - e^{-0.1t}$ and $f_X(t) = 0.1e^{-0.1t}$ for $t \geq 0$.

Let $C = \{5 \leq X \leq 20\}$. Then $P(C) = e^{-0.5} - e^{-2}$ and $E[I_C X] = \int_5^{20} 0.1te^{-0.1t}\, dt = 15e^{-0.5} - 30e^{-2}$.

Thus,

$$E[X|C] = \frac{15e^{-0.5} - 30e^{-2}}{e^{-0.5} - e^{-2}} \approx 10.69.$$

—   □

Example 19.1.2: Reliability of a Standby System with Imperfect Coverage

A system consists of two components, each with life distribution exponential (λ). One component is put into operation; the other is on "cold standby." In the event of failure of the active unit, the standby unit is "brought on line." If the standby unit has not failed while on standby and is brought on line in time, the fault is said to be covered.

Suppose the fault is covered with probability c. Obtain an expression for the life distribution function F and the reliability $R = 1 - F$ for the imperfectly covered system.

Solution

Let C be the event the fault is covered. If C occurs, the system life has distribution gamma $(2, \lambda)$; if C does not occur, the system life has distribution exponential (λ), which is gamma $(1, \lambda)$. We thus take

$$f_X(t|C) = \lambda^2 t e^{-\lambda t} \qquad \text{and} \qquad f_X(t|C^c) = \lambda e^{-\lambda t}.$$

By Theorem 19.1.2,

$$F_X(t) = E[I_{[0,t]}(X)] = E[I_{[0,t]}(X)|C]P(C) + E[I_{[0,t]}(X)|C^c]P(C^c).$$

Use of the conditional densities gives

$$F_X(t) = c \int_0^t \lambda^2 x e^{-\lambda x}\, dx + (1 - c) \int_0^t \lambda e^{-\lambda x}\, dx.$$

Evaluation with the aid of integral formulas in the appendix on mathematical aids gives

$$F_X(t) = c\big(1 - (\lambda t + 1)e^{-\lambda t}\big) + (1 - c)(1 - e^{-\lambda t}) = 1 - (1 + c\lambda t)e^{-\lambda t}.$$

We thus have
$$R(t) = 1 - F_X(t) = (1 + c\lambda t)e^{-\lambda t}.$$

By property (E20b) for expectation, the expected life of the system

$$E[X] = \int_0^\infty R(t)\, dt = \int_0^\infty (1 + c\lambda t)e^{-\lambda t}\, dt = \frac{1 + c}{\lambda}.$$

For $c = 0$ (a single unit system), the expected life is $1/\lambda$. For $c = 1$ (a perfectly covered two-unit system), the expected life is $2/\lambda$. The imperfectly covered system has expected life that varies between these two values in a manner proportional to the probability c of being covered.

— □

19.2. Conditioning by a Random Vector—Special Cases

We consider two simple, but important, special cases, in which the idea of conditional expectation is developed quite intuitively from the notion of a conditional distribution. In each of these cases, we find the conditional expectation to be a random variable $E[g(Y)|X] = e(X)$, where $e(\cdot)$ is a Borel function defined on the range of X. This function satisfies a fundamental equation that

1. Provides a tie with the concept of conditional expectation, given an event.

2. Serves as the basis of the characteristic properties of conditional expectation.

3. Provides the basis for extending the notion of conditional expectation to the general case.

CASE 1—X, Y discrete

Consider $X = \sum_{i=1}^{n} t_i I_{A_i}$ and $Y = \sum_{j=1}^{m} u_j I_{B_j}$ in canonical form, so that

$$A_i = \{X = t_i\} = X^{-1}(\{t_i\}) \qquad \text{and} \qquad B_j = \{Y = u_j\} = Y^{-1}(\{u_j\}).$$

We suppose $P(A_i) > 0$ and $P(B_j) > 0$ for each permissible i, j. By Theorem 19.1.2, the conditional expectation $E[g(Y)|X = t_i] = E[g(Y)|A_i]$ is the expectation of $g(Y)$ with respect to the conditional probability measure $P(\cdot|X = t_i) = P(\cdot|A_i)$, defined by

$$P(Y = u_j|X = t_i) = \frac{P(X = t_i, Y = u_j)}{P(X = t_i)}.$$

Thus

$$E[g(Y)|X = t_i] = \sum_j g(u_j)P(Y = u_j|X = t_i).$$

Since this process determines a value for each t_i in the range of X, it defines a function $e(\cdot)$ on the range of X

$$e(t_i) = E[g(Y)|X = t_i] \qquad \text{for all } t_i \text{ in the range of } X.$$

Now consider any Borel set M in the codomain of X. Using the preceding pattern, we have

$$
\begin{aligned}
E[I_M(X)g(Y)] &= \sum_{i,j} I_M(t_i)g(u_j)P(X = t_i, Y = u_j) \\
&= \sum_{i,j} I_M(t_i)g(u_j)P(Y = u_j|X = t_i)P(X = t_i) \\
&= \sum_i I_M(t_i)\left(\sum_j g(u_j)P(Y = u_j|X = t_i)\right)P(X = t_i) \\
&= \sum_i I_M(t_i)e(t_i)P(X = t_i) = E[I_M(X)e(X)].
\end{aligned}
$$

Hence, the random variable $e(X)$ must satisfy

(A) $$E[I_M(X)g(Y)] = E[I_M(X)e(X)]$$

for all Borel sets M in the codomain of X.

The uniqueness property (E5) for expectations ensures $e(\cdot)$ is unique a.s. $[P_X]$, so that $e(\cdot)$ is determined uniquely on the range of X.

Example 19.2.1:

Suppose the pair of random variables (X,Y) produces the joint distribution that places probability mass 0.1 at points $(1,1)$, $(2,1)$, $(3,1)$, $(4,1)$, $(2,2)$, $(3,2)$, $(4,2)$, $(1,3)$, $(2,3)$, and $(3,3)$. Determine the function $e(\cdot) = E[Y|X = \cdot]$.

Solution

From the joint distribution, we obtain the values:

$$P(X = 1) = 0.2 \quad P(X = 2) = P(X = 3) = 0.3 \quad P(X = 4) = 0.2.$$

From this it follows that

$$P(Y = j|X = 1) = \frac{0.1}{0.2} \quad \text{for } j = 1, 3.$$

Hence
$$e(1) = E[Y|X = 1] = \frac{1}{2}(1 + 3) = 2.$$

Similarly,

$$e(2) = e(3) = \frac{1}{3}(1 + 2 + 3) = 2 \quad \text{and} \quad e(4) = \frac{1}{2}(1 + 2) = \frac{3}{2}.$$

□

INTERPRETATION

The conditional probabilities $P(Y = j|X = t)$, for fixed t, are proportional to the probability masses on the vertical line corresponding to $X = t$. Thus, $E[Y|X = t]$ is the center of mass for that part of the joint distribution that corresponds to $X = t$. In the discussion following Example 14.2.6, we note that the center of mass for a distribution is the best mean-square estimate for the random variable. It seems reasonable to extend this interpretation to the "conditional distribution," given $X = t$, to suppose $E[g(Y)|X = t]$ is the best mean-square estimate of $g(Y)$, given that the value of X is t. We validate this interpretation in Section 19.5.

CASE 2—(X, Y) jointly absolutely continuous

Suppose the joint density function is f_{XY}. Since the event $\{X = t\}$ has zero probability, we cannot begin with the conditional expectation, given the event $\{X = t\}$. However, we appeal to the intuitive notion of a conditional distribution, given $X = t$, by employing the following procedure. Let

$$f_{Y|X}(u|t) = \begin{cases} f_{XY}(t, u)/f_X(t) & \text{for } f_X(t) > 0 \\ 0 & \text{otherwise} \end{cases}.$$

For fixed t such that $f_X(t) > 0$ (i.e., t is in the range of X), the function $f_{Y|X}(\cdot|t)$ has the properties of a probability density function; that is, it is nonnegative and the integral over the whole line is one. It is natural to call $f_{Y|X}(\cdot|t)$ the conditional density function for Y, *given* $X = t$. In part, this terminology is justified by the following development.

Let M be any Borel set on the codomain of X. Then

$$
\begin{aligned}
E[I_M(X)g(Y)] &= \iint I_M(t)g(u)f_{XY}(t, u)\, dt\, du \\
&= \int I_M(t) \left(\int g(u)f_{Y|X}(u|t)\, du \right) f_X(t)\, dt \\
&= \int I_M(t)e(t)f_X(t)\, dt = E[I_M(X)e(X)]
\end{aligned}
$$

where $e(t) = \int g(u)f_{Y|X}(u|t)\, dt$.

Thus $e(X)$ satisfies

(A) $$E[I_M(X)g(Y)] = E[I_M(X)e(X)]$$

for all Borel sets M in the codomain of X.

It seems natural to call $e(t) = \int g(u)f_{X|Y}(u|t)\, du$ the conditional expectation of $g(Y)$, *given* $X = t$. By Theorem 19.1.3 and property (A), we have

(B) $$E[I_M(X)e(X)] = E[g(Y)|X \in M]P(X \in M),$$

provided $P(X \in M) > 0$.

If $e(\cdot)$ is Borel, as it will be in any practical case, the uniqueness property (E5) for expectation ensures that $e(X)$ is a.s. unique or, equivalently, $e(\cdot)$ is unique a.s. $[P_X]$. This means that $e(\cdot)$ is determined essentially on the range of X.

Example 19.2.2:

Suppose (X, Y) produces a joint distribution that is uniform over the triangular region (see Figure 19.1) that has vertices $(0, 0)$, $(1, 0)$, and $(1, 1)$. Then $f_{XY}(t, u) = 2$ over the region (and is zero elsewhere). From this we obtain

$$f_X(t) = \int f_{XY}(t, u)\, du = 2 \int_0^t du = 2t, \qquad 0 \le t \le 1$$

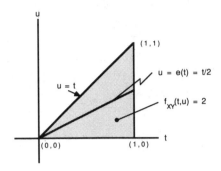

FIGURE 19.1. Joint density for Example 19.2.2.

and

$$f_{Y|X}(u|t) = \frac{1}{t} \qquad \text{for } 0 \le u \le t, \, 0 < t \le 1.$$

Hence

$$e(t) = E[Y|X = t] = \int u f_{Y|X}(u|t) \, du = \frac{1}{t} \int_0^t u \, du = \frac{t}{2} \qquad 0 < t \le 1.$$

———   □

INTERPRETATION

For any t in the range of X, the conditional density describes a "conditional distribution," which may be visualized by considering a very thin strip of the joint distribution. In this case, the conditional distribution is uniform on the interval $(0, t)$. For any t, the center of mass is at the midpoint $t/2$. The graph of $u = e(t)$ is thus the line from the vertex of the triangle at $(0, 0)$ to the midpoint $(1, 1/2)$ of the vertical member (see Figure 19.1). As in the discrete case, we may interpret this center of mass as the best mean-square estimate of Y, given that the value of X is t (see Section 19.5).

The following example has a more complex joint distribution. Although calculations are somewhat more involved, the fundamental pattern is the same as in the previous example.

Example 19.2.3:

The pair $\{X, Y\}$ has joint density function on the unit square (see Figure 19.2)

$$f_{XY}(t, u) = \frac{6}{5}(t^2 + u) \qquad 0 \le t \le 1, \, 0 \le u \le 1.$$

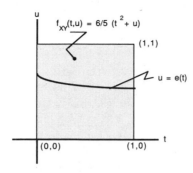

FIGURE 19.2. Joint distribution for Example 19.2.3.

This implies $f_X(t) = \frac{3}{5}(2t^2 + 1)$ for $0 \leq t \leq 1$, and $f_Y(u) = \frac{2}{5}(3u + 1)$ for $0 \leq u \leq 1$.

Determine the conditional expectation $e(t) = E[Y|X = t]$.

Solution

$$f_{Y|X}(u|t) = \frac{2t^2 + 2u}{2t^2 + 1} = 1 + \frac{2u - 1}{2t^2 + 1} \qquad 0 \leq t \leq 1, \quad 0 \leq u \leq 1.$$

Hence,

$$e(t) = E[Y|X = t] = \int u f_{Y|X}(t|u)\, du$$

$$= \int_0^1 u\, du + \frac{1}{2t^2 + 1} \int_0^1 (2u^2 - u)\, du$$

$$= \frac{1}{2} + \frac{1}{6(2t^2 + 1)}.$$

This result may be interpreted as in the previous case. Because of the nonuniformity of the conditional densities, the centers of mass do not lie at the midpoints, so that the graph of $u = e(t)$, shown in Figure 19.2, is not a straight line.

———   □

19.3. Conditioning by a Random Vector—General Case

The treatment of the special cases in the previous section proceeds from the intuitive notion of a conditional distribution. This approach is effective in these simple cases, and leads to a quite useful theory. In many applications involving real random variables, this is the natural way to view and interpret the problem. However, a development along these lines quickly becomes unmanageable in more general cases that involve random vectors of higher dimensions, with mixed distributions. And some of the more powerful features of the theory are not readily obtainable, even in the simple cases.

In each of the special cases considered in Section 19.2, the quantity called the conditional expectation of $g(Y)$, given $X = t$, is the value $e(t)$ of a Borel function $e(\cdot)$, which is defined on the range of X. The corresponding random variable $e(X)$ satisfies

(A) $$E[I_M(X)g(Y)] = E[I_M(X)e(X)]$$

for all Borel sets M in the codomain of X.

Now, in any case, if $e(X)$ satisfies property (A), then by Theorem 19.1.3 we have

(B) $$E[I_M(X)e(X)] = E[g(Y)|X \in M]P(X \in M),$$

provided $P(X \in M) > 0$.

This provides a tie between the concept of $E[g(Y)|X]$, the conditional expectation given the random vector X, and that of $E[g(Y)|X \in M]$, the conditional expectation, given the event $\{X \in M\}$. Subsequent developments show that an extensive and powerful list of properties of $e(X) = E[g(Y)|X]$ may be obtained as a consequence of property (A) and the properties of mathematical expectation discussed in Chapter 14. By the uniqueness property (E5) for expectations, $e(X)$ must be a.s. unique. This is equivalent to the condition that $e(\cdot)$ is unique a.s. $[P_X]$, so that $e(\cdot)$ is effectively determined on the range of X. Its value outside that range is of no consequence in calculations.

The special case of the Radon-Nikodym theorem, property (E19) for expectation, shows that for any integrable $g(Y)$ and any random vector X, there exists a Borel function $e(\cdot)$ such that the random variable $e(X)$ satisfies property (A). By property (E5), the random variable $e(X)$ is a.s. unique. In the light of these facts, it seems reasonable to make the following general definition.

Definition 19.3.1. *Let $e(\cdot)$ be a real-valued Borel function defined on a set that includes the range of random vector X, and let $g(\cdot)$ be a real-valued Borel function such that $g(Y)$ is integrable. Then the random variable $e(X)$*

is called the conditional expectation of $g(Y)$, given X, *denoted* $E[g(Y)|X]$, *iff*

(A)
$$E[I_M(X)g(Y)] = E[I_M(X)e(X)]$$

for all Borel sets M in the codomain of X.

Remark: In spite of the "expectation-style" notation, $E[g(Y)|X]$ is a real-valued function of random vector X, and therefore is a real random variable, unique a.s. The justification for the notation comes from "expectation-like" properties of this random variable, which are discussed in the next section.

Definition 19.3.2. *If $e(\cdot)$ is the Borel function in the definition of conditional expectation, then the quantity $e(t)$ is the* conditional expectation of $g(Y)$, *given $X = t$, denoted $E[g(Y)|X = t]$.*

Remark: We must distinguish between the two symbols:

(a) $E[g(Y)|X = \cdot] = e(\cdot)$, a *Borel function* on the range of X.

(b) $E[g(Y)|X] = e(X)$, a *random variable*. For a given ω we write

$$E[g(Y)|X](\omega).$$

The treatment in this section, so far, is formal and abstract. We have given no real content, as yet, to the extended notion of conditional expectation. It does, however, include the introductory special cases. As a simple, but often useful, extension of the introductory cases, we consider the case that the conditioning random vector X is simple, but the conditioned random vector Y may have any distribution.

Example 19.3.1:

Suppose $X = \sum_{i=1}^{n} t_i I_{A_i}$ is vector-valued and $P(A_i) = P(X = t_i) > 01 \le i \le n$. Let $e(\cdot)$ be defined on the range of X by $e(t_i) = E[g(Y)|X = t_i]$. Then

$$e(X) = \sum_{i=1}^{n} e(t_i)I_{A_i} = \sum_{i=1}^{n} E[g(Y)|X = t_i]I_{\{t_i\}}(X)$$

and

$$e(X) \quad \text{satisfies condition (A)}.$$

Remark: $e(t_i)$ is the conditional expectation of $g(Y)$, given the event $\{X = t_i\}$.

Proof

For any Borel set M on the codomain of X, we have

$$I_M(X) = \sum_{i=1}^{n} I_{\{t_i\}}(X) I_M(t_i),$$

since the ith term in the sum has value one iff $X = t_i$ and $t_i \in M$.

Also, as a special case of Theorem 19.1.3, $E[g(Y)I_{\{t_i\}}(X)] = e(t_i)P(X = t_i)$.

We thus have

$$E[g(Y)I_M(X)] = \sum_{i=1}^{n} I_M(t_i)e(t_i)P(X = t_i).$$

On the other hand,

$$
\begin{aligned}
E[e(X)I_M(X)] &= E\left[\sum_{i=1}^{n} e(t_i)I_M(t_i)I_{\{t_i\}}(X)\right] \\
&= \sum_{i=1}^{n} I_M(t_i)e(t_{ji})P(X = t_i).
\end{aligned}
$$

This establishes proposition (A) and shows that $e(X)$ is the desired conditional expectation.

——   □

A FUNDAMENTAL INTERPRETATION

We have seen in the simple introductory cases that $E[Y|X = t]$ is the center of mass of the conditional distribution, given that X has the value t. As a consequence of the result of Example 14.2.6, we note that for each admissible t,

$$E\left[(Y - r(t))^2\right] \quad \text{is a minimum iff} \quad r(t) = E[Y|X = t].$$

While the notion of conditional distribution, given a value of X, becomes difficult to visualize for mixed distributions and X a random vector, the interpretation that $e(t) = E[Y|X = t]$ is the best mean-square estimate for Y, given $X = t$ is valid in the general case. We establish this carefully in Section 19.5, where we solve the regression problem introduced in Section 16.3.

CONDITIONAL PROBABILITIES AND CONDITIONAL DISTRIBUTIONS

Ordinary, or total, probability may be expressed in terms of expectation by the relationship (E1a), $P(Y \in N) = E[I_N(Y)]$. Use of Theorems 19.1.2 and 19.1.3, enable us to assert

$$P(Y \in N|C) = E[I_N(Y)|C]$$

and

$$P(Y \in N|X \in M) = E[I_N(Y)|X \in M].$$

It seems reasonable to extend the notion of conditional probability to the case of conditioning by a general random vector in a similar manner, as follows.

Definition 19.3.3. $P(Y \in N|X = t) = E[I_N(Y)|X = t]$ a.s. $[P_X]$

Thus, there is no need to develop a separate theory of conditional probability, given a random vector. The essential properties of conditional probability are obtained as special cases of properties of conditional expectation.

Conditional probability, given a random vector, shares with conditional expectation the a.s. stipulation. In developing a general theory, some technical mathematical difficulties may arise because there is an exceptional set of t for each Borel set N, and there is an uncountably infinite class of Borel sets. Although the exceptional set of t for each Borel set N is a Borel set with zero probability assigned by P_X, the union of an uncountable class of such sets may not be Borel, and if so may not have zero probability. For the type of applications we consider, this causes no problem. The nature of the assumptions made generally preclude such difficulties.

We may use conditional probability to obtain a conditional distribution function. Although the notion of a conditional distribution does not serve well as a basis for a general definition of conditional expectation, it turns out to be both natural and useful in many applications.

Definition 19.3.4. $F_{Y|X}(u|t) = P(Y \leq u|X = t) = E[I_{N_u}(Y)|X = t]$ a.s.$[P_X]$ where $N_u = (-\infty, u]$.

Again there is the almost surely stipulation. In practice, we assume there is a suitable version of the conditional expectation for each fixed t in the range of X, so that the ambiguity is removed. If there is a corresponding conditional density function, $f_{Y|X}$, the relationship must be as follows:

$$F_{Y|X}(u|t) = \int I_{N_u}(v) f_{Y|X}(v|t)\, dv = \int_{-\infty}^{u} f_{Y|X}(v|t)\, dv.$$

For a given t, at every point of continuity of $f_{Y|X}$, as a function of u, we must have

$$f_{Y|X}(u|t) = \frac{\partial}{\partial u} F_{Y|X}(u|t).$$

It is more important in applications to be able to go from the conditional probability to the total probability, or from the conditional distribution to the total distribution. The key here is a special case of the defining property (A), which we designate (C). If we let M encompass the entire range of X, so that $I_M(X)$ is one for all ω, then

(C) $$E[g(Y)] = E\big[E[g(Y)|X]\big] = \int E[g(Y)|X = t]F_X(dt).$$

When applied to conditional probability, this becomes

(C*) $$\begin{array}{l} P(Y \in N) = E\big[E[I_N(Y)|X]\big] = \int E[I_N(Y)|X = t]F_X(dt) \\ F_Y(u) = \int F_{Y|X}(u|t)F_X(dt). \end{array}$$

For this reason, condition (C) is often referred to as law of total probability, whether applied to conditional expectation or to conditional probability.

In many investigations, both practical and theoretical, it is expedient first to determine the conditional expectation, given X, then to take the expectation of $e(X)$. Often this involves a natural assumption of a conditional distribution for Y, given $X = t$. This point of view underlies the Bayesian statistics, discussed briefly in Section 22.2. A distribution depends upon a parameter whose value is uncertain. The value of the parameter is taken to be the value of a random variable. For each possible value of the parameter there is a population distribution. The following examples illustrate how such situations arise in a natural way in practical problems, and show how the law of total probability may be used.

Example 19.3.2:

For purposes of analysis, the jobs on a computer may be divided into n classes. Let Y be the execution time for a randomly selected job. If the job is in class i, $1 \le i \le n$, then the execution time is assumed to be exponential (λ_i). Let X represent the job class, and suppose $P(X = i) = p_i$.

Determine $E[Y]$ and the distribution function F_Y.

Solution

(a) The hypothesis may be expressed in terms of the conditional density

$$f_{Y|X}(u|i) = \lambda_i e^{-\lambda_i u} \quad u \ge 0.$$

Then $E[Y|X = i] = \int u f_{Y|X}(u|i)\,du = 1/\lambda_i$, so that $E[Y] = \sum_{i=1}^n p_i/\lambda_i$.

(b) $F_Y(u) = E[I_{[0,u]}(Y)]$, and $E[I_{[0,u]}(Y)|X = i] = \int_0^u f_{Y|X}(s|i)\,ds = 1 - e^{-\lambda_i u}$.

Hence, $F_Y(u) = \sum_{i=1}^n (1 - e^{-\lambda_i u})p_i = 1 - \sum_{i=1}^n e^{-\lambda_i u}p_i$.

—   □

Example 19.3.3:

The time to failure X (in hours) of a manufactured unit is exponential (λ). The parameter λ itself is a random quantity, whose value depends upon the manufacturing process. That is, we suppose λ is the value of a parameter random variable H. Experience indicates the parameter value is somewhere between 0.005 and 0.01. In the absence of further information, it is reasonable to assume H is uniform on the interval $[0.005, 0.01]$. Determine $P(X > 150)$. Then determine $E[X|H = \lambda]$ and from this determine $E[X]$.

Solution

(a) We suppose $F_{X|H}(t|\lambda) = 1 - e^{-\lambda t}$. By assumption, $f_H(\lambda) = 200$ for $0.005 \le \lambda \le 0.01$. Thus

$$P(X > t) \quad = \quad 1 - F_X(t) = 1 - 200 \int_{0.005}^{0.01} (1 - e^{-\lambda t})\, d\lambda$$

$$= \quad 200 \int_{0.005}^{0.01} e^{-\lambda t}\, d\lambda = \frac{200}{t}(e^{-0.005t} - e^{-0.01t}).$$

On substituting $t = 150$, we obtain

$$P(X > 150) = \frac{200}{150}(e^{-0.75} - e^{-1.50}) \approx 0.3323.$$

(b) Under the hypothesis, $E[X|H = \lambda] = 1/\lambda$. Thus, by the law of total probability

$$E[X] \quad = \quad \int E[X|H = \lambda] f_H(\lambda)\, d\lambda = 200 \int_{0.005}^{0.01} \frac{d\lambda}{\lambda}$$

$$= \quad 200 \ln 2 \approx 138.6.$$

—   □

19.4. Properties of Conditional Expectation

The defining condition (A) and properties of mathematical expectation, as developed in Chapter 14, may be used to obtain significant properties of conditional expectation, which play an essential role in many aspects of modern applied probability. The properties listed in the following table have proved to be particularly useful.

The properties numbered (CE1), (CE1a), and (CE1b) in the table are properties (A), (B), and (C) discussed in the previous section. Thus, (CE1) is the defining property, (CE1a) provides the tie with conditioning by an event, and (CE1b) is the law of total probability, illustrated in Example 19.3.2 and 19.3.3. The latter property is used repeatedly in both applications and theoretical developments, in which it is expedient first to condition with respect to a random vector, then take the expectation of the conditional expectation.

The next three properties, linearity (CE2), positivity and monotonicity (CE3), and monotone convergence (CE4), are directly useful in a variety of specific situations. Also, they play a role in developing the character of conditional expectation quite parallel to the fundamental role of the corresponding properties for integrals. This is the basis for the expectation-like character of conditional expectation, which justifies the notational scheme and terminology adopted.

One word of caution. It must be remembered that $E[g(Y)|X]$ is a random variable, and the relations between the random variables in the properties are almost sure relations. This requires some care in handling certain theoretical questions, but usually has little, if any, effect in practical applications.

We provide proofs of several of the properties, in order to illustrate the character of a theoretical treatment. Additional development and discussion is found in the appendix to this chapter.

Proof of Linearity (CE2)

Let $e_1(X) = E[g(Y)|X]$, $e_2(X) = E[h(Z)|X]$, and $e(X) = E[ag(Y) + bh(Z)|X]$.

For any Borel set M on the codomain of X, we have by (CE1)

$$E\big[I_M(X)\big(ag(Y) + bh(Z)\big)\big] = E[I_M(X)e(X)].$$

Also, by linearity (E2) and (CE1), we have

$$
\begin{aligned}
E\big[I_M(X)\big(ag(Y) + bh(Z)\big)\big] \\
&= aE[I_M(X)g(Y)] + bE[I_M(X)h(Z)] \\
&= aE[I_M(X)e_1(X)] + bE[I_M(X)e_2(X)] \\
&= E\big[I_M(X)\big(ae_1(X) + be_2(X)\big)\big].
\end{aligned}
$$

Hence, by the uniqueness property (E5) for expectation,

$$e(X) = ae_1(X) + be_2(X) \quad \text{a.s.}$$

—   □

Property (CE5) provides an additional criterion for independence and provides a simple means of finding conditional expectation when the conditioning and conditioned random quantities are stochastically independent.

Property (CE6) provides an extension of the defining property (CE1). It is apparent that (CE1) is a special case of (CE6). To show that (CE1) implies (CE6), we must show that indicator function I_M may be replaced by a general Borel function h. This very important property provides the key to solving the regression problem in the next section. Again, the proof is illustrative of the types of arguments involved in developing the theory of conditional expectation.

PROPERTIES OF CONDITIONAL EXPECTATION

We suppose, without repeated assertion, that the random variables and functions of random vectors are integrable, as needed.

CE1) $e(X) = E[g(Y)|X]$ a.s. iff $E[I_M(X)g(Y)] = E[I_M(X)e(X)]$ for each Borel set M on the codomain of X.

CE1a) If $P(X \in M) > 0$, then

$$E[I_M(X)e(X)] = E[g(Y)|X \in M]P(X \in M).$$

CE1b) *Law of total probability.* $E[g(Y)] = E\big[E[g(Y)|X]\big]$.

CE2) *Linearity.* For any constants a and b,
$E[ag(Y) + bh(Z)|X] = aE[g(Y)|X] + bE[h(Z)|X]$ a.s.

(extends to any finite linear combination).

CE3) *Positivity; monotonicity.*

    a) $g(Y) \geq 0$ a.s. implies $E[g(Y)|X] \geq 0$ a.s.

    b) $g(Y) \geq h(Z)$ a.s. implies $E[g(Y)|X] \geq E[h(Z)|X]$ a.s.

CE4) *Monotone convergence.* $Y_n \to Y$ a.s. monotonically implies

$$E[Y_n|X] \to E[Y|X] \qquad \text{monotonically a.s.}$$

CE5) *Independence.* $\{X, Y\}$ is an independent pair
iff $E[g(Y)|X] = E[g(Y)]$ a.s. for all Borel functions g.
iff $E[I_N(Y)|X] = E[I_N(Y)]$ a.s. for all Borel sets N on the codomain of Y.

CE6) $e(X) = E[g(Y)|X]$ a.s. iff $E[h(X)g(Y)] = E[h(X)e(X)]$ a.s. for any Borel function h on the codomain of X.

CE7) If $Y = h(X)$, then $E[g(Y)|X] = g(Y)$ a.s. for g, a Borel function.

CE8) $E[h(X)g(Y)|X] = h(X)E[g(Y)|X]$ a.s. for any Borel function h.

CE9) If $X = h(W)$, then $E\big[E[g(Y)|X] \mid W\big] = E\big[E[g(Y)|W] \mid X\big] = E[g(Y)|X]$, a.s.

CE9a) $E\big[E[g(Y)|X] \mid X, Z\big] = E\big[E[g(Y)|X, Z] \mid X\big] = E[g(Y)|X]$ a.s.

CE9b) If $X = h(W)$ and $W = k(X)$, with h, k Borel functions, then $E[g(Y)|X] = E[g(Y)|W]$ a.s.

CE10) If g is a Borel function such that $E[g(t, Y)]$ is finite for all t on the range of X and $E[g(X, Y)]$ is finite, then

    a) $E[g(X, Y)|X = t] = E[g(t, Y)|X = t]$ a.s. $[P_X]$.

    b) If $\{X, Y\}$ is independent, then $E[g(X, Y)|X = t] = E[g(t, Y)]$ a.s. $[P_X]$.

CE11) Suppose $\{X(t) : t \in T\}$ is a real-valued, measurable random process whose parameter set T is a Borel subset of the real line and S is a random variable whose range is a subset of T, so that $X(S)$ is a random variable. If $E[X(t)]$ is finite for all t in T and $E[X(S)]$ is finite, then

    a) $E[X(S)|S = t] = E[X(t)|S = t]$ a.s. $[P_S]$.

    b) If, in addition, $\{S, X_T\}$ is independent, then $E[X(S)|S = t] = E[X(t)]$ a.s. $[P_S]$.

CE12) *Countable additivity and countable sums:*

    a) If Y is integrable on A and $A = \overset{\infty}{\underset{i=1}{\cup}} A_i$ (disjoint union), then

$$E[I_A Y|X] = \sum_{i=1}^{\infty} E[I_{A_i} Y|X] \quad \text{a.s.}$$

    b) If $\sum_{n=1}^{\infty} E[|Y_n|] < \infty$, then $E[\sum_{n=1}^{\infty} Y_n|X] = \sum_{n=1}^{\infty} E[Y_n|X]$ a.s.

CE13) *Triangle inequality:* $|E[g(Y)|X]| \leq E[|g(Y)||X]$ a.s.

CE14) *Jensen's inequality:* If g is a convex function on an interval I that contains the range of a real random variable Y, then $g\{E[Y|X]\} \leq E[g(Y)|X]$ a.s.

CE15) Suppose $E[|Y|^p] < \infty$ and $E[|Z|^p] < \infty$ for $1 \leq p < \infty$.
    Then $E\big[|E[Y|X] - E[Z|X]|^p\big] \leq E[|Y - Z|^p] < \infty$.

Proof of (CE6)

1. If we can establish the property for $g \geq 0$, the general case follows from linearity, since $g = g^+ - g^-$, with both $g^+ \geq 0$ and $g^- > 0$.

2. By the defining condition (CE1), the proposition is true for $h = I_M$, for all Borel sets M.

3. By linearity (CE2), the proposition is true for $h = \sum_{i=1}^n a_i I_{M_i}$

4. For $h \geq 0$, there is a nondecreasing sequence of simple Borel functions $h_n \to h$.

   This means $h_n(X)g(Y)$ increases to $h(X)g(Y)$.

5. By positivity (CE3), $e(X) \geq 0$ a.s., so that $h_n(X)e(X)$ increases to $h(X)e(X)$ a.s.

6. By monotone convergence (E4a), $E[h_n(X)g(Y)] \to E[h(X)g(Y)]$ and $E[h_n(X)e(X)] \to E[h(X)e(X)]$. The two limits must be equal.

7. For general h, consider $h = h^+ - h^-$.

— □

Property (CE7) is essentially a special case of property (CE8). The latter property shows that a random quantity that is a function of the conditioning random vector may be treated as a constant. When combined with property (CE10), the result is particularly intuitive in the Borel function form

$$E[h(X)g(Y)|X = t] = h(t)E[g(Y)|X = t] = E[h(t)g(Y)|X = t].$$

Proof of (CE8)

1. For any Borel set M on the codomain of X, $I_M(X)h(X)$ is a Borel function of X.

2) Set $e(X) = E[g(Y)|X]$ and $e_1(X) = E[h(X)g(Y)|X]$.

2. Now, by (CE6),

$$E[I_M(X)h(X)g(Y)] = E[I_M(X)h(X)e(X)]$$

and

$$E[I_M(X)h(X)g(Y)] = E[I_M(X)e_1(X)].$$

3. By the uniqueness property for expectation, we may assert $e_1(X) = h(X)e(X)$ a.s

— □

Property (CE9) and its variations play an essential role in many theoretical investigations, which lead to results important in applications. For instance, (CE9a) provides a key step in the proof of the interesting and useful Theorem 21.4.2 on Poisson processes. Also, we use it at several crucial places in developing properties of conditional independence in Chapter 22. The concept of conditional independence is essential to the theory of Markov processes and to many topics in decision theory. The phenomenon represented by (CE9) is repeated conditioning. The property ensures that repeated conditioning, in either order, by a random vector and a function of that random vector produces a result that is essentially that of conditioning by the function of the random variable. Conditioning by $X = h(W)$ is "coarser" than conditioning by W itself. The net result of repeated conditioning is the result of the coarser conditioning. In the next section we show that $E[Y|X = t]$ is the best mean-square estimate of Y when the value of X is t. This provides a new point of view for interpreting this and other properties of conditional expectation.

The special form (CE9a) recognizes the fact that the random vector X is a function of the higher dimensional random vector (X, Z). A proof of (CE9) is given in the appendix to this chapter.

We have already referred to property (CE10). This result is highly intuitive, and is rather easy to prove in the simple cases used as introductory examples. For the general case, the proof requires aspects of measure theory beyond the scope of this treatment. The ideas of a proof, however, are sketched in the appendix. Property (CE11) is an extension of (CE10) to a somewhat more general setting.

The countable additivity and the result on integrating countable sums listed as property (CE12) are easy consequences of the corresponding properties for expectation.

As in the case of ordinary expectation, inequalities for conditional expectation are based on inequalities for appropriate random variables and the monotonicity and linearity properties. For proofs, see the appendix to this chapter.

We conclude this section by working some examples that illustrate the use of fundamental properties.

Example 19.4.1:

The pair of random variables $\{X, Y\}$ has joint density function (see Figure 19.3)

$f_{XY}(t, u) = 60t^2 u$ on the triangle Δ bounded by $t = 0$, $u = 0$, $u = 1 - t$,

so that $f_X(t) = 30t^2(1 - t)^2 = 30(t^2 - 2t^3 + t^4)$, $0 \leq t \leq 1$.

Consider

$$Z = \begin{cases} X^2/2Y & \text{for } X \leq 1/2 \\ 2Y & \text{for } X > 1/2 \end{cases} = I_M(X)X^2 + I_N(X)2Y$$

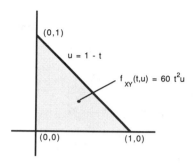

FIGURE 19.3. Joint distribution for Example 19.4.1.

where $M = (0, 1/2]$ and $N = (1/2, 1]$.

Determine $E[Y|X = t]$, $E[Z|X = t]$, and $E[Z]$.

Solution

$$f_{Y|X}(u|t) = \frac{f_{XY}(t, u)}{f_X(t)} = \frac{60t^2 u}{30t^2(1 - t)^2} = \frac{2u}{(1 - t)^2} \quad \text{on} \quad \Delta.$$

For each fixed t on $[0, 1)$, u ranges from 0 to $1 - t$, so that

$$E[Y|X = t] = \int u f_{Y|X}(u|t)\, du = \frac{2}{(1 - t)^2} \int_0^{1-t} u^2\, du = \frac{2}{3}(1 - t)$$

for $0 \le t < 1$. Now by linearity

$$E[Z|X = t] = E[I_M(X)X^2|X = t] + 2E[I_N(X)Y|X = t].$$

By (CE7), the first term on the right-hand side is $I_M(t)t^2$, which has the value t^2 on the interval $M = (0, 1/2]$ and is zero elsewhere. By (CE8), the second term is $I_N(t)2E[Y|X = t]$, which is twice $E[Y|X = t]$ on the interval $(1/2, 1]$ and is zero elsewhere. Thus,

$$E[Z|X = t] = \begin{cases} t^2 & \text{for } 0 \le t \le 1/2 \\ 4/(3(1 - t)) & \text{for } 1/2 < t < 1 \end{cases}.$$

The indicator functions I_M and I_N may be used to separate the two expressions for the different subintervals, so that we may write

$$E[Z|X = t] = I_M(t)t^2 + I_N(t)\frac{4}{3}(1 - t).$$

Warning. We cannot simply add the two expressions to get $t^2 + \frac{4}{3}(1 - t)$. One expression holds for values of t less than or equal to $1/2$; the other

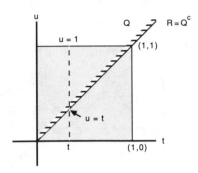

FIGURE 19.4. Joint distribution for Example 19.4.2.

holds only for values of t greater than $1/2$. Simply to add the expressions would give an incorrect value at every t.

By the law of total probability (CE1b),

$$E[Z] = E\big[E[Z|X]\big] = \int I_M(t)t^2 f_X(t)\, dt + \int I_N(t)\frac{4}{3}(1-t)f_X(t)\, dt.$$

The indicator functions effectively determine the limits of integration in each case, so that

$$E[Z] = 30 \int_0^{1/2} (t^4 - 2t^5 + t^6)\, dt + 40 \int_{1/2}^1 (t^2 - 3t^3 + 3t^4 - t^5)\, dt \approx 0.2939.$$

—   □

Example 19.4.2: (See Example 19.2.3 and Figure 19.4)

In Example 19.2.3, the pair $\{X, Y\}$ has joint density function on the unit square

$$f_{XY}(t, u) = \frac{6}{5}(t^2 + u) \qquad 0 \le t \le 1, \quad 0 \le u \le 1.$$

This implies

$$f_X(t) = \frac{3}{5}(1 + 2t^2) \quad 0 \le t \le 1$$

and

$$f_{Y|X}(u|t) = 2\frac{t^2 + u}{(1 + 2t^2)}, \qquad 0 \le t \le 1, \quad 0 \le u \le 1.$$

The conditional expectation is determined in Example 19.2.3 to be

$$e(t) = E[Y|X = t] = \int u f_{Y|X}(u|t)\, du = \frac{1}{2} + \frac{1}{6(1 + 2t^2)} \qquad 0 \le t \le 1.$$

Consider the random variable

$$Z = \begin{cases} 2X^2 & \text{for } X \leq Y \\ 3XY & \text{for } X > Y \end{cases}.$$

Determine $E[Z|X = t]$ and $E[Z]$

Solution

We may use indicator functions to express Z as follows:

$$Z = I_Q(X,Y)2X^2 + I_R(X,Y)3XY$$

where $Q = \{(t,u) : t \leq u\}$ and $R = Q^c$. By linearity (CE2),
$E[Z|X = t] = 2E[I_Q(X,Y)X^2|X = t] + 3E[I_R(X,Y)XY|X = t]$.
By (CE8) and (CE10),

$$
\begin{aligned}
E[Z|X = t] &= 2t^2 E[I_Q(t,Y)|X = t] + 3t E[I_R(t,Y)Y|X = t] \\
&= 2t^2 \int I_Q(t,u) f_{Y|X}(u|t)\, du + 3t \int I_R(t,u) u f_{Y|X}(u|t)\, du.
\end{aligned}
$$

The effective limits of the integrals must be determined for each fixed t in the range of X. Reference to Figure 19.4 shows that for each fixed t, $I_Q(t,u) f_{Y|X}(u|t) > 0$ iff $t \leq u \leq 1$, and $I_R(t,u) f_{Y|X}(u|t) > 0$ iff $0 \leq u \leq t$. Thus

$$E[Z|X = t] = \frac{4t^2}{1 + 2t^2} \int_t^1 (t^2 + u)\, du + \frac{6t}{1 + 2t^2} \int_0^t (t^2 u + u^2)\, du.$$

Evaluation of these integrals is straightforward. The result is

$$E[Z|X = t] = \frac{2t^2 + 4t^4 - t^5}{1 + 2t^2}.$$

Then, by the law of total probability (CE1b),

$$
\begin{aligned}
E[Z] &= E[E[Z|X]] = \int E[Z|X = t] f_X(t)\, dt = \frac{3}{5} \int_0^1 (2t^2 + 4t^4 - t^5)\, dt \\
&= \frac{39}{50} = 0.78.
\end{aligned}
$$

—   □

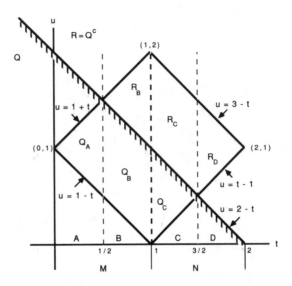

FIGURE 19.5. Joint density and significant regions for Example 19.4.3.

In the next example, both $f_{Y|X}$ and the random variable Z have composite definitions. Use of a combination of indicator functions and geometric representation aid in the formulation of the required integrals.

Example 19.4.3:

The pair $\{X, Y\}$ has joint density function (see Figure 19.5) $f_{XY}(t, u) = \frac{1}{2}tu$ on the square with vertices $(1, 0)$, $(2, 1)$, $(1, 2)$, and $(0, 1)$. This implies $f_X(t) = I_M(t)t^2 + I_N(t)(2t - t^2)$, where $M = (0, 1]$ and $N = (1, 2)$.

Let $Z = \begin{cases} X & \text{for } X + Y \leq 2 \\ X^2 Y & \text{for } X + Y > 2 \end{cases}$.

Determine $E[Z|X = t]$.

Solution

We may write $Z = I_Q(X, Y)X + I_R(X, Y)X^2Y$, where $Q = \{(t, u) : u \leq 2 - t\}$ and $R = Q^c$. Also

$$f_{Y|X}(u|t) = \frac{f_{XY}(t, u)}{f_X(t)} = I_M(t)\frac{u}{2t} + I_N(t)\frac{u}{2(2 - t)}.$$

The regions Q and R and the intervals M and N are shown on Figure 19.5. Use of linearity and property (CE10) shows that

$$E[Z|X = t] \quad = \quad tE[I_Q(t, Y)|X = t] + t^2 E[I_R(t, Y)Y|X = t]$$

$$= \int I_Q(t, u) t f_{Y|X}(u|t)\, du + \int I_R(t, u) t^2 u f_{Y|X}(u|t)\, du.$$

We partition M and N into $M = A \bigcup B$ and $N = C \bigcup D$, as shown in Figure 19.5. Also, we consider the six subregions Q_A, Q_B, Q_C and R_B, R_C, R_D, as shown. Now

$$\int I_Q(t, u) t f_{Y|X}(u|t)\, du$$

$$= \frac{1}{2}\left(I_A(t) \int I_{Q_A}(t, u) u\, du + I_B(t) \int I_{Q_B}(t, u) u\, du \right)$$

$$+ \frac{t}{2(2-t)} I_C(t) \int I_{Q_C}(t, u) u\, du$$

$$= I_A(t) h_A(t) + I_B(t) h_B(t) + I_C(t) h_C(t).$$

Similarly,

$$\int I_R(t, u) t^2 u f_{Y|X}(t|u)\, du$$

$$= \frac{t}{2} I_B(t) \int I_{R_B}(t, u) u^2\, du$$

$$+ \frac{t^2}{2(2-t)}\left(I_C(t) \int I_{R_C}(t, u) u^2\, du + I_D(t) \int I_{R_D}(t, u) u^2\, du \right)$$

$$= I_B(t) g_B(t) + I_C(t) g_C(t) + I_D(t) g_D(t).$$

The limits of integration for each integral are determined from Figure 19.5. Thus,

$$h_A(t) = \frac{1}{2} \int_{1-t}^{1+t} u\, du$$

$$h_B(t) = \frac{1}{2} \int_{1-t}^{2-t} u\, du$$

$$h_C(t) = \frac{t}{2(2-t)} \int_{t-1}^{2-t} u\, du.$$

Similarly,

$$g_B(t) = \frac{t}{2} \int_{2-t}^{1+t} u^2\, du$$

$$g_C(t) = \frac{t^2}{2(2-t)} \int_{2-t}^{3-t} u^2\, du$$

$$g_D(t)$$

$$= \frac{t^2}{2(2-t)} \int_{t-1}^{3-t} u^2\, du.$$

Evaluation is elementary but tedious. The result is

$$E[Z|X = t] \quad = \quad I_A(t)t + I_B(t)\frac{9 - 20t + 30t^2 - 6t^3 + 4t^4}{12}$$

$$+ I_C(t)\frac{9t + 32t^2 - 30t^3 + 6t^4}{12(2 - t)}$$

$$+ I_D(t)\frac{14t^2 - 15t^3 + 6t^4 - t^5}{3(2 - t)}.$$

The indicator functions pick out the correct expression for each value of t. The conditional expectation is assigned the value zero outside the range of the conditioning random variable X.

—   □

In the next example, we illustrate the use of these properties to obtain another important theoretical result.

Example 19.4.4: A Limit Theorem

If $X_n \to X$ a.s., $E[|X_n - X|^p] \to 0$ for some $p \geq 1$ and $\lim_n E[X_n|Z]$ exists a.s. then
$$\lim_n E[X_n|Z] = E[X|Z] \quad \text{a.s.}$$

Proof

By positivity (E3) and Fatou's lemma (E6) for expectation

$$0 \leq E\left[\left|\lim_n E[X_n|Z] - E[X|Z]\right|^p\right] \leq \lim_n E\left[|E[X_n|Z] - E[X|Z]|^p\right].$$

By property (CE15) and hypothesis,

$$\lim_n E\left[|E[X_n|Z] - E[X|Z]|^p\right] \leq \lim_n E[|X_n - X|^p] = 0.$$

By the positivity property (E3a),

$$\left|\lim_n E[X_n|Z] - E[X|Z]\right|^p = 0 \quad \text{a.s.}$$

The conclusion follows immediately.

—   □

19.5.   Regression and Mean-Square Estimation

The development of the previous two sections is rather formal, providing only occasional bits of intuitive insight. We obtain one more formal theoretical result that solves the regression problem introduced in Section 16.3.

The regression problem is that of finding the best mean-square estimator for Y when X is known. That is, we want a rule r, such that $r(t)$ is the best estimate, in the mean-square sense, for $Y(\omega)$ when $X(\omega) = t$. We show that $r(t) = e(t) = E[Y|X = t]$. Not only does this have the advantage of providing a practical solution of the regression problem in many cases, but it also provides a fundamental interpretation of the conditional expectation that does not depend upon the idea of conditional distribution. The formal result is expressed in the following.

Theorem 19.5.1 (Solution of the Regression Problem) *Let Y be a real-valued random variable such that $E[Y^2] < \infty$. For any random vector X, let $\Re_X$ be the class of all Borel functions $r(\cdot)$ such that $E[r^2(X)] < \infty$. Then any function $r(\cdot) \in \Re_X$ for which $E\big[(Y - r(X))^2\big]$ is a minimum satisfies $r(X) = e(X) = E[Y|X]$ a.s.*

Proof

The argument is deceptively simple, using property (CE6) in the key step. For any permissible Borel function r, we take the expectation of the square of the difference as follows:

$$
\begin{aligned}
E\big[(Y - r(X))^2\big] &= E\big[(Y - e(X) + e(X) - r(X))^2\big] \\
&= E\big[(Y - e(X))^2\big] + E\big[(e(X) - r(X))^2\big] \\
&\quad + 2E\big[(Y - e(X))(e(X) - r(X))\big].
\end{aligned}
$$

The first term in the last expression is fixed, since $e(X)$ is unique a.s. The second term is nonnegative by positivity (CE3). We examine the third term. Now $e(X) - r(X)$ is a Borel function of X; call it $h(X)$. Thus

$$
\begin{aligned}
E\big[(Y - e(X))h(X)\big] \\
= E[h(X)Y] - E[h(X)e(X)] = E[h(X)Y] - E\big[h(X)E[Y|X]\big].
\end{aligned}
$$

By (CE6), this expression must be zero. As a result, we have

$$
E\big[(Y - r(X))^2\big] = E\big[(Y - e(X))^2\big] + E\big[(e(X) - r(X))^2\big].
$$

By the positivity property (CE3), this is minimized by setting $r(X) = e(X)$ a.s.

— $\qquad\qquad\qquad\qquad\qquad\qquad\qquad\qquad\qquad\qquad\qquad\qquad$ $\square$

INTERPRETATION

We have a powerful alternative to the limited interpretation of conditional expectation as expectation with respect to a conditional distribution. If $g(Y)$ is an $\mathcal{L}^2$ random variable (i.e., if $E[g^2(Y)] < \infty$), then $e(X) = E[g(Y)|X]$ provides an estimator for the value of $g(Y)$ when the value of

X is known. Thus, if the value of X is observed to be t, we estimate the value of $g(Y)$ to be $e(t)$. This rule for estimation has the property that on the average, the square of the error of estimation is a minimum. We refer to such an estimator as a mean-square estimator for $g(Y)$, given X. Thus

$E[g(Y)|X]$ is a mean-square estimator for $g(Y)$, given X

and $E[g(Y)|X = t]$ is the best mean-square estimate of the value of $g(Y)$, when it is known that X has the value t. We know that $e(X)$ is unique a.s. or, equivalently, $e(\cdot)$ is unique a.s. $[P_X]$.

Suppose that in addition to the information $X = t$ we were to discover that $Z = v$. Then, according to Theorem 19.5.1, our best estimate would be $e^*(t, v) = E[g(Y)|X = t, Z = v]$. That is, we would use a new estimator $e^*(X, Z) = E[g(Y)|X, Z]$. We would expect this estimator to be better (or at least as good) as the estimator based only on the knowledge of the value of X. If this were not so, we should have some serious doubt about the usefulness of the mathematical model. To see that this fact is included in the structure of the model, we first note that $e(X) = r(X, Z)$. Therefore, by Theorem 19.5.1, we must have

$$E\big[(g(Y) - e^*(X, Z))^2\big] \leq E\big[(g(Y) - r(X, Z))^2\big].$$

We cannot rule out equality, as we show in Chapter 22, in the important case of conditional independence, given a random vector. The inequality may be put into an equivalent form in terms of variance. Since by (CE1b),

$$E[g(Y)] = E[e^*(X, Z)] = E[e(X)]$$

we must have

$$\mathrm{Var}[g(Y) - e^*(X, Z)] = E\big[(g(Y) - e^*(X, Z))^2\big]$$

and

$$\mathrm{Var}[g(Y) - e(X)] = E\big[(g(Y) - e(X))^2\big]$$

so that the variance of the error of estimation, given X and Z, is less than, or at most equal to, the variance of the error of estimation, given X. Smaller variance corresponds to smaller uncertainty, in the sense of less variability about the mean. Thus, $e^*(X, Z)$ is at least as good an estimator as $e(X)$.

As an illustration of this interpretation, let us reexamine property (CE9a) in terms of mean-square estimators.

$$E\big[E[g(Y)|X] \mid X, Z\big] = E\big[E[g(Y)|X, Z] \mid X\big] = E[g(Y)|X] \quad \text{a.s.}$$

In words, the following three entities are essentially the same.

1. The mean-square estimate, given X and Z, of the mean-square estimate of $g(Y)$, given X.

2. The mean-square estimate, given X, of the mean-square estimate of $g(Y)$, given X and Z.

3. The mean-square estimate of $g(Y)$, given X.

In each case, we are essentially limited to knowledge of X, so that the best estimate in each case is the same. In a chain of conditionings, the conditioning with the least information controls the overall result.

AN ALTERNATE APPROACH TO CONDITIONAL EXPECTATION

Our starting point in considering conditional expectation is the intuitive notion of averaging with respect to a conditional distribution, which is useful only in simpler cases. We note, in these cases, the relationship to mean-square estimation and find that the relationship survives the transition to the general case. A quite different approach is taken by some authors. One may begin by defining the function $e(\cdot) = E[Y|X = \cdot]$ as the function that minimizes $E\big[(Y - r(X))^2\big]$. Then it may be shown that $e(\cdot)$ has all of the properties of conditional expectation as we have defined it. Property (CE6) and the uniqueness property (E5) for expectation ensure the essential identity of the concepts. One mathematical disadvantage of this alternate approach is that it is limited to $\mathcal{L}^2$ random variables. For most applications, this is not a serious limitation and the mean-square estimate is meaningful only in the $\mathcal{L}^2$ case. The approach taken in Sections 19.2 through 19.4 seems somewhat more intuitive and is not limited to $\mathcal{L}^2$ random variables. In addition, it has the added advantage of tying the general concept to the highly intuitive concept of conditional expectation, given an event. In the $\mathcal{L}^2$ case, we are free to take whichever point of view seems most helpful in a given circumstance.

THE REGRESSION PROBLEM

With Theorem 19.5.1 we have solved the regression problem introduced in Section 16.3. In that section, we solve the problem of finding the affine function $r(\cdot)$ to minimize $E\big[(Y - r(X))^2\big]$. Theorem 19.5.1 shows that the general Borel function to minimize this quantity is $e(\cdot) = E[Y|X = \cdot]$. If $X = t$, the best estimate (in the mean-square sense) for Y is the value $u = e(t) = E[Y|X = t]$. These values could be plotted for each t in the range of X to give a curve.

Definition 19.5.2. *The graph of* $u = e(t) = E[Y|X = t]$ *is known as the regression curve of* Y *on* X.

Remarks

1. The term regression curve is used to distinguish this curve from the regression line introduced in Chapter 16. In many cases the two are the same.

2. If the pair $\{X, Y\}$ is independent, then $E[Y|X] = E[Y]$ and $\text{Cov}[X, Y] = 0$, so that both curves are $u = E[Y]$.

3. The regression line and the regression curve coincide for the joint normal distribution.

Example 19.5.1: The Joint Normal Distribution

Suppose (X, Y) have the joint normal distribution. That is

$$f_{XY}(t, u) = \frac{1}{2\pi\sigma_X\sigma_Y(1 - \rho^2)^{1/2}} e^{-Q(t,u)/2}$$

where

$$Q(t, u) = \frac{1}{1 - \rho^2}\left(\left(\frac{t - \mu_X}{\sigma_X}\right)^2 - 2\rho\left(\frac{t - \mu_X}{\sigma_X}\right)\left(\frac{u - \mu_Y}{\sigma_Y}\right) + \left(\frac{u - \mu_Y}{\sigma_Y}\right)^2\right).$$

Setting $r = (t - \mu_X)/\sigma_X$ and $s = (u - \mu_Y)/\sigma_Y$, we have

$$Q(t, u) = q(r, s) = \frac{r^2 - 2\rho rs + s^2}{1 - \rho^2}.$$

By completing the square, we may write this in the form

$$q(r, s) = r^2 + \frac{(s - \rho r)^2}{1 - \rho^2} = \left(\frac{t - \mu_X}{\sigma_X}\right)^2 + \frac{(u - \alpha(t))^2}{\sigma_Y^2(1 - \rho^2)}$$

where

$$\alpha(t) = \frac{\rho\sigma_Y}{\sigma_X}(t - \mu_X) + \mu_Y.$$

Thus $f_{XY}(t, u) = f_X(t)g(t, u)$ where

$$g(t, u) = \frac{1}{\sigma_Y(1 - \rho^2)^{1/2}\sqrt{2\pi}} \exp(-v^2/2)$$

with $v = u - \alpha(t)\left(\sigma_Y(1 - \rho^2)^{1/2}\right)$.

Hence, $f_{Y|X}(u|t) = g(t, u)$. For each fixed t, $g(t, u)$ is the density for a random variable that is normal with mean value $\alpha(t)$. Thus

$$E[Y|X = t] = \int u f_{Y|X}(u|t)\, du = \alpha(t) = \frac{\rho\sigma_Y}{\sigma_X}(t - \mu_X) + \mu_Y.$$

$\square$

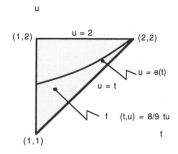

FIGURE 19.6. Joint distribution for Example 19.5.1.

Example 19.5.2:

Consider again the distribution discussed in Example 19.2.2. The joint distribution is uniform over the triangular region with vertices $(0,0)$, $(1,0)$, and $(1,1)$. We have

$$f_X(t) = \int f_{XY}(t,u)\,du = 2\int_0^t du = 2t \quad 0 \le t \le 1$$

and

$$f_{Y|X}(u|t) = \frac{1}{t} \qquad 0 < t \le 1.$$

Hence

$$e(t) = E[Y|X=t] = \int u f_{Y|X}(u|t)\,du = \frac{1}{t}\int_0^t u\,du = \frac{t}{2} \qquad 0 < t \le 1.$$

Thus, the regression curve is the line $u = t/2$, so that the regression curve is the same as the regression line of Y on X.

——   □

Example 19.5.3:

Suppose $f_{XY}(t,u) = 8/9\,tu$, $1 \le t \le u \le 2$. The region is the triangular region (see Figure 19.6) with vertices $(1,1)$, $(2,2)$, and $(1,2)$. For this joint distribution, we find

$$f_X(t) = \frac{4}{9}t(4 - t^2) \quad \text{and} \quad f_{Y|X}(u|t) = \frac{2u}{4 - t^2} \qquad \text{for } 1 \le t \le u < 2.$$

Straightforward calculations give

$$e(t) = \int u f_{Y|X}(u|t)\,du = \frac{2}{4 - t^2}\int_t^2 u^2\,du = \frac{2}{3}\cdot\frac{8 - t^3}{4 - t^2} = \frac{2}{3}\cdot\frac{t^2 + 2t + 4}{t + 2}$$

for $1 \le t < 2$. The graph of $u = e(t)$ is clearly not a straight line in this case.

— $\square$

Example 19.5.4:

Suppose $f_{XY}(t, u) = 1/28\,(4t + 2u + 1)$, $0 \le t \le 20 \le u \le 2$. Then

$$f_X(t) = \frac{1}{14}(4t + 3)f_{Y|X}(u|t) = \frac{4t + 2u + 1}{8t + 6} \qquad 0 \le t \le 20 \le u \le 2.$$

Hence,

$$E[Y|X = t] = \frac{1}{8t + 6}\int_0^2 \left((4t+1)u+2u^2\right)\,du = 1 + \frac{2}{12t + 9} \qquad 0 \le t \le 2.$$

In this case, the regression curve and the regression line are quite different.

— $\square$

19.6. Interpretation in Terms of Hilbert Space $\mathcal{L}^2$

We continue the informal discussion of $\mathcal{L}^2$, begun in Section 13.9. Here we direct attention to the real case, and restrict the measures to probability measures, so that the integrals are mathematical expectations. Since this case is included in the more general system outlined in Section 13.9, $\mathcal{L}^2$ is a vector space over the real numbers and the operator $\langle \cdot, \cdot \rangle$, defined by $\langle X, Y \rangle = E[XY]$ is an inner product. We use this inner product to define a norm or length of a second-order real random variable, viewed as an element of $\mathcal{L}^2$.

Definition 19.6.1. *The* norm *of a second-order random variable X, designated $\|X\|$, is given by*

$$\|X\| = \langle X, X \rangle^{1/2} = E[X^2]^{1/2}.$$

Remark: The norm in $\mathcal{L}^2$ is often called the $\mathcal{L}^2$-norm (read L-two norm).

By standard results of the theory of vector spaces with inner products, it is known that the function $\|\cdot\|$ has all the properties of a norm. That is,

(N1) $\qquad\qquad\qquad\qquad\quad \|X\| \ge 0.$

(N2) $\qquad\qquad\qquad\qquad\quad \|X\| = 0 \qquad$ iff $X = 0$ a.s.

(N3) $\qquad\qquad\qquad\qquad\quad \|aX\| = |a|\|X\|.$

(N4) *Triangle inequality:* $\|X + Y\| \le \|X\| + \|Y\|.$

In turn, the norm can be used to define a metric (distance function) that has all the standard properties.

Definition 19.6.2. *The* distance *between X and Y in $\mathcal{L}^2$ is the $\mathcal{L}^2$-norm of the difference. That is*

$$d(X,Y) = \|X - Y\| = E[(X - Y)^2]^{1/2}.$$

This distance function has the standard properties of a metric, which are derived easily from the properties of the inner product.

(D1) $d(X,Y) \geq 0$.

(D2) $d(X,Y) = 0$ iff $X = Y$ a.s.

(D3) $d(X,Y) = d(Y,X)$.

(D4) *Triangle inequality.* $d(X,Y) \leq d(Z,X) + d(Z,Y)$ for all $Z \in \mathcal{L}^2$.

Notions of convergence and continuity in a metric space are based on the distance derived from the metric. A metric space is complete if every convergent sequence of elements in the space has a limit in the space. It is known from advanced theory that the vector space $\mathcal{L}^2$ is complete in its metric (see Theorem 17.4.2d). Since this metric is based on the inner product, $\mathcal{L}^2$ is a Hilbert space.

Suppose W is an arbitrary random vector. Consider the class $\mathcal{L}^2(W)$ of all real random variables $Z = r(W)$, where $r(\cdot)$ is Borel and $E[r^2(W)] < \infty$, so that Z is an element in $\mathcal{L}^2$. Suppose Z_1, $Z_2 \in \mathcal{L}^2(W)$. Since a linear combination of Borel functions is a Borel function, a linear combination $aZ_1 + bZ_2 = ar_1(W) + br_2(W)$ is also a member of the class. Thus $\mathcal{L}^2(W)$ is a subspace of $\mathcal{L}^2$. Again, we may appeal to advanced results to assert that $\mathcal{L}^2(W)$ is closed (i.e., complete). Thus, $\mathcal{L}^2(W)$ is a closed subspace of $\mathcal{L}^2$.

Suppose X is any random vector and Y is an element of $\mathcal{L}^2$. Let $U = e(X) = E[Y|X]$. By Jensen's inequality (CE13), with $g(t) = t^2$, we have

$$U^2 = E^2[Y|X] \leq E[Y^2|X] \quad \text{a.s.}$$

Thus, by monotonicity (E3)

$$E[U^2] = E\big[E^2[Y|X]\big] \leq E\big[E[Y^2|X]\big].$$

By (CE1b), we have

$$E\big[E[Y^2|X]\big] = E[Y^2] < \infty.$$

Thus, if Y is in $\mathcal{L}^2$, then $U = E[Y|X] = e(X)$ is in $\mathcal{L}^2(X)$

The solution of the regression problem in Section 19.5 shows that $E\big[(Y - e(X))^2\big] \leq E\big[(Y - r(X))^2\big]$, for all Borel functions $r(\cdot)$ such that $E[r^2(X)] < \infty$. Putting $Z = r(X)$, we may express this inequality in terms of the distance function, as follows:

$$d^2(Y,U) \leq d^2(Y,Z) \quad \text{for all } Z \in \mathcal{L}^2(X).$$

If the square of the distance is a minimum, then the distance is a minimum. Thus, the best mean-square estimate for Y, given X is the element of the subspace $\mathcal{L}^2(X)$ that is nearest Y. As is well-known from the theory of linear vector spaces, this is the projection of Y on the subspace $\mathcal{L}^2(X)$. And that projection is the conditional expectation of Y, given X.

This formulation provides a geometric point of view in the analysis of mean-square estimators, and makes available powerful analytical aids from the general theory of vector spaces as developed in linear algebra and functional analysis. These quasi-geometric ideas are used extensively in the study of prediction and control in systems modeled by second-order random processes.

19.7.   Sums of Random Variables and Convolution

Suppose (X, Y) has joint distribution function F_{XY}. Consider $Z = X + Y$. We wish to determine the distribution function F_Z for the sum in terms of the joint distribution function. Now

$$F_Z(v) = P(X + Y \le v) = E[I_{Q_v}(X, Y)]$$

where

$$Q_v = \{(t, u) : t + u \le v\} = \{(t, u) : u \le v - t\}.$$

By the law of total probability (CE1b), we may first condition by X, then take the expectation. Thus,

$$F_Z(v) = E\big[E[I_{Q_v}(X, Y)|X]\big] = E[e(X)] = \int e(t) F_X(dt)$$

where

$$e(t) = E[I_{Q_v}(X, Y)|X = t] = E[I_{Q_v}(t, Y)|X = t].$$

The last equality is justified by property (CE10).

For each fixed v, $(t, u) \in Q_v$ iff $u \in N_{v-t} = (-\infty, v - t]$. Hence, $E[I_{Q_v}(t, Y)|X = t] = E[I_{N_{v-t}}(Y)|X = t] = P(Y \le v - t|X = t) = F_{Y|X}(v - t|t)$ so that

$$F_Z(v) = \int F_{Y|X}(v - t|t) F_X(dt).$$

In the case $\{X, Y\}$ is independent, the result simplifies, as follows. By (CE10b),

$$E[I_{Q_v}(X, Y)|X = t] = E[I_{Q_v}(t, V)] = P(Y \le v - t) = F_Y(v - t)$$

so that

$$F_Z(v) = \int F_Y(v - t) F_X(dt).$$

The latter integral produces the convolution of Y with X. The symmetry of the argument in X and Y shows that these two may be interchanged.

If (X, Y) has joint density function f_{XY}, the preceding argument shows that

$$E[I_{N_{v-t}}(Y)|X = t] = \int_{-\infty}^{v-t} f_{Y|X}(u|t)\, du$$

so that

$$F_Z(v) = \int_{-\infty}^{\infty} \int_{-\infty}^{v-t} f_{Y|X}(u|t) f_X(t)\, du\, dt = \int_{-\infty}^{\infty} \int_{-\infty}^{v-t} f_{XY}(t, u)\, du\, dt.$$

Upon differentiating with respect to v, we obtain

$$f_Z(v) = \int_{-\infty}^{\infty} f_{XY}(t, v - t)\, dt.$$

This latter result may be obtained by more elementary means, using the techniques for distributions of functions of random variables described in Chapter 11. The formula using conditional distributions is not directly obtainable by these elementary arguments, however.

Remark: The notion of conditional probability that $Y \leq v - t$, given $X = t$, and the corresponding conditional distribution function, given $X = t$, appear quite naturally in the preceding discussion. These quantities are expressed as conditional expectations. If we use indicator functions appropriately, it is not necessary to have a separate theory of conditional probability, given a random vector.

Example 19.7.1:

Suppose (X, Y) has joint density function $f_{XY}(t, u) = 4/7\, tu$, $1 \leq t \leq 3$, $0 \leq u \leq 1$.

Let $Z = X + Y$. Determine $f_Z(v)$.

Solution

$f_{XY}(t, u) = I_{[1,3]}(t) I_{[0,1]}(u) tu$ so that $f_{XY}(t, v - t) = I_{[1,3]}(t) I_{[0,1]}(v - t) t(v - t)$

Now, $I_{[0,1]}(v - t) = 1$ iff $0 \leq v - t \leq 1$ iff $v - 1 \leq t \leq v$. Hence, $I_{[0,1]}(v - t) = I_{[v-1,v]}(t)$.

For $v < 1$ or $v > 4$, $f_{XY}(t, v - t) = 0$ for all $t \in [1, 3]$, so that $f_Z(v) = 0$.

For $1 \leq v \leq 2$, $[1, 3] \cap [v-1, v] = [1, v]$, so that $f_Z(v) = 4/7 \int_1^v t(v-t)\, dt = 1/21\, (2v^3 - 6v + 4)$.

For $2 < v \leq 3$, $[1, 3] \cap [v-1, v] = [v-1, v]$, so that $f_Z(v) = 4/7 \int_{v-1}^v t(v - t)\, dt = 1/21\, (2v^3 - 2(v - 1)^3 - 6(v - 1)^2)$.

For $3 < v \le 4$, $[1,3] \cap [v-1, v] = [v-1, 3]$, so that $f_Z(v) = 4/7 \int_{v-1}^{3} t(v-t)\, dt = 1/21 \left(54v - 128 - 2(v-1)^3 - 6(v-1)^2\right)$.

—   □

Computationally, the preceding solution is no easier than the quasi-geometric methods used in Chapter 11. However, it is frequently useful to have the analytic formulation readily available.

19a

Some Theoretical Details

In Section 19.4, we include proofs of several of the properties of conditional expectation, in order to exhibit the character of the underlying theoretical structure. In this appendix, we discuss the proofs of the other properties included in the table properties of conditional expectation. In a few cases, proofs require measure-theoretic results beyond the scope of this book. In those cases, we sketch the ideas of the proofs in a manner that should aid the interested and prepared reader to consult the literature. We refer to the properties by the numbers in the table. As in the table, we assume the indicated integrability of random variables and functions of random vectors.

CE3) By the positivity property (E3) for expectation, the condition $g(Y) \geq 0$ implies $E[I_M(X)g(Y)] \geq 0$ for all Borel sets M. By (CE1), we must therefore have $E[I_M(X)e(X)] \geq 0$ for all Borel sets M. Suppose $e(X) < 0$ for all $\omega \in A$. Then there is a Borel set M_0 such that $A = X^{-1}(M_0)$. Thus, $I_{M_0}(X)e(X) = I_A e(X) \leq 0$. By (E3), we know that $E[I_A e(X)] \leq 0$, with equality iff $I_A e(X) = 0$ a.s. Since $e(X) < 0$ for all $\omega \in A$, the only possibility for the required condition to hold is for $P(A) = 0$. This amounts to the assertion $e(X) \geq 0$ except possibly on a set of probability zero. Positivity is established. Monotonicity follows from positivity and linearity.

— $\qquad\qquad\qquad\qquad\qquad\qquad\qquad\qquad\qquad\qquad\qquad$ $\square$

CE4) Consider the case Y_n increases to Y a.s. Put $e_n(X) = E[Y_n|X]$ and $e(X) = E[Y|X]$.

By monotonicity (CE3)

$$e_n(X) \leq e_{n+1}(X) \leq e(X) \quad \text{a.s.,} \qquad \text{for all } n \geq 1.$$

This means we can neglect an event (set of ω) of zero probability. The indicated relationship holds for all other ω. Thus for all ω not in the exceptional set, we have $e_\infty(X) = \lim_n e_n(X) \leq e(X)$, which means the inequalities hold a.s.

For any Borel set M, $I_M(X)Y_n \to I_M(X)Y$ a.s. and $I_M(X)e_n(X) \to I_M(X)e_\infty(X)$ a.s.

By monotone convergence for expectation,

$$E[I_M(X)e_n(X)] = E[I_M(X)Y_n] \to E[I_M(X)Y] = E[I_M(X)e(X)]$$

and

$$E[I_M(X)e_n(X)] \to E[I_M(X)e_\infty(X)].$$

Hence

$$E[I_M(X)e_\infty(X)] = E[I_M(X)e(X)] \qquad \text{for all Borel sets } M.$$

The uniqueness property (E5) for expectation ensures that $e_\infty(X) = e(X)$ a.s.

The case of decreasing X_n may be reduced to the increasing case by considering the negatives $-X_n$, which form an increasing sequence.

—   □

CE5) We wish to show the equivalence of three conditions:

(a) $\{X, Y\}$ is independent
(b) $E[g(Y)|X] = E[g(Y)]$ a.s. for all Borel functions g
(c) $E[I_N(Y)|X] = E[I_N(Y)]$ a.s. for all Borel sets N.

We show (a) implies (b) implies (c) implies (a).

(b) implies (c), since (c) is a special case of (b).

(a) implies (b).

Since $\{X, Y\}$ is independent, so is $\{g(X), I_M(Y)\}$ for all Borel sets M.

By the product rule (E18) for expectation,

$$E[I_M(X)g(Y)] = E[I_M(X)]E[g(Y)] \qquad \text{for all Borel sets } M.$$

Since $E[g(Y)]$ is a constant, we may move it inside the expectation to get

$$E[I_M(X)g(Y)] = E\big[I_M(X)E[g(Y)]\big] \qquad \text{for all Borel sets } M.$$

By (E5), this is sufficient to ensure that $E[g(Y)] = e(X)$ a.s.

(c) implies (a).

For any permissible Borel sets M, N, we have by hypothesis and (CE1)

$$E[I_M(X)I_N(Y)] = E\big[I_M(X)E[I_N(Y)]\big].$$

Since $E[I_N(Y)]$ is a constant, we may take it out of the expectation to get

$$E[I_M(X)I_N(Y)] = E[I_M(X)]E[I_N(Y)] \qquad \text{for all Borel sets } M, N.$$

By (E18), $\{X, Y\}$ must be independent.

—   □

CE9) We suppose $X = h(W)$. Then $e(X) = E[g(Y)|X] = e(h(W))$. Set $e_0(W) = E[g(Y)|W]$

By (CE7), $E[E[g(Y)|X] \mid W] = E[e(h(W)) \mid W] = e(h(W)) = e(X)$ a.s.

For any Borel set M in the codomain of X, there is a Borel set N such that
$$I_M(X) = I_M[h(W)] = I_N(W).$$
By (CE1), $E[I_M(X)g(Y)] = E[I_M(X)e(X)]$ for all Borel sets M. On the other hand,

$$E[I_M(X)g(Y)] = E[I_N(W)g(Y)] = E[I_N(W)e_0(W)].$$

But $E[I_N(W)e_0(W)] = E[I_M(X)e_0(W)] = E[I_M(X)E[e_0(W)|X]]$. Thus, $e(X) = E[e_0(W)|X] = E[E[g(Y)|W] \mid X]$ a.s., which is the desired result.

—   □

Property (CE10) is so highly intuitive that many writers simply assume it as obvious. It is also quite easy to prove in the case X, Y are real and have a joint density function. Proof of the general case requires some results of measure theory beyond the scope of the present work. Because of the importance of the property, we sketch the following.

IDEA OF A GENERAL PROOF OF (CE10)

If the theorem can be established for $g(t, u) = I_Q(t, u)$, where Q is any Borel set on the codomain of the random vector (X, Y), then a "standard argument" such as the one used in the proof of (CE6) may be used to extend the theorem to any Borel function g such that $E[g(X, Y)]$ is finite. We first consider Borel sets of the form $Q = M \times N$, where M and N are Borel sets on the codomains of X and Y, respectively. In this case

$$I_Q(t, u) = I_M(t)I_N(u).$$

Let $e(t, v) = E[g(v, Y)|X = t] = E[I_M(v)I_N(Y)|X = t] = I_M(v)E[I_N(Y)|X = t]$ and set $e_0(t) = E[g(X, Y)|X = t] = E[I_M(X)I_N(Y)|X = t]$.

Now $e(X, X) = I_M(X)E[I_N(Y)|X]$ a.s., and by (CE8) $e_0(X) = I_M(X)E[I_N(Y)|X]$ a.s.

Hence, $e(X, X) = e_0(X)$ a.s. This ensures that $e(t, t) = e_0(t)$ a.s. $[P_X]$, which is equivalent to the conclusion of the theorem for this case.

By linearity (CE2), the theorem holds for any Borel sets $Q = \bigcup_{i=1}^{n} M_i \times N_i$ (disjoint union), since in this case

$$I_Q = \sum_{i=1}^{n} I_{M_i} I_{N_i}.$$

Hence, equality holds for the class consisting of all finite, disjoint unions of sets of the form $M \times N$.

A number of proofs may be used to show that the equality holds for all Borel sets. We sketch one proof. It is known that the class of sets just described is an algebra of sets and that the minimum sigma algebra that contains it is the class $\mathcal{B}$ of Borel sets. Let $\mathcal{B}^*$ be the class of sets for which the equality holds. If $\{Q_i : 1 \leq i\}$ is a monotone class in $\mathcal{B}^*$, the sequence $\{I_{Q_i} : 1 \leq i\}$ is a monotone sequence of Borel functions. Use of monotone convergence (E4a) for expectation shows that equality holds for I_Q where Q is the limit of the sequence of the Q_i. Thus $\mathcal{B}^*$ is a monotone class. By a well-known theorem of measure theory, this monotone class must contain $\mathcal{B}$. This means that equality holds for every Borel set Q, so that the theorem follows.

If $\{X, Y\}$ is independent, then by (CE5) $e(t, v) = E[g(v, Y)|X = t] = E[g(v, Y)]$ a.s. $[P_X]$.

(CE10b) is the application of $e_0(t) = e(t, t)$ to this case.

— $\qquad\qquad\qquad\qquad\qquad\qquad\qquad\qquad\qquad\qquad$ □

Property (CE11) is an extension of (CE10) to a somewhat more general setting. The extension requires some measure-theoretic considerations and some skill in use of notation to state the situation precisely. We simply borrow this result.

(CE13) Since $-|g(Y)| \leq g(Y) \leq |g(Y)|$, we have by monotonicity (CE3)

$$-E[|g(Y)||X] \leq E[g(Y)|X] \leq E[|g(Y)||X] \quad \text{a.s.,}$$

which is equivalent to

$$|E[g(Y)|X]| \leq E[|g(Y)||X] \quad \text{a.s.}$$

— $\qquad\qquad\qquad\qquad\qquad\qquad\qquad\qquad\qquad\qquad$ □

(CE14) *Jensen's Inequality*

Convex function g satisfies $g(u) \geq g(t) + \lambda(t)(u - t)$, where λ is a nondecreasing function.

Set $e(X) = E[Y|X]$. Then with $u = Y$ and $t = e(X)$, we have

$$g(Y) \geq g(e(X)) + \lambda(e(X))(Y - e(X)).$$

If we take conditional expectations, we have by positivity (CE3) and linearity

$$E[g(Y)|X] \;\geq\; E[g(e(X)) \mid X] + E[\lambda(e(X))Y \mid X]$$
$$- E[\lambda(e(X))e(X) \mid X] \quad \text{a.s.}$$

We may apply (CE8), or the special case (CE7), to each term on the right-hand side to get

$$E[g(Y)|X] \geq g(e(X)) + \lambda(e(X))e(X) - \lambda(e(X))e(X) \quad \text{a.s.}$$

Noting that the last two terms cancel, and replacing $e(X)$ by the conditional expectation notation, we have

$$E[g(Y)|X] \geq g\{E[Y|X]\} \quad \text{a.s.}$$

—   □

(CE15) By Minkowski's inequality (E17) for expectation, we have $E[|X - Y|^p] < \infty$. Now by linearity (CE2)

$$E\big[|E[X|Z] - E[Y|Z]|^p\big] = E\big[|E[X - Y|Z]|^p\big].$$

Since $g(t) = |t|^p$ is convex for $p \geq 1$, Jensen's inequality and monotonicity ensure that the last term is no greater than

$$E\big[E[|X - Y|^p|Z]\big] = E[|X - Y|^p].$$

The last equality is justified by the law of total probability (CE1b).

—   □

PROBLEMS

19–1. Prove propositions (b) and (c) in Theorem 19.1.2

19–2. Show that $E[g(X)|A] = E[g(X)|AB]P(B|A) + E[g(X)|AB^c]P(B^c|A)$.

19–3. Let $X = 0I_{A_1} + 2I_{A_2} + 3I_{A_3}$ and $Y = I_{B_1} + 3I_{B_2}$ (canonical form), with joint probability distribution such that

$$P(A_1 B_1) = 1/6, \qquad P(A_2 B_2) = 1/2, \qquad P(A_3 B_1) = 1/3.$$

Show that $E[X|Y = 1] = E[X|Y = 3] = E[X]$, but that $\{X, Y\}$ is not independent.

19–4. Suppose $\{X, Y\}$ is independent, and each random variable is uniform on [-1,1]. Let $Z = g(X, Y)$ be given by

$$Z = \begin{cases} X & \text{for } X^2 + Y^2 \le 1 \\ c & \text{for } X^2 + Y^2 > 1 \end{cases}.$$

(a) Determine $E[Z|X^2 + Y^2 \le 1]$ and $E[Z|X^2 + Y^2 > 1]$.

(b) Use these results to determine $E[Z]$.

19–5. The pair $\{X, Y\}$ has joint distribution that puts probability mass 0.1 at the points $(0, 1)$, $(1, 0)$, $(1, 3)$, $(3, 2)$, $(3, 4)$, $(4, 3)$, $(4, 4)$, $(4, 5)$ $(5, 3)$, and $(5, 5)$. For this distribution, $E[X] = E[Y] = 3$ and $\text{Var}[X] = 2.8$

(a) Determine the regression line of Y on X.

(b) Determine the regression curve of Y on X.

19–6. The pair $\{X, Y\}$ is independent, with $E[X] = 7$, $\text{Var}[X] = 4$, $E[Y] = 2$, and $\text{Var}[Y] = 1$. If $W = 3X + X^2 Y^2$, what is the regression curve of W on X?

19–7. If X is discrete and Y is absolutely continuous, then the joint distribution can be described by a hybrid mass-density function f_{XY}, such that

$$P(X = t_i, Y \in M) = \int_M f_{XY}(t_i, u)\, du.$$

Develop an expression for $e(u) = E[g(X)|Y = u]$ in this case.

19–8. The pair $\{X, Y\}$ is independent, with $E[X] = 3$, $\text{Var}[X] = 5$, $E[Y] = 3$ and $\text{Var}[Y] = 2$. A value $X(\omega) = 2$ is observed. If $W = X^2 Y + 3Y^2$, what is the mean-square estimate for $W(\omega)$?

19–9. The pair X, Y is independent, with $E[X] = 0$, $\text{Var}[X] = 10$, $E[Y] = 3$, and $\text{Var}[Y] = 2$. Let $W = 2X^2 Y + X$. Determine the regression curve of W on X.

19–10. The pair $\{X, Y\}$ has joint density function $f_{XY}(t, u) = t^2 e^{-t^2 u}$ for $1 \leq t \leq 2, 0 \leq u$. This implies X is uniform on $[1,2]$.

Determine $E[Y|X = t]$ and $E[XY|X = t]$. From these compute $E[Y]$ and $E[XY]$.

19–11. The pair $\{X, Y\}$ has joint density function $f_{XY}(t, u) = 10u^2 t, 0 \leq t \leq u \leq 1$.

(a) Determine the regression curve of Y on X.

(b) If value $X(\omega) = 0.5$ is observed, what is the mean-square estimate of $Y(\omega)$?

19–12. The pair X, Y has joint density function

$$f_{XY}(t, u) = \frac{1}{28}(4t + 2u + 1) \qquad 0 \leq t \leq 2, \ 0 \leq u \leq 2$$

so that $f_X(t) = (4t + 3)/14, 0 \leq t \leq 2$.

(a) Determine the regression curve of Y on X.

(b) Let $W = Y$ for $X \leq 1$, and $W = X$ for $X > 1$. Determine $E[W|X = t]$ and use this to determine $E[W]$.

19–13. The pair X, Y has joint density $f_{XY}(t, u) = \frac{15}{8}t^2 u$ on the triangular region with vertices $(0,0)$, $(2,0)$, and $(0, 2)$.

(a) Determine the conditional density function $f_{Y|X}(u|t)$, noting carefully the (t, u) for which the expression is valid.

(b) Determine the regression curve of Y on X.

(c) Determine $E[X^2 + Y^2|X = t]$.

(d) Let $W = \max\{X, Y\}$. Determine $E[W|X = t]$.

(e) Let $Z = XY$ for $X \leq 1, Y \leq 1$ and $Z = 2$ otherwise. Determine $E[Z|X = t]$.

19–14. Random variables X and Y have joint density function

$$f_{XY}(t, u) = \frac{8}{9}tu \qquad \text{for } 1 \leq t \leq u \leq 2 \quad \text{(and zero elsewhere)}.$$

Then $f_X(t) = 4/9\, t(4 - t^2), 1 \leq t \leq 2$.

(a) Obtain an expression for the regression curve of Y on X.

(b) Determine $E[XY|X = t]$.

(c) Determine $E[X^2 + Y^2|X = t]$.

(d) Determine $E[X^2 Y + Y^2 X|X = t]$.

(e) Determine $E[X|X \le \frac{1}{2}(Y+1)]$.

19-15. The pair $\{X,Y\}$ has joint density function $f_{XY}(t,u) = \frac{20}{3}t^2u$, on the triangle bounded by $t = 1$, $u = 1$, $u = 1 - t$. Calculations show that

$$f_X(t) = \frac{10}{3}(2t^3 - t^4) \quad \text{for } 0 \le t \le 1$$

and

$$f_Y(u) = \frac{20}{9}(3u^2 - 3u^3 + u^4) \quad \text{for } 0 \le u \le 1.$$

Also, $E[X] = 7/9$, $E[X^2] = 40/63$, $E[Y] = 19/27$.

(a) Determine the regression line of Y on X.

(b) Determine the regression curve of Y on X.

(c) Consider $Z = \begin{cases} X^2 & \text{for } X \le 1/2 \\ 2XY & \text{for } X > 1/2 \end{cases} = I_{Q_1}(X)X^2 + I_{Q_2}(X)2XY$

where $Q_1 = (0,1/2]$ and $Q_2 = (1/2,1]$.
Determine $E[Z|X = t]$, and from this determine $E[Z]$.

19-16. The pair $\{X,Y\}$ has joint density function

$$f_{XY}(t,u) = \frac{6}{5}(t^2 + u) \qquad \text{for } 0 \le t \le 1,\, 0 \le u \le 1.$$

This implies $f_X(t) = \frac{6}{5}(t^2 + \frac{1}{2})$, $0 \le t \le 1$, and $f_Y(u) = \frac{6}{5}(u + \frac{1}{3})$, $0 \le u \le 1$.

Also, $E[X] = E[Y] = 3/5$ and $E[X^2] = 11/25$

(a) Determine the regression curve of Y on X.

(b) Let $Z = 3X^2Y + 4XY$. Determine $E[Z|X = t]$.

(c) Let $W = \begin{cases} XY^2 & \text{for } Y \ge X \\ 2X & \text{for } Y < X \end{cases}$. Determine $E[W|X = t]$.

Hint: Use indicator functions to express W and use property (CE10).

19-17. The pair $\{X,Y\}$ has joint density function $f_{XY}(t,u) = t^2e^{-t^2u}$ for $1 \le t \le 2$, $0 \le u$.

This implies that X is uniform on $[1,2]$ (see Problem 19-10).

(a) Obtain a formula for the regression curve of Y on X.

(b) Let $W = XY + 2X^2$. Determine $E[W|X = t]$.

(c) Let $Z = \begin{cases} XY & \text{for } X \le 3/2 \\ X^2 & \text{for } X > 3/2 \end{cases}$. Determine $E[Z|X = t]$ and from this determine $E[Z]$.

19-18. The pair $\{X, Y\}$ has joint density $f_{XY}(t, u) = 24/5\, tu$ on the triangular region bounded by $u = 1$, $t = 1$, $u = 1 - t$. Calculations show that $f_X(t) = f_Y(t) = 12/5\, t(2t - t^2)$, $0 \le t \le 1$. From this, we get $E[X] = 18/25$ and $E[X^2] = 14/25$.

   (a) Determine the regression line of Y on X.

   (b) Determine the regression curve of Y on X.

19-19. The pair $\{X, Y\}$ has joint density function $f_{XY}(t, u) = te^{-2tu}$ for $1 \le t \le 3$, $0 \le u$.

   This implies X is uniform on $[1, 3]$.

   (a) Compute $E[Y|X = t]$ and $E[XY|X = t]$. From these compute $E[X]$ and $E[XY]$.

   (b) Obtain a formula for the regression line of Y on X.

   (c) Obtain a formula for the regression curve of Y on X.

   (d) If $W = 2X + X^2Y$, determine the regression curve of W on X.

19-20. Use the fact that $g(X, Y) = g^*(X, Y, Z)$, with g^* Borel if g is, to establish the following extension of (CE10): $E[g(X, Y)|X = t, Z = v] = E[g(t, Y)|X = t, Z = v]$ a.s. $[P_{XZ}]$.

19-21. Use the result of problem 19-20, (CE9a), and (CE10) to show that

$$E[g(X, Y)|Z = v] = \int E[g(t, Y)|X = t, Z = v]\, dF_{X|Z}(t|v) \text{ a.s. } [P_Z].$$

19-22. The execution time X for jobs input from a terminal connected to a large computer mainframe depend upon both the complexity of the task and the apparent speed of the mainframe because of other tasks being executed. Suppose X is conditionally exponential (βu), given $Y = u$, where Y is a random variable whose value indicates the effective processing speed. That is,

$$F_{X|Y}(t|u) = 1 - e^{-\beta ut}   t \ge 0, \ u \ge 0, \ \beta > 0.$$

   Determine $F_X(t)$ in the case Y is exponential (λ).

19-23. A service facility, which works past nominal closing time to complete jobs on hand, tends to speed up service on any job received during the last hour before closing. Suppose arrival time (in hours) before closing is a random variable T, uniform on $[0,1]$. Service time Y for a unit received in that period is conditionally exponential $\beta(2 - u)$, given $T = u$. Determine the distribution function for the service time Y.

19–24. Unit lifetime of an electronic module is influenced by the power supply voltage. The power supply voltage remains essentially constant, but depends upon which unit is selected from a large lot. Let X be the time to failure (in hours) of a module and let Y be the voltage of the power supply selected. Suppose X is conditionally exponential $(u/5000)$, given $Y = u$, and suppose Y is distributed uniformly on $[4.5, 5.5]$. Determine F_X, and from this determine $P(X \geq 1000)$.

19–25. The time to failure of a manufactured unit is exponential (λ). The parameter λ itself is a random quantity, whose value depends upon the manufacturing process. Let X be the time to failure and let λ be a value of a parameter random variable H. We suppose

$$F_{X|H}(t|\lambda) = P(X \leq t|H = \lambda) = E[I_{[0,t]}(X)|H = \lambda] = 1 - e^{-\lambda t}$$

and
$$H \text{ is uniform on } [0.005, 0.01].$$

 (a) Determine $P(X > 150)$.

 (b) Determine $E[X|H = \lambda]$ and use this to determine $E[X]$.

19–26. A system has n components with independent times to failure $X_i :$ $1 \leq i \leq n$, with X_i exponential (λ_i). The system fails if any one or more of the components fails. What is the probability the failure is due to the ith component? That is, if W is the time to failure of the system, determine $P(W = X_i)$. Note $W = X_i$ iff for all $j \neq i$, we have $X_j > X_i$.

20

Random Selection and Counting Processes

In this chapter, we introduce the problem of selecting, by some chance mechanism, one of a class of random variables. Performance of the basic experiment results in both the selection of the random variable from the class and a determination of the value of the random variable selected.

If we have a sequence $\{Y_i : 0 \leq i\}$, we ordinarily observe the value of a specific random variable, say Y_n; or we observe sequentially the values of a specific number of variables, say $Y_0, Y_1, \ldots, Y_n$. The number n is fixed before the experiment. For example, a sample of size $n+1$ is planned, or we are interested in the maximum value of the first $n+1$ variables in the sequence. Thus, n is "nonrandom" in the sense that it does not depend upon the outcome of the experiment.

There are important situations, however, in which the choice of n is determined by the result of the experiment. In such a case, the number n is taken to be the observed value of a random variable N. The result is a randomly selected random variable. Each outcome ω, corresponding to a trial, results in the selection of a random variable and the determination of its value. By assigning a number to each ω in this way, we determine a new function on the basic space, which we designate Y_N. It seems reasonable to suppose that this new function of the outcomes should itself be a random variable. Although there is a theoretical question of measurability, fundamental theory shows that the new function is indeed a random variable.

In Section 20.1, we set up and illustrate a formal model that serves as a theoretical setting for investigating a variety of important practical problems. In subsequent sections, we analyze several such problems. In Section 20.6, we extend the model by replacing the counting random variable N with a counting process. For each $t \geq 0$, the random variable N_t counts the number of occurrences in the interval $(0, t]$ of a specific phenomenon that take place at discrete instants of time separated by random waiting, or interarrival, times. We consider certain processes that have iid interarrival times, particularly those with exponential interarrival times. This introduces the idea of a Poisson process, which is the subject of the next chapter.

20.1. Introductory Examples and a Formal Representation

The following examples illustrate the situations included in the formal model to be set up.

Example 20.1.1: Total Service Time on a Random Number of Units

A service facility receives a random number N of units to be serviced in a work day. The service time for the ith unit is a random variable Y_i. The total service time for the units received on a given day is the value of the sum of a random number of random variables.

$$X_N = Y_1 + Y_2 + \cdots + Y_N.$$

The problem is to determine the distribution for, or the moments of, the random variable X_N.

———   □

Example 20.1.2: Total Purchases by a Random Number of Customers

The number of orders received on a given day for a certain item of merchandise is a random variable N. The amount of the ith order is a random variable Y_i. It is desired to determine the probability distribution for the total number of items ordered that day. Again, we have the sum of a random number of random variables.

———   □

Example 20.1.3: Extreme Values

Disturbances, such as noise pulses on a power line connected to a computer, arrive on a sporadic basis. The number of pulses in a given time period is a random quantity. The magnitude of each pulse is a random quantity. It is desired to determine the probability that the maximum pulse magnitude does not exceed some critical value.

———   □

Example 20.1.4: Net Gain in a Random Number of Plays of a Game

Let Y_n be the gain on the nth play of a game and let Y_0 be the initial capital, or "bankroll" of the player. Then the net fortune after n plays is

$$X_n = Y_0 + Y_1 + \cdots + Y_n.$$

The number of plays is a random variable. Its value may be determined by an independent chance mechanism, or its value may be determined by a decision rule based on the progress of the game. These situations are different, and we comment on them later.

—   □

Example 20.1.5: Random Sample Size

In taking a random sample, instead of taking a fixed sample size n, it is sometimes more efficient to let the sample size depend upon the outcome of the particular sample. Some decision rule based on the current sampling results determines when to terminate the sampling. The sample size N is a random quantity.

FORMULATION

We wish to formalize systems of the type just illustrated in a manner that enables us to carry out the desired stochastic analysis. To set up the model, we consider two related sequences:

Basic sequence: $\{X_n : 0 \leq n\}$
Incremental sequence: $\{Y_n : 0 \leq n\}$

These are related as follows:

$$X_n = \begin{cases} \sum_{k=0}^{n} Y_k & \text{for } n \geq 0 \\ 0 & \text{for } n < 0 \end{cases}$$

$$Y_n = X_n - X_{n-1} \qquad \text{for all } n.$$

In addition, we consider a nonnegative, integer-valued random variable N, called a counting random variable. Sometimes it is necessary to suppose that N takes on the "value" ∞. For example, if a game has no upper limit on the number of plays, the idealized situation of never stopping is represented by $N = \infty$. In this case, it is necessary to decide on the assignment X_∞. When this is done, we form a new random quantity, $D = X_N$, which has the property

$$D(\omega) = X_n(\omega) = \sum_{k=0}^{n} Y_k(\omega) \qquad \text{whenever } N(\omega) = n.$$

With the aid of the indicator function $I_{\{n\}}(N) = I_{\{N=n\}}$, which has the value one iff $N(\omega) = n$, we may write

$$D = X_N = \sum_{n=0}^{\infty} I_{\{n\}}(N)X_n + I_{\{\infty\}}(N)X_\infty = \sum_{n=0}^{\infty} I_{\{N=n\}}X_n + I_{\{N=\infty\}}X_\infty.$$

When $N < \infty$ a.s., we may also express D in terms of the Y_k as follows:

$$D = X_N = \sum_{k=0}^{N} Y_k \quad \text{a.s.}$$

The counting random variable N "selects" the first $N+1$ members of the Y-process or the Nth member of the X-process. We relate this formal pattern to the introductory examples.

INTERPRETATION AND DISCUSSION

1. In Example 20.1.1, $Y_0 = 0$ and Y_i is the service time for the ith unit received. $D = X_N$ is the service time for N units. In this case, the value of N is determined in a manner that does not depend upon the performance of the service operation. Hence, it is reasonable to assume that N is stochastically independent of the X-process (or, equivalently, of the Y-process).

2. In Example 20.1.2, either $Y_0 = 0$ or it represents "back orders" carried over from previous periods. The variable Y_n represents the demand of the nth customer on the day in question. X_n is the combined demand of the first n customers. Usually, the number N of customers is not influenced by the amounts of the orders, so that N may be considered independent of the X-process. Of course, there are situations in which this is not a good assumption. For example, there may be panic buying or "influenced" buying at a sale, etc.

3. In Example 20.1.3, the values of the disturbances would probably be independent of the number of disturbances. However, in some cases—say a thunderstorm— the intensity of the disturbances might be related to the number that occurs in a given time period.

4. In the game-playing Example 20.1.4, if the value of N is determined by an independent chance mechanism, then we assume N is stochastically independent of the X-process. However, the number of plays could be determined by results of previous plays of the game. The player might have a "system." In this case, the event $\{N = n\}$ that the game be stopped on the nth play would be determined by some decision rule and the values of $(Y_0, Y_1, \ldots, Y_n)$.

5. In the sample of random size in Example 20.1.5, sampling would stop at size n iff the observed values of $(X_1, X_2, \ldots, X_n)$ satisfy the criteria set out by the decision rule. The event $\{N = n\}$, which

means take the sample size to be n, depends upon the decision rule and the "past behavior" of the sampling process.

We have indicated two important cases:

(a) N is independent of the X-process and the Y-process. We refer to this as independent random selection or simply independent selection.

(b) N is determined by a decision rule and the history of the process, so that $\{N = n\}$ is an event determined by $(Y_0, Y_1, \ldots, Y_n)$ or $(X_0, X_1, \ldots, X_n)$. In this case, N is called a Markov time for the process, and we speak of selection at a Markov time.

—   □

As an historically important example that does not fall strictly under either of these cases, we consider Wald's equation. This result is highly intuitive, yet its validation requires a careful examination of the structure of the random variable X_N.

Theorem 20.1.1   (Wald's Equation) *For the incremental sequence, suppose $Y_0 = 0$ and $E[Y_k] = \mu$, for $k \geq 1$. Let N be a counting random variable that satisfies the conditions*

(1) $E[N] < \infty$.

(2) $\{I_{\{N<k\}}, Y_k\}$ is independent for each $k \geq 1$.

Then,
$$E[X_N] = E[N]E[Y] = E[N]\mu.$$

Proof

The condition $E[N] < \infty$ ensures $P(N = \infty) = 0$. Hence, by countable additivity for expectations,

$$
\begin{aligned}
E[X_N] &= \sum_{n=0}^{\infty} E[I_{\{N=n\}} X_n] = \sum_{n=0}^{\infty} \sum_{k=0}^{n} E[I_{\{N=n\}} Y_k] \\
&= \sum_{k=0}^{\infty} \sum_{n=k}^{\infty} E[I_{\{N=n\}} Y_k] = \sum_{k=0}^{\infty} E[I_{\{N\geq k\}} Y_k] \\
&= \sum_{k=1}^{\infty} (E[Y_k] - E[I_{\{N<k\}} Y_k]), \qquad \text{since} \quad Y_0 = 0.
\end{aligned}
$$

By the independence condition (2), $E[I_{\{N<k\}} Y_k] = P(N < k)E[Y_k]$. Thus,

$$E[X_N] = \sum_{k=1}^{\infty} \mu(1 - P(N < k)) = \mu \sum_{k=1}^{\infty} P(N \geq k) = \mu E[N].$$

The last equality is a result of property (E20e), a special form of expectation.

—   □

Example 20.1.6:

Let Y_k be the amount deposited by the kth customer at a savings deposit window in a bank. Suppose $\{Y_k : 1 \leq k\}$ is iid, exponential, with mean value $1/\alpha = 300$ (dollars). The number of customers in a banking day may be represented by a random variable N, approximately Poisson, with mean value $\mu = 120$. What is the average amount deposited in a day?

Solution

If we assume that the number of customers does not depend upon the amounts deposited, we may assume $\{N, Y_k : 1 \leq k\}$ is independent. We have also assumed $Y_0 = 0$. In this case, Wald's equation holds, so that

$$E[D] = E[N]E[Y] = 120 \cdot 300 = 36{,}000.$$

—   □

In the next several sections, we deal with various cases of independent selection. Selection at Markov times is discussed briefly in Section 20.8.

20.2.   Independent Selection from an Iid Sequence

In this section, we develop two very useful theoretical results on independent selection from an iid sequence.

Theorem 20.2.1. *Let $\{Y_k : 0 \leq k\}$ be an incremental sequence and let $\{X_k : 0 \leq k\}$ be the corresponding basic sequence. Suppose*

(1) $Y_0 = X_0 = 0$.

(2) $\{Y_k : 1 \leq k\}$ is iid.

(3) $\{N, Y_k : 1 \leq k\}$ is independent.

The Y_k have common distribution function F_Y, characteristic function ϕ_Y, and moment-generating function M_Y. If the Y_k are nonnegative, integer-valued, then the common generating function is g_Y.
    Then the following propositions are true.

(a) $E[g(D)|N = n] = E[g(X_n)]$.

(b) $\phi_D(u) = g_N(\phi_Y(u))$ $M_D(s) = g_N(M_Y(s))$ $g_D(s) = g_N(g_Y(s))$.

Proof

(a) Since $D = X_N$, proposition (a) is a direct consequence of property (CE11b) for conditional expectation. However, since $P(N = n) > 0$ for all n in the range of N, the result can be established by quite elementary arguments, which may be instructive to observe.

Property (CE1a) for conditional expectation shows that

$$E[I_{\{n\}}(N)g(D)] = E[g(D)|N = n]P(N = n).$$

Since $D = X_n$ whenever $N = n$, we have

$$I_{\{n\}}(N)g(D) = I_{\{n\}}(N)g(X_n).$$

Because of the independence, we have by property (E18) for expectation

$$E[I_{\{n\}}(N)g(X_n)] = E[I_{\{n\}}(N)]E[g(X_n)] = P(N = n)E[g(X_n)].$$

Division by $P(N = n)$ gives the desired result.

(b) By the law of total probability (CE1b), we may assert

$$M_D(s) = E[e^{sD}] = E\big[E[e^{sD}|N]\big].$$

By part (a) and the iid character of the incremental sequence,

$$E[e^{sD}|N = n] = E[e^{sX_n}] = \prod_{i=1}^{n} E[e^{sY_i}] = \big(M_Y(s)\big)^n.$$

Thus, we have

$$M_D(s) = \sum_n \big(M_Y(s)\big)^n P(N = n) = g_N\big(M_Y(s)\big).$$

The other cases are argued similarly, and are left as exercises.

—   □

As an easy consequence of part (a), we may obtain expressions for the mean and variance of the random variable $D = X_N$.

Theorem 20.2.2. *Under the hypothesis of Theorem 20.2.1,*

$$E[D] = E[N]E[Y] \qquad E[D^2] = E[N]\,\mathrm{Var}[Y] + E[N^2]E^2[Y]$$

$$\mathrm{Var}[D] = E[N]\,\mathrm{Var}[Y] + \mathrm{Var}[N]E^2[Y].$$

Proof

By the law of total probability and part (a) of Theorem 20.2.1,

$$E[D] = E\big[E[D|N]\big] = \sum_n E[D|N = n]P(N = n) = \sum_n E[X_n]P(N = n).$$

Since $E[X_n] = nE[Y]$, we have $E[D] = E[Y]\sum_n nP(N = n) = E[Y]E[N]$
The other cases are left as exercises.

—   □

Remark: Theorem 20.2.2 includes Wald's equation for the conditions in the hypothesis.

The following example uses a very simple distribution for the incremental sequence, in order to minimize computational difficulties. It serves to illustrate how the generating function and the moment-generating function automatically handle the task of combining two distributions.

Example 20.2.1: Demand of a Random Number of Customers

Suppose the number N of customers is Poisson with mean value $\mu = 8$. The demand of each customer is a random variable taking on the values 0, 1, and 2 with probabilities 1/4, 1/2, and 1/4, respectively. Under the assumptions of independent selection from an iid sequence, determine $P(D \le 4)$, $E[D]$, and $\text{Var}[D]$.

Solution

(1) $g_D(s) = g_N\big(g_Y(s)\big)$, $g_N(s) = e^{8(s-1)}$, $g_Y(s) = 1/4 + s/4 + s^2/4$. Hence,

$$g_D(s) = \exp\left(8\left(\frac{1}{4} + \frac{1}{2}s + \frac{1}{4}s^2 - 1\right)\right) = e^{4s}e^{2s^2}e^{-6}.$$

Since $e^x = 1 + x + x^2/2! + x^3/3! + \cdots$, we have

$$g_D(s) = e^{-6}\left(1 + 4s + \frac{16s^2}{2} + \frac{64s^3}{6} + \cdots\right)\left(1 + 2s^2 + \frac{4s^4}{2} + \cdots\right)$$

$$= e^{-6}\left(1 + 4s + 10s^2 + \frac{56}{3}s^3 + \frac{86}{3}s^4 + \cdots\right).$$

Thus, $P(D \le 4) = e^{-6}(1 + 4 + 10 + 56/3 + 86/3) = e^{-6}187/3 \approx 0.1545$.

(2) Now from the properties of the Poisson distribution we have $E[N] = \text{Var}[N] = 8$. Also, simple calculations show that $E[Y] = 1$ and $\text{Var}[Y] = 0.5$. Hence

$$E[D] = 8 \cdot 1 = 8 \qquad \text{and} \qquad \text{Var}[D] = 8 \cdot 0.5 + 8 \cdot 1 = 12.$$

—   □

In subsequent sections, we make repeated use of the patterns developed in Theorems 20.2.1 and 20.2.2.

20.3.   A Poisson Decomposition Result—Multinomial Trials

In many applications, we consider a random number of multinomial trials. On each trial in the sequence, the result is one of m kinds. For reasons contained in the theory of Poisson counting processes (see Chapter 21), the counting random variable N is often Poisson. Suppose T_n is the random variable whose value is the kind (or type) of the nth result. If these form an iid sequence, independent of the Poisson counting random variable, then there is a simple but useful decomposition, which shows that the number of results of each kind is Poisson.

We use the notation and terminology of the treatment of multinomial sequences introduced in Section 10.2. Suppose the result on each trial in a sequence is one of m kinds. We let E_{ki} be the event the result on the ith trial is of the kth kind, $1 \le k \le m$. Then the number N_{kn} of results of the kth kind in the first n trials is given by

$$N_{kn} = \sum_{i=1}^{n} I_{E_{ki}} \quad \text{with} \quad \sum_{k=1}^{m} N_{kn} = n, \qquad \text{for each } n \ge 1.$$

We may represent the kind of the ith result by the random variable $T_i = \sum_{k=1}^{m} k I_{E_{ki}}$

Then multinomial sequences are characterized by

$$\{T_i : 1 \le i\} \quad \text{is iid, with } P(T_i = k) = P(E_{ki}) = p_k \quad 1 \le k \le m.$$

If there is a random number N of multinomial trials, the number N_k of results of the kth kind may be expressed in either of two forms.

$$N_k = \sum_{i=1}^{N} I_{E_{ki}} = \sum_{n=1}^{\infty} I_{\{n\}}(N) N_{kn} \quad \text{with} \quad \sum_{k=1}^{m} N_k = N.$$

We suppose that the counting random variable N is independent of the results of the trials. In the important case that N is Poisson, we have the following useful decomposition theorem.

Theorem 20.3.1   (Poisson Decomposition) *Suppose random variable N counts the number of trials in a multinomial sequence, T_i is the kind or type of the ith result, and N_k is the number of results of the kth kind.*

If   (1) N is Poisson (μ).
      (2) $\{T_i : 1 \le i\}$ is iid, with $P(T_i = k) = P(E_{ki}) = p_k 1 \le k \le m$.
      (3) $\{N, T_i : 1 \le i\}$ is independent.
then (a) Each N_k is Poisson (μp_k).
      (b) The class $\{N_k : 1 \le k \le m\}$ is independent.

——   □

Remark: For any fixed n, the class $\{N_{kn} : 1 \le k \le m\}$ has the multinomial distribution and cannot be independent. However, when the number of trials is an independent Poisson random variable, the class $\{N_k : 1 \le k \le m\}$ is independent.

Before proving this theorem, we consider some extensions and make some simple applications. Further results are obtained in Section 21.2 on arrivals of m kinds and compound Poisson processes. The next result follows immediately from Theorem 20.3.1 by elementary properties of moment-generating functions.

Theorem 20.3.2. *In a multinomial sequence we assign to the kth kind of result a "value" a_k. For each k, $1 \le k \le m$, let*

$$Z_k = a_k N_k \quad \text{be the total value of the arrivals of type } k,$$

and

$$Z = \sum_{k=1}^{m} Z_k \quad \text{be the total value of all arrivals.}$$

Then

$$\{Z_k : 1 \le k \le m\} \text{ is independent, and}$$

$$M_Z(s) = \prod_{k=1}^{m} M_{Z_k}(s) = \prod_{k=1}^{m} M_{N_k}(a_k s).$$

If $a_k = 0$, then $M_{Z_k}(s) = 1$. If $a_k = 1$, then $M_{Z_k}(s) = M_{N_k}(s) = e^{\mu p_k (e^s - 1)}$.

——   □

Example 20.3.1:

Experience has shown that the number of parties that arrive at a certain restaurant on Tuesday evenings may be represented by a random variable N that is Poisson (30). The number of persons in the ith party is Y_i. Suppose $\{Y_i : 1 \le i\}$ is iid, with $P(Y_i = k) = p_k$ as follows:

$$p_k = 0.05, \ 0.3, \ 0.1, \ 0.25, \ 0.05, \ 0.1, \ 0.05, \ 0.1$$

for k from 1 to 8, respectively. We assume $\{N, Y_i : 1 \le i\}$ is independent.

(a) What is the expected number of customers?

(b) What is the probability there will be no more than five parties of eight?

(c) What is the expected number of four-person parties?

Solution

(a) For the solution of part (a), we consider the number of customers as a composite demand for a random number of N customers and apply Theorem 20.2.2 to obtain

$$E[D] = E[N]E[Y_k] = 30 \cdot 3.95 = 118.5.$$

To solve parts (b) and (c), we formulate this problem in terms of a multinomial sequence and use Theorem 20.3.1. If we number the kinds of arrivals according the number in the party, then Y_i coincides with the random variable T_i in the development of Theorem 20.3.1, so that the number N_k of parties with k persons is Poisson $(30p_k)$. We thus have the following.

(b) N_8 is Poisson $(30 \cdot 0.1 = 3)$. From a table of summed Poisson probabilities

$$P(N_8 \leq 5) = 1 - P(N_8 \geq 6) = 1 - 0.0839 \approx 0.92.$$

(c) $E[N_4] = 30 \cdot 0.25 = 7.5.$

$\square$

Example 20.3.2:

A mail order house purchases boxes for a thirty-day period. Experience has shown that the number of customer orders in a thirty-day period is a random quantity N that is Poisson (500). Customer orders are shipped in one of five kinds of boxes, depending upon the physical size of the order. The unit costs of the five types of boxes (in dollars) are 0.10, 0.25, 0.35, 0.50, and 0.80, respectively. Customer "demand" for the boxes is iid, with probabilities 0.3, 0.3, 0.2, 0.1, and 0.1 for the five types, respectively.

(a) What is the expected cost of boxes for thirty days?

(b) What is the expected number for each type of box?

Solution

(a) Let Y_i be the cost of the box for the ith customer. Then $E[Y_i] = 0.305$. According to Theorem 20.2.2, the expected cost for the composite demand is

$$E[D] = E[Y]E[N] = 0.305 \cdot 500 = 152.50 \text{ (dollars)}.$$

(b) If N_k is the number of boxes of type k needed, then according to Theorem 20.3.1, N_k is Poisson (μp_k), so that $E[N_k] = \mu p_k$. Thus, the expected numbers of types one through five are 150, 150, 100, 50, and 50. In view of the large variance of the Poisson distribution, the order should probably be somewhat larger for each type, depending on delivery time and cost for reordering.

—   □

Example 20.3.3:

The number of customers in a shop in an eleven-hour day is Poisson (110). The ith customer buys an amount Y_i, in dollars, where Y_i is approximately exponential (0.023). The customers buy independently. For purpose of analysis, the customers are put in categories, according to the amount of purchase:

1. Less than $10.
2. Between $10 and $50.
3. Between $50 and $100.
4. More than $100.

(a) What is the expected number of sales of $50 or more?

(b) What is the probability of 10 or more sales of $100 or greater?

(c) What is the probability of 25 or more sales less than $10 (including customers who buy nothing)?

Solution

The arrivals are of four kinds, with probabilities p_k that the ith arrival is of the kth kind, $1 \le k \le 4$. Thus,

$$p_1 = P(Y_i \le 10) = 1 - e^{-0.023 \cdot 10} = 0.2055$$

$$p_2 = P(10 < Y_i \le 50) = 0.4779$$

$$p_3 = P(50 < Y_i \le 100) = 0.2164 \qquad p_4 = P(Y_i > 100) = 0.1003.$$

(a) Let N_{50}^* be the number who buy \$50 or more. Then $N_{50}^* = N_3 + N_4$. Since $\{N_3, N_4\}$ is an independent pair of Poisson random variables, their sum is Poisson. Hence N_{50}^* is Poisson $(110 \cdot [0.2164 + 0.1003] = 34.84)$, so that $E[N_{50}^*] = 34.84$.

(b) $N_{100}^* = N_4$, which is Poisson $(110 \cdot 0.1003 = 11.03)$. Using $\mu = 11$, we obtain from a table $P(N_4 \geq 10) \approx 0.66$.

(c) N_1 is Poisson $(110 \cdot 0.2055 = 22.6)$. Estimating from a Poisson probability chart, we find $P(N_1 \geq 25) \approx 0.31$. As we show in connection with the homogeneous Poisson process in Chapter 21, the Poisson distribution in this case may be approximated by the normal distribution; i.e., N_1 is approximately $N(22.6, 22.6)$. Use of this approximation gives essentially the same value.

Proof of Theorem 20.3.1

(a) $N_k = \sum_{i=1}^{N} I_{E_{ki}}$. Now hypothesis (1) implies $\{I_{E_{ki}} : 1 \leq i\}$ is iid and (3) implies $\{N, I_{E_{ki}} : 1 \leq i\}$ is independent. By Theorem 20.2.1,

$$g_{N_k}(s) = g_N\big(g_k(s)\big), \qquad \text{where } g_k(s) = q_k + p_k s = 1 + p_k(s - 1).$$

Hence,

$$g_{N_k}(s) = e^{\mu[1 + p_k(s-1) - 1]} = e^{\mu p_k(s-1)}.$$

We conclude that N_k must be Poisson (μp_k).

(b) To show independence, we show that $P(N_1 = n_1, \ldots, N_m = n_m) = \prod_{k=1}^{m} P(N_k = n_k)$ for each m-tuple $(n_1, n_2, \ldots, n_m)$ of nonnegative integers. For any such m-tuple, let $n = n_1 + n_2 + \cdots + n_m$. Then the event

$$\begin{aligned} A &= \{N_1 = n_1, \ldots, N_m = n_m\} \\ &= \{N = n\} \cap \{N_{1n} = n_1, \ldots, N_{mn} = n_m\}. \end{aligned}$$

Since N is independent of the $I_{F_{ki}}$ it must be independent of the N_{kn}. Hence,

$$P(A) = P(N = n)P(N_{1n} = n_1, \ldots, N_{mn} = n_m).$$

Now $P(N = n) = e^{-\mu}\mu^n / n!$ and by the multinomial distribution

$$P(N_{1n} = n_1, \ldots, N_{mn} = n_m) = n! \prod_{k=1}^{m} \frac{p_k^{n_k}}{n_k!}.$$

Since $p_1 + p_2 + \cdots + p_m = 1$ and $n_1 + n_1 + \cdots + n_m = n$, we have

$$e^{-\mu}\mu^n = \prod_{k=1}^{m} e^{-\mu p_k}\mu^{n_k}.$$

Thus

$$P(A) = \prod_{k=1}^{m} e^{-\mu p_k} \frac{(\mu p_k)^{n_k}}{n_k!} = \prod_{k=1}^{m} P(N_k = n_k),$$

which is the desired product rule.

——   □

20.4. Extreme Values

In this section, we consider the problem of determining the distributions for the maximum value and the minimum value of a random number of iid random variables.

Theorem 20.4.1. *Suppose $\{Y_n : 1 \leq n\}$ is an iid sequence of nonnegative random variables with common distribution function F_Y and N is a counting random variable with $\{N, Y_n : 1 \leq n\}$ independent. Set*

$$V_n = \min\{Y_1, Y_2, \ldots, Y_n\} \quad and \quad W_n = \max\{Y_1, Y_2, \ldots, Y_n\}.$$

As a notational convenience, we set $V_0 = W_0 = 0$, and define

$$V_N = \sum_n I_{\{n\}}(N)V_n = \min\{Y_1, Y_2, \ldots, Y_N\}$$

$$W_N = \sum_n I_{\{n\}}(N)W_n = \max\{Y_1, Y_2, \ldots, Y_N\}.$$

If F_V and F_W are the distribution functions for V_N and W_N, respectively, then

(a) $F_V(t) = P(V_N \leq t) = 1 + P(N = 0) - g_N\big(1 - F_Y(t)\big)$ for $t \geq 0$.

(b) $F_W(t) = P(W_N \leq t) = g_N\big(F_Y(t)\big)$ for $t \geq 0$.

Proof

(a) $F_V(t) = E[I_{[0,t]}(V_N)] = E\big[E[I_{[0,t]}(V_N)|N]\big]$. Because of the assumed independence, we have

$$E[I_{[0,t]}(V_N)|N = n] = E[I_{[0,t]}(V_n)] = P(V_n \leq t) = 1 - P(V_n > t).$$

Since for $t \geq 0$, $P(V_0 > t) = 0$ and $P(V_n > t) = P^n(Y > t)$ for $n \geq 1$, we have

$$F_V(t) = \sum_{n=0}^{\infty} \big(1 - P(V_n > t)\big) P(N = n)$$

$$= 1 - \sum_{n=0}^{\infty} P(V_n > t)P(N = n)$$

$$= 1 - \sum_{n=1}^{\infty} P^n(Y > t)P(N = n).$$

We note that $\sum_{n=1}^{\infty} P^n(Y > t)P(N = n) = g_N\big(P(Y > t)\big) - P(N = 0)$, so that

$$F_V(t) = 1 - g_N\big(P(Y > t)\big) + P(N = 0) \qquad \text{for } t \geq 0.$$

Since $P(Y > t) = 1 - F_Y(t)$, the desired result is established.

(b) is an exercise at the end of the chapter.

—   □

Example 20.4.1:

Suppose the number of failures N in a system in a 30 day period is Poisson (3). Let Y_i be the service time in hours for the ith unit that fails. Assume $\{Y_i : 1 \leq i\}$ is iid, with common distribution exponential $(1/2)$. Thus, the mean service time for each unit is two hours. Determine the probability that the longest service time in a 30-day period is no greater than two hours, four hours, eight hours.

Solution

We have $g_N(s) = e^{3(s-1)}$ and $F_Y(t) = 1 - e^{-t/2}$ for $t \geq 0$

Now

$$P(W_N \leq t) = g_N\big(F_Y(t)\big) = e^{3[F_Y(t)-1]} = e^{-3e^{-t/2}}.$$

Hence, $P(W_N \leq 2) = e^{-3e^{-2/2}} \approx 0.3317$. Similarly $P(W_N \leq 4) \approx 0.6663$ and $P(W_N \leq 8) \approx 0.9465$.

—   □

Remark: In the important special case that N is Poisson, as in the previous example, we may obtain the distribution functions F_V and F_W by use of the decomposition theorem 20.3.1. Since the argument is instructive and the result is useful, we develop it in the following theorem.

Theorem 20.4.2. *If the counting function N in Theorem 20.4.1 is Poisson (λ), then*

$$F_V(t) \;=\; P(V_N \leq t) = 1 + e^{-\lambda} - e^{-\lambda F_Y(t)}$$

$$F_W(t) \;=\; P(W_N \leq t) = e^{-\lambda\big(1 - F_Y(t)\big)}.$$

Proof

For a given $t \geq 0$, we consider the arrivals to be of two kinds. The ith arrival is of the first kind iff $Y_i \leq t$; otherwise, it is of the second kind. Let

$$p_1 = P(Y \leq t) = F_Y(t) \qquad \text{and} \qquad p_2 = P(Y > t) = 1 - F_Y(t).$$

Let N_1 be the number of arrivals that are less than or equal to t, and let N_2 be the number of arrivals that are greater than t. Then, by Theorem 20.3.1, N_1 is Poisson (λp_1) and N_2 is Poisson (λp_2). We thus have

$$P(V \leq t) = P(N = 0) + P(N_1 \geq 1) = e^{-\lambda} + 1 - e^{-\lambda p_1} = 1 + e^{-\lambda} - e^{-\lambda F_Y(t)}$$

and

$$P(W \leq t) = P(N_2 = 0) = e^{-\lambda p_2} = e^{-\lambda\left(1 - F_Y(t)\right)}.$$

—   □

Example 20.4.2: Maximum Power Surges

A personal computer is connected to a power line that is subject to electrical noise pulses. Pulses arrive according to a Poisson counting process such that in an eight-hour day the number N of pulses is Poisson (32). Pulse magnitudes (in volts) are iid, exponential with parameter $\alpha = 0.01$. What are the probabilities that in an eight-hour day the maximum pulse will not exceed 300 volts? 500 volts? 750 volts?

Solution

Since the pulses have exponential (0.01) distribution and N is Poisson (32), by Theorem 20.4.2 we have

$$F_W(t) = P(W_N \leq t) = e^{-32e^{-0.01t}}.$$

For $t = 300$, 500, and 750, we obtain the probabilities 0.2033, 0.8060, and 0.9825, respectively.

—   □

Example 20.4.3: Testing an n-transistor Chip

The "microchip" used in many computers has a large number n of transistors on a silicon chip. If a transistor is defective, it has a time to failure that is a random variable with distribution function F_Y. We suppose the defective transistors fail independently of both each other and the number N of defective units, and that each defective transistor has the same life distribution. Nondefective transistors have very long lifetimes, which may be considered infinite for purposes of testing. The system fails iff any one or more of the transistors fails. The probability $R(t)$ that the time to failure of the chip is greater than t is the probability that either no transistors are defective or the minimum failure time of a defective transistor is greater than t.

If the chip operates successfully for time t, we want to determine the probability (conditional) that the number N of defective transistors is zero.

Solution

We assume that each transistor has probability p of being defective, independent of the condition of any other transistor. In this case, the number N of defective transistors is binomial (n, p). However, as shown in Section 11.4, if n is large and p is small enough that np is not too large, we may suppose that N is approximately Poisson (np). In this case, $g_N(s) = e^{np(s-1)}$. If X is the time to failure of the chip, then

$$P(X > t) = P(N = 0) + P(V_N > t).$$

Use of Theorem 20.4.1 shows that

$$P(X > t) = g_N(P(Y > t)) = e^{np(1 - F_Y(t) - 1)} = e^{-np F_Y(t)}.$$

Since $P(X > t | N = 0) = 1$, we have by Bayes' theorem

$$P(N = 0 | X > t) = \frac{P(N = 0)}{P(X > t)} = \frac{e^{-np}}{e^{-np F_Y(t)}} = e^{np(F_Y(t) - 1)}.$$

Suppose $np = 1$ and Y is exponential (1). That is, the average life of a defective transistor is one hour. Then,

$$P(N = 0 | X > t) = e^{-e^{-t}}.$$

For $t = 0, 1, 10$, the probabilities are $0.3769, 0.6922$, and 0.99996, respectively. Thus, a "burn in" of eight to ten hours should identify essentially all of the chips with defective transistors.

—   □

Remark: Examples show (see Problem 20-29) that calculations based on the binomial distribution give results that match these values quite closely.

Example 20.4.4: Lowest Bidder

The number of bids N on a job is a random variable taking on values 1, 2, 3, 4, and 5, with respective probabilities 0.1, 0.2, 0.4, 0.2, and 0.1. The amounts bid (in thousands of dollars) by various contractors are assumed to form an iid class, independent of N, with symmetric triangular distribution on the interval $(100, 140)$. What is the probability there will be at least one bid of not more than 110 thousand dollars.

Solution

Let V_N be the value of the smallest bid. The event there is at least one bid of not more than 110 dollars is the event $\{V_N \leq 110\}$. According to Theorem 20.4.1, if Y_i is the amount of the ith bid, then

$$P(V_N \leq t) = 1 + P(N = 0) - g_N\big(P(Y > t)\big).$$

According to the data given, $P(N = 0) = 0$ and $g_N(s) = 0.1s + 0.2s^2 + 0.4s^3 + 0.2s^4 + 0.1s^5$. Geometrically, we determine from the triangular density function that $P(Y > 110) = 7/8$. Hence,

$$P(V_N \leq 110)$$

$$= 1 - \frac{1}{10}\left(\left(\frac{7}{8}\right) + 2\left(\frac{7}{8}\right)^2 + 4\left(\frac{7}{8}\right)^3 + 2\left(\frac{7}{8}\right)^4 + \left(\frac{7}{8}\right)^5\right)$$

$$\approx 0.3229.$$

The probability that any one contractor will bid 110 or less is 0.125. The effect of opening up for multiple bids is to increase this probability to 0.32.

—   ☐

20.5. Bernoulli Trials with Random Execution Times

In Chapter 6, Bernoulli trials are presented as a special case of repeated trials. In many applications, the trials are performed sequentially in time, so that the number of the trial in the sequence indicates the passage of time. In fact, if each trial takes the same amount of time, then the number of the trial is proportional to the amount of time that has elapsed during the sequence of trials. In this section, we suppose the trials are performed sequentially in time, but the execution times for the trials are iid random quantities. Specifically, we consider the problem of describing the distribution of the time to completion of the first success.

Theorem 20.5.1. *Let N be the number of the first success and Y_i be the time it takes to execute the ith trial, so that*

$$T = Y_1 + Y_2 + \cdots + Y_N \quad \text{is the time to the completion of the first success.}$$

We suppose that

(1) The sequence of trials is Bernoulli, with probability p of success on each trial.

(2) The class $\{Y_i : 1 \leq i\}$ is iid. Then,

   (a) $M_T(s) = pM_Y(s)/(1 - qM_Y(s))$.
   (b) If Y_i is exponential (λ), then T is exponential $(p\lambda)$.
   (c) If $Y_i - 1$ is geometric (p_0), then $T - 1$ is geometric (pp_0).

Proof

(a) Recall that $P(N = k) = pq^{k-1}$ for $k \geq 1$, $q = 1 - p$, so that $N - 1$ is geometric (p).

This implies

$$E[N] = \frac{1}{pg_N(s)} = \frac{ps}{1 - qs}.$$

Hence,

$$M_T(s) = g_N(M_Y(s)) = \frac{pM_Y(s)}{1 - qM_Y(s)}.$$

(b) For Y_i exponential (λ), we have

$$E[Y] = 1/\lambda \quad \text{and} \quad M_Y(s) = \frac{\lambda}{\lambda - s}.$$

Substitution and algebraic manipulations yield the result that $M_T(s) = p\lambda/(p\lambda - s)$.

The form of $M_T(s)$ shows that T must be exponential $(p\lambda)$.

(c) The condition $Y_i - 1$ is geometric (p_0) implies

$$E[Y] = \frac{1}{p_0 g_Y(s)} = \frac{p_0 s}{1 - q_0 s} = \frac{p_0 s}{1 - (1 - p_0)s}.$$

Use of $g_T(s) = g_N(g_Y(s))$ yields, after some algebraic manipulations,

$$g_T(s) = \frac{pp_0 s}{1 - (1 - pp_0)s}.$$

The form of $g_T(s)$ shows that the random variable $T - 1$ is geometric (pp_0).

—   □

Example 20.5.1: Service Shop Performance

A service shop has service time Y_i for the ith unit it services. Its records show that 0.95 of its service jobs are satisfactory, and that failures are independent of other jobs and their service times. Suppose the service times Y_i, in hours, form an iid sequence, exponential (2); that is, the service times are exponential with a mean time of 1/2 hour. Let $T = X_N$ be the service time between completion of unsatisfactory jobs; that is, T is the sum of individual service times for a string of satisfactory service jobs plus the service time of the unsatisfactory job that "breaks the string." Determine $E[T]$ and $P(T > 8)$.

Solution

A "success" is an unsatisfactory job. We have $p = 0.05$. By the general result, T is exponential $(0.05 \cdot 2 = 0.1)$. Thus, $E[T] = 1/0.1 = 10$ and $P(T > 8) = e^{-0.8} \approx 0.449$.

—   □

Example 20.5.2: Job Interviews

Applicants for a job are interviewed sequentially. Interview times in hours are iid, exponential (3). Thus, the average interview time is twenty minutes (one-third hour), and less than five percent take more than one hour. Applicants are selected at random from a pool in which twenty percent are qualified.

(a) What is the probability that a satisfactory candidate will be found in four hours or less?

(b) What is the probability the longest interview will not take more than one hour?

Solution

(a) We have the pattern of Theorem 20.5.1 (b), with $\lambda = 3$ and $p = 0.2$. Accordingly, the total time T is exponential $(3 \cdot 0.2 = 0.6)$. Thus

$$P(T \leq 4) = 1 - e^{-2.4} \approx 0.91.$$

(b) If N is the number of interviews, then $N - 1$ is geometric (0.2) and, under the assumptions, $\{N, Y_i : 1 \leq i\}$ is independent. Let W be the length of the longest interview. We use Theorem 20.4.1 to assert

$$F_W(t) = g_N(F_Y(t)) = \frac{0.2(1 - e^{-3t})}{1 - 0.8(1 - e^{-3t})} = \frac{1 - e^{-3t}}{1 + 4e^{-3t}}.$$

For $t = 1$ we have

$$F_W(1) = \frac{1 - e^{-3}}{1 + 4e^{-3}} \approx 0.792.$$

—   □

20.6. Arrival Times and Counting Processes

We are interested in the occurrences of a given phenomenon that take place at discrete instants of time, separated by random waiting, or interarrival, times. Typical examples are the arrivals of customers or the occurrence of failures in a system. We speak of these occurrences as arrivals and designate the moments of occurrence as arrival times. Such a stream of arrivals may be described in three equivalent ways, which we formulate later.

1. *Arrival times*: $\{S_n : 0 \leq n\}$ with $0 = S_0 < S_1 < \cdots$ a.s. (basic sequence).

2. *Interarrival times*: $\{W_i : 1 \leq i\}$, $W_i > 0$ a.s. for all $i \geq 1$ (incremental sequence).

The strict inequalities incorporate the restriction that, with probability one, there are no simultaneous arrivals. The relationships between the arrival times and the interarrival times are simply:

$$S_0 = 0 \quad S_n = W_1 + W_2 + \cdots + W_n \qquad \text{for all } n \geq 1$$

$$W_n = S_n - S_{n-1} \qquad \text{for all } n \geq 1.$$

The mathematical system is essentially that used for the demand problem in Section 20.2, but with modifications of notation and terminology to correspond to that customarily used in the treatment of arrival and counting processes. We now consider the stream of arrivals in a third way.

1. *Counting process*: $N_t = N(t) =$ the number of arrivals in time interval $(0, t]$.

   Then $N_{t+h} - N_t$ counts the number of arrivals in the time interval $(t, t+h]$. For each $t \geq 0$, $N_t = N(t)$ is a random variable. For a given t, ω, the value is designated $N(t, \omega)$. We thus have an infinite family of counting random variables, one for each nonnegative t. Such an infinite family of random variables is called a random process. In this case, the random process is a counting process. We thus have three equivalent descriptions

$$\{S_n : 0 \leq n\}, \qquad \{W_n : 1 \leq n\}, \qquad \{N_t : 0 \leq t\}.$$

Now $N_0 = 0$ and for any $t > 0$ we have:

$$N_t = \sum_{i=1}^{\infty} I_{(0,t]}(S_i) = \max\{n : S_n \leq t\} = \min\{n : S_{n+1} > t\}.$$

For any given ω, $N(\cdot, \omega)$ is a nondecreasing, right-continuous, integer-valued function on the interval $[0, \infty)$, with $N(0, \omega) = 0$. The following set equivalences are important in formulating various relationships and results.

$$\{N_t \geq n\} = \{S_n \leq t\}\{N_t = n\} = \{S_n \leq t < S_{n+1}\}.$$

Since $\{S_{n+1} \le t\} \subset \{S_n \le t\}$, we have

$$P(N_t = n) = P(S_n \le t) - P(S_{n+1} \le t) = P(S_{n+1} > t) - P(S_n > t).$$

A wide variety of conditions may be assumed on the W-sequence. We assume in this and the following section that

$$\{W_i : 1 \le i\} \text{ is iid, with } W_i > 0 \quad \text{a.s.}$$

Remarks

1. A counting process with iid interarrival times is called a renewal process. The interarrival times are often referred to as renewal times.

2. In the literature on renewal processes, it is common for the random variable N_t to count an arrival or "renewal" at $t = 0$. When this occurs, some slight adjustments must be made in the expressions relating N_t and S_n.

EXPONENTIAL IID INTERARRIVAL TIMES

Several examples considered in previous sections suggest the importance of the case in which the W_i are exponential (λ). Further examples and some theoretical results show why this condition is so commonly assumed in practical situations.

Theorem 20.6.1.

a. $\{W_i : 1 \le i\}$ *is iid exponential* (λ) *implies (b)* S_n *is gamma* (n, λ), *for all* $n \ge 1$.

b. S_n *is gamma* (n, λ), *for all* $n \ge 1$, *and* $S_0 = 0$ *iff (c)* N_t *is Poisson* (λt) *for all* $t > 0$.

Proof

(a) implies (b)

This is established in Section 18.2 as an example of the use of moment-generating functions.

(b) iff (c)

Since $P(S_n \le t) = P(N_t \ge n)$, the result follows immediately from Theorem 8.4.1, on the relationship between the gamma and Poisson distributions.

— □

Remark: Theorem 20.6.1 shows that the counting process for iid exponential interarrival times must be Poisson, in the sense that N_t is Poisson (λt) for each positive t. In Chapter 21, we examine the character of this process in more detail. At this point, we consider some simple examples that illustrate further the usefulness of this process in practical situations.

Example 20.6.1: Job Arrivals During a Down Time

Jobs arrive at a computer center with interarrival times, in hours, iid exponential (5). Thus, the average interarrival time is $1/5$ of an hour (twelve minutes). If the machine is down for a period of two hours, what is the probability that the number of jobs arriving during the period is no greater than twelve?

Solution

N_2 is Poisson $(5 \cdot 2 = 10)$. From a table of summed Poisson distribution,

$$P(N_2 \leq 12) = 1 - P(N_2 \geq 13) = 1 - 0.2084 \approx 0.79.$$

—   □

We develop next a simple but useful computational result for processes with arrival times S_n that are gamma (n, λ).

Theorem 20.6.2. *Consider* $\{S_n : 1 \leq n\}$, *with* S_n *gamma* (n, λ) *for all* $n \geq 1$.
   Suppose $\int_0^\infty |g| < \infty$ *and* $E\left[\sum_{n=1}^\infty |g(S_n)|\right] < \infty$. *Then* $E\left[\sum_{n=1}^\infty g(S_n)\right] = \lambda \int_0^\infty g$.

Proof

We use the countable sums property (E8b) for expectation and the corresponding property for integrals to assert

$$E\left[\sum_{n=1}^\infty g(S_n)\right] = \sum_{n=1}^\infty E[g(S_n)] = \sum_{n=1}^\infty \int g(t) f_n(t) \, dt$$

where

$$f_n(t) = \frac{\lambda e^{-\lambda t} (\lambda t)^{n-1}}{(n-1)!} \qquad \text{for } t \geq 0.$$

It is easy to show that the hypotheses for property (E8b) holds, so that

$$\sum_{n=1}^\infty \int g f_n = \int g \sum_{n=1}^\infty f_n.$$

Since

$$\sum_{n=1}^{\infty} f_n(t) = \lambda e^{-\lambda t} \sum_{n=1}^{\infty} \frac{(\lambda t)^{n-1}}{(n-1)!} = \lambda e^{-\lambda t} e^{\lambda t} = \lambda$$

the desired conclusion follows.

— □

Example 20.6.2: Discounted Replacement Cost

Suppose the cost of replacement of a failed unit in a system is c dollars. If money is discounted at a rate α, then a dollar spent t units of time in the future has current value $e^{-\alpha t}$. Suppose the life of a unit is exponential (λ), that the units fail independently, and units are replaced immediately upon failure. If S_n is the time of failure of the nth unit, then the present cost C of all future replacements is

$$C = \sum_{n=1}^{\infty} ce^{-\alpha S_n}.$$

The expected replacement cost is

$$E[C] = E\left[\sum_{n=1}^{\infty} g(S_n)\right] \qquad \text{where} \quad g(t) = ce^{-\alpha t}.$$

Since S_n is gamma (n, λ), we have by Theorem 20.6.2

$$E[C] = \lambda \int_0^{\infty} ce^{-\alpha t}\,dt = \frac{\lambda c}{\alpha}.$$

Suppose unit replacement cost $c = 100$ dollars, average time in years to failure $1/\lambda = 1/3$, and discount rate per year $\alpha = 0.1$ (10 percent). Then

$$E[C] = 100 \cdot \frac{3}{0.1} = 3000.$$

— □

Remark: Instead of a fixed cost c of unit replacement, we could consider the cost C_n of replacing the nth unit to be a random variable. If each C_n has common expected value c, and for each n the pair $\{C_n, S_n\}$ is independent, the result is the same.

We may wish to consider replacement for only a finite period. In this case, we use the following modification of Theorem 20.6.2:

Theorem 20.6.3. *Consider* $\{S_n : 1 \leq n\}$, *with* S_n *gamma* (n, λ) *for all* $n \geq 1$. *Suppose* $\int_0^{\infty} |g| < \infty$ *and* $E\left[\sum_{n=1}^{\infty} |g(S_n)|\right] < \infty$. *Let* N_t *be the counting random variable for the arrivals in the interval* $(0, t]$, *and let* $Z_t = \sum_{n=1}^{N_t} g(S_n)$. *Then*

$$E[Z_t] = \lambda \int_0^t g(u)\,du.$$

Proof

Since $n \leq N_t$ iff $S_n \leq t$, we have $\sum_{n=1}^{N_t} g(S_n) = \sum_{n=1}^{\infty} I_{[0,t]}(S_n) g(S_n)$

In Theorem 20.6.2, replace g by $I_{[0,t]} g$, and note that $\int_0^{\infty} I_{[0,t]}(u) g(u) \, du = \int_0^t g(u) \, du$.

$\square$

Remark: In Section 21.4, we obtain more general results of this type. As a special case of Theorem 21.4.2, we obtain the moment-generating function for Z_t of Theorem 20.6.3.

Example 20.6.3: Replacement Costs, Finite Horizon

Under the conditions in Example 20.6.2, consider the replacement costs over a two-year period.

Solution

$$E[C] = \lambda c \int_0^t e^{-\alpha u} \, du = \frac{\lambda c}{\alpha} [1 - e^{-\alpha t}].$$

The expected cost figure in Example 20.6.2 is reduced by the factor $[1 - e^{-0.2}] = 0.1813$ to give $E[C] \approx 543.81$.

$\square$

Example 20.6.4:

A shipment of spare parts is delayed twenty days past the time the last part in stock is used. Intervals between breakdowns of machines requiring this part in the shop are exponential with mean value $1/\lambda = 2$ days. Breakdowns occur independently. It costs \$100 per day of idle time for a machine. What is the expected downtime cost until the spare parts arrive?

Solution

The arrival times S_n for breakdowns are gamma $(n, 1/2)$. Time $t = 20$. Unit cost $c = 100$. The idle time for a breakdown at S_k is $20 - S_k$. Hence $g(u) = 100(20 - u)$. According to Theorem 20.6.3,

$$E[C] = 100 \cdot \frac{1}{2} \int_0^{20} (20 - u) \, du = 50 \cdot 200 = 10{,}000.$$

$\square$

Example 20.6.5:

Items arrive at a processing plant with arrival times S_n gamma (n, λ). The units are processed immediately. All must be shipped not later than a fixed time $T > 0$. The processed units are to be shipped in two batches: the first at time $A < T$ and the second at time T. Determine A to minimize the total storage time of the units processed.

Solution

Total waiting time is $Z_T = \sum_{n-1}^{N_T} g(S_n)$ where $g(u) = I_{[0,A]}(u)(A - u) + I_{(A,T]}(u)(T - u)$.

Thus

$$E[Z_T] = \lambda \int_0^A (A - u)\, du + \lambda \int_A^T (T - u)\, du = \lambda \frac{A^2}{2} + \lambda \frac{(T - A)^2}{2}.$$

The minimum occurs for $A = T/2$.

—   □

In the important special case that $g(u) = ce^{-\alpha u}$, the expression for $E[\sum_{n=1}^{\infty} g(S_n)]$ may be put into an alternate useful form that does not require the interarrival times to be exponential. We can generalize slightly by replacing the constant c with independent random coefficients having the same mean value.

Theorem 20.6.4. *Suppose $S_0 = 0$ and $S_n = \sum_{i=1}^{n} W_n$ for all $n \geq 1$, where $\{W_i : 1 \leq i\}$ is iid.*

Also, suppose $\{V_n : 1 \leq n\}$ is such that for each positive n, $\{V_n, S_n\}$ is independent and $E[V_n]$ has the common value v.

Let $M_W(s)$ be the common moment-generating function for the W_i. Then, for $\alpha > 0$,

$$E[C] = E\left[\sum_{n=1}^{\infty} V_n e^{-\alpha S_n}\right] = v\frac{M_W(-\alpha)}{1 - M_W(-\alpha)}.$$

Proof

First, we note that

$$E[V_n e^{-\alpha S_n}] = vM_{S_n}(-\alpha).$$

By fundamental properties of the moment-generating function,

$$M_{S_n}(s) = M_W^n(s).$$

Hence

$$E[C] = v\sum_{n=1}^{\infty} M_W^n(-\alpha).$$

The last sum is a geometric series, which has the value

$$\sum_{n=1}^{\infty} u^n = \frac{u}{1-u} \qquad \text{for } |u| < 1.$$

Now

$$M_W(-\alpha) = \int_0^{\infty} e^{-\alpha t} f_W(t)\, dt < \int_0^{\infty} f_W(t)\, dt = 1 \qquad \text{for all } \alpha > 0.$$

Putting $u = M_W(-\alpha)$ yields the desired result.

— $\square$

Example 20.6.6:

Suppose the interarrival times are uniformly distributed on $[a, b]$. Then

$$M_W(-\alpha) = \frac{1}{b-a} \int_a^b e^{-\alpha t}\, dt = \frac{e^{-a\alpha} - e^{-b\alpha}}{\alpha(b-a)}$$

so that

$$E[C] = v \frac{e^{-a\alpha} - e^{-b\alpha}}{\alpha(b-a) - (e^{-a\alpha} - e^{-b\alpha})}.$$

Suppose $a = 1$, $b = 3$, and $\alpha = 0.05$. Substitution and calculations show that $E[C] \approx 9.55v$. A numerical check also shows $M_W(-0.05) \approx 0.9052 < 1$.

— $\square$

20.7.   Arrivals and Demand in an Independent Random Time Period

For each specific $t \geq 0$, the random variable N_t in the counting process N determines the number of arrivals in the time interval $(0, t]$. This can be the counting random variable for "customers" that arrive according to the counting process.

ARRIVALS IN A RANDOM TIME PERIOD

We now consider the case that the length of the time interval is a random variable T that is independent of the arrival process.

Theorem 20.7.1   (Arrivals in an Independent Random Time)
Suppose

   1. The interarrival times W_i form an iid class.

2. T is a nonnegative random variable with continuous distribution function F_T.

3. The class $\{T, W_i : 1 \leq i\}$ is independent.

Let N_T be the number of arrivals in the time period $(0, T]$. Then

$$P(N_T = n) = E[F_T(S_{n+1})] - E[F_T(S_n)].$$

Proof

Use of the law of total probability (CE1b) and property (CE11) for conditional expectation shows that

$$P(N_T = n) = E[I_{\{n\}}(N_T)] = E\big[E[I_{\{n\}}(N_T)|T]\big] = \int E[I_{\{n\}}(N_t)]F_T(dt).$$

Now, as noted in Section 20.6,

$$E[I_{\{n\}}(N_t)] = P(N_t = n) = P(S_{n+1} > t) - P(S_n > t).$$

By the special form of expectation (E20a),

$$\int_0^\infty \big(P(S_{n+1} > t) - P(S_n > t)\big)F_T(dt) = E[F_T(S_{n+1})] - E[F_T(S_n)].$$

$\square$

In the important special case that T is exponential, we have the following result

Theorem 20.7.2   (Arrivals in an Exponential Time Period)
Suppose

1. *The interarrival times W_i form an iid class.*

2. *T is exponential (ν).*

3. *The class $\{T, W_i : 1 \leq i\}$ is independent.*

Let $N_T = N(T)$ be the number of arrivals in the time period $(0, T]$. Then

$$P(N_T = n) = \big(1 - M_W(-\nu)\big)\big(M_W(-\nu)\big)^n \qquad \text{for all } n \geq 0$$

so that N_T is geometric, with $p = 1 - M_W(-\nu)$ and has generating function

$$g_{N_T}(s) = \frac{1 - M_W(-\nu)}{1 - sM_W(-\nu)}.$$

Proof

By Theorem 20.7.1,

$$
\begin{aligned}
P(N_T = n) &= E[1 - e^{-\nu S_{n+1}} - 1 + e^{-\nu S_n}] = E[e^{-\nu S_n}] - E[e^{-\nu S_{n+1}}] \\
&= M_W^n(-\nu) - M_W^{n+1}(-\nu) = \big(1 - M_W(-\nu)\big)M_W^n(-\nu).
\end{aligned}
$$

$\square$

Example 20.7.1:

Periodically, the mainframe in a computer center is taken out of service for preventive maintenance. Jobs arriving during shutdown are backlogged, unless the number of jobs reaches the value m. In the latter case, excess jobs are sent out for processing (at extra cost). Times are measured in minutes. We suppose

1. Interarrival times W_i are exponential (λ).

2. Service time T is exponential (ν).

3. Job times Y_i are independent, exponential (γ).

4. $\{T, W_i, Y_j : 1 \leq i, 1 \leq j\}$ is an independent class.

On any given maintenance period:

a. What is the probability that jobs must be processed externally?

b. What is the expected backlog of jobs to be done internally, after maintenance?

c. What is the expected machine time for units sent out for processing?

Solution

1. N_T has geometric distribution with $p = 1 - \lambda/(\lambda + \nu) = \nu/(\lambda + \nu)$, so that $E[N_T] = \lambda/\nu$.

   Jobs are sent out iff $N_T > m$. Now, $P(N_T > m) = q^{m+1} = (\lambda/(\lambda + \nu))^{m+1}$.

2. The backlog $B = I_{[0,m-1]}(N_T)N_T + I_{[m,\infty)}(N_T)m$, so that

$$E[B] = \sum_{k=0}^{m-1} kpq^k + mP(N_T > m - 1) = \sum_{k=1}^{m-1} kpq^k + mq^m.$$

The last term comes from part (a). Now the first term on the right-hand side is

$$pq \sum_{k=1}^{m-1} kq^{k-1} = pq\frac{d}{dq}\left(\frac{1 - q^m}{1 - q}\right) = \frac{q}{p}(1 - q^m) - mq^m.$$

Thus,

$$E[B] = \frac{q}{p}(1 - q^m) = \frac{\lambda}{\nu}\left(1 - \left(\frac{\lambda}{\lambda + \nu}\right)^m\right).$$

3. The counting random variable for the jobs sent out is

$$C = N_T - B = I_{[m+1,\infty)}(N_T)(N_T - m).$$

The expected machine time for units sent out is

$$E[C]E[Y] = \left(E[N_T] - E[B]\right)E[Y] = \frac{\lambda}{\gamma\nu}\left(\frac{\lambda}{\lambda+\nu}\right)^m.$$

Suppose average down time is 30 minutes, average interarrival time is 5 minutes, average job time is 12 minutes, and m is 10. This means

$$\nu = \frac{1}{30}, \qquad \lambda = \frac{1}{5}, \qquad \gamma = \frac{1}{12}, \qquad \text{and} \qquad \frac{\lambda}{\lambda+\nu} = \frac{6}{7}.$$

Thus $E[N_T] = 30/5 = 6$, $E[B] = 6\left(1-(6/7)^{10}\right) \approx 4.72$, $E[C]E[Y] = 6 \cdot 12 \cdot (6/7)^{10} \approx 15.4$

$\square$

COMPOSITE DEMAND IN RANDOM TIME PERIOD

We may use N_T to count arrivals (e.g., failures, customers, etc.), each of which has a demand (service time, order size, etc.). Let Y_i be the demand of the ith arrival. We consider only the case T is exponential (ν).

Theorem 20.7.3 (Demand in Random Time Period) *Suppose*

1. *The interarrival times W_i form an iid class, with common distribution function F_W and moment-generating function M_W.*

2. *Period length T is exponential (ν).*

3. *Individual demands Y_i form an iid class with common distribution function F_Y and moment-generating function M_Y.*

4. *$\{T, W_i, Y_j : 1 \le i, 1 \le j\}$ is an independent class.*

Let $D = \sum_{n=1}^{N_T} Y_i$ be the composite demand. Then

$$M_D(s) = g_{N_T}\left(M_Y(s)\right) = \frac{1 - M_W(-\nu)}{1 - M_Y(s)M_W(-\nu)}.$$

Proof

Substitute $M_Y(s)$ for s in the formula for $g_{N_T}(s)$ determined in Theorem 20.7.2.

$\square$

In the following special case, the distribution function F_D may be obtained explicitly.

Theorem 20.7.4 (Demand for Exponential Interarrival Times, Exponential Period) *Suppose*

1. *The interarrival times W_i form an iid class, exponential (λ).*

2. *Period length T is exponential (ν).*

3. *Individual demands Y_i form an iid class, exponential (γ).*

4. *$\{T, W_i, Y_j : 1 \le i, 1 \le j\}$ is an independent class.*

Let $D = \sum_{n=1}^{N_T} Y_i$ be the composite demand. Then

$$F_D(t) = u(t)\left(1 - \frac{\lambda}{\lambda + \nu}e^{-\beta t}\right)$$

where

$$\beta = \frac{\gamma\nu}{\lambda + \nu} \qquad and \qquad M_D(s) = \frac{\gamma\nu - s\nu}{\gamma\nu - s(\lambda + \nu)}.$$

Also

$$E[D] = \frac{\lambda}{\gamma\nu} \qquad and \qquad \mathrm{Var}[D] = \frac{\lambda^2 + 2\lambda\nu}{\gamma^2\nu^2}.$$

Proof

1. We first obtain the moment-generating function M_D. Use of

$$M_W(-\nu) = \frac{\lambda}{\lambda + \nu} \qquad and \qquad M_Y(s) = \frac{\gamma}{\gamma - s}$$

   leads by simple algebra to the result

$$M_D(s) = \frac{\gamma\nu - s\nu}{\gamma\nu - s(\lambda + \nu)}.$$

2. To recover the distribution function F_D, we may use Laplace transforms. As shown in Section 18.4, the Laplace transform of F_D is given by $\frac{1}{s}M_D(-s)$. Use of the algebraic technique of the partial fraction expansion (as used in the integration of rational functions) yields the result

$$\frac{1}{s}M_D(-s) = \frac{1}{s} - \frac{\lambda}{\lambda + \nu}\frac{1}{s + \beta} \qquad \beta = \frac{\gamma\nu}{\lambda + \nu}.$$

   From a table of Laplace transforms, we find that F_D has the desired form.

3. To obtain the moments, we may either use derivatives of M_D or use the density function for the absolutely continuous part

$$f_D(t) = \frac{\lambda}{\lambda + \nu} \beta e^{-\beta t} \qquad \text{for } t > 0.$$

Note that there is positive probability mass $F_D(0) = \nu/(\lambda + \nu)$ at $t = 0$. This does not add to the values of $E[D]$ and $E[D^2]$ obtained from the density function alone.

—   □

Example 20.7.2:

For the computer maintenance problem in Example 20.7.1, determine the distribution function for the random variable D whose value is the total machine time accumulated during shutdown, using $\nu = 1/30$, $\lambda = 1/5$, and $\gamma = 1/12$. Use this to determine the probability the backlog for the mainframe will not exceed three hours (180 minutes).

Solution

We obtain by simple arithmetic the values $\lambda/(\lambda + \nu) = 6/7$, as before, and $\beta = \gamma \nu/(\lambda + \nu) = 1/84$. Hence, by Theorem 20.7.4,

$$F_D(t) = 1 - \frac{6}{7} e^{-t/84} \qquad \text{so that} \quad F_D(180) \approx 0.90.$$

—   □

A second special case may be obtained as follows.

Theorem 20.7.5. *Suppose*

1. *The interarrival times W_i form an iid class, exponential (λ).*

2. *Period length T is exponential (ν).*

3. *Individual demands Y_i form an iid class, with $Y_i - 1$ geometric (p).*

4. *$\{T, W_i, Y_j : 1 \le i, 1 \le j\}$ is an independent class.*

Let $D = \sum_{n=1}^{N_T} Y_i$ be the composite demand. Set

$$p_0 = 1 - M_W(-\nu) = \frac{\nu}{\lambda + \nu} \qquad and \qquad q_0 = M_W(-\nu) = \frac{\lambda}{\lambda + \nu}.$$

Then,

$$P(D = 0) = p_0 \qquad and \qquad P(D \le m) = p_0 + q_0(1 - a^m) = \frac{(1 - a^m)\lambda + \nu}{\lambda + \nu}$$

for $m \geq 1$, where

$$a = q + q_0 p = 1 - pp_0 = \frac{\lambda + q\nu}{\lambda + \nu}.$$

Proof

$$g_D(s) = g_{N_T}\big(g_Y(s)\big) = \frac{p_0}{1 - q_0 ps/(1 - qs)} = p_0 \frac{1 - qs}{1 - as}$$

where $a = q + q_0 p = 1 - pp_0$. By algebraic division,

$$
\begin{aligned}
g_D(s) &= p_0\left(1 + \frac{(a-q)s}{1-as}\right) = p_0\left(1 + \frac{q_0 ps}{1 - as}\right) \\
&= p_0\big(1 + q_0 ps(1 + as + a^2 s^2 + \cdots)\big).
\end{aligned}
$$

Using coefficients of powers of s, we find $P(D = 0) = p_0$ and $P(D = k) = p_0 q_0 p a^{k-1}$ for $k \geq 1$.

Summing, we have

$$
\begin{aligned}
P(D \leq m) &= p_0 + p_0 q_0 p(1 + a + \cdots + a^{m-1}) = p_0 + p_0 q_0 p\left(\frac{1 - a^m}{1 - a}\right) \\
&= p_0 + q_0(1 - a^m).
\end{aligned}
$$

□

Example 20.7.3:

A merchant stocks a certain popular item. Customers interested in this item arrive according to a counting process with iid interarrival times (in days) each exponential ($\lambda = 3$). The demands Y_i of individual customers form an iid class, with $Y_i - 1$ geometric ($p = 1/3$). The merchant receives a shipment of 100 units. There is some uncertainty when the next shipment will arrive. The time to the arrival of the next shipment may be considered a random variable T exponential ($\nu = 1/10$). Thus, an average of three customers per day, with an average purchase of 3 units per customer. The average interarrival time between shipments is 10 days. What is the probability that the 100 items will not be sold out before the next shipment?

Solution

In order to use the results of Theorem 20.7.5, we calculate

$$q_0 = \frac{\lambda}{\lambda + \nu} = \frac{3}{3.1} \qquad p_0 = \frac{0.1}{3.1} \qquad \text{and} \qquad a = \frac{2}{3} + \frac{3}{3.1} \cdot \frac{1}{3} = \frac{92}{93}.$$

This yields the result

$$P(D \leq 100) = \frac{1}{31} + \frac{30}{31}(1 - (92/93)^{100}) \approx 0.672.$$

□

20.8. Decision Schemes and Markov Times

In the preceding sections, we suppose the counting random variable N is independent of the basic and incremental sequences. There are many practical problems in which this is a good assumption. However, in the introductory examples in Section 20.1, we encounter simple situations in which the value of N is determined by a decision rule and the history of the process. Thus, N is not independent of the basic sequence, but is conditioned by it (or, equivalently, by the incremental sequence) in a special way. In this section, we look more carefully at that case.

We consider a sequence $Z_\mathcal{N} = \{Z_n : 0 \leq n\}$ of random vectors. In various applications, this could be the basic X-sequence or the incremental Y-sequence of random variables, or it could be a quite general sequence of random vectors. In particular, the Z-sequence may be a decision system to which the basic and incremental sequences are adapted, as described at the end of this section.

We suppose the Z-sequence is evolving in time. In applications, the values of the sequence represent the successive states of a physical or behavioral system evolving in time. The vector Z_0 represents the initial value or state of the system. At discrete instants of time t_1, t_2, ..., the system makes a transition from one value to the succeeding value. Thus

At $t_1 \sim n = 1$, the value changes from Z_0 to Z_1.
At $t_2 \sim n = 2$, the value changes from Z_1 to Z_2.

If n represents the present time (i.e., t is in the interval $[t_n, t_{n+1})$), then the vector $W_n = (Z_0, Z_1, \ldots, Z_n)$ represents the past behavior, at n, of the Z-sequence, and the class $\{W_n : 0 \leq n\}$ constitutes the history of the Z-sequence. We call the Z-sequence a decision sequence.

On the basis of the "performance" of the decision sequence, a decision is made to select Z_n (or X_n in some suitably related sequence). This decision is expressed by setting the counting random variable N to the value $N = n$. The counting random variable is thus nonnegative, integer-valued. In some situations, as in the independent case considered in previous sections, we need to consider the possibility that no Z_n is selected. This is usually done by assigning the "value" $N = \infty$. The quotation marks are a reminder that ∞ is not a number; it is a symbol that can be treated formally as a number if certain precautions and limitations are observed.

If, as we suppose in this treatment, the Z_n are observed sequentially in time, then a decision must be based on past behavior of the decision sequence. The future is simply not available at the time of the decision. We ordinarily include the present as part of the past in this characterization, since we have available both the strict past and the present. To determine whether or not $N = n$, we examine the pattern of values of Z_0, Z_1, ..., Z_n and check whether or not they take on prescribed patterns of values. These prescribed patterns constitute the decision rule. Now a pattern of values

$(t_0, t_1, \ldots, t_n)$ may be considered a point in $(n+1)$-dimensional space. The set of all patterns satisfying the decision rule to select Z_n constitute a subset Q_n in that space. This means that

$$\{N = n\} = \{W_n \in Q_n\} \qquad \text{for a certain Borel set } Q_n.$$

Formally, then, the decision rule specifies a Borel set Q_n in each $(n+1)$-dimensional space, $n \geq 0$.

The following elementary examples illustrate these ideas.

1. Suppose Z_n is selected the first time the Z-process takes on a value in a fixed set M. We speak of a first visit to set M. Thus,

$$N(\omega) = n \qquad \text{iff both} \quad Z_i(\omega) \in M^c \;\; 0 \leq i \leq n - 1 \text{ and } Z_n(\omega) \in M.$$

In the notation of events,

$$
\begin{aligned}
\{N = n\} &= \bigcap_{k=0}^{n-1} \{Z_k \in M^c\}\{Z_n \in M\} \\
&= \{W_n \in M^c \times M^c \times \cdots \times M^c \times M\}.
\end{aligned}
$$

Thus,

$$Q_n = M^c \times M^c \times \cdots \times M^c \times M.$$

That is, Q_n is the set of those $(n+1)$-tuples such that the first n coordinates fail to be in M and the nth coordinate is in M.

2. Suppose each Z_n is real-valued. Select $\epsilon > 0$. The decision rule is: select Z_n iff n is the smallest integer such that $|Z_n - Z_{n-1}| \leq \epsilon$. Thus

$$N = n \quad \text{iff} \quad |Z_i - Z_{i-1}| > \epsilon, \; 1 \leq i \leq n - 1 \quad \text{and} \quad |Z_n - Z_{n-1}| \leq \epsilon.$$

The inequalities $|x_i - x_{i-1}| > \epsilon$, $1 \leq i \leq i - 1$, and $|x_n - x_{n-1}| \leq \epsilon$ determine a subset Q_n in the $(n+1)$-dimensional space $\Re^{n+1}$.

The condition $\{N = n\} = \{W \in Q_n\}$ may be expressed in the language of measurability. Thus, $\{N = n\}$ is an event determined by W_n, or, equivalently, $\{N = n\} \in \sigma(W_n)$.

We adopt the following terminology.

Definition 20.8.1.   *A nonnegative, integer-valued random variable N (which may take on the "value" ∞) is called a Markov time for a decision sequence $\{Z_n : 0 \leq n\}$ iff for each $n \geq 0$ there exists a Borel set Q_n such that $\{N = n\} = \{W_n \in Q_n\}$. If $P(N = \infty) = 0$, then N is called a stopping time for the decision sequence.*

Remarks

1. Although this terminology is not adopted universally, it is rapidly becoming the conventional choice. Some authors do not distinguish between Markov times and stopping times. Some use other terminology entirely.

2. In some situations, the event $\{N = n\}$ should be determined by the strict past W_{n-1}. This may be handled by requiring $N - 1$ to be a Markov time for the Z-sequence, since in that case

$$\{N = n\} = \{N - 1 = n - 1\} = \{W_{n-1} \in Q_n\}.$$

Some Structure and Properties of Markov Times

In the following, we consider a fixed decision sequence $\{Z_n : 0 \le n\}$ and suppose N and N_k are Markov times for this sequence. We state a number of important facts about Markov times. Although some of the proofs are rather complicated in detail, they involve elementary ideas. We illustrate by giving some of the simpler proofs.

MT1) $\{N = n\} \in \sigma(W_n)$ for each $n \ge 0$, iff $\{N \le n\} \in \sigma(W_n)$ for each $n \ge 0$.

MT2) The event $\{N = \infty\} = \{N < \infty\}^c$ is a set determined by the entire Z-sequence, and thus by the infinite dimensional random vector W_∞.

MT3) The class $\{\{N = n\} : 0 \le n \le \infty\}$ is a partition of the basic space Ω.

The following class of events plays an important role in the theory of Markov times.

Definition 20.8.2. *An event A determined by the decision sequence is said to be* prior to Markov time N *iff for each $n \ge 0$, the event $\{N = n\} \cap A$ is an event determined by W_n.*

Remark: If A is a prior event, then $\{N = \infty\} \cap A$ is an event determined by W_∞.

Definition 20.8.3. *The class $\mathcal{F}_N$ of all events prior to Markov time N is called the* pre-N sigma algebra.

MT4) $\mathcal{F}_N$ is a sigma algebra of events determined by W_∞ and every event determined by N is in $\mathcal{F}_N$.

MT5) Event A determined by W_∞ is prior to N iff for each $n \ge 0$, $\{N \le n\} \cap A$ is an event determined by W_n.

MT6) Event A is prior to N iff $A = \bigcup\limits_{n=0}^{\infty} A_n \cup A_{\infty}$, with each A_n determined by W_n for $0 \le n \le \infty$.

MT7) $\min\{N_1, N_2\}$, $\max\{N_1, N_2\}$, and $N_1 + N_2$ are Markov times.

MT8) $\min\{N, n\}$ is a Markov time and is a function of W_n.

MT9) $N_1 \le N_2$ implies $\mathcal{F}_{N_1} \subset \mathcal{F}_{N_2}$.

Proof that Every Event Determined by N is in $\mathcal{F}_N$

Since $\{N \in M\} = \bigcup\limits_{n \in M} \{N = n\}$, we must have $\{N \in M\}\{N = n\}$ is either $\emptyset$ or $\{N = n\}$. In either case, this is an event determined by W_n.

$\qquad\qquad\qquad\qquad\qquad\qquad\qquad\qquad\qquad\qquad\qquad\qquad \square$

Proof that $N = \min\{N_1, N_2\}$ is a Markov Time

Since $N \le n$ iff either $N_1 \le n$ or $N_2 \le n$, we have

$$\{N \le n\} = \{N_1 \le n\} \cup \{N_2 \le n\}.$$

By (MT1) each of the two sets on the right is in $\sigma(W_n)$; hence, their union must be. Since the latter holds for any $n \ge 0$, property (MT1) ensures that N is a Markov time.

$\qquad\qquad\qquad\qquad\qquad\qquad\qquad\qquad\qquad\qquad\qquad\qquad \square$

In Theorem 20.1.1, Wald's equation is obtained under a general set of conditions. We consider next an important case in which the counting random variable N is a Markov time.

Theorem 20.8.4 (Wald's Equation for a Markov Time) *Suppose $Y_0 = 0$ and $\{Y_n : 1 \le n\}$ is independent with common mean $E[Y_n] = \mu$. Set $X_n = \sum_{i=0}^{n} Y_i$.*

Let N be a Markov time for the Y-sequence (or, equivalently, for the X-sequence), with $E[N]$ finite.

Then

$$E[X_N] = \mu E[N], \qquad E[Y_{N+k}] = \mu, \qquad and \qquad E[X_{N+k}] = \mu(E[N] + k)$$

for all $k \ge 1$.

Proof

1. We note that the event $\{N < k\}$ is an event determined by Y_1, Y_2, $\ldots$, Y_{k-1}, which is independent of the events determined by Y_k. The first equation follows from Theorem 20.1.1.

2. For any integer $k \ge 1$, $Y_{N+k} = \sum_n I_{\{N=n\}} Y_{n+k}$.

Since $\{N = n\}$ in an event determined by $(Y_1, Y_2, \ldots, Y_n)$, we have $\{I_{\{N=n\}}, Y_{n+k}\}$ independent. Thus,

$$E[Y_{N+k}] = \sum_n E[I_{\{N=n\}}Y_{n+k}] = \mu \sum_n P(N = n) = \mu.$$

3. Since $X_{N+k} = X_N + Y_{N+1} + \cdots + Y_{N+k}$, the third result follows from linearity and the first two results.

—   □

Remark: Examination of the proof for Theorem 20.8.4 shows that if N is independent of the Y sequence, then $E[Y_N] = \mu$. However, if N is a Markov time, this claim cannot be made. For example, suppose N is the first time that $Y_n(\omega) \geq 2\mu$. Then $I_{\{N=n\}}Y_n \geq 2\mu I_{\{N=n\}}$ and if $P(N < \infty) = 1$

$$E[Y_N] = \sum_n E[I_{\{N=n\}}Y_n] \geq 2\mu \sum_n E[I_{\{N=n\}}] = 2\mu.$$

Example 20.8.1:

Suppose a store has a stock of M items of a certain kind. Let Y_k be the demand of the kth customer. Assume $\{Y_k : 1 \leq k\}$ is independent, with common mean value μ. Let N be the number of customers who must be served to sell the M units of merchandise. We must have

$$X_N - Y_N < M \leq X_N.$$

Now N is the "first visit" of the X-sequence to the interval $[M, \infty)$, so that N is a Markov time for the X-sequence. By Wald's equation, we have,

$$\mu E[N] - E[Y_N] < M \leq \mu E[N].$$

It follows, by simple algebra, that

$$\frac{M}{\mu} \leq E[N] < \frac{M + E[Y_N]}{\mu}.$$

Although we cannot say that $E[Y_N] = \mu$, it is reasonable to suppose that $E[Y_N] \approx \mu$. Since N is the first customer to push the total demand to M, we should ordinarily suppose that with high probability the demand Y_N of that customer is considerably smaller than M.

—   □

We next establish a simple result that shows that no matter how one decides where to begin observation of an iid sequence, the result is still an iid sequence. For example, suppose Y_i represents the gain on the ith play in a repetitive game. If the Y-sequence is iid, then no system of deciding when to play is going to alter the probability structure for succeeding plays. This result may be extended to the more general situation in which the Y_i form the incremental sequence of a basic sequence known as a submartingale. Martingales and submartingales play a significant role in many theoretical and practical investigations in modern probability. While the theory is somewhat more complex for submartingales, the following treatment illustrates essential features of the role of Markov times.

Theorem 20.8.5. *Suppose $\{Y_i : 1 \leq i\}$ is an iid class and Y is a random variable with the common distribution. Let N be a counting random variable. If either*

1. *$\{N, Y_i : 1 \leq i\}$ is independent or*

2. *N is a Markov time for the Y-sequence,*

then $\{X_N, Y_{N+i} : 1 \leq i\}$ is independent and each Y_{N+i} has the common distribution for Y.

Proof

1. For any Borel set M, consider

$$
\begin{aligned}
P(Y_{N+i} \in M) &= E[I_M(Y_{N+i})] = \sum_n E[I_{\{N=n\}} I_M(Y_{n+i})] \\
&= \sum_n P(N = n) P(Y \in M) = P(Y \in M).
\end{aligned}
$$

2. To show independence of the complete class, it is sufficient to show independence for $1 \leq i \leq m$ for all $m \geq 2$. We write out the case $m = 2$, which demonstrates the essential structure of the proof for the general case. For any Borel sets M, Q_1, Q_2

$$
\begin{aligned}
\mathcal{P} &= P(X_N \in M, Y_{N+1} \in Q_1, Y_{N+2} \in Q_2) \\
&= E[I_M(X_N) I_{Q_1}(Y_{N+1}) I_{Q_2}(Y_{N+2})] \\
&= \sum_n E[I_{\{N=n\}} I_M(X_n) I_{Q_1}(Y_{n+1}) I_{Q_2}(Y_{n+2})]
\end{aligned}
$$

Now either (a) $\{N = n\}$ is independent of the Y-sequence, or (b) $\{N = n\}$ is an event determined by $(Y_1, Y_2, \ldots, Y_n)$. In either case, $\{I_{\{N=n\}} I_M(X_n), Y_{n+i} : 1 \leq i\}$ is independent. Hence,

$$
\mathcal{P} = \sum_n E[I_{\{N=n\}} I_M(X_n)] E[I_{Q_1}(Y_{n+1})] E[I_{Q_2}(Y_{n+2})]
$$

$$
\begin{aligned}
&= \sum_n E[I_{\{N=n\}} I_M(X_n)] E[I_{Q_1}(Y)] E[I_{Q_2}(Y)] \\
&= E[I_M(X_N)] E[I_{Q_1}(Y)] E[I_{Q_2}(Y)] \\
&= P(X_N \in M) P(Y_{N+1} \in Q_1) P(Y_{N+2} \in Q_2).
\end{aligned}
$$

— $\square$

Example 20.8.2: Betting on the Basis of a Run of Heads

A coin is flipped repeatedly. Bets are place on heads or tails. A player has a system. He waits until there is a "run" of five consecutive heads, then bets that the next flip will result in a tail. How effective is his system?

Discussion

Let Y_i represent the ith play, with value one corresponding to a head and value zero corresponding to a tail. The number N of the play on which the fifth consecutive head turns up is a Markov time for the sequence. Thus Y_{N+1} is the random variable that designates the play on which the bet is made. Now Y_{N+1} is independent of the class of previous random variables and has the same distribution. Thus, the "system" gains the player no advantage.

— $\square$

Remark: This result is not surprising. It is usually anticipated by arguing that since the sequence is independent it is obvious that the next play must be independent of the previous ones. However, it is not known before play begins which will be the play on which the bet is placed. And if there were another trial, in all likelihood a different play would be selected. Thus we are dealing with Y_{N+1}, not Y_{n+1} for a specified n.

Frequently, we do not observe directly either the basic X-sequence or the incremental Y-sequence; instead, we observe an auxiliary Z-sequence whose "behavior" provides information about the behavior of the basic or incremental sequence. A variety of relationships may exist between the Z-sequence, taken as the decision sequence, and the basic or incremental sequence. One of the most common and most useful in terms of analytical studies is the case that each event determined by X_n is an event determined by the past W_n for the Z-sequence. In the language of measurability, for each Borel set M in the codomain of X_n there is a corresponding Borel set Q in the codomain of W_n such that

$$
X_n^{-1}(M) = W_n^{-1}(Q).
$$

Theorem 11.1.2 ensures there is a Borel function ϕ_n such that $X_n = \phi_n(W_n)$.

We adopt the following terminology.

Definition 20.8.6. *We say the sequence $\{X_n : 0 \leq n\}$ is* adapted *to the sequence $\{Z_n : 0 \leq n\}$ iff every event determined by X_n is an event determined by W_n, so that for each n there exists a Borel function ϕ_n such that $X_n = \phi_n(W_n)$. A Z-sequence to which the X-sequence (equivalently, the Y-sequence) is adapted is called a* decision sequence *for the process.*

Remark: It should be apparent that the X-sequence is adapted to the Z-sequence iff the Y-sequence is so adapted.

The results in Theorem 20.8.5 are limited in scope and value, being quite intuitive, although they illustrate important ideas and make precise ideas usually arrived at quite informally (and imprecisely). We make further, more significant, use of Markov times for sequences in Chapter 23, in connection with the strong Markov property for sequences. Also, certain problems for Poisson processes use Markov times in the more general setting of continuous-parameter random processes. In Chapter 21, we note without proof some results for Poisson processes.

PROBLEMS

20–1. Suppose the number N of automobile accidents in a city in a day is Poisson (300). The amount of time, in hours, spent by city police on each accident is exponential (2). These times are independent. What is the expected number of police hours spent on accidents each day?

20–2. Prove the following propositions about the model for independent selection from an iid sequence.

(a) $g_D(s) = g_N(g_Y(s))$.

(b) $E[D^2] = E[N]\,\mathrm{Var}[Y] + E[N^2]E^2[Y]$ and $\mathrm{Var}[D] = E[N]\,\mathrm{Var}[Y] + \mathrm{Var}[N]E^2[Y]$.

20–3. For the demand random variable $D = X_N$, suppose N is uniform on the integers 0, 1, 2 and each Y_i has the distribution 0.1, 0.2, 0.3, 0.2, 0.2 on 0, 1, 2, 3, 4. Determine the generating function g_D and from this determine the distribution for D.

20–4. Consider $D = \sum_{i=0}^{N} Y_i$, with $\{N, Y_1, Y_2, Y_3\}$ independent, $Y_0 = 0$, $P(N = k) = 0.2,\ 0.3,\ 0.4,\ 0.1$ for $k = 0, 1, 2, 3$, respectively, and $\{Y_1, Y_2, Y_3\}$ iid, with $P(Y = j) = \frac{1}{3}, \frac{1}{2}, \frac{1}{6}$, for $j = 0, 1, 2$, respectively. Determine the generating function g_D and use this to determine the distribution for D.

20–5. The number of customers in a major appliance store is equally likely to be 1, 2, or 3. Each customer buys 0, 1, or 2 items, with probabilities 0.5, 0.4, and 0.1 respectively. Customers buy independently, regardless of the number of customers.

(a) What is the expected number of items sold?

(b) What is the probability of three or more items sold?

20–6. The number N of trials in a Bernoulli sequence is a random variable, independent of the sequence, which takes on values 1, 2, or 3, with probabilities 0.3, 0.4, and 0.3, respectively. The probability of success on any trial is $p = 0.4$. Determine the distribution for the number of successes S.

20–7. The number of lightning strikes during a severe thunderstorm over a national forest area is a random quantity N having the Poisson (100) distribution. Experience shows that the strikes may be considered to form a Bernoulli sequence, independent of N, with probability $p = 0.005$ that any strike will start a fire. Let F be the number of fires started during a storm. Determine the distribution for F.

20–8. The number of customers in a store on a given day is Poisson (100). The amount purchased by the ith customer is a random quantity Y_i. The Y_i form an iid class with the common density function

$$f_Y(t) = \frac{100 - t}{5000}, \qquad 0 \leq t \leq 100.$$

Let D be the total amount purchased by the customers on a given day. Determine $E[D]$ and $\text{Var}[D]$.

20–9. A warranty service facility for a certain type of photographic equipment charges the manufacturer $10, $20, or $30, depending upon the category of service required. In an eight-hour day, the number N of items received is a random variable that takes on values 0, 1, 2, 3, and 4 with equal probability 0.2. The job categories for the various arrivals may be represented as an iid class $\{Y_i : 1 \leq i \leq 4\}$, independent of the arrival process. Suppose

$$P(Y_i = 10) = 0.5, \quad P(Y_i = 20) = 0.3, \quad \text{and} \quad P(Y_i = 30) = 0.2.$$

Let C be the total service charge for a given day.

(a) Determine $E[C]$.

(b) Calculate the first four terms of the generating function $g_C(s)$. From this, determine

$$P(C \geq 40) = 1 - P(C \leq 30).$$

20–10. Suppose X is conditionally Poisson (λt), given $Y = t$.

(a) Show that $g_X(s) = M_Y(\lambda(s - 1))$.

(b) If Y is gamma (α, β), show that X is negative binomial $(\alpha, \beta/(\beta + \lambda))$.

20–11. The number of messages N input to a communications channel per unit time is Poisson (λ). The number of characters in the ith message is Y_i. Assume the Y_i form an iid class, independent of the arrival process, with $Y_i - 1$ geometric (p). Let D be the total number of characters handled by the channel per unit time. Let M_k be the number of messages with k or more characters

(a) Determine $g_D(s)$, $E[D]$, and $\text{Var}[D]$.

(b) If $E[Y] = \dfrac{1}{p} = 100$ and $E[N] = \lambda = 100$, determine $P(M_{200} \geq 5)$.

20–12. The number N of orders in a day forwarded to the shipping department of a book dealer is Poisson (100). The individual orders constitute an iid class $\{Y_i : 1 \le i\}$, independent of the arrival process, with each Y_i having values 1, 2, 5, 10, 20, 50, or 100, with probabilities 0.1, 0.15, 0.15, 0.25, 0.20, 0.10, and 0.05, respectively.

   (a) What is the expected number of books shipped in a day?

   (b) What is the probability there will be no more than six orders of 100?

   (c) What is the expected number of 50-book orders?

20–13. One car in twenty on a certain highway is a Mercedes. If the number of cars passing a given point in an hour is Poisson (60), what is the probability that five or more will be Mercedes?

20–14. A service center on an interstate highway experiences customers during a one-hour period as follows.

   Northbound traffic: Total vehicles, Poisson (100). Twenty percent are trucks.
   Southbound traffic: Total vehicles, Poisson (120). Twenty-five percent are trucks.

   Ten percent of the vehicles stop. Each truck has one or two persons, with probabilities 0.7 and 0.3, respectively. Each car has 1, 2, 3, 4, or 5 persons, with probabilities 0.3, 0.3, 0.2, 0.1, and 0.1, respectively. Assume independence of the arrival processes and the numbers of persons in the vehicles. Let D be the total number of persons to be served. Determine $E[D]$ and $g_D(s)$.

20–15. The number of passenger cars through a toll gate in an eight-hour period is approximately Poisson (5000). The number of persons in the ith car is a simple random variable Y_i with distribution

$k =$	1	2	3	4	5	6
$P(Y = k)$	0.35	0.25	0.20	0.10	0.05	0.05

Suppose the Y_i form an independent class, independent of the counting random variable N. Let Z be the total number of passengers and let N_2 be the number of cars with two passengers.

   (a) Determine $E[Z]$ and $\text{Var}[Z]$.

   (b) What is the distribution for N_2?

   (c) Use the decomposition theorem to determine whether, on the average, there is a greater total number of individuals in two-person cars, three-person cars, or four-person cars. Verify your answer by obtaining the expected values in each case.

20–16. The number N of customers in a shop during a given day is Poisson (100). Customers pay with cash, MasterCard, or Visa charge cards. Suppose the type of payment by the ith customer is indicated by a random variable Y_i, with values 1, 2, and 3 signifying cash, Master-Card, and Visa, respectively. Experience indicates the Y_i form an iid class, with common distribution 0.40, 0.35, and 0.25.

    (a) What is the probability that 25 or fewer persons will pay cash? Justify your answer by reference to the appropriate results in the text. Put your answer in a form that would enable you to go to a table and tell what table is needed.

    (b) What is the expected number of credit card transactions in a day?

20–17. In Theorem 20.4.1 on extreme values, prove that $F_W(t) = g_N[F_Y(t)]$ for $t \geq 0$.

20–18. Suppose a teacher of applied probability is equally likely to have 0, 1, 2, or 3 students come in during office hours on a given day. If the visit times, in minutes, of the individual students are iid, exponential (0.1), what is the probability that no visit will last more than 20 minutes?

20–19. Noise pulses arrive on a data phone line according to an arrival process such that for each $t > 0$, N_t, the number of arrivals in the time interval $(0, t]$, in hours, is Poisson $(5t)$. The ith pulse has an "intensity" Y_i, such that the Y_i form an iid class, independent of the arrival process, with common distribution function $F_Y(u) = 1 - e^{-2u^2}$, $u \geq 0$. Determine the probability that in an eight-hour day the intensity will not exceed two.

20–20. The number N of noise bursts on a data transmission line in an hour of operation is Poisson (μ). The number of digit errors caused by the ith burst is a random variable Y_i. Assume the Y_i form an iid class, independent of the arrival process, with $Y_i - 1$ geometric (p). An error correction system is capable of correcting five or fewer errors in a burst. Suppose $\mu = 10$ and $p = 0.5$. What is the probability of no uncorrected error in an hour of operation?

Hint: There are no uncorrected errors iff the maximum number of errors in a burst does not exceed five.

20–21. Four percent of the cars on a highway stop at a station for gasoline. Individual purchases, in gallons, may be considered iid, exponential $(\lambda = 0.1)$. In the period from 6:00 A.M. to 11:00 P.M., the number N of cars passing a point on this highway is Poisson (3000). Under the appropriate independence assumptions, determine the probability that the maximum gasoline purchase is no more than 50 gallons.

20–22. Five solid-state modules are installed in a computer system. If the modules are satisfactory, they have practically unlimited life. However, with probability $p = 0.05$, any unit could have a defect that results in a lifetime, in hours, which is exponential (0.0025). The system fails if any one or more module fails. If no module is defective, the system lasts indefinitely. Note that the number N of defective modules is binomial $(5, 0.05)$. What is the probability the system does not fail because of a defective module in the first 500 hours after installation?

Note. If V_N is the minimum life of the N defective modules and V_S is the life of the system with the modules, then $P(V_S > t) = P(N = 0) + P(V_N > t)$.

20–23. During a one-hour wind storm, the number of wind gusts causing a force of 100 pounds or more that strikes a television antenna is a random variable N that is Poisson (200). Suppose the ith gust has force $Y_i = 100 + X_i$, with the X_i iid, exponential (0.02), and independent of N. Thus, $F_Y(t) = 1 - e^{-0.02(t-100)}$ for $t \geq 100$. What is the probability the maximum gust will not exceed 500 pounds?

20–24. A truck stop and restaurant on an interstate highway welcomes tour groups. During a summer day, the number of arriving groups is approximately Poisson (12). The number of members in a group (although integer-valued) may be approximated by a random variable exponential (1/30). That is, the average is 30, and the probability of more than 100 in any group is only about 0.036. Suppose the numbers in the various groups are iid. What is the probability that the maximum group on a given day will exceed 100?

20–25. A personal computer is connected to a power line subject to electrical noise pulses. Pulses arrive according to a Poisson counting variable. In an eight-hour day, the number of pulses is Poisson (30). Pulse magnitudes, in volts, are exponential ($\alpha = 0.02$). What are the probabilities that in such a period the maximum pulse will not exceed 200, 300, or 500 volts?

20–26. Digital signals are transmitted at 9,600 baud (bits per second) in "bytes" of ten bits each, or 960 bytes per second. If Y_1 is the number of the first byte with one or more errors, $Y_1 + Y_2$ is the number of the second byte with one or more errors, etc., we may assume the class $\{Y_i : 1 \leq i\}$ is iid, with $Y_i - 1$ geometric (0.001). Error-correcting devices have probability 0.0001 of failing to correct the errors in any byte. This may be analyzed as a sequence of Bernoulli trials with random execution time. The execution time is the time between two bytes that have errors to be corrected. A "success" is an uncorrected error. What is the average time (in hours) between uncorrected errors?

20–27. As a sales inducement, a video electronics store offers each customer who makes a purchase of 100 dollars or more a free chance at a prize. This is in the form of a "drawing," so that the prize is available immediately. Suppose the times, in hours, between sales (hence drawings) is exponential, with $\lambda = 4$. The probability of winning on a draw is 0.01. If the times between sales are independent, and the results of the draws are independent, what is the probability of a winner in an eight-hour day? What is the expected time between winning drawings?

Hint: Consider the process to be a Bernoulli trial with random execution times.

20–28. In Example 20.4.3, the number N of defective transistors is binomial (n, p), but the calculations are carried out under the approximation N is Poisson (np). For $n = 1000$, $p = 0.001$ and Y exponential (1), determine $P(N = 0|X > t)$ for $t = 0, 1, 10$ under the binomial distribution. Compare with the results in Example 20.4.3.

20–29. The number N of incoming and outgoing calls in a university residential hall, initiated between 8:00 P.M. and 9:00 P.M. on a Friday evening, is Poisson, with mean value 100. The duration of individual calls is exponential, with mean length two minutes. If Y_i is the duration of the ith call, the Y_i form an iid class, independent of the arrival process. Let W be the maximum duration of any call initiated during that period. Determine $P(W \leq 10)$ and $P(W \leq 15)$.

20–30. Interarrival times, in hours, for customers in a computer store are iid exponential (2). One customer in 10 buys a computer. Customers buy independently. Let T be the time between customers who purchase a computer. Determine $E[T]$ and $P(T \leq 2)$.

20–31. A memory module in a computer produces occasional errors. Error-correcting hardware is added with the result that the probability of an uncorrected error is p. Successive interarrival times, in hours, of the errors from the module are an iid sequence $\{Y_i : 1 \leq i\}$. The error-correcting hardware operates independently of the module. Suppose the Y_i are exponential, with mean time to failure 30 minutes and the probability p of an uncorrected error is 0.01. Let T be the time to the first uncorrected error.

Determine $E[T]$ and $P(T \geq 24)$.

20–32. The process of discovering the fault in an electronics system is one of testing components sequentially, until a defective element is discovered. Suppose any of the components is defective with probability $p = 0.05$. Let the time in minutes to test the ith component be Y_i. These form an iid sequence, exponential with mean value 0.1. Let D be the time to discover the first defective unit (i.e., time to completion of the test on the first defective unit). Determine $E[D]$ and $P(D > 3)$.

20–33. Statistics show that about one call in twelve to a police switchboard requires dispatching a car to the scene. The number N of calls in an hour is Poisson (30). We may suppose the calls are independent.

(a) What is the expected number of dispatches in an hour?

(b) What is the probability that no more than four cars will be dispatched in an hour?

20–34. Customers arrive according to a Poisson counting process with $E[N_t] = t$. This means that $\lambda = 1$. Individual demands form an iid class, independent of the arrival process. Each Y_i takes on values 0, 1, and 2 with probabilities 0.25, 0.5, and 0.25, respectively.

(a) For any $t > 0$, determine $E[D_t]$ and $g_{D_t}(s)$.

(b) Use g_{D_t} to determine $P(D_8 \geq 5)$.

20–35. A service shop has service time Y_i for the ith unit it services. Its service record shows that 95 percent of its service jobs are satisfactory, and that failures are independent of other jobs and of service times. Suppose the service times Y_i in hours form an iid sequence, exponential (2). Let T be the service time in the shop between completion of unsatisfactory jobs; that is, T is the sum of the individual service times for the string of satisfactory jobs before the next unsatisfactory job plus the service time for the unsatisfactory job. Determine $E[T]$ and $P(T > 8)$.

20–36. A repair shop finds that repair times Y_i on units of a certain type of equipment are iid exponential (λ). It decides to put a limit T on the amount of time it will spend on any one unit. Thus, if $Y_i \leq T$, the job is completed successfully; if $Y_i > T$, the job is abandoned as a failure. Find the distribution of the time W to the first failure.

Hints: Consider the process as a Bernoulli sequence with random execution times. The probability of success on any trial is

$$p = P(Y \leq T) = 1 - e^{-\lambda T} M_Y(s) = \frac{\lambda}{\lambda - s}.$$

Let N be the number of successes before the first failure and let D be the time to complete N successful jobs. Then $W = D + T$ is the time to the first failure. We have

$$N \text{ is geometric}(q) \qquad \text{and} \qquad g_N(s) = \frac{q}{1 - ps}.$$

Determine the moment-generating function for D and use Laplace transforms to obtain the distribution function F_D. Then shift F_D by T units to obtain F_W.

20–37. A company recruiter interviews students in a university placement center. The time in hours for each interview is a random quantity exponential (3). If the interview times may be considered iid, what is the probability the interviewer will see at least fifteen students in six hours of interviewing? What is the average number seen in six hours of interviewing? What is the probability no interview will last for more than one hour?

20–38. A customer brings her camera to a repair shop for cleaning and adjustment. She is told there are four cameras ahead of hers. Suppose the jobs are done sequentially, with job times in hours iid exponential (2). What is the probability her camera (the fifth to be worked on) will be ready when she returns four hours later?

20–39. Consider the discounted replacement costs in Example 20.6.2. Suppose costs are inflated on a "straight line" basis, so that the cost at time t in the future is $c(t) = c_0 + c_1 t$. Then the discounted cost will be

$$C = \sum_{n=1}^{\infty} (c_0 + c_1 S_n) e^{-\alpha S_n}.$$

Determine $E[C]$.

20–40. It is desired to budget for a two-year supply of cartridges for a laser printer. The units cost \$90 each, with immediate delivery of individual units available. Money is invested at eight percent interest, with immediate withdrawal possibility. This means for one dollar to be spent t years from now, $e^{-0.08t}$ dollars must be invested now. The life in years of the ith unit is a random quantity W_i, exponential $(\lambda = 15)$. This means that the average life of a unit is $1/15$ year. The present value of the units is

$$C = \sum_{n=1}^{N_2} 90 e^{-0.08 S_n}$$

where S_n is the time of replacement of the nth unit and N_2 is the number of units replaced in the two-year period. Determine the expected discounted cost $E[C]$.

20–41. A device is subject to shocks that occur with iid exponential (λ) interarrival times. The ith shock gives rise to damage D_i. The D_i form an iid class, independent of the arrival process. The device is self-healing, so that the effect of a shock occurring at S_i is reduced exponentially with time. The effects are additive. Thus, at time t, the net residual effect is

$$D(t) = \sum_{i=1}^{N_t} D_i e^{-\alpha(t-S_i)}.$$

Determine $E[D(t)]$ as a function of $E[D_i]$, λ, α, and t.

21

Poisson Processes

In the treatment of arrival processes in Section 20.6, we show that if the interarrival times are iid exponential (λ), then the random variable N_t that counts the arrivals in the interval $(0, t]$ is Poisson (λt) for each $t \geq 0$. Such a counting process is known as a homogeneous Poisson process. In this chapter, we examine the structure of these processes in more detail. After introducing the notion of increments for a process, we examine several sets of conditions that characterize homogeneous Poisson processes. These axiom sets show why the Poisson process is a suitable model for so many practical systems. The exponential interarrival times have the property of "starting over," beginning anew, at any time $t \geq 0$. We show that the Poisson process preserves this property. We establish the fact that under the condition of n arrivals in an interval $(0, t]$, the arrivals are distributed as if they were values of an iid class of variables uniformly distributed on that interval. Previous results on arrivals of m kinds are sharpened to take advantage of the independence and stationarity of the increments, making possible several significant applications. In addition, we note that this formulation characterizes compound Poisson processes, in which the condition of only single arrivals is relaxed. We show that independent Poisson processes can be combined additively to obtain a new Poisson process. The conditional order statistics property is exploited to study an important class of problems. Finally, nonhomogeneous Poisson processes are considered briefly.

21.1. The Homogeneous Poisson Process

As a first step toward completing the characterization of homogeneous Poisson processes, we examine several equivalent axiom sets for these processes. Identification of these axiom systems is important for modeling, for they show what assumptions must hold in order for a homogeneous Poisson process to be a good model.

INCREMENTS OF A RANDOM PROCESS

To extend and sharpen our characterization of Poisson counting processes, we need the concept of an increment of a process. Since N_t counts the arrivals in the time interval $(0, t]$, the difference or the increment $N_{t+h} - N_t$ counts the arrivals in the interval $(t, t + h]$. For each fixed $t \geq 0$, $h > 0$, the increment $N_{t+h} - N_t$ is a random variable. We say the process has independent increments iff the increments for any class of nonoverlapping time intervals form an independent class of random variables. The process has stationary increments iff the probability of a given number of arrivals in any interval is determined by the length of that interval, but does not depend upon its location. Thus $N_{t+h} - N_t$ has the same distribution as does N_h, for any $t \geq 0$, $h > 0$.

The following notation is frequently useful in dealing with increments.

$$p_k(t, t + h) = P(N_{t+h} - N_t = k) \quad \text{and} \quad \pi_k(t, t + h) = P(N_{t+h} - N_t \geq k).$$

We generally shorten $p_k(0, 0 + h)$ to $p_k(h)$, and similarly for $\pi_k(h)$.

AXIOM SETS FOR THE POISSON PROCESS

In the appendix to this chapter, we establish the following theorem. Although the proofs given there appear quite formidable, they involve elementary ideas and, for the most part, use concepts and techniques developed in this book. It is not necessary to examine the proofs in order to use the theorem. However, the results are quite important, and the proofs provide insight into the structure of the process and the nature of the arguments employed to establish this structure.

Theorem 21.1.1 (Axiom Sets for the Homogeneous Poisson Process) *For a counting process $N = \{N_t : 0 \leq t\}$, the following four sets of conditions are equivalent, in the sense that if any one set holds, so do the others.*

AXIOM SET I

1. *The interarrival times $\{W_i : 1 \leq i\}$ are iid exponential (λ).*

AXIOM SET II

1. *$N_0 = 0$ a.s. and $0 < E[N_t] < \infty$ for all $t > 0$.*

2. *There are no multiple arrivals (i.e., $W_i > 0$ a.s. for all $i \geq 1$).*

3. *The increments are independent.*

4. *The increments are stationary.*

AXIOM SET III

1. *There is a positive number λ such that N_t is Poisson (λt) for all $t > 0$.*

2. *The increments are independent.*

3. *The increments are stationary.*

AXIOM SET IV

1. *$p_1(t, t+h) = \lambda h + o(h)$ as $h \to 0+$, uniformly in $t \geq 0$.*

2. *$\pi_2(t, t+h) = o(h)$ as $h \to 0+$, uniformly in $t \geq 0$.*

3. *The increments are independent.*

4. *The increments are stationary.*

We take the third axiom system as the basis for the definition, since this is most closely related to the name and the important properties for working with the process.

—   □

Definition 21.1.2. *A counting process N is a* homogeneous Poisson process *with* rate λ *iff N_t is Poisson (λt) for each $t \geq 0$ and N has independent, stationary increments.*

The parameter λ is called the *rate*, since the average number of arrivals per unit time is given by

$$\frac{1}{t} E[N_t] = \lambda.$$

REGULARITY OF COUNTING PROCESSES

A number of properties of the Poisson counting process are developed in Section 20.6. The principal new feature treated in this section is the class of increments and their properties. The homogeneous Poisson process satisfies a condition on the increments known as regularity, defined as follows.

Definition 21.1.3. *A counting process N is* regular *iff*

$$\frac{\pi_2(t, t+h)}{\pi_1(t, t+h)} \to 0 \qquad \text{as } h \to 0+, \text{ uniformly in } t \geq 0.$$

Theorem 21.1.4. *If $p_1(t, t+h) = \lambda h + o(h)$ as $h \to 0+$, uniformly in $t \geq 0$, then N is regular iff both*

a. *$\dfrac{\pi_2(t, t+h)}{p_1(t, t+h)} \to 0$ as $h \to 0+$, uniformly in $t \geq 0$.*

b. *$\pi_2(t, t+h) = o(h)$ as $h \to 0+$, uniformly in $t \geq 0$.*

Proof

$$\frac{\pi_2}{\pi_1} = \frac{\pi_2}{p_1 + \pi_2} = \frac{\pi_2/p_1}{1 + \pi_2/p_1} \to 0$$

$$\text{iff} \quad \frac{\pi_2}{p_1} \to 0 \quad \text{iff} \quad \frac{\pi_2}{\lambda h + o(h)} \to 0 \quad \text{iff} \quad \frac{\pi_2}{h} \to 0.$$

□

Axiom Set IV is often stated as:

1. $p_1(t, t + h) = \lambda h + o(h)$ as $h \to 0+$, uniformly in $t \geq 0$.

2. N is regular.

3. The increments are independent.

4. The increments are stationary.

WAITING TIMES AFTER t

The interarrival times for the homogeneous Poisson process are exponential. An essential characteristic of an exponential random variable is that it "starts over" at any time. That is,

$$P(W_i > t + h | W_i > t) = P(W_i > h).$$

This property is maintained in the behavior of the homogeneous Poisson process, as the following theorem shows. The complications in the proof (see the appendix to this chapter) arise from the fact that at a given t it is not known which interarrival time is in effect.

Theorem 21.1.5   (Waiting Times after t) *Suppose $\{W_i : 1 \leq i\}$ is iid exponential (λ).*

For any fixed $t \geq 0$, let W_{t_1}, W_{t_2}, ... be the waiting times (interarrival times) after t. Then

a. *For any $t \geq 0$, $\{W_{ti} : 1 \leq i\}$ is iid exponential (λ).*

b. *For $0 < t_1 < \cdots < t_m = t$,*

$$\{(N_{t_1}, N_{t_2} - N_{t_1}, \ldots N_t - N_{t_{m-1}}), W_{t_1}, W_{t_2}, \ldots\} \quad \text{is independent.}$$

□

Remark: In applications of this result, the time of observation is frequently a random time T. Usually, T is either independent of the process N or is dependent in some manner on the past of the process.

Example 21.1.1:

Traffic passes a point on a highway according to a Poisson process with rate 5 (per minute).

1. Observation begins at a random time T, independent of the process. Let V_T be the elapsed time before the next vehicle arrives. In the light of Theorem 21.1.5, it seems reasonable to suppose that V_T is exponential (5).

2. Observation begins at the arrival of the next vehicle. The random variable T is thus dependent upon the past of the process. Again, it seems reasonable to suppose that V_T is exponential (5).

—   □

The pertinent general assertion is made in the following theorem.

Theorem 21.1.6. *Consider a homogeneous Poisson process with rate (λ). Suppose T is a nonnegative random variable and let $\{W_{Ti} : 1 \le i\}$ be the interarrival times after T. If*

(1) T is independent of the process N, or

(2) T is a Markov time for N (see Section 20.8), then

$\{W_{Ti} : 1 \le i\}$ *is iid exponential*(λ) *and* $\{T, W_{Ti} : 1 \le i\}$ *is independent.*

Proof of Case 1

1. In this case, $\{T, W_{ti} : 1 \le i\}$ is an independent class. To simplify writing, we let

$$Z = \prod_{i=1}^{n} g_i(W_i), \qquad Z_t = \prod_{i=1}^{n} g_i(W_{ti}) \quad \text{and} \quad Z_T = \prod_{i=1}^{n} g_i(W_{Ti}).$$

Use of the law of total probability (CE1b), property (CE11) for conditional expectation, and Theorem 21.1.5 enables us to assert

$$E[Z_T] = E\big[E[Z_T|T]\big] \qquad E[Z_T|T = t] = E[Z_t] = E[Z].$$

Now by the independence of $\{W_i : 1 \le i\}$, we have

$$E[Z] = \prod_{i=1}^{n} E[g_i(W_i)].$$

In a similar manner, $E[g_i(W_{Ti})] = E[g_i(W_i)]$. Since these results hold for any n, any i, and any choices of the g_i, we have the desired independence and exponential distributions.

2. The case for T, a Markov time, is established by a similar argument. The details are somewhat more complicated, and the argument requires a more general formulation of conditional expectation.

— $\qquad\qquad\qquad\qquad\qquad\qquad\qquad\qquad\qquad\qquad\qquad\quad$ $\square$

Example 21.1.2:

A laser printer uses cartridges that must be replaced after 3000 to 5000 sheets are printed. Use of the machine varies from day to day. Experience indicates the times, in days, between replacements may be considered exponential with mean value ten (i.e., $\lambda = 1/10$). It is desired to order a year's supply. The cartridges are packed two to a box. How many boxes should be ordered to achieve probability 0.95 that the supply will last a year?

Solution

The counting process is homogeneous Poisson with rate 0.1. Since it does not matter when the count begins, we are concerned for a subsequent period of 365 days. The number needed will be Poisson (36.5). Use of tables give 45 for $\mu = 35$ and 51 for $\mu = 40$. Thus, the order should be for approximately 48 cartridges, or 24 boxes.

— $\qquad\qquad\qquad\qquad\qquad\qquad\qquad\qquad\qquad\qquad\qquad\quad$ $\square$

CONDITIONAL ORDER STATISTICS

The next theorem provides an important characterization of the homogeneous Poisson process. Not only does it provide insight into the behavior of the process, but it also provides a powerful resource for analyzing certain classes of problems.

Theorem 21.1.7. *Suppose N is a homogeneous Poisson process with rate λ. Consider a disjoint class $\{E_i : 1 \le i \le n\}$, with each E_i an interval of the form $(t_i, t_i + c_i]$. Let $N(E_i)$ be the number of arrivals in the interval E_i. Put*

$$E = \bigcup_{i=1}^{n} E_i, \qquad and \qquad c = \sum_{i=1}^{n} c_i \text{ the total length of } E.$$

For $k_1 + k_2 + \cdots + k_n = k$, we have

$$P\big(N(E_1) = k_1, N(E_2) = k_2, \ldots, N(E_n) = k_n | N(E) = k\big)$$

$$= \frac{k!}{k_1! \, k_2! \ldots k_n!} \prod_{i=1}^{n} \left(\frac{c_i}{c}\right)^{k_i}.$$

In particular, for $E = (0, t]$, $E_1 = (0, u]$, $E_2 = (u, t]$, and $k = n$, we have

$$P(N_u = j | N_t = n) = C(n, j) \left(\frac{u}{t}\right)^{j} \left(\frac{t - u}{t}\right)^{n-j}.$$

Proof

1. Since the E_i are disjoint, the $N(E_i)$ form an independent class.

2. Let $A = \{N(E_1) = k_1, N(E_2) = k_2, \ldots, N(E_n) = k_n\}$ and $B = \{N(E) = k\}$.

   Since $A \subset B$, we have $P(A|B) = P(A)/P(B)$. We note also that $\prod_{i=1}^{n} e^{-\lambda c_i} = e^{-\lambda c}$. Thus,

$$
\begin{aligned}
P(A|B) &= \prod_{i=1}^{n} e^{-\lambda c_i} \frac{(\lambda c_i)^{k_i}}{k_i!} \cdot \frac{k!}{e^{-\lambda c}(\lambda c)^k} \\
&= \frac{k!}{k_1! \, k_2! \ldots k_n!} \cdot \left(\frac{c_1}{c}\right)^{k_1} \left(\frac{c_2}{c}\right)^{k_2} \cdots \left(\frac{c_n}{c}\right)^{k_n}.
\end{aligned}
$$

$\square$

Interpretation

We interpret this result in terms of order statistics, developed in Example 11.2.17. The central facts are these.

Suppose $\{X_i : 1 \leq i \leq n\}$ is iid, with common distribution function F. Let

$$
\begin{aligned}
Y_1 &= \text{smallest of} \quad X_1, X_2, \ldots, X_n. \\
Y_2 &= \text{next larger of} \quad X_1, X_2, \ldots, X_n. \\
&\vdots \\
Y_n &= \text{largest of} \quad X_1, X_2, \ldots, X_n.
\end{aligned}
$$

Then Y_k is called the kth order statistic for the class $\{X_i : 1 \leq i \leq n\}$. Now,

$$
\begin{aligned}
P(Y_k \leq t) &= P(k \text{ or more of the } X_i \text{ lie in } (-\infty, t]) \\
&= \sum_{j=k}^{n} C(n, j) F^j(t) \big(1 - F(t)\big)^{n-j}.
\end{aligned}
$$

We are interested in the special case that $X_i = U_i$, uniform on $(0, t]$, for which the kth order statistic Y_k has the distribution

$$P(Y_k \leq u) = \sum_{j=k}^{n} C(n, j) \left(\frac{u}{t}\right)^{j} \left(\frac{t - u}{t}\right)^{n-j} \qquad \text{for } 0 < u < t.$$

If there are n arrivals in the interval $(0, t]$, then S_k is the kth largest arrival time, so that S_k is the kth order statistic for the n arrival times. Also, we recall that $\{S_k \leq u\} = \{N_u \geq k\}$. By the special case of Theorem 21.1.7,

$$P(S_k \leq u | N_t = n) = P(N_u \geq k | N_t = n) = \sum_{j=k}^{n} C(n, j) \left(\frac{u}{t}\right)^j \left(\frac{t-u}{t}\right)^{n-j}.$$

Thus, conditionally, given $N_t = n$, the arrival time S_k has the same distribution as the kth order statistic for a class $\{U_i : 1 \leq i \leq n\}$ that is iid uniform on $(0, t]$. This means the arrivals tend to be spread uniformly over the interval $(0, t]$, and provides partial support for the adjective homogeneous in the name of the process. It also provides the basis for testing whether or not a process is Poisson: do the observed arrivals in a given interval have the appropriate statistical features to satisfy the uniform distribution? Standard methods of hypothesis testing may be used. In Section 21.4, we show some important consequences of this property.

BEHAVIOR FOR LARGE t

The following limiting behavior is important in many applications.

Theorem 21.1.8 (Behavior for Large t) *If N is a homogeneous Poisson process with rate λ, then*

a. $N_t/t \to \lambda$ *a.s. as $t \to \infty$.*

b. N_t *is approximately $N(\lambda t, \lambda t)$ for large λt.*

Proof

1. Because of independent, stationary increments, N_n is the sum of n independent random variable, each Poisson (λ). By the strong law of large numbers,

$$\frac{1}{n} N_n \to \lambda = E[N_1] \quad \text{a.s.} \qquad \text{as } n \to \infty.$$

It is simple, but somewhat tedious, to show that this implies $N_t/t \to \lambda$ a.s.

2. $E[N_t] = \text{Var}[N_t] = \lambda t$. By the central limit theorem

$$P\left(\frac{N_n - \lambda n}{\sqrt{\lambda n}} \leq u\right) \approx \Phi(u) \qquad \text{for } \lambda n \text{ sufficiently large.}$$

For large t,

$$\frac{N_t - \lambda t}{\sqrt{\lambda t}} \approx \frac{N_{\lfloor t \rfloor} - \lambda \lfloor t \rfloor}{\sqrt{\lambda \lfloor t \rfloor}} \qquad \text{where } \lfloor t \rfloor \text{ is the integer part of } t.$$

$\square$

Example 21.1.3:

Suppose N is a homogeneous Poisson process with rate $\lambda = 20$. Determine

$$P\left(\left|\frac{1}{30}N_{30} - 20\right| \le 0.1\right).$$

Solution

Since N_{30} is Poisson $(20 \cdot 30 = 600)$, we do not expect to find a table of probabilities. However, we may use the normal approximation.

$$P\left(\left|\frac{1}{30}N_{30} - 20\right| \le 0.1\right) = P\left(\left|\frac{N_{30} - 600}{\sqrt{600}}\right| \le \frac{0.1 \cdot 30}{\sqrt{600}}\right)$$
$$\approx 2\Phi(0.12) - 1 \approx 0.78.$$

$\square$

21.2. Arrivals of m kinds and compound Poisson processes

We now adapt the Poisson decomposition Theorem 20.3.1 to the case of a Poisson counting process N with rate λ. We assume that the kind of each arrival is determined by a trial in a multinomial sequence, so that each arrival is one of m kinds or types. Once more, we use the notation for multinomial trials as introduced in Sections 10.2 and 20.3. That is, we let

E_{ki} be the event the result on the ith trial is of the kth kind, $1 \le k \le m$.

Then the number N_{kn} of results of the kth kind in the first n trials is given by

$$N_{kn} = \sum_{i=1}^{n} I_{E_{ki}} \quad \text{with} \quad \sum_{k=1}^{m} N_{kn} = n, \quad \text{for each } n \ge 1$$

and the kind of the ith result is represented by $T_i = \sum_{k=1}^{m} kI_{E_{ki}}$

In this context, the ith result is the kind of the ith arrival. If the counting process is $N = \{N_t : 0 \le t\}$, then N_t counts the total number of arrivals in the interval $(0, t]$. We introduce m counting processes N^k, $1 \le k \le m$, such that N_t^k is the number of arrivals of the kth kind in the interval $(0, t]$. Then

$$N_t^k = \sum_{i=1}^{N_t} I_{E_{ki}} = \sum_{n=1}^{\infty} I_{\{n\}}(N_t)N_{kn}.$$

It is a direct result of Theorem 20.3.1 that for each $t > 0$, the class $\{N_t^k : 1 \le k \le m\}$ is an independent class with N_t^k Poisson $(p_k\lambda t)$. To show that the processes N^k are homogeneous Poisson, we must show that they have stationary, independent increments.

We first obtain a result for the case N has stationary, independent increments, but is not necessarily Poisson.

Theorem 21.2.1 (Arrivals of m Kinds) *Suppose the counting process $N = \{N_t : 0 \leq t\}$ counts the number of arrivals, T_i is the kind of the ith arrival, and N^k is the counting process for the arrivals of the kth kind.*
   If

1. $\{T_i : 1 \leq i\}$ *is iid, with* $P(T_i = k) = P(E_{ki}) = p_k 1 \leq k \leq m$,
2. $\{N, T_i : 1 \leq i\}$ *is independent, and*
3. N *has stationary, independent increments,*

then each process N^k, $1 \leq k \leq m$ has stationary, independent increments.

Proof

We carry out the proof for the intervals $(0, t]$ and $(u, u+c]$, $t \leq u$. Extension to the general case should then be apparent. Let

$$A = \{N_t^j = m\} \qquad \text{and} \qquad B = \{N^k(u, u+c] = n\}, \qquad t \leq u.$$

Then

$$P(A) = E[I_A], \qquad P(B) = E[I_B], \qquad \text{and} \qquad P(AB) = E[I_A I_B].$$

1. By the law of total probability (CE1b), $E[I_A] = E\big[E[I_A|N_t]\big]$. Also, by virtue of property (CE10b), we have

$$
\begin{aligned}
E[I_A|N_t = v] &= E\left[I_{\{m\}}\left(\sum_{r=1}^{N_t} I_{E_{jkr}}\right)\,\Big|\,N_t = v\right]\\[2mm]
&= E\left[I_{\{m\}}\left(\sum_{r=1}^{v} I_{E_{jkr}}\right)\right] = e_m(v)
\end{aligned}
$$

since $\{N_t, I_{E_{jks}} : 1 \leq s\}$ is independent. Thus, $E[I_A] = E[e_m(N_t)]$. Similarly, $E[I_B] = E\big[E[I_B|N_u]\big]$, with

$$
\begin{aligned}
E[I_B|N_u = v] &= E\left[I_{\{n\}}\left(\sum_{s=1}^{N(u,u+c]} I_{E_{k,s+v}}\right)\right]\\[2mm]
&= E\left[I_{\{n\}}\left(\sum_{s=1}^{N(u,u+c]} I_{E_{ks}}\right)\right],
\end{aligned}
$$

invariant with v, since $\{N_u, N(u, u + c], I_{E_{ks}}, : 1 \leq s\}$ is independent. Because of the stationary increments for N,

$$
E\left[I_{\{n\}}\left(\sum_{s=1}^{N(u,u+c]} I_{E_{ks}}\right)\right] = E\left[I_{\{n\}}\left(\sum_{s=1}^{N_c} I_{E_{ks}}\right)\right]
$$

$$= \quad e_n\big(N(u, u+c]\big) = e_n(N_c).$$

Since $E[I_B] = E[e_n(N_c)]$ depends only upon c, the increments of N^k are stationary.

2. Again by the law of total probability,

$$E[I_A I_B] = E\big[E[I_A I_B | N_t, N(t, u]]\big],$$

with

$$E[I_A I_B | N_t = v, N(t, u] = w]$$

$$= \quad E\left[I_{\{m\}}\left(\sum_{r=1}^{v} I_{E_{jkr}}\right) I_{\{n\}}\left(\sum_{s=1}^{N(u,u+c]} I_{E_{k,s+v+w}}\right)\right]$$

$$= \quad E\left[I_{\{m\}}\left(\sum_{r=1}^{v} I_{E_{jkr}}\right)\right] E\left[I_{\{n\}}\left(\sum_{s=1}^{N(u,u+c]} I_{E_{ks}}\right)\right]$$

$$= \quad e_m(v) e_n\big(N(u, u+c]\big)$$

since $\{N_t, N(t, u], N(u, u+c], I_{E_{ks}} : 1 \le s\}$ is independent. Now,

$$\begin{aligned}
E[I_A I_B] &= E\big[e_m(N_t) e_n\big(N(u, u+c]\big)\big] \\
&= E[e_m(N_t)] E\big[e_n\big(N(u, u+c]\big)\big] \\
&= E[I_A] E[I_B].
\end{aligned}$$

Hence the pair $\{N_t, N(u, u+c]\}$ is independent.

—   ☐

We may assign to each kind k of arrival a "value" a_k. We let

$$Z_t^k = \sum_{i=1}^{N_t} a_k I_{E_{ki}} = \text{the total value of arrivals of the } k\text{th kind in } (0, t]$$

and

$$Z_t = \sum_{k=1}^{m} Z_t^k = \text{the total value of all arrivals in } (0, t].$$

Since the Z^k are constants times the N^k, it is clear that the properties of increments of N are maintained, so that we have the following.

Corollary 21.2.2 (Arrivals of m Kinds with Values) *Suppose the counting process $N = \{N_t : 0 \le t\}$ counts the number of arrivals, T_i is the kind of the ith arrival, a_k is the value for an arrival of the kth kind, and N^k is the counting process for the arrivals of the kth kind. Suppose, further,*

1. $\{T_i : 1 \leq i\}$ is iid, with $P(T_i = k) = P(E_{ki}) = p_k$. $1 \leq k \leq m$,

2. $\{N, T_i : 1 \leq i\}$ is independent.

3. N has stationary, independent increments.

If we set

$$Z_t^k = \sum_{i=1}^{N_t} a_k I_{E_{ki}} \qquad and \qquad Z_t = \sum_{k=1}^{m} Z_t^k$$

then Z and each Z^k, $1 \leq k \leq m$ has stationary, independent increments.

We may now apply these results to Poisson arrival processes.

Theorem 21.2.3 (Poisson Arrivals of m Kinds) *Suppose the homogeneous Poisson process $N = \{N_t : 0 \leq t\}$ with rate λ counts the number of arrivals, T_i is the kind of the ith arrival, and N^k is the counting process for the arrivals of the kth kind. If*

1. $\{T_i : 1 \leq i\}$ is iid, with $P(T_i = k) = P(E_{ki}) = p_k$, $1 \leq k \leq m$ and

2. $\{N, T_i : 1 \leq i\}$ is independent,

then each N^k is a homogeneous Poisson process with rate (λp_k), and the class $\{N_t^k : 1 \leq k \leq m\}$ is independent for each $t > 0$.
—   □

Theorem 21.1.5 on waiting times after t shows that a homogeneous Poisson process starts over after any time t. Thus, the statistics of arrivals of various kinds in an interval $(t, t + c]$ must have the same properties as the statistics of arrivals in $(0, c]$. We may thus assert the following.

Theorem 21.2.4. *Suppose the homogeneous Poisson process $N = \{N_t : 0 \leq t\}$ with rate λ counts the number of arrivals, T_i is the kind of the ith arrival, and N^k is the counting process for the arrivals of the kth kind. If*

1. $\{T_i : 1 \leq i\}$ is iid, with $P(T_i = k) = P(E_{ki}) = p_k$, $1 \leq k \leq m$ and

2. $\{N, T_i : 1 \leq i\}$ is independent,

then each $N^k(t, t + c]$ is Poisson (λc) and $\{N^k(t, t + c] : 1 \leq k \leq m\}$ is independent.
—   □

Example 21.2.1: Bernoulli Trials with Exponential Execution Times

Theorem 21.2.3 provides an alternate approach to the problem of Bernoulli trials with exponential execution times. The execution times become interarrival times for a Poisson process with rate λ. The arrivals (results of the trials) are of two kinds: successful and unsuccessful. The counting process for successful arrivals is homogeneous Poisson with rate (λp). The interarrival times T for successful arrivals are therefore exponential (λp). This agrees with the result in Theorem 20.5.1 (b).

— $\square$

Remark: This example points to another way of viewing the counting processes N^k. We may view N^k as the result of a sequence of Bernoulli trials with execution times iid, exponential (λ) and probability p_k of success on each trial. By Theorem 20.5.1, the time to the first arrival of the kth kind is exponential (λp_k). By Theorem 21.1.6, the system starts over at each arrival time, with iid interarrival times, independent of previous behavior. Thus, the time to each succeeding arrival of kth kind is exponential (λp_k), independent of the previous interarrival times. Hence, N^k is characterized by iid interarrival times, each exponential (λp_k). By Axiom Set I, N^k is Poisson with rate (λp_k).

Example 21.2.2: Burst Errors on a Digital Data Line

Errors on a digital data line come in bursts or packets. That is, a burst of one or more errors occurs in a very short time interval; then the line is free of errors until another burst arrives. The system has an error-correcting scheme that will correct the errors in any packet or burst of two or fewer errors. The arrival of bursts is Poisson with rate $\lambda = 3.5$ (per hour). The bursts are independent, identically distributed. Thus, if Y_i is the number of errors in the ith burst, then $\{Y_i : 1 \leq i\}$ is iid. Suppose

$$P(Y_i = 1) = 0.900, \qquad P(Y_i = 2) = 0.099, \qquad P(Y_i \geq 3) = 0.001.$$

What is the probability of an uncorrected error in an eight-hour period of operation? What is the expected number of uncorrected error bursts in that period?

Solution

We may consider the bursts to be of two kinds: correctable ones and uncorrectable ones, with $p_1 = 0.999$ and $p_2 = 0.001$. The arrival process N^2 for uncorrected bursts is Poisson with rate $\lambda p_2 = 0.0035$. The probability of an uncorrected error in an eight-hour period is the probability the first arrival time is no greater than eight. Now this interarrival time is exponential (0.0035). Thus,

$$P(S_1 \le 8) = 1 - e^{-8 \times 0.0035} \approx 0.0276.$$

Also, N_8^2 is Poisson (0.028), so that $E[N_8^2] = 0.028$.

—   □

Example 21.2.3:

The problem of job interviews in Example 20.5.2 may be viewed in terms of an arrival process. Let the completion of an interview represent an arrival. The interview times are then interarrival times. In that problem, these times were assumed iid, exponential (3). The counting process for arrivals (interview completions) is Poisson with rate 3. Now the arrivals are of two kinds: qualified and unqualified, with $p_1 = 0.2$ and $p_2 = 0.8$. By Theorem 21.2.1, the counting process for arrivals of qualified candidates is Poisson with rate $\lambda p_1 = 3 \times 0.2 = 0.6$. The time to the first qualified arrival is therefore exponential (0.6). Thus, as before,

$$P(T \le 4) = 1 - e^{-0.6} \approx 0.91.$$

—   □

COMPOUND POISSON PROCESSES

The axiom systems for homogeneous Poisson processes require that (with probability one) there be no multiple arrivals. If this restriction is removed, the resulting processes are called compound Poisson processes. The theory of processes with Poisson arrivals of several kinds contains the essential character of compound Poisson processes.

Suppose N is a homogeneous Poisson process with rate λ, whose arrivals constitute a multinomial sequence. Let $Y_n = \sum_{k=1}^{m} k I_{E_{kn}}$ be the number of arrivals at arrival time S_n. Then $\{Y_i : 1 \le i\}$ is an iid family of nonnegative, integer-valued random variables with common generating function $g_Y(s)$. The random process Z such that Z_t counts the total number of arrivals in $(0, t]$ may be expressed

$$Z_t = \sum_{i=1}^{N_t} Y_i \quad t \ge 0 \text{ with generating function } g(t, s) = E[s^{Z_t}] = g_{N_t}\big(g_Y(s)\big).$$

Corollary 21.2.2 shows that the Z process has independent, stationary increments.

Some classical developments show that the generating function for a compound Poisson process must have the form $g_{N_t}\big(g(s)\big)$ for a suitable function *g*. Thus, a compound Poisson process can always be represented as an ordinary homogeneous Poisson process, with various numbers of arrivals. The only new feature is the possibility of an arbitrary number of arrivals; that is, the Y_k are not necessarily bounded. The proof that establishes the form of the generating function uses some lemmas that combine probabilistic and analytical methods. The ideas are simple, but the details are tedious.

21.3. Superposition of Poisson Processes

.

Not only can we decompose Poisson processes with arrivals of several kinds into independent Poisson processes, one for each kind, but also we can combine additively several independent Poisson processes to form a new Poisson process. To establish the fundamental proposition, Theorem 21.3.2, we use the following result, which is also useful in other contexts.

Theorem 21.3.1. *Consider counting processes* $M = \{M_t : 0 \le t\}$ *and* $N = \{N_t : 0 \le t\}$. *Suppose* $\{M_t, N_t\}$ *is an independent pair for each* $t \ge 0$. *Define the process* Z *by* $Z_t = M_t + N_t$ *for any* $t \ge 0$.

1. *If* M *and* N *both have independent increments, so does* Z.

2. *If* M *and* N *both have independent, stationary increments, so does* Z.

Proof

1. $Z_{t+h} - Z_t = (M_{t+h} - M_t) + (N_{t+h} - N_t)$ and $Z_t = M_t + N_t$. By hypothesis, the class $\{M_t, N_t, M_{t+h} - M_t, N_{t+h} - N_t\}$ is independent. By elementary properties of functions of independent random variables, we conclude that $\{Z_t, Z_{t+h} - Z_t\}$ is independent. This argument can be extended to any finite set of increments for nonoverlapping time intervals.

2. For any $t > 0$, we have $P(Z_t = k) = \sum_{j=0}^{k} P(M_t = j)P(N_t = k - j)$. For any $t \ge 0$, $h > 0$, we have

$$
\begin{aligned}
P(Z_{t+h} - Z_t = k) &= \sum_{j=0}^{k} P(M_{t+h} - M_t = j)P(N_{t+h} - N_t = k - j) \\
&= \sum_{j=0}^{k} P(M_h = j)P(N_h = k - j) = P(Z_h = k).
\end{aligned}
$$

—

$\square$

Theorem 21.3.2. *Let M and N be homogeneous Poisson processes, with rates μ and λ, respectively. Suppose $\{M_t, N_t\}$ is independent for each $t \geq 0$. Define the process Z by $Z_t = M_t + N_t$, for any $t \geq 0$. Then Z is a homogeneous Poisson process with rate $\mu + \lambda$.*

Proof

The theorem follows immediately from Theorem 21.3.1 and the fact that the sum of independent Poisson random variables is Poisson.

— $\qquad\qquad\qquad\qquad\qquad\qquad\qquad\qquad\qquad\qquad$ □

Remark: It is apparent that the result extends immediately to the sum of any finite number of independent Poisson processes.

The following example uses both decomposition and superposition.

Example 21.3.1:

Passenger cars on an interstate highway reach a service center according to a Poisson process with rate $\lambda = 200$ (per hour). Trucks arrive according to a Poisson process with rate $\mu = 50$. The arrival processes are independent. If there is probability $p_1 = 0.1$ that any passenger car will stop and probability $p_2 = 0.15$ that any truck will stop, describe the process that counts the vehicles that stop at the center.

Solution

The process N^P that counts the passenger cars that stop is Poisson with rate $200 \cdot 0.1 = 20$, and the process N^T that counts the trucks that stop is Poisson with rate $50 \cdot 0.15 = 7.5$. The pair $\{N^P, N^T\}$ is independent. The process $N^P + N^T$ that counts both trucks and passenger cars is Poisson with rate $20 + 7.5 = 27.5$.

— $\qquad\qquad\qquad\qquad\qquad\qquad\qquad\qquad\qquad\qquad$ □

21.4. Conditional Order Statistics

In Section 21.1, we show that conditionally, given $N_t = n$, the arrival time S_k has the same distribution as the kth order statistic for a class $\{U_i : 1 \leq i \leq n\}$ that is iid uniform on $(0, t]$. The following consequence of Theorem 21.1.7, made apparent by this interpretative development, is frequently useful. A function ϕ of n variables is called a symmetric function iff interchanging the values of its arguments does not change its values. For example,

$$\phi(t_1, t_2, \ldots, t_n) = \sum_{k=1}^{n} g(t_k)$$

is symmetric, since it does not matter in what order the terms are added.

Theorem 21.4.1 (Symmetric Function of the Arrival Times) *Suppose N is a homogeneous Poisson process and ϕ is a symmetric function of n variables. Then*

$$E[\phi(S_1, S_2, \ldots, S_n)|N_t = n] = E[\phi(U_1, U_2, \ldots, U_n)]$$

where $\{U_1, U_2, \ldots, U_n\}$ is iid, uniform on $(0, t]$.

—   □

Remarks

1. The class $\{U_1, U_2, \ldots, U_n\}$ is not only iid, but can be independent of the counting process and any auxiliary random variables.

2. This result may be used, for example, to obtain an alternate proof of Theorems 20.6.2 and 20.6.3. We use it to obtain a generalization of Theorem 20.6.3, which we use later.

An important class of problems has the following features:

Arrivals are according to a Poisson process with rate λ and the ith arrival "survives" or "persists" or "remains active" for a time whose value is a random variable Y_i. Usually, we may assume the class $\{Y_i : 1 \leq i\}$ is iid and independent of the arrival process.

It is desired either (i) to count the number of active individuals at time t or (ii) to determine the distribution for the total remaining time for those active at time t.

Example 21.4.1: Active Individuals at Time t

A. Customers arrive in a store according to a Poisson process. The amount of time the ith person spends in the store (including "looking around" as well as "being served") is a random quantity Y_i. Then Z_t is the number of persons in the store at time t. Note that we have a question of the validity of the independence assumption. If there are many persons in the store, browsing times may be shorter and service times longer. We suppose these variations compensate.

B. Automobile accidents occur according to a Poisson process. Police investigation time, including time to reach the scene, for the ith accident is a random quantity Y_i. Then Z_t is the number of accidents under investigation at time t.

C. Emergency patients arrive in a hospital trauma center according to a Poisson process. The ith patient is in the facility (before discharge or before being sent to a hospital room) for a time Y_i. The number of patients in the center at time t is the random variable Z_t.

D. Customers arrive at an infinite-server queue according to a Poisson process. Each customer is served immediately. The customer arriving at time S_n has service time Y_n. The number of customers at time t is Z_t.

These problems exhibit a common pattern: The nth arrival is counted at time t iff both

1. That individual arrives before t; i.e., $t - S_n \geq 0$.

2. That individual is still "active" at t; i.e., $S_n + Y_n \geq t$, or $t - S_n \leq Y_n$.

   Thus,

$$Z_t = \sum_{n=1}^{N_t} g(t, S_n, Y_n), \quad \text{where } g(t, S_n, Y_n) = \begin{cases} 1 & \text{if } 0 \leq t - S_n \leq Y_n \\ 0 & \text{otherwise} \end{cases}.$$

$\square$

Example 21.4.2: Remaining Time

E. Automobiles arrive at a repair shop during working hours according to a Poisson process with rate λ. The shop time for the nth unit is a random variable Y_n. Let Z_t be the total remaining shop time for the automobiles in the shop at time t, say at closing time.

F. In a large company, employee accidents causing lost work time occur according to a Poisson counting process. The nth accident occurs at time S_n and results in Y_n days lost from work. Let Z_t be the total remaining days of lost time for employees out at time t.

G. Orders to be filled in a mail order house arrive according to a Poisson process with rate λ. The nth order takes time Y_n to complete. Let Z_t be the backlog of time required to complete processing of orders pending at time t.

The nth individual is active iff $S_n \leq t \leq S_n + Y_n$. Now $S_n \leq t$ iff $N_t \geq n$, and if $S_n \leq t$, the time remaining for the nth individual is

$$g(t, S_n, Y_n) = \max\{0, S_n + Y_n - t\}.$$

Hence, the total remaining time is $Z_t = \sum_{n=1}^{N_t} g(t, S_n, Y_n)$.

$\square$

We now state a general result that includes these and other important cases. Theorem 21.4.1 plays an essential role in the proof, provided at the end of this section.

Theorem 21.4.2. *Suppose $N = \{N_t : t \geq 0\}$ is homogeneous Poisson with rate λ and $\{Y_i : 1 \leq i\}$ is iid, with $\{N, Y_i : 1 \leq i\}$ independent. Consider*

$$Z_t = \sum_{n=1}^{N_t} g(t, S_n, Y_n) \quad \text{for } N_t \geq 1 \qquad \text{and} \qquad Z_t = 0 \quad \text{for } N_t = 0$$

with suitable integrability conditions on g. Let

$$M_t(s) = E[e^{sZ_t}] \qquad \text{and} \qquad M_Y(t, u, s) = E[e^{sg(t,u,Y)}]$$

where Y has the common distribution for the Y_i. Then

$$M_t(s) = \exp\left(\lambda \int_0^t \left(M_Y(t, u, s) - 1 \right) du \right)$$

$$E[Z_t] = \lambda \int_0^t E[g(t, u, Y)] \, du \qquad \text{and} \qquad \text{Var}[Z_t] = \lambda \int_0^t E[g^2(t, u, Y)] \, du.$$

$\square$

Before proving this theorem, we apply it to several special cases.

Theorem 21.4.3 (Special Cases of Theorem 21.4.2 for Active Individuals at Time t)

1. *For $Y_i = c$ (a constant) and g invariant with t, $E[g(t, u, Y)] = g(t, u, c)$. If we replace $g(t, u, c)$ by $g(u)$, we have*

$$M_t(s) = \exp\left(\lambda \int_0^t [e^{sg(u)} - 1] \, du \right)$$

and

$$E[Z_t] = \lambda \int_0^t g(u) \, du.$$

Thus, we have extended Theorem 20.6.3 to include an expression for the moment-generating function.

2. *If $g(t, u, Y)$ has the form $Y g(u)$, invariant with t, we have*

$$M_t(s) = \exp\left(\lambda \int_0^t E[e^{sg(u)Y} - 1] \, du \right) \quad \text{and} \quad E[Z_t] = \lambda E[Y] \int_0^t g(u) \, du.$$

The latter expression confirms the remark after Example 20.6.2.

3. *For $g(t, S_n, Y_n) = 1$ iff $0 \leq t - S_n \leq Y_n$ and is 0 otherwise, as in Example 21.4.1, and $0 \leq S_n \leq t$, we have*

$$e^{sg(t,u,Y)} - 1 = \begin{cases} e^s - 1 & \text{for } t - u \leq Y \\ 0 & \text{for } t - u > Y \end{cases}$$

so that $E\left[e^{sg(t,u,Y)} - 1\right] = (e^s - 1)P(Y \geq t - u)$. By a standard change of variable technique, $\int_0^t P(Y \geq t - u)\, du = \int_0^t P(Y \geq u)\, du$. According to Theorem 21.4.2, $M_t(s)$ is given by

$$M_t(s) = \exp\left(\lambda \int_0^t P(Y > u)\, du \cdot (e^s - 1)\right).$$

This result shows that, Z_t is Poisson $(\lambda \int_0^t P(Y > u)\, du)$. Since $\int_0^\infty P(Y > u)\, du = E[Y]$, by property (E20), it follows that for large t, Z_t is approximately Poisson $(\lambda E[Y])$. If $0 \leq Y \leq M$, then Z_t is Poisson $(\lambda E[Y])$ for all $t \geq M$.

— □

Example 21.4.3:

For the emergency room case, Example 21.4.1C, suppose the arrival rate λ is 3 (per hour) and the time Y_i (in hours) for the ith patient is uniform $[0, 3]$. Now

$$F_Y(u) = \frac{u}{3} \quad \text{for } 0 \leq u \leq 3 \quad \text{and} \quad \int_0^t (1 - u/3)\, du = t - \frac{t^2}{6} \quad \text{for } 0 \leq t \leq 3.$$

For $t \geq 3$, the integral is $E[Y] = 3/2$. Thus, the number Z_t is Poisson $(3t - t^2/2)$ for $0 \leq t \leq 3$ and is Poisson $(9/2)$ for $t \geq 3$.

— □

For the case of remaining time, as in Example 21.4.2, we do not have a general result to match that for the number in the system at time t. However, for Y exponential (α), a simple form is obtained for $M_t(s)$.

Theorem 21.4.4 (The Remaining Time Case when Y is Exponential) *If Y is exponential (α), then*

$$M_t(s) = \exp\left(\frac{\lambda}{\alpha} \frac{s}{\alpha - s}(1 - e^{-\alpha t})\right) \quad \text{and} \quad E[Z_t] = \frac{\lambda}{\alpha^2}\left(1 - e^{-\alpha t}\right).$$

Proof

Since $g(t, u, Y) = 0$ for $Y \leq t - u$ and $g(t, u, Y) = u + Y - t$ for $Y \geq t - u$, we have

$$e^{sg(t,u,Y)} - 1 = \begin{cases} 0 & \text{for } Y \leq t - u \\ e^{s(u-t)}e^{sY} - 1 & \text{for } Y > t - u \end{cases}.$$

For Y is exponential (α),

$$E[e^{sg(t,u,Y)} - 1] = e^{s(u-t)} \int_{t-u}^{\infty} \alpha e^{-(\alpha-s)v} \, dv - \int_{t-u}^{\infty} \alpha e^{-\alpha v} \, dv.$$

Elementary calculations show this to be $(s/(\alpha - s)) \, e^{-\alpha(t-u)}$. Hence,

$$M_t(s) = \exp\left(\lambda \cdot \frac{s}{\alpha - s} \int_0^t \alpha e^{-\alpha(t-u)} \, du\right) = \exp\left(\frac{\lambda}{\alpha}\frac{s}{\alpha - s}(1 - e^{-\alpha t})\right).$$

Either determination of $M_t'(0)$ or direct calculation of $\lambda \int_0^t E[g(t,u,Y)] \, du$ shows that

$$E[Z_t] = \frac{\lambda}{\alpha^2}\left(1 - e^{-\alpha t}\right).$$

$\square$

Example 21.4.4:

A shop begins the day with work caught up. Jobs come in during the day according to a Poisson process with rate $\lambda = 4$ (per hour). Shop time for jobs received are iid exponential $(\alpha = 1/2)$. Jobs are begun immediately. At closing time, after ten hours, what is the expected total backlog of shop time to be carried over to the next day?

Solution

According to the preceding result,

$$E[Z_{10}] = 4 \cdot 2^2 [1 - e^{-10/2}] \approx 16 \qquad \text{since} \quad [1 - e^{-5}] \approx 0.9933.$$

The number of unfinished jobs on hand is Poisson with rate

$$4 \int_0^{10} e^{-u/2} \, du \approx 8.$$

$\square$

A variety of other situations can be treated by proper choice of function g. Parzen (see the bibliographical note) has considered a number of these under the general heading of filtered Poisson processes.

We conclude this treatment with a proof of Theorem 21.4.2

1. Use of the law of total probability (CE1b) shows that $M_t(s) = E[e^{sZ_t}] = E\left[E[e^{sZ_t}|N_t]\right]$.

   By property (CE11) for conditional expectation, we have

$$E[e^{sZ_t}|N_t = n] = E[e^{sZ_n}|N_t = n] \qquad \text{where} \quad Z_n = \sum_{k=1}^{n} g(t, S_k, Y_k).$$

Hence,

$$E[e^{sZ_t}] = \sum_{n=0}^{\infty} E[e^{sZ_n}|N_t = n]P(N_t = n).$$

2. By the rule on iterated conditioning (CE9a),

$$E[e^{sZ_n}|N_t = n] = E\big[E[e^{sZ_n}|N_t, S_1, S_2, \ldots, S_n] \,\big|\, N_t = n\big].$$

Since the class $\{S_k, Y_i : 1 \leq k, 1 \leq i\}$ is independent, for $0 < t_1 < t_2 < \cdots < t_n \leq t$, we have by (CE11b),

$$E[e^{sZ_n}|N_t = n, S_1 = t_1, \ldots, S_n = t_n] = E\left[\prod_{k=1}^{n} e^{sg(t,t_k,Y_k)}\right]$$

$$= \phi_n(t, t_1, t_2, \ldots, t_n).$$

Hence

$$E[e^{sZ_n}|N_t = n] = E[\phi_n(t, S_1, \ldots, S_n)|N_t = n].$$

Since for each fixed t, ϕ_n is a symmetric function in the other variables, we have by Theorem 21.4.1,

$$E[\phi_n(t, S_1, \ldots, S_n)|N_t = n] = E[\phi_n(t, U_1, \ldots, U_n)]$$

$$= E\left[\prod_{k=1}^{n} e^{sg(t,U_k,Y_k)}\right]$$

where $\{U_i : 1 \leq i \leq n\}$ is iid, uniform on $(0, t]$.

3. By independence and the rule of exponents,

$$E\left[\prod_{k=1}^{n} e^{sg(t,U_k,Y_k)}\right] = \prod_{k=1}^{n} E[e^{sg(t,U_k,Y_k)}].$$

By the law of total probability (CE1b),

$$E[e^{sg(t,U_k,Y_k)}] = E\big[E[e^{sg(t,U_k,Y_k)}|U_k]\big] = \frac{1}{t}\int_0^t E[e^{sg(t,u,Y)}]\, du.$$

Thus, if we use $P(N_t = n) = e^{-\lambda t}(\lambda t)^n/n!$, we have

$$M_t(s) = e^{-\lambda t}\sum_{n=0}^{\infty}\frac{1}{n!}\left(\frac{\lambda t}{t}\int_0^t E[e^{sg(t,u,Y)}]\, du\right)^n$$

$$= e^{-\lambda t}\exp\left(\lambda\int_0^t E[e^{sg(t,u,Y)}]\, du\right).$$

Since $\int_0^t 1\, du = t$, we have $e^{-\lambda t} = \exp(\lambda\int_0^t 1\, du)$. Upon adding exponents, we obtain the desired result for the moment-generating function $M_t(s)$.

4. Evaluation of $M_t'(0)$ gives the desired expression for $E[Z_t]$. Similarly, $M_t''(0)$ may be used to obtain $E[Z_t^2]$, which then leads to the desired expression for $\mathrm{Var}[Z_t]$.

—   □

21.5.   Nonhomogeneous Poisson Processes

The homogeneous Poisson process has increments that are both independent and stationary. Stationarity means that the arrival statistics do not change with time. In many systems, this condition is not met. In the following treatment, we consider Poisson processes whose increments need not be stationary.

Definition 21.5.1. *If N is a counting process, the* mean-value function m *is given by*

$$m(t) = E[N_t], \quad for\ 0 \le t.$$

Now, m is nonnegative, nondecreasing on $[0, \infty)$ and $E[N_b - N_a] = m(b) - m(a)$ for all $0 \le a \le b$.

Definition 21.5.2. *A counting process N is* nonhomogeneous Poisson *iff*

1. m is continuous on $[0, \infty)$ and $m(0) = 0$.

2. N_t is Poisson $(m(t))$ for all $t \ge 0$.

3. The increments of N are independent.

As in the homogeneous case, various axiom sets are possible. The following is often useful. For definitions of $\pi_1(t, t+h)$ and regularity, see Section 21.1.

An alternate Axiom Set for nonhomogeneous Poisson processes is:

1. $0 < m(t) < \infty$ for all $t > 0$.

2. $\pi_1(t, t+h) \to 0$ as $h \to 0+$, uniformly in t.

3. N is regular.

4. N has independent increments.

Proof

In the appendix, see the first remark after the proof that Axiom Set II implies Axiom Set III.

—   □

Theorem 21.5.3. *If N is a nonhomogeneous Poisson process with mean-value function m, then*

$$N_{t+h} - N_t \quad \text{is Poisson } \big(m(t+h) - m(t)\big).$$

Proof

We noted earlier that $E[N_{t+h} - N_t] = m(t+h) - m(t)$. It is an easy exercise with generating functions (see Problem 18-26) to show that if $\{X, Y\}$ is independent, Y is nonnegative and both X and $X + Y$ are Poisson, then Y is also Poisson. Apply this to $\{N_t, N_{t+h} - N_t\}$.

— $\square$

CHANGE OF TIME SCALE

Our next result shows that by an appropriate nonlinear change of time scale a nonhomogeneous process can be transformed into a homogeneous one. This leads to an important simulation technique.

Definition 21.5.4. *If N is a Poisson counting process with mean-value function m, we define the function τ by*

$$\tau(t) = \inf\{s : m(s) > t\} \qquad \text{for all } t \geq 0.$$

Note the similarity of this function to the quantile function introduced in Section 11.4. The difference is the use of $>$ in the definition of τ rather than the $\geq$ used for the quantile function. An analysis similar to that for the quantile function shows that (1) τ is right-continuous, strictly increasing, and (2) $m\big(\tau(t)\big) = t$ for all $t \geq 0$.

The desired transformation is displayed in the following

Theorem 21.5.5. *Suppose N is Poisson with mean-value function m. Define the counting process N^* by*

$$N_t^* = N_{\tau(t)}, \qquad \text{where} \quad \tau(t) = \inf\{s : m(s) > t\}.$$

Then $S_n^* = m(S_n)$ is the nth arrival time and N^* is homogeneous, Poisson with rate $\lambda = 1$.

Proof

1. The nth arrival S_n^* for N^* has the value t iff the nth arrival S_n for N has the value $\tau(t)$. By the second property of τ,

$$S_n^* = t \quad \text{iff} \quad S_n = \tau(t) \quad \text{iff} \quad m(S_n) = t.$$

2. Since τ is right-continuous, strictly increasing, the interval $(a, b]$ is transformed into the interval $(\tau(a), \tau(b)]$ and disjoint intervals go into disjoint intervals. Hence, intervals of N^* are independent and

$$N_b^* - N_a^* = N_{\tau(b)} - N_{\tau(a)} \quad \text{is Poisson } \big(m(\tau(b)) - m(\tau(a))\big) = b - a.$$

— $\square$

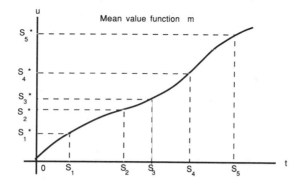

FIGURE 21.1. Construction for arrival times for nonhomogeneous Poisson process.

SIMULATION

Suppose it is desired to simulate the arrivals for a nonhomogeneous Poisson process with mean-value function m. We suppose that m is strictly increasing, as is usually the case. This implies that m is one-one, so that $\tau = m^{-1}$, the inverse of m. If we obtain a set of arrival times S_k^* for N^*, then the quantities $S_k = \tau(S_k^*)$ are the corresponding arrival times for N.

To obtain a sample of n arrival times for N^*, we obtain a sample of interarrival times and combine them additively. That is, we take a sample of size n from an iid set of random variables exponential (1). To do this, we do the following.

1. Take a sample $u_1, u_2, \ldots, u_n$ from a uniform distribution by the use of a random-number generator (see Example 11.4.1).

2. Then, $w_i^* = -\ln(1 - u_i)$, $1 \le i \le n$ is a corresponding sample from a population with distribution exponential (1).

3. Set $s_i^* = \sum_{k=1}^{i} w_k^*$. Then $s_1^*, s_2^*, \ldots, s_n^*$ are the first n arrival times for N^*, Poisson with rate 1.

4. Finally, we transform the s_i^* into $s_i = \tau(s_i^*) = m^{-1}(s_i^*)$. This may be done graphically, as indicated in Figure 21.1, or analytically, as in the following example.

Example 21.5.1:

A nonhomogeneous Poisson process has mean-value function

$$m(t) = \begin{cases} \lambda t^{3/2} & \text{for } 0 \le t \le 4 \\ \lambda\big(3(t-4) + 8\big) & \text{for } 4 < t \end{cases}.$$

Suppose $\lambda = 1/2$. For the following sample, size $n = 8$, of random numbers, uniform on $(0, 1)$, determine the corresponding arrival times s_i for N.

0.8095, 0.7025, 0.0654, 0.1461, 0.5312, 0.9241, 0.7508, 0.4901.

Solution

We assume the random numbers u_1, u_2, ..., u_8 constitute a sample of size eight from a population uniform on $(0, 1)$. The numbers $w_i^* = -\ln(1 - u_i)$ form a corresponding sample from a population exponential (1). The sums $s_i^* = \sum_{k=1}^i w_k^*$ are the corresponding arrival times for N^*, Poisson with rate 1. To complete the problem, we must have the inverse function for m.

If $u = m(t)$, we note that $m(0) = 0$ and $m(4) = 4$. Thus

$$t = (2u)^{2/3} \quad \text{for } 0 \le u \le 4 \quad \text{and} \quad t = \frac{2}{3}(u + 2) \quad \text{for } 4 \le u.$$

The results of the indicated calculations are tabulated as follows:

$i =$	1	2	3	4	5	6	7	8
$u_i =$	0.8095	0.7025	0.0654	0.1461	0.5312	0.9241	0.7508	0.4901
$w_i^* =$	1.6581	1.2123	0.0676	0.1579	0.7576	2.5783	1.3895	0.6735
$s_i^* =$	1.6581	2.8704	2.9380	3.0959	3.8535	6.4318	7.8213	8.4948
$s_i =$	2.224	3.206	3.256	3.372	3.902	5.621	6.548	6.997

The numbers in the final row constitute the sample of eight arrival times for N.

—   ☐

For the homogeneous Poisson process, the rate, or arrival rate, is a constant. The notion of arrival rate can be extended to the nonhomogeneous case as follows:

Definition 21.5.6. *If the mean-value function m is differentiable, then $r(t) = m'(t)$ is the* arrival rate *at time t.*

Remark: . If the rate is specified, the mean-value function may be determined by integration.

Example 21.5.2:

Suppose $r(t) = 5000te^{-3t}$. The maximum value is $(5000/3e) \approx 613$ at $t = 1/3$. $m(t) = E[N_t] = \int_0^t r(u)\,du = 5000 \int_0^t ue^{-3u}\,du = (5000/9)\big(1 - e^{-3t}(3t + 1)\big)$. For large values of t, we have $m(t) \approx m(\infty) = 5000/9 \approx 556$.

For large t, N_t is approximately Poisson (556). Also, for large t, N_t is approximately normal $N\big(m(t), m(t)\big)$. For large n, the time of the nth arrival is S_n. Now

$$P(S_n \leq t) = P(N_t \geq n) \approx 1 - \Phi\left(\frac{n - m(t)}{\sqrt{m(t)}}\right).$$

□

21.6. Bibliographic Note

Because of its importance, the Poisson counting process is treated in many places, and fundamental results have been derived in a wide variety of ways. The theoretical treatments and applications in this and the preceding chapter draw from many sources. We note specifically only two of these sources that have contributed significantly.

The proof in the appendix that Axiom Set II for the homogeneous Poisson process implies Axiom Set III uses Lemma 11.4.2 (the coupling lemma) and Theorem 11.4.5. This treatment is patterned after a similar development by

> Timothy C. Brown: Poisson approximations and the definition of the Poisson process, *The American Mathematical Monthly*, 92 (2), Feb., 1984, 116–23.

This elegant treatment uses essentially probabilistic arguments to show how the Poisson distribution reflects certain simple structural features of the process.

A number of decomposition and superposition results are found in the applications literature. Among the more sophisticated are the patterns developed in Theorem 21.4.2 and its applications to the examples in Section 21.4. Our principal source for these particular results is

> Emanuel Parzen: *Stochastic processes*, Holden-Day, San Francisco, Calif., 1962.

We have adapted certain parts of Section 21.4–5 on "filtered Poisson processes." In particular, the proof of Theorem 21.4.2 uses the essential ideas of Parzen's proof of his theorem 5A. Parzen develops the topic more fully, with a variety of extensions and applications.

21a

21a.1. Independent Increments

The condition of independent increments is a reasonable assumption in a wide variety of systems modeled by random processes. As sketched in Section 21a.1, the notion is a simple one. Consider a random process $X_T = \{X_t : t \in T\}$. If $t, u \in T$ and $t < u$, then the random variable $X_u - X_t$ is an increment for the process. For simplicity of exposition, we suppose T is an interval on the real line. The notion of independent increments is the notion that for nonoverlapping subintervals of the parameter set the increments form an independent class.

There are some ambiguities in the literature because of various formulations of the concept. We consider the usual formulations and examine their relationships. To facilitate the formulation, we adopt the following terminology.

Definition 21a.1.1. *We say a finite subset $T_n = \{t_0, t_1, \ldots, t_n\}$ of the parameter set T is* strictly ordered *iff $t_0 < t_1 < \cdots < t_n$.*

We now consider four alternative definitions found in the literature. We suppose T is an interval and that subscripts refer to elements of T.

Definition 21a.1.2. *The random process $X_T = \{X_t : t \in T\}$ has independent increments iff for each strictly ordered, finite subset $T_n \subset T$ we have*

(A) $\{X_{t_1} - X_{t_0},\ X_{t_2} - X_{t_1}, \ldots,\ X_{t_n} - X_{t_{n-1}}\}$ is independent.

(B) $\{X_{t_0},\ X_{t_1} - X_{t_0},\ X_{t_2} - X_{t_1},\ \ldots,\ X_{t_n} - X_{t_{n-1}}\}$ is independent.

(C) $\{X_t - X_s, U_s\}$ is independent for all $s \leq t$, where $U_s = \{X_u : u \leq s\}$.

(D) If T has a first element t^* and $T_n = \{t^*, t_1, t_2, \ldots, t_n\}$, then $\{X_{t^*}, X_{t_1} - X_{t^*}, \ldots, X_{t_n} - X_{t_{n-1}}\}$ is independent.

The relationships between these conditions is as follows:

1. (B) and (C) are equivalent and they imply (A).

2. If T has a first element, then (B), (C), and (D) are equivalent, and they imply (A).

3. If T has a first element, with $X_{t^*} = 0$ a.s., then all four are equivalent.

Remarks

1. In many important cases, such as the Poisson counting process, it is assumed the first element is $t = 0$ and $X_0 = 0$. Thus, all of these characterizations are equivalent. That is, if any one holds, so do the others.

2. For many purposes, condition (C) is the most useful, and it is customary to take it as the defining condition.

3. Proofs of the relationships are rather easy consequences of properties of independent classes of random variables, although they are somewhat cumbersome to write out. We forgo the proofs.

21a.2.   Axiom Systems for the Homogeneous Poisson Process

The principal objective of this section is to establish the equivalence of four sets of conditions, any one of which ensures that a counting process is a homogeneous Poisson process. That is, we want to prove Theorem 21.1.1. As an initial step, we obtain an important result, Theorem 21.1.5, which shows that the homogeneous process starts over at any given time $t \geq 0$. For convenience, we restate these theorems.

Theorem 21a.2.1   (Axiom Sets for the Homogeneous Poisson Process) *For a counting process $N = \{N_t : 0 \leq t\}$, the following four sets of conditions are equivalent, in the sense that if any one set holds, so do the others.*

Axiom Set I

1. *The interarrival times $\{W_i : 1 \leq i\}$ are iid exponential (λ).*

Axiom Set II

1. *$N_0 = 0$ a.s. and $0 < E[N_t] < \infty$ for all $t > 0$.*

2. *There are no multiple arrivals (i.e., $W_i > 0$ a.s. for all $i \geq 1$).*

3. *The increments are independent.*

4. *The increments are stationary.*

Axiom Set III

1. *There is a positive number λ such that N_t is Poisson (λt) for all $t > 0$.*

2. *The increments are independent.*

3. *The increments are stationary.*

Axiom set IV

1. $p_1(t, t+h) = \lambda h + o(h)$ *as* $h \to 0+$.

2. $\pi_2(t, t+h) = o(h)$ *as* $h \to 0+$.

3. *The increments are independent.*

4. *The increments are stationary.*

— $\qquad\qquad\qquad\qquad\qquad\qquad\qquad\qquad\qquad\qquad$ □

To establish this theorem, we first establish Theorem 21a.2.2, which has Axiom set I as hypothesis. Then we show that

Theorem 21a.2.2 implies II implies III implies I, and III iff IV

Theorem 21a.2.2   (Waiting Times After t) *Suppose* $\{W_i : 1 \le i\}$ *is iid exponential* (λ).
For any fixed $t \ge 0$, *let* W_{t1}, W_{t2}, ... *be the waiting times (interarrival times) after* t. *Then*

a. *For any* $t \ge 0$, $\{W_{ti} : 1 \le i\}$ *is iid exponential* (λ).

b. *For* $0 < t_1 < \cdots, < t_m = t$,

$$\{(N_{t_1}, N_{t_2} - N_{t_1}, \ldots N_t - N_{t_{m-1}}), W_{t1}, W_{t2}, \ldots\} \quad \text{is independent.}$$

Proof
We first prove the theorem in the case $m = 2$.
    Consider any $0 < s < t$, any $j \ge 0$, $k \ge 0$, and any $r \ge 1$. Set $n = j + k$. Then the event

$$A = \{N_s = j, N_t - N_s = k, W_{t1} \le y_1, \ldots, W_{tr} \le y_r\}$$

may be expressed in terms of S_j, S_{j+1}, S_n, W_{n+1}, $\ldots$, W_{n+r}, as follows:

$$
\begin{aligned}
A \;=\; & \{(S_j, S_{j+1}, S_n) \in Q, \\
& \quad t < S_n + W_{n+1} \le t + y_1, W_{n+2} \le y_2, \ldots, W_{n+r} \le y_r\} \\
=\; & B \cap \{W_{n+2} \le y_2, \ldots, W_{n+r} \le y_r\}
\end{aligned}
$$

where $Q = \{(v, w, x) : v \le s < w, x \le t\}$. Also we have

$$C = \{N_s = j, N_t - N_s = k\} = \{(S_j, S_{j+1}, S_n) \in Q, t < S_n + W_{n+1}\}.$$

Because of the iid character of the interarrival times, if F_W is the common distribution function,

$$P(A) = P(B)F_W(y_2)F_W(y_3)\dots F_W(y_r).$$

Since r is arbitrary, we will have established (a) and (b) if we can show that $P(B) = P(C)F_W(y_1)$. If the joint density for (S_j, S_{j+1}, S_n) is f, then the joint density for $(S_j, S_{j+1}, S_n, W_{n+1})$ is ff_W. The probabilities $P(B)$ and $P(C)$ are the integrals of ff_W over the appropriate regions in four-space. If (v, w, x, y) represents a point in this space, the first three coordinates are restricted to Q.

- For $P(B)$, for each x we have $t < x+y \le t+y_1$ or $t-x < y \le t-x+y_1$.

- For $P(C)$, for each x we have $t < x + y$ or $t - x < y$

Thus,

$$P(B) \quad = \quad \iiint_Q f \int_{t-x}^{t-x+y_1} f_W \, dy \, dv \, dw \, dx$$

$$P(C) \quad = \quad \iiint_Q f \int_{t-x}^{\infty} f_W \, dy \, dv \, dw \, dx.$$

Elementary integration and use of properties of the exponential show that

$$\int_{t-x}^{t-x+y_1} \lambda e^{-\lambda y} \, dy = (1-e^{-\lambda y_1})e^{-\lambda(t-x)} \quad \text{and} \quad \int_{t-x}^{\infty} \lambda e^{-\lambda y} \, dy = e^{-\lambda(t-x)}.$$

Since the factor $(1 - e^{-\lambda y_1})$ may be treated as a constant, we have

$$P(C) = \iiint_Q f e^{-\lambda(t-x)} \, dv \, dw \, dx \quad \text{and} \quad P(B) = (1 - e^{-\lambda y_1})P(C).$$

The case $m = 1$ and the general case are proved in a parallel manner. In each case, the difference between the integral for $P(B)$ and that for $P(C)$ is the restriction on the last coordinate. This is the coordinate for the density for the waiting time W_{n+1}. In each case Q is the same for both $P(B)$ and $P(C)$.

—   □

Theorem 21a.2.2 implies Axiom Set II

1. By Theorem 20.6.1, N_t is Poisson (λt), so that the finite expectation is assured. Also, the a.s. positivity of the interarrival times ensures no multiple arrivals. We need only establish the fact that the increments are independent and stationary.

2. For any $0 < s < t$, $N_t - N_s = k$ is the same function of W_{s1}, W_{s2}, ... as $N_{t-s} = k$ is of W_1, W_2, ... Thus, Theorem 21a.2.2 ensures that $N_t - N_s$ is Poisson $(\lambda(t - s))$, and that $\{N_s, N_t - N_s\}$ is independent.

3. With the aid of Theorem 21a.2.2, we set up an inductive argument to establish the independence of the increments. Suppose

$$0 < t_2 < \cdots < t_k = t < u$$

and

$$\{N_{t_1}, N_{t_2} - N_{t_1}, \ldots, N_t - N_{t_{k-1}}\} \text{ is independent.}$$

As in 2,

$$\{(N_{t_1}, N_{t_2} - N_{t_1}, \ldots, N_t - N_{t_{k-1}}), N_s - N_t\} \qquad \text{is independent.}$$

An elementary fact about independence ensures that the entire set of $k + 1$ increments must be independent. The independence of any finite number of increments follows by mathematical induction.

Axiom Set II implies Axiom Set III

1. The condition $E[N_t] < \infty$ implies $N_t(\omega) < \infty$ for all ω except possibly a negligible set having probability zero. This is usually written $N_t < \infty$ a.s. This condition, along with condition (2), ensures that for almost every outcome ω there will be a finite number of discrete arrivals in the interval $(0, t]$. For each such ω, there will be a minimum interarrival time among this finite set of interarrival times. If D_t is the random variable whose value for any outcome ω is the minimum interarrival time in the interval $(0, t]$, then condition (2) requires that $D_t > 0$ a.s., so that $P(D_t < t/k) \to 0$ as $k \to \infty$.

2. We consider a sequence of partitions of the interval $(0, t]$ in which the nth partition subdivides the interval into 2^n equal subintervals $I_{in} = (t_{i-1,n}, t_{i,n}]$, each of length $t/2^n$.

   Let A_{in} = the event of one or more arrivals in I_{in}, and set $p_{in} = P(A_{in})$.

3. Consider $Z_n = \sum_{i=1}^{2^n} I_{A_{in}}$ = the number of subintervals in which there is at least one arrival.

   For any outcome ω yielding a finite number of arrivals, the number of subintervals having one or more arrivals may increase, but can never decrease, as the subdivisions are made finer by increasing n. Also, for large enough n, there will be at most one arrival in any subinterval. This means that the number of subintervals having arrivals is equal to the total number of arrivals. We may express these facts symbolically by asserting

$$Z_n \leq Z_{n+1} \quad \text{for all } n \geq 1 \qquad \text{and} \qquad Z_n \to N_t \quad \text{a.s.}$$

By the monotone convergence theorem

$$E[Z_n] = \sum_{i=1}^{2^n} p_{in} \to E[N_t] = m(t).$$

We note that $m(t) = E[N_t]$ must be nondecreasing with increasing t.

4. Let Y_t be Poisson $(m(t))$. Then, by the triangle inequality for absolute value

$$|P(N_t \in M) - P(Y_t \in M)| \leq |P(N_t \in M) - P(Z_n \in M)| \\ + |P(Z_n \in M) - P(Y_t \in M)|.$$

5. By the coupling lemma 11.4.2,

$$|P(N_t \in M) - P(Z_n \in M)| \leq P(N_t \neq Z_n).$$

Now $N_t \neq Z_n$ iff at least one subinterval has more than one arrival. This requires that the minimum distance between arrivals be less than the subinterval length. Hence,

$$P(N_t \neq Z_n) \leq P(D_t < t/2^n) \to 0 \qquad \text{as } n \to \infty.$$

6. Because of the independent increments, the class $\{A_{in} : 1 \leq i \leq 2^n\}$ is independent for each $n \geq 1$. By Theorem 11.4.5, with $\mu = m(t)$ and $\lambda = \sum_{i=1}^{2^n} p_{in} = E[Z_n]$, we have

$$|P(Z_n \in M) - P(Y_t \in M)| \leq |m(t) - E[Z_n]| + \sum_{i=1}^{2^n} p_{in}^2.$$

7. By stationary increments, $p_{in} = p_n$ for all i. By 3, $E[Z_n] = 2^n p_n \to m(t)$. Hence,

$$\sum_{i=1}^{2^n} p_{in}^2 = 2^n p_n^2 = 2^{-n} E^2[Z_n] \to 0 \cdot m^2(t) = 0.$$

We have thus established that N_t is Poisson $(m(t))$.

8. By stationary increments

$$m(t + u) = E[N_{t+u}] = E[N_{t+u} - N_t] + E[N_t] = m(u) + m(t).$$

By Cauchy's theorem (see the appendix on mathematical aids), we must have

$$m(t) = m(1)t = \lambda t \qquad \text{for } t > 0.$$

$\square$

Remarks

1. If we make the reasonable assumption that for each $t > 0$,

$$P(N_{u+h} - N_u \geq 1) \to 0 \qquad \text{as } h \to 0, \text{ uniformly in } u \text{ on } [0, t]$$

then $p_n = \max\{p_{in} : 1 \leq i \leq 2^n\} \to 0$ as $n \to \infty$, so that we do not need stationary increments in step 7 to ensure that $\sum_i p_{in}^2 \to 0$. Without stationary increments, we do not have $m(t) = \lambda t$. In this case, the Poisson process is nonhomogeneous.

2. At the expense of some refinement of the argument showing $2^n p_n$ is bounded, the condition $E[N_t] < \infty$ may be replaced by the slightly less restrictive assumption $N_t < \infty$ a.s.

Axiom Set III implies Axiom Set I

1. First, we obtain the joint density for $(S_1, S_2, \ldots, S_n)$. We identify the region Q in which the probability mass must lie. Then we show that for any n-dimensional rectangle R in Q, the probability is the integral of a certain function over that rectangle. From this, we infer that the function must be the desired density function (an intuitive fact established rigorously by the use of fundamental measure theory). Since $0 < S_1 < S_2 \ldots$, the density must be zero outside the region Q in euclidean n-space, where

$$Q = \{t : 0 < t_1 < t_2 < \cdots < t_n\}.$$

Let $R = \{\mathbf{t} : a_i < t_i \leq b_i, 1 \leq i \leq n\}$ be a rectangle. In order for R to be in Q, we must have

$$0 < a_1 < b_1 < a_2 < b_2 < \cdots < a_n < b_n.$$

A sketch in two dimensions will illustrate the pattern (see Figure 21a.2). Now $(S_1, \ldots, S_n) \in R$ iff $a_i < S_i \leq b_i$ for $1 \leq i \leq n$. These conditions hold iff

$$N_{a_1} = 0, \ N_{b_1} - N_{a_1} = 1, \ \ldots, \ N_{a_n} - N_{b_{n-1}} = 0, \ N_{b_n} - N_{a_n} \geq 1.$$

To establish this assertion, we argue as follows

$N_{a_1} = 0$ requires $S_1 > a_1$.
$N_{b_1} - N_{a_1} = 1$ requires $a_1 < S_1 \leq b_1 < S_2$.
$N_{a_2} - N_{b_1} = 0$ requires $S_2 > a_2$.
$N_{b_2} - N_{a_2} = 1$ requires $a_2 < S_2 \leq b_2 < S_3$.

$\vdots$

$N_{a_n} - N_{b_{n-1}} = 0$ requires $S_n > a_n$.
$N_{b_n} - N_{a_n} \geq 1$ requires $a_n < S_n \leq b_n$.

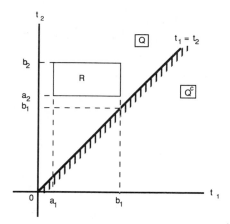

FIGURE 21a.2. Conditions for rectangle R to be a subset of region Q $(n = 2)$.

We may, of course, have $a_n < S_{n+k} \leq b_n$ for one or more positive k, since this does not affect the condition

$$(S_1, S_2, \ldots, S_n) \in R.$$

By the independence and the Poisson distribution of the increments

$$
\begin{aligned}
P\big((S_1, \ldots, S_n) \in R\big) \\
= \quad & e^{-\lambda a_1} \lambda (b_1 - a_1) e^{-\lambda(b_1 - a_1)} e^{-\lambda(a_2 - b_1)} \\
& \ldots e^{-\lambda(a_n - b_{n-1})} \big(1 - e^{-\lambda(b_n - a_n)}\big) \\
= \quad & \lambda^{n-1} \prod_{i=1}^{n-1} (b_i - a_i)(e^{-\lambda a_n} - e^{-\lambda b_n})
\end{aligned}
$$

Repeated integration shows this last expression is equal to

$$\int_{a_1}^{b_1} \int_{a_2}^{b_2} \cdots \int_{a_n}^{b_n} \lambda^n e^{-\lambda t_n} \, dt_n \ldots dt_1.$$

2. We may now use standard techniques for functions of random variables to obtain the joint distribution for $(W_1, \ldots, W_n)$.

$$P^* = P(W_1 \leq a_1, \ldots, W_n \leq a_n) = P\big((S_1, \ldots, S_n) \in Q^*\big)$$

where Q^* is the subset of Q such that

$$Q^* = \{ \boldsymbol{t} : 0 < t_1 \leq a_1, t_1 < t_2 \leq t_1 + a_2, \ldots, t_{n-1} < t_n \leq t_{n-1} + a_n \}.$$

Thus

$$P^* = \int_0^{a_1} \int_{t_1}^{t_1+a_2} \cdots \int_{t_{n-1}}^{t_{n-1}+a_n} \lambda^n e^{-\lambda t_n} \, dt_n \cdots dt_1.$$

Integrating with respect to t_n gives

$$P^* = (1 - e^{-\lambda a_n}) \int_0^{a_1} \cdots \int_{t_{n-2}}^{t_{n-2}+a_{n-1}} \lambda^{n-1} e^{-\lambda t_{n-1}} \, dt_{n-1} \cdots dt_1.$$

Repeated integration gives

$$P^* = \prod_{i=1}^{n} (1 - e^{-\lambda a_i}).$$

Since this probability is the value of the joint distribution function at $(a_1, a_2, \ldots, a_n)$, we can assert that the interarrival times form an iid, exponential (λ) class.

□

Axiom Set III is equivalent to Axiom Set IV

1. Examination of the probabilities $P(N_t = k)$ shows that Axiom Set III implies Axiom set IV. We leave the details to an exercise.

2. Stationarity of the increments implies that (1) and (2) hold uniformly in t.

3. For any interval $E = (t, t+h]$, we have $p_0(E) = 1 - p_1(E) - \pi_2(E) = 1 - \lambda h + o(h)$.

4. The generating function g_E for $N(E) = N(t, t+h] = N_{t+h} - N_t$ is

$$g_E(s) = 1 - \lambda h + \lambda h s + o(h) = 1 + \lambda h(s - 1) + o(h) = 1 + z.$$

5. For any positive integer n, let

$$E = (t, t+h] = \bigcup_{i=1}^{n} E_i \quad \text{where} \quad E_i = (t_{i-1}, t_i], \quad \text{with} \quad t_i - t_{i-1} = \frac{h}{n}.$$

By independent increments,

$$\ln g_E(s) = \sum_{i=1}^{n} \ln g_{E_i}(s) = \sum_{i=1}^{n} \ln(1 + z_i),$$

where

$$z_i = \lambda \frac{h}{n}(s - 1) + o\left(\frac{h}{n}\right).$$

6. For any $\epsilon : 0 < \epsilon < 1/4$, there is an $n = n(h)$ such that

$$\lambda \frac{h}{n} < \epsilon < \frac{1}{4} \quad \text{and} \quad \left| o\left(\frac{h}{n}\right) \right| < \epsilon \lambda \frac{h}{n} < \epsilon^2 < \frac{1}{16}.$$

7. Now $\ln(1 + z_i) = z_i + R_i$, so that

$$\ln g_E(s) = \sum_{i=1}^{n} \lambda \frac{h}{n}(s - 1) + \sum_{i=1}^{n} o\left(\frac{h}{n}\right) + \sum_{i=1}^{n} R_i.$$

8. For $|s| < 1/2$, we have $|s - 1| < 3/2$ and

$$|z_i| = \left| \lambda \frac{h}{n}(s - 1) + o(h) \right| < \frac{3\lambda}{2} \cdot \frac{h}{n} + \frac{\lambda}{4} \cdot \frac{h}{n} < 2\lambda \frac{h}{n} < \frac{1}{2}.$$

By a simple analytical lemma (see the appendix on mathematical aids),

$$|R_i| \leq |z_i|^2 < 4\lambda^2 \frac{h^2}{n^2} < 4\epsilon\lambda \frac{h}{n}.$$

We thus have

$$|\ln g_E(s) - \lambda h(s - 1)| \leq \epsilon\lambda h + 4\epsilon\lambda h = 5\epsilon\lambda h.$$

9. Since the latter expression holds for any ϵ with $0 < \epsilon < 1/4$, we conclude that

$$\ln g_E(s) = \lambda h(s - 1) \quad \text{or, equivalently,} \quad g_E(s) = e^{\lambda h(s-1)}.$$

The form of g_E shows that $N(E)$ is Poisson (λh), which is sufficient to establish Axiom Set III.

□

PROBLEMS

21–1. Show that Axiom Set III implies Axiom Set IV.

21–2. A homogeneous Poisson process N has rate $\lambda = 3$. Because of independent increments, for large λt, N_t is approximately normal $N(\lambda t, \lambda t)$. Use the central limit theorem to approximate $P(50 \leq N_{20} \leq 60)$.

21–3. Show that if N_t is Poisson (λt) for all $t > 0$, and T is independent of the arrival process, then

$$g_{N_T}(s) = M_T\big(\lambda(s - 1)\big).$$

Compare with the result of Theorem 20.7.2 when W_i is exponential (λ) and T is exponential (ν).

21–4. Interarrival times W_i (in minutes) of passengers at a bus stop form an iid class, exponential with mean value 1, so that N_t is Poisson (t). A bus leaves at $t = 0$; the next arrives at time T minutes later.

Suppose T is uniformly distributed $[6, 8]$, independent of the arrivals. Determine

$$E[N_T], \qquad \text{Var}[N_T], \qquad P(8 \leq N_T \leq 10), \qquad \text{and} \qquad g_{N_T}(s).$$

21–5. Customers arrive in a shop according to a Poisson arrival process with rate $\lambda = 10$ (per hour). The customers buy independently. The ith customer buys an amount Y_i, in dollars, where Y_i is approximately exponential (0.023). The shop is open from 10:00 A.M. to 9:00 P.M. (eleven hours). For purpose of analysis, the customers are put in categories, according to the amount of purchase:

1. Less than $10.
2. Between $10 and $50.
3. Between $50 and $100.
4. More than $100.

(a) What is the expected number of sales of $50 or more?

(b) What is the probability of four or more sales of $100 or greater?

(c) What is the probability of 25 or more sales less than $10 (including customers who buy nothing)?

21–6. Customers arrive at a department store according to a Poisson process N with rate $\lambda = 100$ (per hour). The probability that any customer is female is 0.6. The probability that any female customer makes a purchase is 0.8; the probability that any male customer makes a purchase is 0.7. Use the model for arrivals of m kinds, with $m = 3$: nonbuyers, female buyers, and male buyers, with

$$p_1 = 0.3 \cdot 0.4 + 0.2 \cdot 0.6 = 0.24, \qquad p_2 = 0.8 \cdot 0.6 = 0.048,$$

and

$$p_3 = 0.7 \cdot 0.4 = 0.28.$$

Let Y_i be the amount purchased by the ith male buyer and X_j be the amount purchased by the jth female buyer. Suppose $\{Y_i : 1 \le i\}$ iid uniform on $(10, 100)$ and $\{X_j : 1 \le j\}$ iid uniform on $(5, 75)$, with $\{N, Y_i, X_j : 1 \le i, 1 \le j\}$ independent.

(a) What is the probability that in any given fifteen-minute period at least 12 buyers will be women and not more than 8 buyers will be men?

(b) What is the expected total purchase in a ten-hour day?

(c) What is the probability the minimum purchase by both men and women buyers in any one-hour period will be at least fifteen dollars?

21–7. Passengers arrive at a bus stop with iid exponential (λ) interarrival times. The sum of the passenger waiting times until time t is $D(t) = \sum_{i=1}^{N_t}(t - S_i)$. If the arrival time of the bus is T, independent of the passenger arrival times, determine the expected total waiting time $E[D(T)]$.

21–8. For the shop in problem 21-5, the time (in hours) the ith individual spends in the store is a random quantity T_i. The T_i form an iid class, each distributed uniformly on $[0, 1]$. Let Z_t be the number of customers in the store at time t (in hours after opening). Determine $E[Z_3]$ and $P(Z_t \ge 3)$ for $t = 1/2$, 1, and 5.

21–9. Customers arrive in a bookstore according to a Poisson arrival process with rate $\lambda = 9$ (per hour). The time (in hours) the ith individual spends in the store is a random quantity Y_i. The Y_i form an iid class, with common density function $f_Y(t) = 2(1 - t)$ for $0 \le t \le 1$. Let Z_t be the number of customers in the store at time t (in hours after opening). Determine $E[Z_2]$ and $P(Z_t \ge 5)$ for $t \ge 1$.

21–10. A large minicomputer has numerous terminals. On occasion, a large, top-priority job ties up the CPU for a time T (seconds). During this time, other jobs are queued up for processing. Suppose jobs arrive according to a Poisson process with rate $\lambda = 0.5$. This means inter-arrival times are iid, exponential (0.5). The individual job times may be considered exponential (10). Suppose T is exponential (0.05) and

$\{T, W_i, Y_j : 1 \le i, 1 \le j\}$ is an independent class. Let D be total backlog demand time accumulated during time T.

(a) Use the moment-generating function $M_D(s)$ to determine $E[D]$ and $\text{Var}[D]$.

(b) What is the probability D is no greater than $2E[D]$?

21–11. Individual suppliers of a perishable food item arrive at a collection center according to a Poisson process with rate λ (per day). Individual lots are stored until a wholesale buyer arrives to take the entire stock. However, the stock deteriorates in a manner to make the value decrease exponentially in time. Let D_i be the value of the ith lot at the time of receipt. If the buyer arrives at time t, then the value of the accumulated food is

$$D(t) = \sum_{i=1}^{N_t} D_i e^{-\alpha(t - S_i)}.$$

Let T be the arrival time of the buyer. We suppose $\{D_i : 1 \le i\}$ is i.i.d., and the class $\{D_i, N_t, T : 1 \le i, t \ge 0\}$ is independent.

(a) Show that the expected value $E[D(T)]$ of the produce at the time of arrival of the buyer is given by

$$E[D(T)] = E[D]\frac{\lambda}{\alpha}\left(1 - M_T(-\alpha)\right)$$

where M_T is the moment-generating function for T.

(b) Suppose $E[D] = 250$ (dollars), $\alpha = 0.1$, and $\lambda = 12$. If the arrival time of the buyer is uniform on 1, 2, 3, 4 (days), determine the expected value $E[D(T)]$ of the produce at the time of arrival of the buyer. Compare this with $E[D]E[N_T]$.

21–12. Consider a nonhomogeneous Poisson process with $\lambda(t) = m'(t) > 0$ for all $t > 0$. Define the new process N^* by $N_t^* = N\left(m^{-1}(t)\right)$. Show that N^* is homogeneous with rate $\lambda = 1$.

21–13. Let N be a nonhomogenous Poisson process with mean-value function $m(\cdot)$. Show that the conditional distribution of arrival times, given n arrivals in the time interval $(0, t]$ is iid, with common distribution function

$$F(u) = \begin{cases} m(u)/m(t) & \text{for } 0 \le u \le t \\ 1 & \text{for } t < u \end{cases}.$$

That is

$$P\left(\bigcap_{k=1}^{n}\{S_k \le t_k\} \,\middle|\, N_t = n\right) = \prod_{k=1}^{n} P\left(U_k \le \frac{m(t_k)}{m(t)}\right) = \prod_{k=1}^{n} P(V_k \le t_k)$$

where $\{V_i : 1 \leq i \leq n\}$ is iid with common distribution function F defined earlier.

21–14. Show that $N_t + 1$ is a Markov time for the arrival process.

22

Conditional Independence, Given a Random Vector

In Chapter 5, we extend the concept of stochastic independence to include conditional independence, given an event. A variety of examples demonstrate the usefulness of this concept. In Chapter 19, we move from conditioning by an event to conditioning by a random vector, to obtain the powerful concept of conditional expectation, given a random vector. Conditional expectation includes, as a special case, conditional probability, given a random vector. In the present chapter, we take the additional step of formulating the notion of conditional independence, given a random vector.

Historically, conditional independence has been associated with the study of Markov processes. Such random processes are characterized by conditional independence of the past and future, given the present. This conditional independence models the fact that the system represented is memoryless. The likelihood of any event in the future depends only on the present state of the system and not how the present state was attained. The future is affected stochastically by the past only through the present. Modern treatments of Markov processes examine the consequences of conditional independence for system behavior. Such treatments are usually quite abstract, and conditional independence is expressed in terms of conditional expectation, given a sigma algebra of events. In Chapter 23, we reformulate some of the essential features of that theory in terms of the more readily accessible concept of conditional independence, given a random vector.

Recent investigations show that the condition is encountered in many other applications of probability. For example, the analysis of many decision processes assumes a Markov condition, although this assumption is often a tacit one. In the present chapter, we illustrate the importance of conditional independence by formulating the Bayesian approach to statistics and by setting up some elementary decision models. In order to do this, we first introduce the concept and establish some of its more useful properties.

22.1. The Concept and Some Basic Properties

Although there seems to be no historical connection, it may be instructive to see how conditional independence, given a random vector, can be viewed as an extension of the simpler concept of conditional independence, given an event. Suppose $\{A, B\}$ ci$|$ C. If $A = X^{-1}(M)$ and $B = Y^{-1}(N)$, then the product rule

$$P(AB|C) = P(A|C)P(B|C)$$

may be expressed

$$E[I_M(X)I_N(Y)|C] = E[I_M(X)|C]E[I_N(Y)|C].$$

If this rule holds for all Borel sets M and N on the respective codomains of X and Y, we should be inclined to say the pair of random vectors $\{X, Y\}$ is conditionally independent, given the event C.

Now suppose Z is a simple random variable and z_k is a possible value. Let $C_k = \{Z = z_k\}$, the event that Z takes on the value z_k. If the product rule $E[I_M(X)I_N(Y)|C_k] = E[I_M(X)|C_k]E[I_N(Y)|C_k]$ holds for all Borel sets M and N in the respective codomains of X and Y and for all z_k in the range of Z, we should then be inclined to say the pair $\{X, Y\}$ is conditionally independent, given Z. To examine this possibility further, we write $Z = \sum_k z_k I_{C_k}$ (canonical form). Using the result of Example 19.3.1, we have

$$E[I_M(X)I_N(Y)|Z] = \sum_k E[I_M(X)I_N(Y)|Z = z_k]I_{C_k},$$

with similar expressions for $E[I_M(X)|Z]$ and $E[I_N(Y)|Z]$. Since $I_{C_j}I_{C_k} = 0$ for $j \neq k$ and $I_{C_k}^2 = I_{C_k}$, we have

$$E[I_M(X)|Z]E[I_N(Y)|Z] = \sum_k E[I_M(X)|Z = z_k]E[I_N(Y)|Z = z_k]I_{C_k}.$$

We thus have

(A) $E[I_M(X)I_N(Y)|Z] = E[I_M(X)|Z]E[I_N(Y)|Z]$ iff

(B) $E[I_M(X)I_N(Y)|Z = z_k] = E[I_M(X)|Z = z_k]E[I_N(Y)|Z = z_k]$ for all k

This argument shows that condition (B) is a reasonable basis for the notion that $\{X, Y\}$ is conditionally independent, given simple random vector Z. The equivalent condition (A) proves to be the more useful way to characterize the condition, since it may be used in the general case. Subsequent development shows the following definition to be appropriate.

Definition 22.1.1. *A pair $\{X, Y\}$ of random vectors is said to be conditionally independent, given random vector Z, denoted $\{X, Y\}$ ci$|$ Z, iff the following product rule holds:*

$$E[I_M(X)I_N(Y)|Z] = E[I_M(X)|Z]E[I_N(Y)|Z] \text{ a.s. for all Borel } M \text{ and } N.$$

An arbitrary class $\{X_i : i \in J\}$ of random vectors is conditionally independent, given Z, iff such a product rule holds for each finite subclass of two or more members from the class.

Since X, Y, and Z are random vectors of arbitrary dimension, we are usually interested in conditional independence of pairs.

A FUNDAMENTAL INTERPRETATION

In the case of conditional independence of events, various equivalent conditions provide insight into the nature of the concept and clues to using it in applications. A similar situation holds for this extension to conditioning by random vectors. In the following table of properties, we have the property (special case of (CI6))

$$\{X,Y\} \text{ ci}|\ Z \qquad \text{iff} \qquad E[g(X)|Z,Y] = E[g(X)|Z] \text{ a.s. for all Borel } g.$$

We read this in terms of conditional expectation as a mean-square estimator. The best estimate of $g(X)$, given the value of Z, is not changed by additional knowledge of the value of Y. Conditioning by (Y, Z) produces essentially the same result as conditioning solely by Z. We say "essentially," since the equality holds almost surely, that is, with probability one.

A list of properties is provided, for ready reference, in the following table. Proofs, or discussions of the proofs, are found in the appendix to this chapter. The reader may want to peruse one of more of the proofs to see how these properties are based on the properties of expectation and of conditional expectation, developed in Chapters 14 and 19.

Any of the first eight properties, (CI1) through (CI8), could be used as the defining condition. The simplicity and symmetry of (CI1), as well as its similarity to the defining condition for independence of classes of random vectors, leads naturally to its choice.

A cursory examination of the table of properties shows that (CI5) is an extension of (CI1), since the latter is a special case of the former. Similarly, (CI6) is an extension of (CI3), which is an extension of (CI2), and (CI8) is an extension of (CI4). Property (CI9) is essentially a restatement of (CI5) in alternate, but often convenient, form. It is desirable to retain the simpler forms in the list of equivalents, since these are usually the easier to establish, either by assumption of reasonable conditions in a model or by deduction from other conditions. The fact that these simpler forms imply the others makes it possible to move directly to the stronger statements, once the simpler ones are established.

Properties (CI1) and (CI5) display the fundamental product rule. Properties (CI2), (CI3), and (CI6) exhibit the fact that if $\{X,Y\}$ ci$|\ Z$, then conditioning of a function of (X, Z) by Z is not enhanced by further conditioning by Y. This conditioning may be interpreted by recalling that $E[g(X, Z)|Z = u]$ is the best mean-square estimate of $g(X, Z)$ based on the knowledge that $Z = u$ (see Section 19.5). If we know the value of Z, we cannot improve our estimate by the additional knowledge of the value of Y. Given the value of Z, the likelihood of any event determined by X is not altered by knowledge of the value of Y. Because the condition is symmetric in X, Y, it is also true that if the value of Z is known, the best estimate of $h(Y, Z)$ will not be changed by obtaining knowledge of the value of X.

CONDITIONAL INDEPENDENCE, GIVEN A RANDOM VECTOR

Definition 22.1.2. *The pair* $\{X, Y\}$ *is conditionally independent, given* Z, *denoted* $\{X, Y\}$ *ci*$|$ Z, *iff the following product rule holds:*

$$E[I_M(X)I_N(Y)|Z] = E[I_M(X)|Z]E[I_N(Y)|Z] \text{ a.s. for all Borel sets } M \text{ and } N.$$

An arbitrary class $\{X_i : i \in J\}$ *of random vectors is conditionally independent, given* Z, *iff such a product rule holds for each finite subclass of two or more members from the class.*

Remark: The expression "for all Borel sets M and N," here and elsewhere, implies the sets are on the appropriate codomains. Also, the expressions "for all Borel functions g," etc. imply that the functions are real-valued, such that the indicated expectations are finite.

The following conditions are equivalent, in the sense that if any one of them holds, so do all of the others. Each is a necessary and sufficient condition that $\{X, Y\}$ ci$|$ Z.

CI1) $E[I_M(X)I_N(Y)|Z] = E[I_M(X)|Z]E[I_N(Y)|Z]$ a.s. for all Borel sets M and N.

CI2) $E[I_M(X)|Z, Y] = E[I_M(X)|Z]$ a.s. for all Borel sets M.

CI3) $E[I_M(X)I_Q(Z)|Z, Y] = E[I_M(X)I_Q(Z)|Z]$ a.s. for all Borel sets M and Q.

CI4) $E[I_M(X)I_Q(Z)|Y] = E\big[E[I_M(X)I_Q(Z)|Z] \big| Y\big]$ a.s. for all Borel sets M and Q.

_____

CI5) $E[g(X, Z)h(Y, Z)|Z] = E[g(X, Z)|Z]E[h(Y, Z)|Z]$ a.s. for all Borel functions g and h.

CI6) $E[g(X, Z)|Z, Y] = E[g(X, Z)|Z]$ a.s. for all Borel functions g.

CI7) For any Borel function g there exists a Borel function e_g such that

$$E[g(X, Z)|Z, Y] = e_g(Z) \quad \text{a.s.}$$

CI8) $E[g(X, Z)|Y] = E\big[E[g(X, Z)|Z] \big| Y\big]$ a.s. for all Borel functions g.

_____

CI9) $\{U, V\}$ ci$|$ Z for all $U = g(X, Z)$, $V = h(Y, Z)$, for any Borel functions g and h.

ADDITIONAL PROPERTIES OF CONDITIONAL INDEPENDENCE

CI10) $\{X, Y\}$ ci| Z implies $\{X, Y\}$ ci| $(Z, U), \{X, Y\}$ ci| (Z, V), and $\{X, Y\}$ ci| (Z, U, V), where $U = h(X)$ and $V = k(Y)$, with h and k Borel.

CI11) $\{X, Z\}$ ci| Y and $\{X, W\}$ ci| (Y, Z) iff $\{X, (Z, W)\}$ ci| Y.

CI12) $\{X, Z\}$ ci| Y and $\{(X, Y), W\}$ ci| Z implies $\{X, (Z, W)\}$ ci| Y.

CI13) $\{X, Y\}$ is independent and $\{X, Z\}$ ci| Y iff $\{X, (Y, Z)\}$ is independent.

CI14) $\{X, Y\}$ ci| Z implies $E[g(X, Y)|Y = u, Z = v] = E[g(X, u)|Z = v]$ a.s. $[P_{YZ}]$.

CI15) $\{X, Y\}$ ci| Z implies

$$
\text{(a)} \quad E[g(X, Z)h(Y, Z)] \quad = \quad E\big[E[g(X, Z)|Z]E[h(Y, Z)|Z]\big] \quad = \\
E[e_1(Z)e_2(Z)]
$$

$$
\text{(b)} \quad E[g(Y)|X \in M]P(X \in M) = E\big[E[I_M(X)|Z]E[g(Y)|Z]\big].
$$

CI16) $\{(X, Y), Z\}$ ci| W
iff $E[I_M(X)I_N(Y)I_Q(Z)|W] = E[I_M(X)I_N(Y)|W]E[I_Q(Z)|W]$ a.s. for all Borel sets M, N, and Q.

The pair (CI4) and (CI8) shows that the best estimate of $g(X, Z)$ based on the knowledge of Y is the same as the best estimate of $E[g(X, Z)|Z]$ based on the knowledge of Y. This property plays an essential role in the development of the theory of Markov processes.

The additional patterns (CI10) through (CI15) have been found useful in a variety of contexts. Some of these properties are used in subsequent developments and various exercises. They are collected in the table for ready reference. Should some variation be needed, the reader may find it helpful to examine the proofs in the appendix. These proofs may suggest how a given theorem may be extended or otherwise modified.

Before turning to applications, we use a simple example to show that conditional independence is not the same as independence. The example also demonstrates some typical calculations.

Example 22.1.1:

Suppose $\{X, Y\}$ ci| H, with

$$
f_{X|H}(t|u) = f_{Y|H}(t|u) = ue^{-ut} \quad \text{and} \quad f_H(u) = \lambda e^{-\lambda u} \quad t \geq 0, \ u \geq 0.
$$

Determine (a) $F_X(t)$ and (b) $F_{X|Y}(t|v) = P(X \leq t|Y = v)$.

Solution

To simplify writing, let $I_t = I_{(-\infty, t]}$. Then

$$
F_X(t) = E[I_t(X)] = E\big[E[I_t(X)|H]\big].
$$

Since

$$E[I_t(X)|H = u] = 1 - e^{-ut}, \qquad t \geq 0, \quad u \geq 0$$

we must have

$$F_X(t) = \int_0^\infty (1 - e^{-ut})\lambda e^{-\lambda u}\, du = 1 - \frac{\lambda}{\lambda + t} \qquad t \geq 0.$$

Upon differentiating by t, we get

$$f_X(t) = f_Y(t) = \frac{\lambda}{(\lambda + t)^2} \qquad t \geq 0.$$

We use (CI8) to assert

$$
\begin{aligned}
F_{X|Y}(t|v) &= E[I_t(X)|Y = v] = E\big[E[I_t(X)|H]\,\big|\,Y = v\big] \\
&= \int_0^\infty [1 - e^{-ut}] f_{H|Y}(u|v)\, du.
\end{aligned}
$$

Now

$$f_{H|Y}(u|v) = \frac{f_{Y|H}(v|u) f_H(u)}{f_Y(v)} = u(\lambda + v)^2 e^{-(\lambda+v)u}.$$

From this it follows that

$$F_{X|Y}(t|v) = 1 - (\lambda + v)^2 \int_0^\infty u e^{-(\lambda+v+t)u}\, du = 1 - \left(\frac{\lambda + v}{\lambda + v + t}\right)^2$$

for $t \geq 0$, $v \geq 0$. Since we do not have $E[I_t(X)|Y] = E[I_t(X)]$ a.s., it follows from (CE5) that $\{X, Y\}$ cannot be independent. Although conditionally independent, given H, the pair $\{X, Y\}$ is not independent. This is a consequence of the fact that both random variables are tied stochastically to H.

—   □

22.2.   The Bayesian Approach to Statistics

Classical statistics begins with the assumption of a population distribution, which describes how some attribute varies over a set of entities known as the population. The goal is to obtain information about this population distribution, by sampling, or some other suitable mode of experimentation. Typically, the distribution is assumed to belong to a specified class of distributions. Thus, it may be normal, exponential, binomial, Poisson, etc. These standard distributions are characterized by certain parameters. If the values of the parameters are determined, then the entire distribution is determined. Since it is rarely possible to evaluate the parameters exactly, the problem becomes one of estimating the values of the parameters on the basis of experimental evidence. A finite set of parameters can be viewed as the coordinates of a single vector-valued parameter θ. The value of θ is assumed fixed, although there is uncertainty about its value.

Suppose there is reason to believe the population has the normal, or Gaussian, distribution. The parameters are the mean value μ, and the standard deviation σ. There is uncertainty about the values; i.e., there is uncertainty about the value of the vector (μ, σ). Since μ is not a random variable, we cannot make statements such as "With probability p, μ lies between a and b." This leads to conceptual and practical difficulties in many important situations.

One attempt to solve the dilemma is the so-called "Bayesian" approach. In this view, the uncertainty about the population is modeled by considering it to be randomly selected from a class of populations whose distributions are characterized by the parameters under investigation.

1. The value of the parameter θ for the distribution of the selected population is a realization or experienced value of a parameter random variable (or random vector) H.

2. The population distribution is determined by the parameter value $H(\omega)$. This means that if random variable X has the distribution for the population selected, then (X, H) has a joint distribution, and for each possible value θ of the parameter random variable H, the population random variable X has a conditional distribution, given $H(\omega) = \theta$.

To implement a statistical analysis, we must characterize appropriately the joint distribution for (X, H). Further, we must clarify the notion of a random sample. Recall from Section 18.7 that a random sample of size n in classical statistics is an observation of an iid class $\{X_1, X_2, \ldots, X_n\}$ whose common distribution is the population distribution.

THE USUAL BAYESIAN APPROACH

1. Assume a conditional distribution for X, given $H = \theta$.

2. Assume a prior distribution for the parameter random vector H, using any available information, based on experiment or previous knowledge and experience.

3. Extend the notion of a random sampling process by assuming that the class $\{X_i : 1 \leq i\}$ is conditionally independent, given H, with all X_i having the same conditional distribution.

Since the last assumption involves conditional independence, we examine it more closely. Under a given value of H (i.e., for a given population distribution), the result of taking an observation of X_i does not affect and is not affected by the results of observing any combination of the other random variables. We have an "operational independence" in the sampling procedure; but the variables are not independent, for their distributions depend upon the parameter. The stochastic tie, however, is only through the random quantity H. This is precisely the relationship modeled by conditional independence. We find it convenient to adopt the following terminology:

Definition 22.2.1. *A class of random variables is* ciid, *given H, denoted $\{X_i : i \in J\}$ ciid| H, iff the class is conditionally independent, given H, and each random variable in the class has the same conditional distribution, given H.*

A random sampling process, given H, is a finite class of random variables that is ciid, given H.

Let us consider the conditional distribution function, given H. To simplify writing, put

$$W = (X_1, X_2, \ldots, X_n) \qquad \text{and} \qquad I_t = I_{(-\infty, t]}$$

$$
\begin{aligned}
F_{W|H}(t_1, t_2, \ldots, t_n|u) &= P(X_1 \leq t_1, X_2 \leq t_2, \ldots, X_n \leq t_n|H = u) \\
&= E[I_{t_1}(X_1)I_{t_2}(X_2)\ldots I_{t_n}(X_n)|H = u].
\end{aligned}
$$

Conditional independence yields the product rule

$$F_{W|H}(t_1, t_2, \ldots, t_n|u) = \prod_{i=1}^{n} E[I_{t_i}(X_i)|H = u] = \prod_{i=1}^{n} F_{X|H}(t_i|u).$$

Thus the conditional distribution function obeys a product rule analogous to that in the case of ordinary independence. It is easy to show (see Section 19.3) that the conditional density, if it exists, satisfies a corresponding product rule.

$$f_{W|H}(t_1, t_2, \ldots, t_n|u) = \prod_{i=1}^{n} f_{X|H}(t_i|u).$$

We illustrate these ideas by considering the important case of Bernoulli trials.

BERNOULLI TRIALS, GIVEN H

A sequence of "identical trials" is performed in an operationally independent manner.

Let E_i be the event of a "success" on the ith trial in the sequence. Set $X_i = I_{E_i}$.

Then X_i takes on the value 1 if event E_i occurs and takes on the value 0 if E_i fails to occur. In statistics, this represents the problem of determining a population proportion. A proportion p of a population has a given attribute or property. A sample of size n is taken, in an attempt to discover this proportion p. In a random sample, the probability of observing this attribute in any selected member of the population is p. Now, in many situations, the population proportion p may itself be a random quantity.

Example 22.2.1: The Population Proportion as a Random Quantity

A. The fraction of the voting population who favors a given candidate may be different at different times and in different circumstances.

B. The fraction of items coming off a production line that meets specifications depends upon factors that may vary in a "random" manner with time.

C. The fraction of potential customers who prefer a certain brand or model will vary with conditions.

D. The ratio of men to women students in a university may vary from year to year, so that the fraction of women students is not fixed.

— $\square$

In order to carry out a statistical analysis, we suppose

- The entire sampling operation constitutes one composite trial.

- The population proportion p is an unknown quantity, but its value does not change over the time the sample is taken. Thus, the probability of success on any component trial in the sample is the same.

This probability of success on any trial is a parameter representing the state of nature (i.e., the actual population proportion). In the Bayesian approach, we model the population proportion as the value of a parameter random variable H whose range is $(0, 1)$. The presumed "operational independence", which characterizes a random sample, is modeled as conditional independence, given H. During a run, with the value of H fixed at u, the result on any trial does not affect, nor is it affected by, the result of any other trial. We thus make the fundamental assumptions

$$\{X_i : 1 \leq i\} \text{ ciid} | H$$

with $P(E_i|H = u) = E[X_i|H = u] = e(u) = u$ a.s. $[P_H]$.

Under such assumptions for Bernoulli trials, given H, we may carry out an analysis parallel to that for an ordinary Bernoulli sequence. We must have $P(E_i^c|H = u) = 1 - u$.

If we let F_i be either E_i or E_i^c, the choice made arbitrarily for each i, then

$$P\left(\bigcap_{i=1}^{n} F_i \,\Big|\, H = u\right) = \prod_{i=1}^{n} P(F_i | H = u).$$

Thus, for a given u, the Bernoulli analysis goes through as in the ordinary Bernoulli case, except that we write u for p. To obtain the conditional probability, given $H = u$, include a factor u for each uncomplemented E_i and a factor $1 - u$ for each E_i^c.

The random variable $S_n = \sum_{i=1}^{n} X_i$ is the number of successes in n trials. We have

$$P(S_n = k | H = u) = C(n, k) u^k (1 - u)^{n-k}, \quad 0 \le k \le n, \quad [\text{Bionomial}(n, u)]$$

$$E[I_{\{k\}}(S_n) | H] = C(n, k) H^k (1 - H)^{n-k} \quad \text{a.s.}$$

$$E[S_n | H = u] = nu \quad \text{or} \quad E[S_n | H] = nH \quad \text{a.s.}$$

If Y is the number of failures before the first success, then

$$P(Y = k | H = u) = u(1 - u)^k, \quad 0 \le k, \quad [\text{Geometric}(u)]$$

$$E[Y | H = u] = \frac{1 - u}{u} \quad \text{or} \quad E[Y | H] = \frac{1 - H}{H} \quad \text{a.s.}$$

The following numerical example illustrates the type of calculations required.

Example 22.2.2: A Numerical Example

In a sequence of Bernoulli trials, the probability of success is fixed, but unknown. It is represented as the value of a random variable H. Suppose H is taken to be uniform on the interval $[1/2, 3/4]$. Let random variable Y be the number of failures before the first success.

Determine $P(Y = 2)$ and $E[Y]$.

Solution

The preceding discussion indicates that Y must be geometric, given $H = u$, where u is the operative probability. Using known facts about the geometric distribution, we assume

$$P(Y = 2 | H = u) = E[I_{\{2\}}(Y) | H = u] = u(1 - u)^2$$

and

$$E[Y | H = u] = \frac{1 - u}{u}.$$

Then

$$P(Y = 2) = E\big[E[I_{\{2\}}(Y)|H]\big] = \int E[I_{\{2\}}(Y)|H = u]f_H(u)\,du$$

$$= 4\int_{1/2}^{3/4}(u - 2u^2 + u^3)\,du = \frac{67}{768} \approx 0.0872.$$

In a similar manner,

$$E[Y] = E\big[E[Y|H]\big] = \int E[Y|H = u]f_H(u)\,du = 4\int_{1/2}^{3/4}\left(\frac{1}{u} - 1\right)du$$

$$= 4\ln(3/2) - 1 \approx 0.6219.$$

$\square$

A Posterior Distribution for H

We model the sampling process as a sequence of Bernoulli trials, given H, and proceed as follows:

- We use all available information to determine a prior distribution for H.

  It is frequently reasonable to assume a Beta distribution and use the fact that
  $$\int_0^1 u^{r-1}(1 - u)^{s-1}\,du = \frac{\Gamma(r)\Gamma(s)}{\Gamma(r + s)}.$$

  The random variable H is Beta (r, s) iff $f_H(t) = \frac{\Gamma(r+s)}{\Gamma(r)\Gamma(s)}t^{r-1}(1-t)^{s-1}$, $0 < t < 1$.

  Then, as shown in Section 12.2,

  $$E[H] = \frac{r}{r + s} \quad \text{and} \quad E[H^m] = \frac{r(r + 1)\ldots(r + m - 1)}{(r + s)(r + s + 1)\ldots(r + s + m - 1)}.$$

  By proper choice of the parameters r, s we may "fit" the density to the assumed distribution for H.

- Observe a sample of size n; if there are k successes, then $S_n = k$. This determines a conditional distribution for S_n, given H.

- We use the fact $S_n = k$ and perform a Bayesian reversal to update the distribution for H. We thus determine a posterior distribution for H.

THE BAYESIAN REVERSAL

We have calculated $P(S_n = k | H = u)$ and $E[S_n | H = u]$. At this point, we need to reverse the direction of conditioning, since what we want is the probability distribution for H, given that $S_n = k$. It is this reversal of the direction of conditioning that gives rise to the designation Bayesian, since this is the role of Bayes' theorem in elementary probability. The key idea, however, is the modeling of the population parameter, the state of nature, as a random quantity. To simplify writing, put S for S_n, and I_u for $I_{(-\infty,u]}$, as before. Then

$$F_{H|S}(u|k) = P(H \leq u | S = k) = E[I_u(H) | S = k].$$

Since $P(S = k) > 0$, we may use properties (CE1) and (CE1a) for conditional expectation to assert

$$E[I_u(H) | S = k] P(S = k) = E[I_u(H) I_{\{k\}}(S)]$$

so that

$$E[I_u(H) | S = k] = \frac{E[I_u(H) I_{\{k\}}(S)]}{P(S = k)} = \frac{E[I_u(H) I_{\{k\}}(S)]}{E[I_{\{k\}}(S)]}.$$

By properties (CE1b) and (CE8),

$$E[I_u(H) | S = k] = \frac{E\big[I_u(H) E[I_{\{k\}}(S) | H]\big]}{E\big[E[I_{\{k\}}(S) | H]\big]}.$$

As shown earlier,

$$E[I_{\{k\}}(S) | H = u] = P(S = k | H = u) = C(n,k) u^k (1 - u)^{n-k}.$$

For the case H has Beta (r, s) distribution, the conditional distribution function becomes

$$F_{H|S}(u|k) = \frac{A \int_0^u t^{r+k-1}(1-t)^{s+n-k-1}\, dt}{A \int_0^1 t^{r+k-1}(1-t)^{s+n-k-1}\, dt}, \quad 0 < t < 1$$

where the common factor A is $C(n,k)\frac{\Gamma(r+s)}{\Gamma(r)\Gamma(s)}$. Use of the basic integral shows the denominator to be $\Gamma(r+k)\Gamma(s+n-k)/\Gamma(r+s+n)$. Using the fundamental theorem of calculus, we differentiate with respect to u to obtain the conditional density function

$$f_{H|S}(u|k) = \frac{\Gamma(r+s+n)}{\Gamma(r+k)\Gamma(s+n-k)} u^{r+k-1}(1-u)^{s+n-k-1}, \quad 0 < u < 1.$$

It is convenient to summarize this important result for ready reference.

Theorem 22.2.2   (Posterior Distribution for Beta Prior) *In a conditional Bernoulli sequence, if the prior distribution for H is Beta (r, s), then the conditional distribution for H, given $S_n = k$, is Beta $(r + k, s + n - k)$, so that*

$$E[H|S_n = k] = \frac{r + k}{r + s + n} \quad \text{or} \quad E[H|S_n] = \frac{S_n + r}{r + s + n}.$$

—   □

The solution of the *regression problem* in Section 19.5 shows that the best estimate (in the mean-square sense) of the value of H, given $S_n = k$ is

$$E[H|S_n = k] = \frac{r + k}{r + s + n} \quad \text{so that} \quad E[H|S_n] = \frac{S_n + r}{r + s + n}.$$

Such an estimator is frequently used for small sample analysis, since classical statistics is formulated in such a manner that large samples are highly desirable.

It may be instructive to compare the Bayesian result with the estimator obtained in classical analysis. In the classical case, the estimator for H is

$$A_n = \frac{1}{n} S_n.$$

Since the classical analysis presumes no prior knowledge of the parameter p, we should compare with the Bayesian case of no prior knowledge. If there is no prior knowledge, values in one part of the interval are as likely as in any other comparable part. Thus, it is reasonable to assume the uniform prior distribution, so that H is Beta $(1, 1)$. In this case, the estimator is

$$E[H|S_n] = \frac{S_n + 1}{n + 2}.$$

For large sample size n, the two values are quite close. To see how the estimates may differ, let us consider a simple numerical example. Now A_n is just the fraction of successes in n trials. Suppose we keep this fraction constant at $2/10$ and compare with the Bayesian estimate for $n = 10$, 20, 50, and 100. The corresponding Bayesian estimates are $3/12 = 0.250$, $5/22 = 0.227$, $11/52 = 0.212$, and $21/102 = 0.206$, respectively.

This development illustrates how the very reasonable assumption of conditional independence, given H, makes it possible to use experience to modify estimates of probabilities or other parameters that control population distributions. Conditional independence is assumed frequently in applications of Bayesian statistics. Unfortunately, many presentations do not make this assumption explicit. As a consequence, explanations are not complete and frequently do not exploit some of the useful properties of conditional independence, such as those summarized in the table. Also, failure to note the assumption increases the risk that a particular model may be used in situations in which the presumed conditional independence does not hold.

In some recent work, A.P. Dawid has shown that conditional independence provides a conceptual framework for much of the theory of statistical inference. While he casts his work in a somewhat different theoretical framework, the patterns he identifies and uses can be formulated in terms of conditional expectation, as is done in this chapter. For references, see the bibliographic note at the end of this chapter.

22.3.   Elementary Decision Models

In making decisions, it is usually not enough to analyze the probabilities of alternatives under consideration. Decisions are based on both estimates of likelihoods and costs or rewards associated with actions and outcomes. The decision maker is apt to proceed more cautiously (i.e., to require higher probabilities for favorable outcomes) if the costs of failure are high. If rewards for success are great, this tends to tip the balance in the other direction. This is exhibited in elementary life situations. The probability of winning in a state lottery is extremely small. Yet the cost of losing (the price of a lottery ticket) is so affordable and the rewards for winning are so great, that large numbers of citizens are willing to take the risk. The probability of a disastrous nuclear power plant failure is likewise extremely small. But the potential cost of failure in money, lives, and health is so great, while the rewards are seen as so minimal (at least in the US), that there is strong resistance to locating such a plant anywhere near a populated area. Only if the energy situation is so severe that the rewards in terms of quality of life tend to balance out the risks is there incentive for accepting the risks. And always there is pressure to reduce the probabilities of failure, even though the costs of marginal increases in reliability may be quite high.

A wide variety of decision models have been used in the literature on decision theory, especially in the field of operations research. In this section, we explicate one of the standard models, which exhibits many essential features of statistical decision problems. We consider two cases: (a) without experimentation, and (b) with experimentation. The first case does not ordinarily involve a conditional independence assumption; the second case does.

The transition from probable inference (determining the most likely alternatives) to decision (determining a course of action to be chosen from among a set of alternatives) turns on the idea of a loss function (or a gain function, with a loss a negative gain). The loss function uses some symbol of value, such as monetary units. But its specification may require quite subtle, often subjective, judgments of worth or utility. At this point, the mathematical modeler must work closely with the decision maker in order to "build into the model" a valid representation of the appropriate loss function.

The decision maker is faced with the task of choosing one of a set of available or feasible alternative actions, under a condition of uncertainty about the result of the action, but with the goal of achieving some optimum return or of minimizing the loss in a suitable fashion.

ELEMENTS OF A DECISION MODEL

1. There is a set $A = \{a_1, a_2, \ldots, a_m\}$ of possible actions.

2. There is a set of possible outcomes. Usually, there is uncertainty about which of these outcomes will occur. In such a case, the outcome is represented as the value of a random variable Y. The distribution for Y is controlled by a state of nature (and perhaps by the action chosen).

3. The state of nature is uncertain. In the Bayesian model, this state is the value of a parameter random variable H.

4. There is a loss function L that has the property that

$$L(a, y) = \text{the loss when the action is } a \text{ and the outcome is } y.$$

In the analysis of this model, we also use the risk function $r(\cdot, \cdot)$ defined by

$$r(a, u) = E[L(a, Y)|H = u].$$

We note that in some problems $Y = H$, so that in this case

$$r(a, u) = E[L(a, H)|H = u] = L(a, u).$$

5. The usual *objective* is to select an action a that *minimizes the expected loss*

$$R(a) = E[L(a, Y)] = E\big[E[L(a, Y)|H]\big] = E[r(a, H)].$$

CASE A. WITHOUT EXPERIMENTATION

Assume $\{Y, H\}$ have a joint distribution. The objective is to select action a to minimize

$$R(a) = E[L(a, Y)] = E[r(a, H)].$$

In this case of no experimentation, no conditional independence assumptions are needed.

Example 22.3.1:

The manager of a university computer center is planning to order a supply of printout paper for the fall semester. It is presumed that by the beginning of the spring semester the printers will be replaced with a type that uses a different size paper. The order can be for 5, 10, or 20 boxes, at $20 per box. If too much is ordered, the excess can be returned for $10 per box. If too little is ordered, individual boxes, as needed, may be special-ordered for immediate delivery at $25 per box. Demand is assumed to be a random quantity with Poisson distribution. The parameter λ is unknown. However, on the basis of past experience the manager assumes λ to be the value of a random variable H with possible values $\{10, 15, 20\}$, taken on with probabilities $\{0.3, 0.5, 0.2\}$, respectively. From the point of view of Bayesian statistics, what amount should the manager order?

Solution

The set of actions is $A = \{5, 10, 20\}$. Let Y be the demand random variable. If $B = \{Y \le a\}$, then the loss function is given by

$$L(a, Y) = 10(a - Y)I_B + 5(Y - a)(1 - I_B) = 5(Y - a) - 15I_B Y + 15a I_B.$$

It follows that

$$
\begin{aligned}
r(a, \lambda) &= E[L(a, Y)|H = \lambda] \\
&= 5E[Y - a|H = \lambda] - 15E[I_B Y|H = \lambda] + 15aE[I_B|H = \lambda].
\end{aligned}
$$

Now, $E[Y - a|H = \lambda] = \lambda - a$, $E[I_B|H = \lambda] = P(Y \le a|H = \lambda)$, and

$$E[I_B Y|H = \lambda] = \sum_{k=0}^{a} k\frac{\lambda^k}{k!}e^{-\lambda} = \lambda\sum_{j=0}^{a-1}\frac{\lambda^j}{j!}e^{-\lambda} = \lambda P(Y \le a - 1|H = \lambda).$$

Using these, we find that

$$r(a, \lambda) = 5(\lambda - a) + 15aP(Y \le a|H = \lambda) - 15\lambda P(Y \le a - 1|H = \lambda).$$

This expression may be put in the form

$$r(a, \lambda) = 5(\lambda - a) - 15(\lambda - a)P(Y \le a - 1|H = \lambda) + 15aP(Y = a|H = \lambda).$$

Values of $r(a, \lambda)$ may be calculated with the aid of tables of summed and cumulative terms for the Poisson distribution. For example, for $a = 10$, $\lambda = 15$, we have

$$r(10, 15) = 5 \cdot 5 - 15 \cdot 5(1 - 0.9301) + 15 \cdot 10 \cdot 0.0486 = 27.0.$$

The other cases are calculated similarly, and the results are tabulated here.

Risk Function $r(a, \lambda)$			
$\lambda =$	10	15	20
$a = 5$	25.6	50.0	75.0
10	18.8	27.0	50.1
20	100.0	53.2	26.6

The problem reduces to selecting the action a that yields the smallest value of

$$R(a) = E[L(a, Y)] = E[r(a, H)].$$

Using the tabulated values of $r(a, \lambda)$ and the prior distribution for H given in the problem statement, we determine $R(5) = 0.3 \cdot 25.6 + 0.5 \cdot 50.0 + 0.2 \cdot 75.0 = 47.7$.

Similarly, we obtain: $R(10) = 29.2$ and $R(20) = 61.9$. The optimum action is thus $a = 10$.

—   □

CASE B. WITH EXPERIMENTATION

The value of a test random vector X is observed. We suppose that $\{X, Y, H\}$ has a joint distribution. The objective is to determine an optimal decision rule d for selecting an action $a = d(x)$ when $X(\omega) = x$, such that the Bayesian risk

$$B(d) = E\big[L(d(X), Y)\big]$$

is a minimum. We obtain a solution in terms of the function

$$R(a, x) = E[L(a, Y)|X = x].$$

Now, by (CE10),

$$E[L(d(X), Y)|X = x] = E[L(d(x), Y)|X = x] = R(d(x), x)$$

so that

$$B(d) = E\big[E[L(d(X), Y)|X]\big] = E[R(d(X), X)].$$

For each x in the range of test random vector X, let $d^*(x)$ be an action that makes

$$R(d^*(x), x) \le R(a, x) \qquad \text{for all } a \in A.$$

Then $d^*(\cdot)$ is defined on the range of X and

$$B(d^*) = E[R(d^*(X), X)] \le E[R(d(X), X)] = B(d) \quad \text{for all permissible } d.$$

The rule $d^*(\cdot)$ is an optimal decision rule. We say "an optimal" rule, since $d^*(\cdot)$ may not be unique. We can thus solve the problem if we can determine $R(a, x) = E[L(a, Y)|X = x]$.

If we have this function R, then for any observed value $X(\omega) = x$, we select an action that minimizes $R(a, x)$. At this point, we make a conditional independence assumption. Ordinarily, the outcome following an action is affected only by the value of the state-of-nature parameter, and not by the result of experimentation. In fact, it is usually immaterial whether or not the decision maker obtains experimental data. The experimental evidence may be in the form of previously available data. In such cases, it is appropriate to assume that $\{X, Y\}$ ci$|$ H. We may use this conditional independence further to reduce the problem to manageable form. By property (CI8), we have

$$
\begin{aligned}
R(a, x) &= E[L(a, Y)|X = x] = E\big[E[L(a, Y)|H] \,\big|\, X = x\big] \\
&= E[r(a, H)|X = x].
\end{aligned}
$$

In order to evaluate the last expectation, we must have the conditional distribution for H, given $X = x$. Since we begin with the conditional distribution for X, given H, we must use the Bayesian procedure for reversing the direction of conditioning. The details depend upon the joint distribution for (X, H).

Example 22.3.2:

Consider, again, the problem of ordering computer paper, in Example 22.3.1. The manager checks the amount of paper used the previous semester, and finds that seventeen boxes were required. There is reason to believe that the conditions of use are essentially unchanged (e.g., approximately the same enrollment, the same assignments, etc.). Now there is operational independence, since the same contingencies remain and no new policies for paper usage are initiated. If X is the random variable whose value is the demand in the previous semester, it is reasonable under these conditions to suppose that $\{X, Y\}$ ci$|$ H. In fact, we suppose that $\{X, Y\}$ ciid$|$ H.

Solution

With these assumptions and this datum, the task is to select $a = d^*(17)$ to minimize

$$
R(a, 17) = E[r(a, H)|X = 17] = \sum_\lambda r(a, \lambda) p_{H|X}(\lambda|17).
$$

Now

$$
p_{H|X}(\lambda|17) = \frac{p_{X|H}(17|\lambda) p_H(\lambda)}{p_X(17)}
$$

where

$$
p_X(17) = p_{X|H}(17|10) p_H(10) + p_{X|H}(17|15) p_H(15) + p_{X|H}(17|20) p_H(20).
$$

Using assumed values for p_H and values from a table for $p_{X|H}$, we get

$$p_X(17) = 0.0128 \cdot 0.3 + 0.0874 \cdot 0.5 + 0.0760 \cdot 0.2 = 0.0614 = \frac{1}{16.29}.$$

Thus,

$$\begin{aligned} R(5, 17) &= 16.29(25.6 \cdot 0.0128 \cdot 0.3 + 50.0 \cdot 0.0847 \cdot 0.5 \\ &\quad + 75.0 \cdot 0.0760 \cdot 0.2) \\ &= 54.7. \end{aligned}$$

Similarly,

$$R(10, 17) = 32.2 \qquad \text{and} \qquad R(20, 17) = 49.5.$$

Again, the indicated action is to order ten boxes, even though use during the previous semester was greater than ten. The penalty for returning unused boxes is greater than the penalty for special orders. Thus, the model calls for conservative ordering.

—   □

The cost structure and range of parameter values in these examples are quite simple, so that the results may seem somewhat artificial. In practice, for instance, there would be additional factors such as price breaks for larger orders. These would increase the computational complexity without changing the essential structure of the problem or the strategy of solution.

The following simple example gives another illustration of the role of conditional independence in decision making and demonstrates some technique in the Bayesian reversal of the direction of conditioning.

Example 22.3.3: A Prediction Strategy

An electronic game provides conditional Bernoulli trials as follows. It selects randomly a probability of success, then performs a Bernoulli sequence of trials based on that value of the parameter random variable H. The player has no information about the distribution for H, nor is he told the value of H selected on the initial trial. He is allowed to observe the number of successes in m trials; then he is to predict the number of successes in the next n trials. If he guesses within one of the correct number, he gains one dollar (loses -1); otherwise, he loses one dollar.

Suppose $m = 3$ and $n = 10$. On the trial run, there are two successes out of three. What number a of successes should he guess will occur on the following ten trials? He seeks a strategy that will minimize his expected loss.

Solution

Let $X = T_m$ = the number of successes on the initial run of m trials. The action a is the number ($0 \leq a \leq n$) of successes predicted in the n trials. The outcome $Y = S_n$ = the number of successes in the n trials. The loss function $L(\cdot, \cdot)$ is defined by

$$L(a, y) = \begin{cases} -1 & \text{for } a - 1 \leq y \leq a + 1 \\ 1 & \text{otherwise} \end{cases}.$$

We seek a strategy to minimize $E[L(a, Y)]$. By the law of total probability (CE1b) for the first equality and conditional independence and (CI8) for the second equality, we have

$$E[L(a, Y)] = E\big[E[L(a, Y)|T_m]\big] = E\Big[E\big[E[L(a, Y)|H] \mid T_m\big]\Big].$$

From the analysis of conditional Bernoulli sequences in Section 22.2, we know that

$$P(Y = k|H = u) = C(n, k)u^k(1 - u)^{n-k}$$

so that

$$
\begin{aligned}
r(a, u) &= E[L(a, Y)|H = u] = \sum_{k=0}^{n} L(a, k)P(Y = k|H = u) \\
&= \sum_{k=0}^{n} L(a, k)C(n, k)u^k(1 - u)^{n-k}.
\end{aligned}
$$

Use of the specific values 1 or -1 for L and the fact that the probabilities $P(S_n = k|H = u)$ sum to one for fixed n and u gives the result

$$r(a, u) = 1 - 2p(a, u)$$

where

$$
\begin{aligned}
p(a, u) &= C(n, a - 1)u^{a-1}(1 - u)^{n-a+1} + C(n, a)u^a(1 - u)^{n-a} \\
&\quad + C(n, a + 1)u^{a+1}(1 - u)^{n-a-1}.
\end{aligned}
$$

For each x, $0 \leq x \leq m$, we seek a value $a = d(x)$ such that

$$R(a, x) = E[r(a, H)|X = x] = E[r(d(x), H)|X = x] \qquad \text{is a minimum.}$$

Since we have the conditional distribution for $X = T_m$, given H, we are faced again with the Bayesian problem of reversing the direction of conditioning. Since $P(X = x) > 0$, we have

$$
\begin{aligned}
E[r(a, H)|X = x]P(X = x) &= E[I_{\{x\}}(X)r(a, H)] \\
&= E\big[E[I_{\{x\}}(X)r(a, H)|H]\big].
\end{aligned}
$$

The player has no knowledge of the distribution for H; he, therefore, assumes that H is uniformly distributed. Use of fundamental properties shows that

$$
\begin{aligned}
P(X = x) &= E[I_{\{x\}}(X)] \\
&= E\big[E[I_{\{x\}}(X)|H]\big] = \int P(X = x|H = u)F_H(du).
\end{aligned}
$$

Since $X = T_m$ is conditionally binomial, given $H = u$,

$$
P(X = x) = C(m, x) \int_0^1 u^x (1 - u)^{m-x}\, du.
$$

Use of the integrals associated with the Beta distribution shows that

$$
\int_0^1 u^x (1 - u)^{m-x}\, du = \frac{1}{(m+1)C(m, x)} \quad \text{so that} \quad P(X = x) = \frac{1}{m+1}.
$$

Now, in a similar fashion we have

$$
E\big[E[I_{\{x\}}(X)r(a, H)|H]\big] = \int_0^1 P(X = x|H = u)r(a, u)\, du.
$$

Examination of the expression for $r(a, u)$ shows that we must evaluate integrals of the form

$$
C(m, x)C(n, r) \int_0^1 u^{x+r}(1 - u)^{m+n-x-r}\, du
$$

$$
= \frac{C(m, x)C(n, r)}{(m+n+1)C(m+n, x+r)}.
$$

Using these, we find that

$$
R(a, x) = E[r(a, H)|X = x] = 1 - \frac{2(m+1)C(m, x)}{m+n+1}K(a, x)
$$

where

$$
\begin{aligned}
K(a, x) &= \frac{C(n, a-1)}{C(m+n, x+a-1)} + \frac{C(n, a)}{C(m+n, x+a)} \\
&\quad + \frac{C(n, a+1)}{C(m+n, x+a+1)}.
\end{aligned}
$$

To minimize $R(a, x)$ for a given x, we seek an a that maximizes $K(a, x)$. Use of the values $m = 3$, $n = 10$, and $x = 2$ yields the values:

$$
K(5, 2) = 0.4324 \quad K(6, 2) = 0.4779 \quad K(7, 2) = 0.4883
$$

$$K(8,2) = 0.4534 \quad K(9,2) = 0.3625.$$

Since, for $x = 2$, the largest value of $K(a,2)$ corresponds to $a = 7$, the best prediction is $d(2) = 7$. This is consistent with the fact that two thirds of the results on the calibration run were successes.

——   □

Bibliographic Note

Until recently, the systematic development and use of conditional independence has been found principally in advanced theoretical treatments of Markov processes. In these, conditional independence is formulated in terms of conditional expectation, given a sigma algebra of events.

Some treatments of Bayesian statistics in standard textbooks tacitly use conditional independence in the form of the product rule for the conditional distribution function. However, the concept is often not identified and developed, and it is not related to the theory of Markov processes. The result has been restricted insight and failure to exploit powerful theoretical resources for dealing with important problems.

The present chapter presents an updated version of certain aspects of the treatment in the author's UMAP expository monograph,

Paul E. Pfeiffer: *Conditional independence in applied probability*, COMAP, Inc., Lexington, Mass., 1978.

In a series of seminal papers, A.P. Dawid has used conditional independence "to form a conceptual framework for much of the theory of statistical inference." Probably the most helpful introduction for the reader interested in applications to statistics is

A.P. Dawid: Conditional independence in statistical theory, *Journal of the Royal Statistical Society*, Series B, 41 (1), 1979, 1–31.

In a more theoretical paper, Dawid presents an abstract formalization of conditional independence. In that paper and the applications paper cited here, he identifies several useful patterns, that he uses to illuminate a wide variety of topics in statistics. Although Dawid uses an abstract formulation that does not employ conditional expectation, we have expressed the patterns of interest in terms of conditional expectation, given a random vector. These have been incorporated into the exposition in the present chapter. It has seemed preferable to remain within the framework of the standard probability model, since nothing essential is lost thereby. This is consistent with Dawid's goal of unifying the theory underlying statistics. In addition, it makes clear the relationships to other topics. Frequently referred to as the Markov property, conditional independence is an appropriate assumption in many models in decision theory and applied random processes.

22a

Proofs of Properties

We wish to establish the properties in the table on conditional independence, given a random vector. We first show the equivalence of (CI1) through (CI4), then consider the extensions to (CI5) through (CI9).

We begin by showing (CI1) holds iff (CI2) holds. Then we establish the chain of implications: (CI2) implies (CI3) implies (CI4) implies (CI2). To simplify writing, we drop the "a.s." designation in many of the step-by-step arguments.

(CI1) implies (CI2) Set $e_1(Z,Y) = E[I_M(X)|Z,Y]$ and $e_2(Z,Y) = E[I_M(X)|Z]$. If we can show that

$$E[I_N(Y)I_Q(Z)e_1(Z,Y)] = E[I_N(Y)I_Q(Z)e_2(Z,Y)]$$

for all Borel sets N and Q, then by the uniqueness property (E5b) for expectation, we may assert the almost sure equality of $e_1(Z,Y)$ and $e_2(Z,Y)$. Using, in order, properties (CE8), (CI1), (CE9a), and (CE8), we have

$$
\begin{aligned}
E\big[I_N(Y)E[I_M(X)|Z] \mid Z\big] &= E[I_M(X)|Z]E[I_N(Y)|Z] \\
&= E[I_M(X)I_N(Y)|Z] \\
&= E\big[E[I_M(X)I_N(Y)|Z,Y] \mid Z\big] \\
&= E\big[I_N(Y)E[I_M(X)|Z,Y] \mid Z\big].
\end{aligned}
$$

By property (CE1)

$$E\Big[I_Q(Z)E\big[I_N(Y)E[I_M(X)|Z] \mid Z\big]\Big] = E\big[I_Q(Z)I_N(Y)E[I_M(X)|Z]\big]$$

for all Borel Q. Similarly,

$$
\begin{aligned}
E\big(I_Q(Z)E\big[I_N(Y)E[I_M(X)|Z,Y] \mid Z\big]\big) \\
= E\big(I_Q(Z)I_N(Y)E[I_M(X)|Z,Y]\big)
\end{aligned}
$$

for all Borel Q. Hence,

$$E\big(I_Q(Z)I_N(Y)E[I_M(X)|Z]\big) = E\big(I_Q(Z)I_N(Y)E[I_M(X)|Z,Y]\big)$$

for all Borel N and Q. We may therefore assert that

$$E[I_M(X)|X] = E[I_M(X)|Z,Y] \quad \text{a.s.}$$

—   □

(CI2) implies (CI1) By properties (CE9a), (CE8), (CI2), and (CE8), in that order,

$$
\begin{aligned}
E[I_M(X)I_N(Y)|Z] &= E\big[E[I_M(X)I_N(Y)|Z,Y] \mid Z\big] \\
&= E\big[I_N(Y)E[I_M(X)|Z,Y] \mid Z\big] \\
&= E\big[I_N(Y)E[I_M(X)|Z] \mid Z\big] \\
&= E[I_M(X)|Z]E[I_N(Y)|Z]
\end{aligned}
$$

—   □

(CI2) implies (CI3) By (CE8) and (CI2), then (CE8),

$$
\begin{aligned}
E[I_M(X)I_Q(Z)|Z,Y] &= I_Q(Z)E[I_M(X)|Z,Y] = I_Q(Z)E[I_M(X)|Z] \\
&= E[I_M(X)I_Q(Z)|Z].
\end{aligned}
$$

—   □

(CI3) implies (CI4) By (CE9) and (CI3),

$$
\begin{aligned}
E[I_M(X)I_Q(Z)|Y] &= E\big[E[I_M(X)I_Q(Z)|Z,Y] \mid Y\big] \\
&= E\big[E[I_M(X)I_Q(Z)|Z] \mid Y\big].
\end{aligned}
$$

—   □

(CI4) By (CI4) then (CE9),

$$
\begin{aligned}
E\big[E[I_M(X)I_Q(Z)|Z] \mid Y\big] &= E[I_M(X)I_Q(Z)|Y] \\
&= E\big[E[I_M(X)I_Q(Z)|Z,Y] \mid Y\big].
\end{aligned}
$$

By (CE1), this ensures that for all Borel sets N on the codomain of Y,

$$
E\big[I_N(Y)E[I_M(X)I_Q(Z)|Z]\big] = E\big[I_N(Y)E[I_M(X)I_Z(Z)|Z,Y]\big].
$$

But by (CE8), we must have

$$
E\big[I_N(Y)I_Q(Z)E[I_M(X)|Z]\big] = E\big[I_N(Y)I_Q(Z)E[I_M(X)|Z,Y]\big].
$$

By the uniqueness property (E5b), we conclude that

$$
E[I_M(X)|Z] = E[I_M(X)|Z,Y] \quad \text{a.s.}
$$

—   □

It is clear that (CI5) implies (CI1), (CI6) implies (CI2), and (CI8) implies (CI4), since the second property in each pair is a special case of the first property in that pair.

(CI1) implies (CI5) The proof involves three phases:

1. Establish the theorem for $g(X,Y) = I_M(X)I_Q(Z)$ and $h(Y,Z) = I_N(Y)I_R(Z)$.

2. Show that this implies the theorem for $g(X,Z) = I_B(X,Z)$ and $h(Y,Z) = I_C(Y,Z)$.

3. Use a "standard argument" to extend to general $g(X,Z)$ and $h(Y,Z)$.

First, by (CE8) and (CI1), then (CE8),

$$
\begin{aligned}
E[I_M(X)&I_Q(Z)I_N(Y)I_R(Z)|Z] \\
&= I_Q(Z)I_R(Z)E[I_M(X)|Z]E[I_N(Y)|Z] \\
&= E[I_M(X)I_Q(Z)|Z]E[I_N(Y)I_R(Z)|Z]
\end{aligned}
$$

The second step requires an argument similar to that for the proof of (CE10), sketched in appendix 19a, to show that

$$
E[I_B(X,Z)I_C(Y,Z)] = E[I_B(X,Z)|Z]E[I_C(Y,Z)]
$$

for all Borel sets B and C. We omit the details, which require measure theory beyond the scope of this work.

Third, a standard argument

(i) Uses linearity to show the statement is true for all simple functions g and h.

(ii) Uses monotone convergence to show true for all nonnegative integrable functions $g(X,Z)$ and $h(Y,Z)$.

(iii) Establishes the general case by considering $g = g^+ - g^-$ and $h = h^+ - h^-$.

—   ☐

Similar arguments establish the extension from (CI3) to (CI6) and from (CI4) to (CI8).

(CI6) implies (CI7) Since $E[g(X,Z)|Z] = e_g(Z)$ a.s., by definition.

—   ☐

(CI7) implies (CI6) By (CE9a), (CI7), and (CE7), then (CI7), we have

$$
\begin{aligned}
E[g(X,Z)|Z] &= E\big[E[g(X,Z)|Z,Y]\,\big|\,Z\big] \\
&= E[e_g(Z)|Z] = e_g(Z) = E[g(X,Z)|Z,Y].
\end{aligned}
$$

—   □

(CI9) implies $\{X,Y\}$ ci| Z since $X = g(X,Z)$ and $Y = h(Y,Z)$.

—   □

(CI5) implies (CI9) since $\phi(U) = \phi[g(X,Z)] = g_1(X,Z)$ and $\psi[h(Y,Z)] = h_1(Y,Z)$.

—   □

CI10) $(X,Z,U) = \phi(X,Z)$ and $(X,Z) = \psi(X,Z,U)$ with ϕ, ψ Borel. Now by (CE9b), then the hypothesis and (CI6), then (CI6) and (CI9), we have

$$
E[h(Y)|X,Z,U] = E[h(Y)|X,Z] = E[h(Y)|Z] = E[h(Y)|Z,U].
$$

Thus, by (CI6) we have $\{X,Y\}$ ci| (Z,U).

A parallel argument, interchanging X for Y and U for V shows that $\{X,Y\}$ ci| (Z,V).

Let $W = (Z,U)$. We have $\{X,Y\}$ ci| W from the first result. By the second result, we have $\{X,Y\}$ ci| (W,V), which is equivalent to $\{X,Y\}$ ci| (Z,U,V)

—   □

CI11) First we apply (CE9a), then second hypothesis for (CI11) and (CI6) to get

$$
\begin{aligned}
E[g(Y,Z,W)|X,Y] &= E\big[E[g(Y,Z,W)|X,Y,Z]\,\big|\,X,Y\big] \\
&= E\big[E[g(Y,Z,W)|Y,Z]\,\big|\,X,Y\big].
\end{aligned}
$$

Next, we apply the first hypothesis and (CI6), then (CE9a) to establish

$$
\begin{aligned}
E\big[E[g(Y,Z,W)|Y,Z]\,\big|\,X,Y\big] &= E\big[E[g(Y,Z,W)|Y,Z]\,\big|\,Y\big] \\
&= E[g(Y,Z,W)|Y]
\end{aligned}
$$

so that by (CI6), $\{X,(Z,W)\}$ ci| Y.

Conversely, by (CI9), $\{X,(Z,W)\}$ ci| Y implies $\{X,Y\}$ ci| Y and by (CI10) it also implies $\{X,(Z,W)\}$ ci| (Y,Z).

—   □

CI12) By (CI10), $\{(X,Y),W\}$ ci| Z implies $\{(X,Y),W\}$ ci| (Y,Z), which by (CI9) implies $\{X,W\}$ ci| (Y,Z). The desired result then follows from (CI11).

—   ⊔

CI13) Suppose the first two conditions hold. Then

$$
\begin{aligned}
E[g(Y,Z)|X] &= E\big[E[g(Y,Z)|X,Y]\mid X\big] \\
&= E\big[E[g(Y,Z)|Y]\mid X\big] \\
&= E\big[E[g(Y,Z)|Y]\big] = E[g(Y,Z)].
\end{aligned}
$$

The first equality follows by (CE9a), the second by the second hypothesis and (CI6), the third by the first hypothesis and (CE5), and the fourth by (CE1b). By (CE5), we may assert that $\{X,(Y,Z)\}$ is independent.

To show the converse, we first note that $\{X,(Y,Z)\}$ independent implies $\{X,Y\}$ independent. For any Borel sets M, N, and Q, consider

$$
E[I_M(X)I_N(Y)I_Q(Z)] = E\big[I_N(Y)E[I_M(X)I_Q(Z)|Y]\big].
$$

Also, by (E18), (CE1), (E2), and (CE5), we have

$$
\begin{aligned}
E[I_M(X)I_N(Y)I_Q(Z)] &= E[I_M(X)]E[I_N(Y)I_Q(Z)] \\
&= E[I_M(X)]E\big[I_N(Y)E[I_Q(Z)|Y]\big] \\
&= E\big[I_N(Y)E[I_M(X)]E[I_Q(Z)|Y]\big] \\
&= E\big[I_N(Y)E[I_M(X)|Y]E[I_Q(Z)|Y]\big].
\end{aligned}
$$

By the uniqueness property (E5) for expectation, we conclude that

$$
E[I_M(X)I_Q(Z)|Y] = E[I_M(X)|Y]E[I_Q(Z)|Y] \quad \text{a.s.}
$$

for all Borel sets M and Q so that $\{X,Z\}$ ci| Y.

—   □

CI14) As in the proof of (CE10) for conditional expectation, it suffices to establish the theorem for $g = I_{M \times N} = I_M I_N$. Now by (CE8), (CI6), and (CE2), we have

$$
\begin{aligned}
E[I_M(X)I_N(Y)|Y=u,Z=v] &= I_N(u)E[I_M(X)|Y=u,Z=v] \\
&= I_N(u)E[I_M(X)|Z=v] \\
&= E[I_M(X)I_N(u)|Z=v].
\end{aligned}
$$

—   □

CI15) (a) Follows from (CE1b) and (CI5).

(b) Follows from the definition of $E[g(Y)|X \in M]$, (CE1a), and (CI5).

— $\qquad\qquad\qquad\qquad\qquad\qquad\qquad\qquad\qquad\qquad$ □

CI16) As in the proof of (CE10), use $I_M I_N = I_{M \times N}$. Then

$$E[I_{M \times N}(X,Y)I_Q(Z)|W] = E[I_{M \times N}(X,Y)|W]E[I_Q(Z)|W] \quad \text{a.s.}$$

for all Borel M, N, Q iff

$$E[I_R(X,Y)I_Q(Z)|W] = E[I_R(X,Y)|W]E[I_Q(Z)|W] \quad \text{a.s.}$$

for all Borel R and Q.

— $\qquad\qquad\qquad\qquad\qquad\qquad\qquad\qquad\qquad\qquad$ □

PROBLEMS

22–1. Show that if $\{X, (Y, Z)\}$ is independent, then
$$E[g(X)h(Y)|Z] = E[g(X)]E[h(Y)|Z] \quad \text{a.s.}$$

22–2. Suppose $\{N, H\}$ is independent and $\{N, Y\}$ ci| H. Show that

$$E[g(N)h(Y)|H] = E[g(N)]E[h(Y)|H] \quad \text{a.s.}$$

22–3. Let $\{X_i : 1 \leq i \leq n\}$ be a random sample, given H. Set $W = (X_1, X_2, \ldots, X_n)$.

Determine the best mean-square estimator for H, given W, for each of the following cases:

(1) X is delayed exponential, given $H = u : f_{X|H}(t|u) = e^{-(t-u)}$ for $t \geq u$ and H is exponential (1).

(2) X is Poisson (u), given $H = u : p_{X|H}(k|u) = e^{-u}u^k/k!$ for $k = 0$, 1, 2, $\ldots$, and H is gamma (m, λ).

(3) X is geometric (u), given $H = u : p_{X|H}(k|u) = u(1 - u)^k$ for all $k \geq 0$ and H is uniform $[0, 1]$.

22–4. Consider the following "electronic model" of a conditional Bernoulli sequence. We use a random-number generator (see Section 11.4) to produce a sequence of "observed values" $(u_0, u_1, \ldots, u_n, \ldots)$ from an iid sequence $\{U_0, U_1, \ldots, U_n, \ldots\}$ with each U_i uniform on $[0, 1)$. We then define a sequence $\{H, X_1, X_2, \ldots, X_n, \ldots\}$ as follows.

(a) Suppose F_H is the distribution function for a probability distribution whose range is included in the interval $(0, 1)$. Let Q_H be the corresponding quantile function. Then $H = Q_H(U_0)$ has the distribution function F_H.

(b) For any $i \geq 1$, set $X_i = 1$ iff $U_i \leq H$ and $X_i = 0$ otherwise.

The selection of $p = Q_H(u_0)$ determines the value of the "random parameter" H for the sequence of observations of the X_i. Given that $H = p$, for any i the probability is p that X_i has the value one. Once the value of $H = Q_H(U_0)$ is determined, the values of the X_i are determined in an operationally independent manner.

Show that (1) $\{X_i : 1 \leq i\}$ ciid| H and (2) $P(X_i = 1|H = u) = u$.

Hint: Note that $X_i = f(U_i, H)$. Since $\{H, U_i, U_j\}$ is independent, for any Borel g and h, it follows by property (CE10b) for conditional expectation that

$$
\begin{aligned}
E[g(X_i)h(X_j)|H = u] &= E\big[g[f(U_i, H)]h[f(U_j, H)] \mid H = u\big] \\
&= E\big[g[f(U_i, u)]h[f(U_j, u)]\big] \quad \text{a.s. } [P_H].
\end{aligned}
$$

Similarly, establish

$$E[g(X_i)|H = u] = E\big[g[f(U_i u)]\big]$$

and

$$E[h(X_j)|H = u] = E\big[h[f(U_j u)]\big] \quad \text{a.s. } [P_H].$$

Now by the independence of $\{U_i : 1 \leq i\}$, for any fixed u,

$$E\big[g[f(U_i, u)]h[f(U_j, u)]\big] = E\big[g[f(U_i u)]\big] E\big[h[f(U_j u)]\big]$$

so that

$$E[g(X_i)h(X_j)|H = u] = E[g(X_i)|H = u]E[h(X_j)|H = u] \quad \text{a.s. } [P_H].$$

It is apparent that this product rule extends to any finite number of the X_i.

22–5. In a sequence of Bernoulli trials, the probability of success is fixed but unknown. It is represented by a random variable H. Let random variable Y be the number of failures before the first success.

(a) Suppose H is Beta $(2, 3)$. Determine an expression for $P(Y = k)$, $0 \leq k$, and determine $E[Y]$.

(b) Suppose H is Beta (r, s). A sequence of trials is observed. Show that if $Y = k$, then H is conditionally Beta $(r+1, s+k)$, so that $E[H|Y = k] = (r + 1)/(r + s + k + 1)$.

22–6. Consider the composite demand random variable $D = X_N$ introduced in Section 20.1. Suppose $\{N, (H, X_1, X_2, \ldots, X_n)\}$ is independent for each $n \geq 1$, and $E[Y_i|H = u] = e(u)$, invariant with i. Use the result of Problem 2 to show that

$$E[D|H] = E[N]e(H) = E[N]E[Y|H] \quad \text{a.s.}$$

22–7. Consider a conditional Bernoulli sequence, given $H = u$. Thus, in the system of Problem 22-6, $P(Y_i = 1|H = u) = u$ for $0 < u < 1$, and D is the number of successes in a random number N of trials. Suppose N is Poisson (10) and H is Beta $(2, 3)$. Determine $E[D]$.

22–8. Let $S_N = \sum_{i=1}^{N} X_i = \sum_{i-1}^{N} I_{E_i}$ be the number of successes in a random number N of trials in a Bernoulli sequence, given H. Suppose $\{N, H\}$ is independent and $\{N, (X_1, X_2, \ldots, X_n)\}$ ci$|$ H for each $n \geq 1$. Recall that $\{X_i : 1 \leq i\}$ ciid$|$ H. Use the result of previous problems to determine $E[S_N]$.

22-9. Suppose $W = D/V$, with $V > 0$ a.s., N is nonnegative, integer valued, and $\{V, (N, D)\}$ is independent. Show that

$$P(W \leq t|N = n) = \int P(D \leq ut|N = n) F_V(du).$$

Hint: Calculate $P(W \leq t, N = n)$. Use CI13), CE1b), and CI14). Note that

$$E[h(V, N)] = \sum_k \int h(u, k) F_V(du) P(N = k)$$

22-10. It is desired to study the waiting time for the arrival of an ambulance after reporting an accident. Direct statistical data are difficult to obtain. The following approach has been used. Let H be the number of ambulances in service, D be the distance traveled by the ambulance dispatched, and V be the average velocity of the ambulance for the trip. By considering the geometry of the deployment scheme, it is possible to make reasonable assumptions about $P(D \leq t|N = n)$. Also, from statistical data for traffic conditions, etc., it is possible to make reasonable assumptions about the distributions for V and N. We may reasonably suppose that $\{V, (D, N)\}$ is independent. Let A be the area served in square miles. Suppose D is in miles and V is in miles per hour.

   (a) Use the result of Problem 22-9 and the following assumptions to calculate $P(W \leq t|N = n)$. (1) $P(D \leq s|N = n) = rs$, $0 \leq s \leq 1/r$ where $r^2 = n\pi/A$ and (2) V is uniform on $(20, 35)$.

   (b) Repeat part (a) with condition (1) replaced by the condition (1′) $P(D \leq s|N = n) = 1 - e^{-rs}$, $s \geq 0$ where $r^2 = n\pi/A$.

22-11. A company is considering whether or not to make a certain investment. Thus, there are two actions: a_0 =do not buy, and a_1 = buy. The venture could be classified as unsuccessful, moderately successful, or successful. Success is influenced by market conditions, which are categorized as doubtful, mildly favorable, or favorable. We thus have two uncertain conditions that may be represented by a performance random variable Y and a market-condition random variable H. We assign values:

   $Y = -1$ (unsuccessful), 0 (moderately successful), 1 (successful).
   $H = -1$ (doubtful), 0 (mildly favorable), 1 (favorable).

   (a) Company executives determine a loss function $L(a, k)$ (in hundreds of thousands of dollars), where a is the action and k is the value of the performance function. Also, they estimate the pertinent probabilities.

$L(a,k)$				$P(Y = k\|H = j)$				$P(H = j)$
$k =$	-1	0	1	$k =$	-1	0	1	
a_0	1	0	-2	$j = -1$	0.5	0.3	0.2	0.3
a_1	3	-1	-4	0	0.2	0.5	0.3	0.2
				1	0.1	0.3	0.6	0.5

Which action, based on "expected loss," should be taken?

(b) A market survey is considered. Let X be the random variable that represents the result of the survey. Then $X = -1$ indicates a doubtful market, etc.

Past experience indicates the following conditional probabilities $P(X = j | H = i)$

	$j =$	-1	0	1
$i = -1$		0.7	0.2	0.1
0		0.2	0.6	0.2
1		0.1	0.1	0.8

The result of the market survey is that the market is doubtful. Thus, $X = 1$. What action should be taken?

23

Markov Sequences

We wish to set up a model for a system characterized by a sequence of states taken on at discrete instants that we call transition times. At each transition time, there is either a change to a new state or a renewal of the state immediately before the transition. Each state is maintained unchanged during the period between transitions. At any transition time, the move to the next state is characterized by a conditional transition probability distribution. We suppose that the system is memoryless, in the sense that the transition probabilities are dependent upon the current state (and perhaps the period number), but not upon the manner in which that state was reached. The past influences the future only through the present. This is the Markov property, which we model in terms of conditional independence.

23.1.   The Markov Property for Sequences

We are interested in sequences $X_{\mathcal{N}} = \{X_n : 0 \leq n\}$ of random variables that take values, called states, in a set $\mathcal{E}$, known as the state space. For a discrete state space, we identify states by integers. In general, state space $\mathcal{E}$ can be any complete metric space; usually it is the real line $\Re$ or Euclidean n-space $\Re^n$. For a given outcome or observation ω, $\{X_n(\omega) : 0 \leq n\}$ is a sequence of elements of $\mathcal{E}$. This sequence is known as a realization or trajectory of the process. We suppose the X-sequence is evolving in time. At discrete instants of time $t_1, t_2, \ldots$, the system makes a transition from one value or state to the succeeding one.

Initial period: $n = 0$, $t \in [0, t_1)$, state is $X_0(\omega)$; at t_1 the transition is to $X_1(\omega)$.

Period one: $n = 1$, $t \in [t_1, t_2)$, state is $X_1(\omega)$; at t_2 the transition is to $X_2(\omega)$.

$\vdots$

Period k: $n = k$, $t \in [t_k, t_{k+1})$, state is $X_k(\omega)$; at t_{k+1} the transition is to $X_{k+1}(\omega)$.

$\vdots$

The parameter n indicates the period $t \in [t_n, t_{n+1})$. If the periods are of unit length, then $t_n = n$. At t_{n+1}, there is a transition from the state $X_n(\omega)$ to the state $X_{n+1}(\omega)$ for the next period. We shall generally suppose the state space $\mathcal{E}$ is finite or countably infinite, in which case the states are numbered and represented as integers.

In some cases, we assume a decision sequence $Z_{\mathcal{N}} = \{Z_n : 0 \leq n\}$. Each Z_n may be vector-valued.

Definition 23.1.1. *The sequence $X_{\mathcal{N}}$ is adapted to $Z_{\mathcal{N}}$, denoted $X_{\mathcal{N}} \sim Z_{\mathcal{N}}$, iff for each $n \geq 0$, every event determined by X_n is an event determined by the vector $W_n = (Z_0, Z_1, \ldots, Z_n)$.*

According to Theorem 11.1.2, this is equivalent to the condition that there is a Borel function ϕ_n such that

$$X_n = \phi_n(Z_0, Z_1, \ldots, Z_n).$$

To simplify writing, we adopt the following convention:

$$U_n = (X_0, X_1, \ldots, X_n), \; U_{m,n} = (X_m, \ldots, X_n), \; \text{and} \; U^n = (X_n, X_{n+1}, \ldots).$$

Similarly, $W_n = (Z_0, Z_1, \ldots, Z_n)$, etc. If the Z_k are vector-valued, we understand W_n to be the vector whose coordinates consist of the collection of coordinates of the Z_k in some preassigned order. We note that we could have each $Z_k = X_k$ as a special case. The random vector U_n is called the past at n of the sequence $X_{\mathcal{N}}$ and W_n is the past at n of $Z_{\mathcal{N}}$. Each past of $Z_{\mathcal{N}}$ determines the corresponding past of $X_{\mathcal{N}}$. The sequence $\{U_n : 0 \leq n\}$ is the history of $X_{\mathcal{N}}$ and $\{W_n : 0 \leq n\}$ is the history of $Z_{\mathcal{N}}$.

In order to capture the notion that the system is without memory, so that the future is affected by the present, but not by how the present is reached, we use the notion of conditional independence, given a random vector, as formulated in Chapter 22.

Definition 23.1.2. *The process $X_{\mathcal{N}}$ is Markov relative to $Z_{\mathcal{N}}$ iff*

(1) $X_{\mathcal{N}} \sim Z_{\mathcal{N}}$ and (2) (M) $\{X_{n+1}, W_n\}$ ci| X_n for all $n \geq 0$

In the next section, we examine several equivalent conditions for the Markov property. For the present, we note that this condition is equivalent to

$$P(X_{n+1} = k | X_n = j, W_{n-1} \in Q_n) = P(X_{n+1} = k | X_n = j)$$

for each $n \geq 0$, j, $k \in \mathcal{E}$, and Q_n Borel. The state in the next period is conditioned by the past only through the present state, and not by the manner in which the present state is reached. The statistics of the process are determined by the initial state probabilities and the transition probabilities

$$P(X_{n+1} = k | X_n = j) \qquad \text{for all } j, k \in \mathcal{E}, n \geq 0.$$

We examine these matters more closely in Section 23.3.

Remarks

1. When $X_{\mathcal{N}} = Z_{\mathcal{N}}$, we simply say $X_{\mathcal{N}}$ is Markov.

2. If $X_{\mathcal{N}}$ is Markov relative to $Z_{\mathcal{N}}$, with $X_{\mathcal{N}} \sim H_{\mathcal{N}}$ and $H_{\mathcal{N}} \sim Z_{\mathcal{N}}$, then $X_{\mathcal{N}}$ is Markov relative to $H_{\mathcal{N}}$. This is a simple consequence of property (CI9) for conditional independence.

3. By setting $H_{\mathcal{N}} = X_{\mathcal{N}}$, we see that if $X_{\mathcal{N}}$ is Markov relative to some $Z_{\mathcal{N}}$, then it is Markov.

The following examples exhibit a pattern that implies the Markov condition and that can be exploited to obtain the transition probabilities.

Example 23.1.1: One-Dimensional Random Walk

An object starts at a given initial position. At discrete instants t_1, t_2, ... the object moves a random distance along a line. The various moves are independent of each other. Let

Y_0 be the initial position.
Y_k be the amount the object moves at time $t = t_k$ $\{Y_k : 1 \leq k\}$ iid.
$X_n = \sum_{k=0}^{n} Y_k$ be the position after n moves.

We note that $X_{n+1} = g(X_n, Y_{n+1})$. Since the position after the transition at t_{n+1} is affected by the past only by the value of the position X_n and not by the sequence of positions that led to this position, it is reasonable to suppose that the process $X_{\mathcal{N}}$ is Markov. We verify this later.

— $\qquad\qquad\qquad\qquad\qquad\qquad\qquad\qquad\qquad\qquad\qquad$ □

Example 23.1.2: A Class of Branching Processes

Each member of a population is able to reproduce. For simplicity, we suppose that at certain discrete instants the entire next generation is produced. Some mechanism limits each generation to a maximum population of M members.

Let

Z_{in} be the number produced by the ith member of the nth generation.
$X_{n+1} = \min\{M, \sum_{i=1}^{X_n} Z_{in}\} = g(X_n, Y_{n+1})$, where
$Y_{n+1} = (Z_{1n}, Z_{2n}, \ldots, Z_{Mn})$.

If $\{Z_{in} : 1 \le i \le M, 0 \le n\}$ is independent, then $\{Y_{n+1}, U_n\}$ is independent for any $n \ge 0$. Again, it is reasonable to suppose the process X_N is Markov.

—   □

Example 23.1.3: An Inventory Problem

A certain item is stocked according to an (m, M) inventory policy, as follows:

- If stock at the end of a period is less than m, order up to M.
- If stock at the end of a period is m or greater, do not order.

Let X_0 be the initial stock, and X_n be the stock at the end of the nth period (before restocking), and let D_n be the demand during the nth period. Then for $n \ge 0$,

$$X_{n+1} = \left\{ \begin{array}{ll} \max\{M - D_{n+1}, 0\} & \text{if } 0 \le X_n < m \\ \max\{X_n - D_{n+1}, 0\} & \text{if } m \le X_n \end{array} \right\} = g(X_n, D_{n+1}).$$

If we suppose $\{D_n : 1 \le n\}$ is independent, then $\{D_{n+1}, U_n\}$ is independent for each $n \ge 0$ and the Markov condition seems to be indicated.

—   □

Remark: In this case, the actual transition takes place throughout the period. However, for purposes of analysis, we examine the state only at the end of the period (before restocking). Thus, the transitions are dispersed in time, but the observations are at discrete instants.

Example 23.1.4: Remaining Lifetime

A piece of equipment has lifetime that is an integral number of units of time. When a unit fails, it is replaced immediately with another unit of the same type.

Suppose

X_n is the remaining lifetime of the unit in service at time n.
Y_{n+1} is the lifetime of the unit installed at time n, with $\{Y_n : 1 \le n\}$ iid.

Then

$$X_{n+1} = \left\{ \begin{array}{ll} X_n - 1 & \text{if } X_n \ge 1 \\ Y_{n+1} - 1 & \text{if } X_n = 0 \end{array} \right\} = g(X_n, Y_{n+1}).$$

—   □

Remark: Each of these examples exhibits the pattern

(i) $\{X_0, Y_n : 1 \leq n\}$ is independent.

(ii) $X_{n+1} = g_{n+1}(X_n, Y_{n+1})$, $n \geq 0$.

We now verify the Markov condition and obtain a method for determining the transition probabilities.

Theorem 23.1.3 **(A Pattern Yielding Markov Sequences)** *Suppose* $\{Y_n : 0 \leq n\}$ *is independent. Set*

$$X_0 = g_0(Y_0) \qquad and \qquad X_{n+1} = g_{n+1}(X_n, Y_{n+1}) \qquad for\ all\ n \geq 0.$$

Then

a. $X_{\mathcal{N}}$ *is Markov.*

b. $P(X_{n+1} \in Q | X_n = u) = P(g_{n+1}(u, Y_{n+1}) \in Q)$ *for all* n, u *and any Borel set* Q.

Proof

1. It is apparent that if $Y_0, Y_1, \ldots, Y_n$ are known, then U_n is known.

   Thus $U_n = h_n(Y_0, Y_1, \ldots, Y_n)$, which ensures $\{Y_{n+1}, U_n\}$ is independent.

   By property (CI13), with $X = Y_{n+1}$, $Y = X_n$, and $Z = U_{n-1}$, we have $\{Y_{n+1}, U_{n-1}\}$ ci$|$ X_n.

   Since $X_{n+1} = g_{n+1}(Y_{n+1}, X_n)$ and $U_n = h_n(X_n, U_{n-1})$, property (CI9) ensures $\{X_{n+1}, U_n\}$ ci$|$ X_n for all $n \geq 0$, which is the Markov property.

2. $P(X_{n+1} \in Q | X_n = u) = E\big[I_Q(g_{n+1}(X_n, Y_{n+1}))\ \big|\ X_n = u\big]$ a.s. $= E\big[I_Q(g_{n+1}(u, Y_{n+1}))\big]$ a.s. $[P_X]$ by (CE10b) $= P(g_{n+1}(u, Y_{n+1}) \in Q)$ by (E1a).

   $\square$

HOMOGENEOUS MARKOV SEQUENCES

The application of Theorem 23.1.3 to the previous examples shows that the transition probabilities are invariant with n. This case is important enough to warrant separate classification.

Definition 23.1.4. *If* $P(X_{n+1} \in Q | X_n = u)$ *is invariant with* n, *for all Borel sets* Q, *all* $u \in \mathcal{E}$, *the Markov process* $X_{\mathcal{N}}$ *is said to be* homogeneous.

In this regard, we note the following special case of Theorem 23.1.3.

Corollary 23.1.5. *If* $\{Y_n : 1 \leq n\}$ *is iid and* $g_{n+1} = g$ *for all* n, *then the process is a homogeneous Markov process, and*

$$P(X_{n+1} \in Q | X_n = u) = P(g(u, Y_{n+1}) \in Q), \qquad invariant\ with\ n.$$

Remarks

1. Theorem 23.1.3 and Corollary 23.1.5 are not limited to discrete state spaces.

2. If $\mathcal{E}$ is discrete, the stochastic behavior is determined by the distribution for X_0 and the transition probabilities $P(X_{n+1} = k | X_n = j)$. These are also called the one-step transition probabilities.

3. In the homogeneous case, the transition probabilities are invariant with n. In this case, we write

$$P(X_{n+1} = k | X_n = j) = p(j, k) \qquad \text{or} \qquad p_{jk} \quad \text{invariant with } n.$$

We show in Theorem 23.3.2 that the stochastic behavior of the sequence is determined by the initial distribution and the transition probability matrix, or simply the transition matrix,

$$\boldsymbol{P} = [p(j, k)].$$

We examine some of the consequences of this fact in subsequent sections.

We return to the examples. From Theorem 23.1.3, it is apparent that each is Markov. In fact, since the function g is the same for all n and the random variables corresponding to the Y_i form an iid class, the processes must be homogeneous. We may use part (b) of Theorem 23.1.3 or Corollary 23.1.5 to obtain the one-step transition probabilities.

Example 23.1.5: Random Walk (Continued)

$g_n(u, Y_{n+1}) = u + Y_{n+1}$, so that g_n is invariant with n. Since $\{Y_n : 1 \leq n\}$ is iid, $P(X_{n+1} = k | X_n = j) = P(j + Y = k) = P(Y = k - j) = p_{k-j}$, where $p_k = P(Y = k)$.

—   □

Example 23.1.6: Branching Process (Continued)

$g(j, Y_{n+1}) = \min\{M, \sum_{i=1}^{j} Z_{in}\}$ and $\mathcal{E} = \{0, 1, \ldots, M\}$. If $\{Z_{in} : 1 \leq i \leq M, 1 \leq n\}$ is iid, then

$$W_{jn} = \sum_{i=1}^{j} Z_{in} \qquad \text{ensures} \qquad \{W_{jn} : 1 \leq n\} \quad \text{iid for each } j \in \mathcal{E}.$$

We thus have

$$P(X_{n+1} = k | X_n = j) = \begin{cases} P(W_{jn} = k) & \text{for } 0 \leq k < M \\ P(W_{jn} \geq M) & \text{for } k = M \end{cases}, \qquad 0 \leq j \leq M.$$

With the aid of moment-generating functions, one may determine distributions for

$$W_1 = Z_1, \ W_2 = Z_1 + Z_2, \ \ldots, \ W_M = Z_1 + \cdots + Z_M.$$

□

Example 23.1.7: Inventory Problem (Continued)

In this case,

$$g(j, D_{n+1}) = \begin{cases} \max\{M - D_{n+1}, 0\} & \text{for } 0 \le j < m \\ \max\{j - D_{n+1}, 0\} & \text{for } m \le j \le M \end{cases}.$$

Numerical example

$$m = 1 \qquad M = 3 \qquad D_n \text{ is Poisson}(1).$$

Use D for D_n. Because of the invariance with n, set $P(X_{n+1} = k | X_n = j) = p(j, k)$

$$g(0, D) = \max\{3 - D, 0\}$$

$g(0, D) = 0$ iff $D \ge 3$ implies $p(0, 0) = P(D \ge 3)$
$g(0, D) = 1$ iff $D = 2$ implies $p(0, 1) = P(D = 2)$
$g(0, D) = 2$ iff $D = 1$ implies $p(0, 2) = P(D = 1)$
$g(0, D) = 3$ iff $D = 0$ implies $p(0, 3) = P(D = 0)$

$$g(1, D) = \max\{1 - D, 0\}$$

$g(1, D) = 0$ iff $D \ge 1$ implies $p(1, 0) = P(D \ge 1)$
$g(1, D) = 1$ iff $D = 0$ implies $p(1, 1) = P(D = 0)$
$g(1, D) = 2, 3$ is impossible

$$g(2, D) = \max\{2 - D, 0\}$$

$g(2, D) = 0$ iff $D \ge 2$ implies $p(2, 0) = P(D \ge 2)$
$g(2, D) = 1$ iff $D = 1$ implies $p(2, 1) = P(D = 1)$
$g(2, D) = 2$ iff $D = 0$ implies $p(2, 2) = P(D = 0)$
$g(2, D) = 3$ is impossible

$g(3, D) = \max\{3 - D, 0\} = g(0, D)$ so that $p(3, k) = p(0, k)$.

The various probabilities for D may be obtained from a table (or may be calculated easily) to give the transition probability matrix

$$P = \begin{bmatrix} 0.0803 & 0.1839 & 0.3679 & 0.3679 \\ 0.6321 & 0.3679 & 0 & 0 \\ 0.2642 & 0.3679 & 0.3679 & 0 \\ 0.0803 & 0.1839 & 0.3679 & 0.3679 \end{bmatrix}.$$

— $\square$

Example 23.1.8: Remaining Lifetime (Continued)

$g(0, Y) = Y - 1$, so that $p(0, k) = P(Y - 1 = k) = P(Y = k + 1)$.
$g(j, Y) = j - 1$ for $j \geq 1$, so that $p(j, k) = \delta_{j-1,k}$ for $j \geq 1$.

The resulting transition probability matrix is

$$P = \begin{bmatrix} p_1 & p_2 & p_3 & \cdots \\ 1 & 0 & 0 & \\ 0 & 1 & 0 & \\ \vdots & & & \ddots \end{bmatrix}.$$

The matrix is an infinite matrix, unless Y is simple. If the range of Y is $\{1, 2, \ldots, M\}$ then the state space $\mathcal{E}$ is $\{0, 1, \ldots, M - 1\}$.

— $\square$

23.2.   Some Further Patterns and Examples

Properties of conditional independence provide a number of equivalent conditions for the Markov property (M), under the common condition that $X_{\mathcal{N}}$ is adapted to $Z_{\mathcal{N}}$, denoted $X_{\mathcal{N}} \sim Z_{\mathcal{N}}$. For example,

$$E[g(X_n, X_{n+1})|W_n] = E[g(X_n, X_{n+1})|X_n] \quad \text{a.s.} \qquad \text{for all } n \geq 0.$$
((M'))
In the appendix to this chapter, we use various properties of conditional independence, particularly (CI9), (CI10), and (CI12), to establish the following facts. The immediate future X_{n+1} may be replaced by any finite future $U_{n,n+p}$ and the present X_n may be replaced by any extended present $U_{m,n}$. Some results of abstract measure theory show that the finite future $U_{n,n+p}$ may be replaced by the entire future U^n. Thus, we may assert the following.

Theorem 23.2.1   (Extended Markov Property) *$X_{\mathcal{N}}$ is Markov relative to $Z_{\mathcal{N}}$ iff*

(1) $X_{\mathcal{N}} \sim Z_{\mathcal{N}}$.

(2) (M^*) $\{U^n, W_m\}$ ci$|$ $U_{m,n}$ for all $0 \leq m \leq n$.

— $\square$

The following property is frequently of interest.

Theorem 23.2.2   (Nearest-Neighbor Property) *If $X_\mathcal{N}$ is Markov relative to $Z_\mathcal{N}$, then*

(N) $E[g(U_{m,n})|W_{m-1}, U^{n+1}] = E[g(U_{m,n})|X_{m-1}, X_{n+1}]$ a.s.

for $1 \le m \le n$, or, equivalently

(N*) $\{U_{m,n}, (W_{m-1}, U^{n+1})\}$ ci| (X_{m-1}, X_{n+1}) $1 \le m \le n$.

——   □

Thus, an extended present is affected only by the immediate past and the immediate future. In the appendix to this chapter, we construct a proof with the aid of properties (CI9), (CI10), and (CI11).

SPACE-TIME PROCESSES

The following general pattern is useful here and elsewhere. To follow the proof, consult the table of properties of conditional independence.

Theorem 23.2.3. *If $\{W, (X, Y, Z)\}$ is independent and $\{X, Y\}$ ci| Z, then*

$$\{(X, W), Y\}\text{ ci| }Z,\quad \{X, Y\}\text{ ci| }(Z, W),\quad \text{and}\quad \{(X, W), (Y, W)\}\text{ ci| }(Z, W).$$

Proof

By hypothesis and (CI13), we have $\{W, (X, Y)\}$ ci| Z. By (CI9), we have also $\{W, X\}$ ci| Z. For any Borel sets M, N, and Q, we have

$$
\begin{aligned}
E[I_Q(W)I_M(X)I_N(Y)|Z] &= E[I_Q(W)|Z]E[I_M(X)I_N(Y)|Z] \quad \text{a.s.}\\
&= E[I_Q(W)|Z]E[I_M(X)|Z]E[I_N(Y)|Z] \quad \text{a.s.}\\
&= E[I_Q(W)I_M(X)|Z]E[I_N(Y)|Z] \quad \text{a.s.}
\end{aligned}
$$

By (CI16), we have $\{(X, W), Y\}$ ci| Z. An application of (CI10) then (CI9) gives $\{X, Y\}$ ci| (Z, W).

A further application of (CI9) gives $\{(X, W), (Y, W)\}$ ci| (Z, W).

——   □

Example 23.2.1: Space-Time Processes

Suppose $X_\mathcal{N}$ is a Markov process and N is a nonnegative, integer-valued random variable with $\{X_\mathcal{N}, N\}$ independent. Set $N_n = N + n$ for $n \ge 0$ and put

$$Z_n = (X_n, N_n) = (X_n, N + n) : \Omega \to \mathcal{E} \times \mathcal{N}.$$

Then $W_n = (Z_0, Z_1, \ldots, Z_n) = h_n(U_n, N)$ and $(U_n, N) = g_n(W_n)$, where both g_n and h_n are Borel functions, for each $n \ge 0$.

We want to show that $Z_{\mathcal{N}}$ is a vector-valued Markov process. That is, we want to show that

$$\{Z_{n+1}, W_n\} \text{ ci} |\ Z_n \qquad \text{for all } n \geq 0.$$

Proof

To establish this, we note that for each $n \geq 0$, $\{N_n, (X_{n+1}, U_n, X_n)\}$ is independent and $\{X_{n+1}, U_n\}$ ci$|\ X_n$. By Theorem 23.2.3.

$$\{(X_{n+1}, N_n), (U_n, N_n)\} \text{ ci} |\ (X_n, N_n).$$

Now $Z_n = (X_n, N_n), Z_{n+1} = k_{n+1}(X_{n+1}, N_n)$, and $W_n = f_n(U_n, N_n)$. By (CI9), we may assert $\{Z_{n+1}, W_n\}$ ci$|\ Z_n$.

—   □

In many applications, it is reasonable to assume the process has independent increments. As discussed in Appendix 21a, this means that $\{X_{n+1} - X_n, U_n\}$ is independent for each $n \geq 0$. If there is a decision sequence $Z_{\mathcal{N}}$, we may well have $\{X_{n+1} - X_n, W_n\}$ is independent for each $n \geq 0$. The next theorem shows such systems are Markov sequences.

Theorem 23.2.4 (Independent Increments) *If $X_{\mathcal{N}} \sim Z_{\mathcal{N}}$ and $\{X_{n+1} - X_n, W_n\}$ is independent for each $n \geq 0$, then $X_{\mathcal{N}}$ is Markov relative to $Z_{\mathcal{N}}$.*

Proof

The hypothesis implies $\{X_{n+1} - X_n, (X_n, W_n)\}$ is independent. By (CI13), $\{X_{n+1} - X_n, W_n\}$ ci$|\ X_n$.
    Since $X_{n+1} = g(X_{n+1} - X_n, X_n)$, we have $\{X_{n+1}, W_n\}$ ci$|\ X_n$, by (CI9).
—   □

Example 23.2.2: Random Walk (23.1.1)

It is apparent that this process has independent increments, hence is Markov. The advantage of using Theorem 23.1.3 is that it shows how to compute the transition probabilities.

—   □

Remarks

1. Theorem 23.2.4 extends to processes with continuous parameters, so that the Poisson process, which has independent increments, is Markov. A number of other commonly used processes, such as the Wiener process, have independent increments, and thus are Markov processes.

2. The theory of Markov processes with continuous parameters is quite important. However, at certain key points in the development there are technical complications not found in the discrete-parameter case. In this brief, elementary introduction, we limit consideration to the discrete-parameter case.

23.3. The Chapman-Kolmogorov Equation

The Markov property is

(M) $\{X_{n+1}, W_n\}$ ci$|$ X_n for all $n \geq 0$.

By Theorem 23.2.1 and property (CI9) on conditional independence, we have

$$\{X_{n+k+m}, X_n\} \text{ ci}| X_{n+k} \quad \text{for all } n \geq 0,\ k,\ m \geq 1.$$

By the iterated conditioning rule (CI8) for conditional independence, it follows that

(CK) $E[g(X_{n+k+m})|X_n] = E\big[E[g(X_{n+k+m})|X_{n+k}] \,\big|\, X_n\big]$ a.s.

for all $n \geq 0$, k, $m \geq 1$

This is the Chapman-Kolmogorov equation, which plays a significant role in the study of Markov sequences. For a discrete subspace $\mathcal{E}$, with $P(X_n = j|X_m = i) = p_{m,n}(i,j)$, this equation takes the form

(CK') $p_{m,q}(i,k) = \sum_{j \in \mathcal{E}} p_{m,n}(i,j) p_{n,q}(j,k) \qquad 0 \leq m < n < q.$

To see that this is so, consider

$$
\begin{aligned}
P(X_q = k|X_m = i) &= E[I_{\{k\}}(X_q)|X_m = i] \\
&= E\big[E[I_{\{k\}}(X_q)|X_n] \,\big|\, X_m = i\big] \\
&= \sum_j E[I_{\{k\}}(X_q)|X_n = j] p_{m,n}(i,j) \\
&= \sum_j p_{n,q}(j,k) p_{m,n}(i,j)
\end{aligned}
$$

Homogeneous Case

For this case, we may put (CK') in a useful matrix form. The conditional probabilities p^m of the form

$$p^m(i,k) = P(X_{n+m} = k|X_n = i) \qquad \text{invariant in } n$$

are known as the m-step transition probabilities. The Chapman-Kolmogorov equation in this case becomes

$$(CK'') \qquad p^{m+n}(i,k) = \sum_{j \in \mathcal{E}} p^m(i,j) p^n(j,k) \qquad \text{for all } i, k \in \mathcal{E}.$$

In terms of the m-step transition matrix $\boldsymbol{P}^{(m)} = [p^m(i,k)]$, this set of sums is equivalent to the matrix product

$$(CK'') \qquad\qquad \boldsymbol{P}^{(m+n)} = \boldsymbol{P}^{(m)} \boldsymbol{P}^{(n)}.$$

A simple inductive argument based on this fact establishes the following.

Theorem 23.3.1. *The m-step probability matrix* $\boldsymbol{P}^{(m)} = \boldsymbol{P}^m$, *the mth power of the transition matrix* $\boldsymbol{P}$.

—   □

Example 23.3.1: Inventory Problem (Continued)

For the inventory problem in Example 23.1.3, the three-step transition probability matrix $\boldsymbol{P}^{(3)}$ is obtained by raising $\boldsymbol{P}$ to the third power to get

$$\boldsymbol{P}^{(3)} = \boldsymbol{P}^3 = \begin{bmatrix} 0.2930 & 0.2917 & 0.2629 & 0.1524 \\ 0.2619 & 0.2730 & 0.2753 & 0.1898 \\ 0.2992 & 0.2854 & 0.2504 & 0.1649 \\ 0.2930 & 0.2917 & 0.2629 & 0.1524 \end{bmatrix}.$$

—   □

We consider next the state probabilities for the various stages. That is, we examine the distributions for the various X_n, letting $p_k(n) = P(X_n = k)$ for each k in $\mathcal{E}$. To simplify writing, we consider a finite state space $\mathcal{E} = \{0, 1, \ldots, M\}$. We use $\boldsymbol{\pi}(n)$ for the row matrix

$$\boldsymbol{\pi}(n) \equiv [p_0(n) p_1(n) \ldots p_M(n)].$$

As a consequence of Theorem 23.3.1, we have the following.

Theorem 23.3.2. *For a homogeneous Markov sequence, the probability distribution for any* X_n *is determined by the initial distribution (i.e., the distribution for* X_0) *and the transition probability matrix* $\boldsymbol{P}$.

Proof

Suppose that the homogeneous sequence X_N has finite state-space $\mathcal{E} = \{0, 1, \ldots, m\}$. For any $n \geq 0$, let $p_j(n) = P(X_n = j)$ for all $j \in \mathcal{E}$. Put

$$\boldsymbol{\pi}(n) = [p_0(n) p_1(n) \ldots p_M(n)].$$

Then

$\pi(0) = $ the initial probability distribution.
$\pi(1) = \pi(0)\boldsymbol{P}$.

$\vdots$

$\pi(n) = \pi(n-1)\boldsymbol{P} = \pi(0)\boldsymbol{P}^{(n)} = \pi(0)\boldsymbol{P}^n$ the nth-period distribution.

The last expression is an immediate consequence of Theorem 23.3.1.

— $\square$

Example 23.3.2: Inventory Problem (Continued)

In the inventory system for Example 23.1.3 and 23.3.1, suppose the initial stock is $M = 3$. This means that

$$\pi(0) = [0\ 0\ 0\ 1].$$

The product of $\pi(0)$ and $\boldsymbol{P}^3$ is the fourth row of $\boldsymbol{P}^3$, so that the distribution for X_3 is

$$\pi(3) = [p_0(3)\ p_1(3)\ p_2(3)\ p_3(3)] = [0.2930\ 0.2917\ 0.2629\ 0.1524].$$

Thus, given a stock of $M = 3$ at startup, the probability is 0.2917 that $X_3 = 1$. This is the probability of one unit in stock at the end of period number three.

— $\square$

Remarks

1. A similar treatment shows that for the nonhomogeneous case the distribution at any stage is determined by the initial distribution and the class of one-step transition matrices. In the nonhomogeneous case, transition probabilities $p_{n,n+1}(i,j)$ depend on the stage n.

2. A discrete-parameter Markov process, or Markov sequence, is characterized by the fact that each member X_{n+1} of the sequence is conditioned by the value of the previous member of the sequence. This one-step stochastic linkage has made it customary to refer to a Markov sequence as a Markov chain. In the discrete-parameter Markov case, we use the terms process, sequence, or chain interchangeably.

We conclude this section with an extension of the Markov property that is useful in the study of Markov sequences with associated costs and rewards.

Theorem 23.3.3. *Suppose X_N is Markov and $\{D_n : 1 \le n\}$ is such that for each $n \ge 0$ the pair $\{D_{n+1}, U_n\}$ is independent. For a fixed Borel function g, put $G_{n+1} = g(X_n, D_{n+1})$ for all $n \ge 0$. Then*

$$\{G_{n+k+1}, U_n\} \text{ ci} |\ X_{n+j} \qquad and \qquad \{G_{n+k+1}, X_n\} \text{ ci} |\ X_{n+j}$$

*for all $n \geq 0$, all j, k, $0 \leq j \leq k$. More generally, if $V^n = (U^n, G_n, G_{n+1}, \ldots)$,
then*

$$\{V^{n+k+1}, U_n\} \text{ ci} | \ X_{n+j} \qquad and \qquad \{V^{n+k+1}, X_n\} \text{ ci} | \ X_{n+j}$$

for all $n \geq 0$, all j, k, $0 \leq j \leq k$.

Proof

For any permissible n, j, and k, we have $\{D_{n+k+1}, (X_{n+k}, U_n, X_{n+j})\}$ is independent. By the Markov property, we have we have $\{X_{n+k}, U_n\}$ ci| X_{n+j}.

By Theorem 23.2.3, $\{(X_{n+k}, D_{n+k+1}), U_n\}$ ci| X_{n+j}.

Use of (CI9) shows that $\{G_{n+k+1}, U_n\}$ ci| X_{n+j} and $\{G_{n+k+1}, X_n\}$ ci| X_{n+j}.

A parallel argument establishes the more general case.

—   □

23.4.   The Transition Diagram and the Transition Matrix

The examples in previous sections suggest that a Markov chain is a dynamic system, evolving in time. On the other hand, the stochastic behavior of a homogeneous chain is determined completely by the probability distribution for the initial state and the one-step transition probabilities $p(i, j)$ as presented in the transition matrix $\boldsymbol{P}$. The time-invariant transition matrix may convey a static impression of the system. However, a simple geometric representation, known as the transition diagram, makes it possible to link the unchanging structure, represented by the transition matrix, with the dynamics of the evolving system behavior.

Definition 23.4.1. *A transition diagram for a homogeneous Markov chain is a linear graph with one node for each state and one directed edge for each possible one-step transition between states (nodes).*

We ignore, as essentially impossible, any transition that has zero transition probability. Thus, the edges on the diagram correspond to positive one-step transition probabilities between the nodes connected. Since for some pair (i, j) of states, we may have $p(i, j) > 0$ but $p(j, i) = 0$, we may have a connecting edge between two nodes in one direction, but none in the other. The system can be viewed as an object jumping from state to state (node to node) at the successive transition times. As we follow the trajectory of this object, we achieve a sense of the evolution of the system.

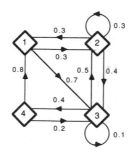

FIGURE 23.1. Transition diagram for Example 23.4.1.

Example 23.4.1: (See Figure 23.1)

Determine the transition diagram for the homogeneous chain with transition probability

$$P = \begin{bmatrix} 0 & 0.3 & 0.7 & 0 \\ 0.3 & 0.3 & 0.4 & 0 \\ 0 & 0.5 & 0.1 & 0.4 \\ 0.8 & 0 & 0.2 & 0 \end{bmatrix}.$$

SOLUTION AND DISCUSSION

The transition diagram is shown in Figure 23.1. It has four nodes, corresponding to four possible states of the system, which are numbered one through four.

- The edges leaving node i correspond to the entries on row i in P. Thus the total probability "leaving" a node must be one. This provides a consistency check for the diagram.

- Nodes (states) 2 and 3 have self-loops; this corresponds to positive $p(2,2)$ and $p(3,3)$. These are entries on the main diagonal of P.

- Nodes 1 and 4 are connected by a single edge, directed from 4 to 1. This corresponds to the fact $p(1,4) = 0$ (the element on the first row and fourth column of P) but $p(4,1) = 0.8 > 0$ (the element on the fourth row and first column).

- There is no edge connecting node 2 with node 4. This corresponds to the fact that both $p(2,4) = 0$ and $p(4,2) = 0$.

- Each of the pairs $(1,2)$, $(2,3)$, and $(3,4)$ has a bilateral connection, as shown on the diagram. This corresponds to the fact that in each case, both $p(i,j) > 0$ and $p(j,i) > 0$.

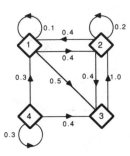

FIGURE 23.2. Transition diagram for Example 23.4.2.

- Every state for the chain can be reached from every other state in the chain, for there is a path with positive probability joining any two nodes. However, not every state can be reached in one step from every other state. For example, it takes at least two steps to go from state four to state two.

—   □

Example 23.4.2: (See Figure 23.2)

The diagram in Figure 23.2 illustrates some possibilities not encountered in the previous example. Its transition matrix P is

$$P = \begin{bmatrix} 0.1 & 0.4 & 0.5 & 0 \\ 0.4 & 0.2 & 0.4 & 0 \\ 0 & 1 & 0 & 0 \\ 0.3 & 0 & 0.4 & 0.3 \end{bmatrix}.$$

Discussion

- If the system is in state 4 initially, it will remain in that state on the next transition with probability 0.3. With probability 0.3 it will go to state 1, and with probability 0.4 it will go to state 3. The system can remain in state 4 indefinitely, but always with the same positive probability it will leave on the next transition. Once it leaves state 4, however, it can never return. As we see in Section 23.6, such a state is called transient.

- If the system reaches state 3, it must go to state 2 at the next transition. However, state 3 is not transient, since it can be visited again at a later stage.

- Once the system reaches any of the states 1, 2, or 3, it never leaves that subset of states. Once it reaches any of these states, it then has positive probability of going to any other state in the subset. As the system continues to evolve in time, each state will be reached an unlimited number of times. We see in Section 23.6 that such states are called recurrent, and that the class $\{1, 2, 3\}$ is called a recurrent class. This class is closed in the sense that no state outside the class can be reached from any state in the class.

- As a result of the manner in which the states are numbered, the element in the fourth column for each recurrent state is zero. Consider the submatrix P_1 obtained by removing the fourth column and the fourth row from P.

$$P_1 = \begin{bmatrix} 0.1 & 0.4 & 0.5 \\ 0.4 & 0.2 & 0.4 \\ 0 & 1 & 0 \end{bmatrix}.$$

The submatrix P_1 is a stochastic matrix and is thus the transition matrix of a homogeneous Markov chain. Removal of the fourth column and the fourth row from P effectively removed state four from the chain. The subchain with states $\{1, 2, 3\}$ is a homogeneous Markov chain in its own right. This sort of reduction or decomposition of chains is discussed in Section 23.8 (see Example 23.8.1 and the accompanying discussion).

———   □

The previous examples are constructed artificially for illustrative purposes. It may be instructive to consider the diagram for a chain arising from an application.

Example 23.4.3: Inventory Problem (See Figure 23.3)

The inventory problem described in Example 23.1.3 gives rise to a homogeneous chain with transition matrix (which we round here to two places)

$$P = \begin{bmatrix} 0.08 & 0.18 & 0.37 & 0.37 \\ 0.63 & 0.37 & 0 & 0 \\ 0.26 & 0.37 & 0.37 & 0 \\ 0.08 & 0.18 & 0.37 & 0.37 \end{bmatrix}.$$

DISCUSSION OF TRANSITION DIAGRAM

In keeping with the development in Example 23.1.3, we number the states 0 through 3.

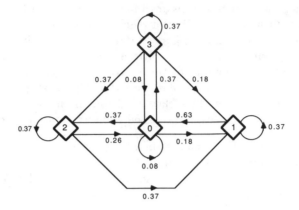

FIGURE 23.3. Transition diagram for Example 23.4.3.

- Each state has a self-loop, since there is a positive probability of remaining in any state. For example, no sales in states 1, 2, or 3 result in remaining in that state. Since state 0 corresponds to zero stock at the end of a period, with subsequent restocking, three sales (i.e., demand of three or more) are required to have state 0 at the end of the next period. The probability of remaining in state 0 is therefore of lower probability than that of staying in one of the other states.

- Every state may be reached from every other state, though not necessary in one period.

- For this system, the more interesting questions concern certain limiting or averaging properties, which the transition diagram does not address directly.

—   □

23.5.   Visits to a Given State in a Homogeneous Chain

In this section, we obtain various probabilities related to the number of visits to a given state in a homogeneous chain. These results are usually obtained by intuitive arguments that are quite plausible and persuasive. However, such arguments lack the precision desired for careful development of the model. We consider an arbitrarily chosen but fixed state j. We let

T_1 be the time of the first arrival or visit to state j, after the initial state, which may or may not be j. The event

$$\{T_1 = n\} = \{X_1 \neq j, X_2 \neq j, \ldots, X_{n-1} \neq j, X_n = j\}.$$

Definition 23.5.1. *The* first-passage-time probabilities *from state i to state j are given by*

$$F_k(i,j) = P(T_1 = k | X_0 = i) = E[I_k(T_1) | X_0 = i] \qquad for\ k \geq 1.$$

If $i = j$, then the $F_k(j,j)$, $k \geq 1$ are the recurrence-time probabilities *for state j.*

Remark: We use here, and elsewhere, I_k instead of the more complete symbol $I_{\{k\}}$ to designate the indicator function for the single-element set $\{k\}$.

The first pattern we identify is quite intuitive. The probability of the first passage from state i to state j in k steps is the sum over $b \neq j$ of the products of the probability of one-step passage from i to b and the first passage from b to j in $k - 1$ steps. While this result is quite plausible, an argument such as the following is needed to establish it carefully.

Theorem 23.5.2.

$$F_k(i,j) = \begin{cases} p(i,j) & for\ k = 1 \\ \sum_{b \neq j} p(i,b) F_{k-1}(b,j) & for\ k > 1 \end{cases} \quad \text{a.s.}$$

Proof

1. For $k = 1$, $T_1 = 1$ iff $X_1 = j$. Hence $F_1(i,j) = P(X_1 = j | X_0 = i) = p(i,j)$.

2. For $k > 1$,

$$
\begin{aligned}
F_k(i,j) &= \sum_{b \neq j} P(X_1 = b, X_2 \neq j, \ldots, X_{k-1} \neq j, X_k = j | X_0 = i) \\
&= E[I_{Q_1}(X_1) I_{Q_2}(X_2) \ldots I_{Q_k}(X_k) | X_0 = i],
\end{aligned}
$$

where $Q_r = \mathcal{E} - \{j\}$ for $1 \leq r \leq k - 1$ and $Q_k = \{j\}$. We use first (CE9), then (CE8) and the Markov property (M) to assert

$$
\begin{aligned}
F_k(i,j) &= E\big[E[I_{Q_1}(X_1) \ldots I_{Q_k}(X_k) | X_0, X_1] \,\big|\, X_0 = i\big] \\
&= E\big[I_{Q_1}(X_1) E[I_{Q_2}(X_2) \ldots I_{Q_k}(X_k) | X_1] \,\big|\, X_0 = i\big] \\
&= \sum_{b \in \mathcal{E}} I_{Q_1}(b) E[I_{Q_2}(X_2) \ldots I_{Q_k}(X_k) | X_1 = b] p(i,b) \quad \text{a.s.}
\end{aligned}
$$

Now $I_{Q_1}(b) = 1$ iff $b \neq j$. Also, for $b \neq j$ we have

$$E[I_{Q_2}(X_2) \ldots I_{Q_k}(X_k) | X_1 = b] = F_{k-1}(b,j) \quad \text{a.s.}$$

Thus

$$F_k(i,j) = \sum_{b \neq j} F_{k-1}(b,j)p(i,b).$$

□

It is sometimes useful to have this in matrix form. As before, let $\boldsymbol{P} = [p(i,j)]$, and let $\boldsymbol{P}(\cdot, j)$ be the jth column and $\boldsymbol{P}(i, \cdot)$ be the ith row. Further, let $\boldsymbol{Q}_j$ be the matrix obtained from $\boldsymbol{P}$ by replacing column j with zeros. Then

$$\boldsymbol{F}_k(\cdot, j) = \begin{cases} \boldsymbol{P}(\cdot, j) & \text{for } k = 1 \\ \boldsymbol{Q}_j \mathcal{F}_{k-1}(\cdot, j) & \text{for } k > 1 \end{cases} \qquad \text{[column matrices]}.$$

The event that state j is reached from state i is the event that the first passage time is k for some finite k. Thus, we make the following.

Definition 23.5.3. $F(i,j) = P(T_1 < \infty | X_0 = i) = E[I_{\mathcal{N}_1}(T_1)|X_0 = i] = \sum_{k=1}^{\infty} F_k(i,j)$, where $\mathcal{N}_1$ is the set of positive integers.

The next result expresses $F(i,j)$ as the probability of getting from i to j in one step plus the sum of the products for each $b \neq j$ of the probability of going from i to b in one step times the probability of getting from b to j. This quite intuitive result may be established as follows.

Theorem 23.5.4.

$$F(i,j) = \sum_{k=1}^{\infty} F_k(i,j) = p(i,j) + \sum_{b \neq j} p(i,b)F(b,j).$$

Proof

$$
\begin{aligned}
E[I_{\mathcal{N}_1}(T_1)|X_0 = i] &= \sum_{k=1}^{\infty} E[I_k(T_1)|X_0 = i] = \sum_{k=1}^{\infty} F_k(i,j) \\
&= p(i,j) + \sum_{k=2}^{\infty}\sum_{b \neq j} p(i,b)F_{k-1}(b,j) \qquad \text{by Thm 23.5.2} \\
&= p(i,j) + \sum_{b \neq j} p(i,b)\sum_{k=2}^{\infty} F_{k-1}(b,j) \\
&= p(i,j) + \sum_{b \neq j} p(i,b)F(b,j).
\end{aligned}
$$

□

THE NUMBER OF VISITS TO STATE j

The number of visits to state j, including the initial state if it is state j, is designated N_j. Since we have a visit to state j at stage n iff $X_n(\omega) = j$ iff $I_j(X_n(\omega)) = 1$, this random variable may be expressed in the following useful fashion

$$N_j = \sum_{n=0}^{\infty} I_j(X_n).$$

The next theorem states a very basic result concerning N_j. Since a key step in the proof uses a consequence of the strong Markov property, we defer the proof to Section 23.10.

Theorem 23.5.5.

$$P(N_j = m | X_0 = j) = (1 - F(j,j)) F(j,j)^{m-1}$$

$$P(N_j = m | X_0 = i) = \begin{cases} 1 - F(i,j) & \text{for } m = 0, \ i \neq j \\ F(i,j)(1 - F(j,j)) F(j,j)^{m-1} & \text{for } m > 0, \ i \neq j \end{cases}.$$

—   □

As a consequence, we have the following zero-one law for the probability that N_j is finite, given that the system starts in state j.

Theorem 23.5.6 (A Zero-One Law)

$$P(N_j < \infty | X_0 = j) = \begin{cases} 1 & \text{if } F(j,j) < 1 \\ 0 & \text{if } F(j,j) = 1 \end{cases}.$$

Proof

Since $P(N_j < \infty | X_0 = j) = \sum_{m-1}^{\infty} P(N_j = m | X_0 = j)$, we have by Theorem 23.5.5

$$P(N_j < \infty | X_0 = j) = \begin{cases} 0 & \text{for } F(j,j) = 1 \\ (1 - F(j,j)) \sum_{m=1}^{\infty} F(j,j)^{m-1} & \text{for } F(j,j) < 1 \end{cases}.$$

The geometric series shows the value for $F(j,j) < 1$ must be one, as asserted.

—   □

We next introduce a concept that is useful in the classification of states in a homogeneous Markov chain.

Definition 23.5.7. *For each pair $\{i, j\}$ of states in a time-homogeneous Markov chain with discrete state space, let*

$$R(i, j) = E[N_j | X_0 = i].$$

The matrix $\boldsymbol{R} = [R(i, j)]$ is called the potential matrix *for the chain.*

Theorem 23.5.8.

$$R(j,j) \;=\; \begin{cases} \infty & \text{if } F(j,j)=1 \\ 1/\big(1-F(j,j)\big) & \text{if } F(j,j)<1 \end{cases}$$
$$R(i,j) \;=\; F(i,j)R(j,j) \quad \text{for } i \neq j.$$

Proof

Put $p = 1 - F(j,j)$ and $q = F(j,j)$.

1. For $i = j$, we have by Theorem 23.5.5 $P(N_j = m | X_0 = j) = pq^{m-1}$ so that $R(i,j) = E[N_j | X_0 = i] = p\sum_{m=1}^{\infty} mq^{m-1}$. The series diverges for $q = 1$ and converges to $1/(1-q)^2 = 1/p^2$ for $q < 1$, from which the desired result is immediate.

2. For $i \neq j$, we have $P(N_j = m | X_0 = i) = F(i,j)pq^{m-1}$, so that $R(i,j) = F(i,j)R(j,j)$.

— $\qquad\qquad\qquad\qquad\qquad\qquad\qquad\qquad\qquad\qquad\qquad\qquad$ $\square$

We obtain an alternate series expansion for $R(i,j)$ that yields an important matrix expression.

Theorem 23.5.9.

$$R(i,j) = \sum_{n=0}^{\infty} p^n(i,j), \qquad \text{where} \quad p^0(i,j) = \delta_{ij}$$

$$R = I + PR = I + RP.$$

Proof

$R(i,j) = \sum_{n=0}^{\infty} E[I_j(X_n) | X_0 = i] = \sum_{n=0}^{\infty} p^n(i,j)$.
Using the definition of a matrix product, we have

$$RP \;=\; \left(\sum_{k} R(i,k)p(k,j) \right) = \left(\sum_{k} \sum_{n=0}^{\infty} p^n(i,k)p(k,j) \right)$$
$$=\; \left(\sum_{n=0}^{\infty} \sum_{k} p^n(i,k)p(k,j) \right).$$

Applying the Chapman-Kolmogorov equation to each element of the matrix, we get

$$RP = \left(\sum_{n=0}^{\infty} p^{n+1}(i,j) \right) = \left(\sum_{n=1}^{\infty} p^n(i,j) \right) = R - I.$$

A similar argument shows $PR = R - I$.

— $\qquad\qquad\qquad\qquad\qquad\qquad\qquad\qquad\qquad\qquad\qquad\qquad$ $\square$

23.6.   Classification of States in Homogeneous Chains

The concepts and theorems of Section 23.5 make possible some important classifications of the states in discrete-state-space Markov chains. The first classification is as follows:

Definition 23.6.1. *State j in a homogeneous Markov chain with discrete state space $\mathcal{E}$ is called:*

Transient *iff $F(j,j) < 1$ and* Recurrent *iff $F(j,j) = 1$.*

State j is transient if, starting in j, there is a positive probability that the system will not return to state j. This situation can be visualized in terms of a transition diagram. If there is only a finite number of states, the system can reach a part of the diagram from which there are no paths returning to the node for state j. On the other hand, if state j is recurrent, then the system can move only to nodes from which there is a return path to the node for state j.

If the state space $\mathcal{E}$ is discrete, but infinite, some complications may arise for recurrent states. We may have $F(j,j) = P(T_1 < \infty | X_0 = j) = 1$, yet have $E[T_1 | X_0 = j]$ infinite. The conditional probability is one that the recurrence time is finite, yet the conditional expectation of that time is infinite.

Definition 23.6.2. *A recurrent state j in a homogeneous Markov chain is*

Positive *iff $E[T_1 | X_0 = j] < \infty$ and* Null *iff $E[T_1 | X_0 = j] = \infty$.*

The recurrent null condition can occur only if the state space is infinite. In this introductory treatment, we can ignore this case.

Sometimes a chain has a structure (which may be apparent from the transition diagram) that makes possible returns to a state only on multiples of some integer δ. This condition may be characterized as follows.

Definition 23.6.3. *For recurrent state j in a homogeneous Markov chain with discrete state space, let*

$$\delta \ \text{be the largest integer such that} \ P\left(\bigcup_{n=1}^{\infty} \{T_1 = n\delta\} \,\Big|\, X_0 = j\right) = 1.$$

Then state j is periodic *with period δ if $\delta \geq 2$ and is* aperiodic *if $\delta = 1$.*

Figure 23.4 displays a simple example of a chain in which state 1 has period two. If any state i had a self-loop [i.e., $p(i,i) > 0$], then periodicity would not be possible. An alternate characterization of periodicity is given in the following theorem, which we state without proof.

FIGURE 23.4. A chain with period two.

Theorem 23.6.4. *For recurrent state j in a homogeneous Markov chain with discrete state space, let δ be the greatest common divisor of $A = \{n : p^n(j,j) > 0\}$*

   a. If $\delta > 1$, then state j is periodic with period δ.

   b. If $\delta = 1$, then state j is aperiodic.

—   □

As a result of Theorems 23.5.6 and 23.5.8, in the previous section, we may state the following.

Theorem 23.6.5 (Conditions for Recurrent and Transient States)
Consider a state j in a homogeneous Markov chain with discrete state space.

   a. State j is recurrent iff $F(j,j) = 1$ iff $R(j,j) = \infty$ iff $P(N_j = \infty | X_0 = j) = 1$.

   b. State j is transient iff $F(j,j) < 1$ iff $R(j,j) < \infty$ iff $P(N_j < \infty | X_0 = j) = 1$.

—   □

We may use Theorems 23.5.8 and 23.5.9 to establish the following.

Theorem 23.6.6. *Suppose state j is a transient state in a homogeneous Markov chain with discrete state space. Then*

   (a) $R(i,j) < \infty$ for all $i \in \mathcal{E}$.

   (b) $p^n(i,j) \to 0$ as $n \to \infty$ for all $i \in \mathcal{E}$.

Proof

   a. By Theorem 23.5.8, $R(i,j) = F(i,j)R(j,j) \le R(j,j) < \infty.$

   b. By Theorem 23.5.9, $R(i,j) = \sum_{n=0}^{\infty} p^n(i,j) < \infty$ which implies $p^n(i,j) \to 0$.

—   □

Next we state a very important limit theorem that is useful in characterizing the long-run behavior of many chains arising in practice. We omit the proof of the general theorem. In Section 23.9, we examine the evolution of the process in a very important special case, in a manner that sheds light on the nature of the convergence.

Theorem 23.6.7. *Suppose j is a recurrent state in a homogeneous Markov chain with discrete state space.*

a. *If state j is null, then $p^n(i,j) \to 0$ as $n \to \infty$ for any $i \in \mathcal{E}$.*

b. *If j is positive, aperiodic, then*

$$p^n(j,j) \to \pi_j > 0 \text{ and } p^n(i,j) \to F(i,j)\pi_j \quad \text{as } n \to \infty \text{ for all } i \in \mathcal{E}.$$

—   □

The next concepts we introduce have a simple interpretation in terms of the transition diagram. They provide the basis for important analysis of homogeneous chains.

Definition 23.6.8. *In a homogeneous Markov chain with discrete state space, we say*

State j is accessible to state i, designated $i \to j$, iff there is an $n \geq 0$ such that $p^n(i,j) > 0$.

States i,j communicate, designated $i \leftrightarrow j$ or $j \leftrightarrow i$, iff both $i \to j$ and $j \to i$.

Remarks

1. The condition $i \to j$ holds iff there is a path in the transition diagram from node i to node j, since if there is a path of length n edges then $p^n(i,j)$ is at least as great as the product of the transition probabilities on these edges.

2. Since we take $p^0(i,j) = \delta_{ij}$, the condition $p^0(j,j) = 1$ implies that each state j communicates with itself.

The following theorem is an easy consequence of these facts and of Theorems 23.5.8 and 23.5.9.

Theorem 23.6.9 (Communication of States as an Equivalence Relation)

a. *$i \to j$ iff $F(i,j) > 0$ iff $R(i,j) > 0$.*

b. *$\to$ is a reflexive, transitive relation on $\mathcal{E}$. [$j \to j$ and $i \to j$, $j \to k$ implies $i \to k$.]*

    c. ↔ *is an equivalence relation on* $\mathcal{E}$ *[reflexive, transitive and symmetric].*

—   □

Remark: As an equivalence relation, ↔ determines equivalence classes (i.e., subsets whose members communicate with one another).

We consider some important types of subsets of $\mathcal{E}$.

Definition 23.6.10. *Let $\mathcal{E}$ be the state space for a homogeneous Markov chain and let C be a subset of $\mathcal{E}$. Then*

1. C *is* closed *iff no state outside of C is accessible to any state in C.*

2. *If the single-element set $C = \{j\}$ is closed, the state j is an* absorbing *state.*

3. *Closed set C is* irreducible *iff no proper subset of C is closed.*

4. *The chain is an* irreducible *chain iff the only closed set in $\mathcal{E}$ is $\mathcal{E}$ itself.*

5. C *is* communicating *iff each state in C communicates with every state in C.*

6. *The* accessible set A_i *for i is the set of those states j that are accessible to state i.*

We summarize several significant facts in numbered theorems for ease of reference.

Theorem 23.6.11. *For a homogeneous Markov chain with discrete state space $\mathcal{E}$*

1. *State j is absorbing iff $p(j,j) = 1$.*

2. *The accessible set A_i for state i is closed.*

3. *A closed set C is communicating iff it is irreducible.*

Proof

For the most part, proofs are based on elementary properties of → and ↔ and elementary properties of the transition probabilities. The essential patterns can be visualized in terms of the transition diagram. The total probability on edges leaving a node must add to one.

1. $p(j,j) = 1$ iff the only state accessible to state j is state j.

2. If any state can be reached from A_i that state is accessible to state i and hence belongs to A_i.

3. If C is closed and communicating, then no proper subset D could be closed, for any set in C is accessible to every state in D.

On the other hand, if C is closed and irreducible, then the accessible set A_i for any state in C cannot be a proper subset and hence must be C itself. Thus every state is accessible to every state, so that C is communicating.

$\square$

Theorem 23.6.12. *Let i and j be states in a homogeneous Markov chain with discrete state space.*

a. *If $i \leftrightarrow j$, then i is recurrent iff j is recurrent.*

b. *If $i \rightarrow j$ and i is recurrent, then $i \leftrightarrow j$.*

c. *If $i \rightarrow j$ and i is recurrent, then j is recurrent.*

Proof

1. Given $i \leftrightarrow j$. To show $R(i,i) = \infty$ iff $R(j,j) = \infty$. By Theorem 23.5.9, it is sufficient to show that

$$\sum_{n=0}^{\infty} p^n(i,i) = \infty \qquad \text{iff} \qquad \sum_{n=0}^{\infty} p^n(j,j) = \infty.$$

To show that $\sum_{n=0}^{\infty} p(j,j) = \infty$, it is sufficient to show that

$$\sum_{n=a}^{\infty} p(j,j) = \infty \qquad \text{for some } a \geq 0.$$

Suppose $R(i,i) = \infty$. Since $i \leftrightarrow j$, there exist integers N and M such that $A = p^N(i,j) > 0$ and $B = p^M(j,i) > 0$. By the Chapman-Kolmogorov equation,

$$p^{N+M+k}(j,j) \geq p^M(j,i)p^k(i,i)p^N(i,j) = ABp^k(i,i).$$

Thus

$$\sum_{n=N+M}^{N+M+s} p^n(j,j) \geq AB\sum_{k=0}^{s} p^k(i,i) \rightarrow \infty \qquad \text{as } s \rightarrow \infty.$$

This implies $\sum_{n=N+M}^{\infty} p^n(j,j) = \infty$, so that $R(j,j) = \infty$.

The same argument may be carried out with i and j interchanged.

2. Given $i \rightarrow j$ and i is recurrent. To show $j \rightarrow i$, it is sufficient to show that $F(j,i) > 0$.

Let B_n be the set of states that may be reached from i in n steps.

Then each state in B_{n+1} may be reached in one step from some state in B_n. If we can show $F(c, i) = 1$ for each state in B_n for each n, then proposition (b) will be proved, since each state accessible to i will be in some B_n. We use the pattern provided by Theorem 23.5.4, and set up an inductive argument. Since $p(i, b) = 0$ for b not in B_1, we have

$$1 = F(i, i) = p(i, i) + \sum_{b \in B_1} p(i, b)F(b, i).$$

Since $p(i, i) + \sum_{b \in B_1} p(i, b) = 1$, we must have $F(b, i) = 1$ for all $b \in B_1$.

Suppose $F(b, i) = 1$ for all $b \in B_n$. Then for each such b,

$$1 = F(b, i) = p(b, i) + \sum_{c \in B_{n+1}} p(b, c)F(c, i).$$

Since $p(b, i) + \sum_{c \in B_{n+1}} p(b, c) = 1$, we must have $F(c, i) = 1$ for all $c \in B_{n+1}$.

The desired consequence follows by the principle of mathematical induction.

3. If $i \to j$ and i is recurrent, then by (b) $i \leftrightarrow j$. By (a), it follows that j is recurrent.

—   □

We close this section by gathering, for ready reference, a number of useful propositions. Some of them are easy consequences of previous theorems. Others, principally (a) and (c), are stated without proof.

Theorem 23.6.13. *Consider a homogeneous Markov chain with discrete state space $\mathcal{E}$.*

a. *If $\mathcal{E}$ is irreducible, then all states are either (1) transient, (2) recurrent null, or (3) recurrent positive.*

b. *If $\mathcal{E}$ is finite, then no state is recurrent null and not all states are transient.*

c. *If class C is irreducible, all states in the class are either (1) aperiodic or (2) periodic with the same period.*

d. *If class C is finite irreducible, then all its states are recurrent positive.*

e. *If class C is a closed, irreducible set of recurrent states and i is a transient state, then $F(i, j) = F(i, k)$ for all $j, k \in C$.*

—   □

23.7.   Recurrent States and Limit Probabilities

In this section we consider, without proof, several important limit theorems for homogeneous Markov chains with discrete state space $\mathcal{E}$. In particular, we are interested in states that are recurrent, positive, aperiodic. As in Section 23.3, we let $\pi(n) = [p_0(n)p_1(n)\ldots]$ where $p_j(n) = P(X_n = j)$.

We let $\pi = [\pi_0\ \pi_1 \ldots]$ and $\mathbf{1} = [1\ 1\ldots]^T$ (a column matrix).

Theorem 23.7.1. *Suppose a homogeneous Markov chain with discrete state space $\mathcal{E}$ is irreducible aperiodic. Then all states are recurrent positive iff the following system has a solution:*

$$\pi_j = \sum_{i\in\mathcal{E}} \pi_i p(i,j) \qquad \sum_{i\in\mathcal{E}} \pi_i = 1.$$

In matrix form, this system is $\pi P = \pi$, $\pi \mathbf{1} = 1$.

If there is a solution π, it satisfies the conditions

(a) It is unique.

(b) It is strictly positive.

(c) $\lim_n p^n(i,j) = \pi_j$ *for all i, $j \in \mathcal{E}$.*

$\hspace{10cm}\Box$

Corollary 23.7.2. *Suppose a homogeneous Markov chain with finite state space $\mathcal{E}$ is irreducible aperiodic. Then*

$$\pi P = \pi \qquad \text{with} \quad \pi \mathbf{1} = 1$$

has an unique solution and $\lim_n p^n(i,j) = \lim_n 1/n \sum_{k=1}^{n} p^k(i,j) = \pi_j$ *for all i, $j \in \mathcal{E}$.*

The limiting distribution is referred to variously as the long-run distribution or as the invariant distribution. The term invariant distribution comes from the fact that

If $\pi(n) = \pi$, then $\pi(n+1) = \pi(n)P = \pi P = \pi$.

The term long-run distribution comes from the fact that

$$\lim_n \pi(n) = \lim_n \pi(0)P^n = \pi(0) \begin{bmatrix} \pi_0 & \pi_1 & \cdots & \pi_M \\ \pi_0 & \pi_1 & \cdots & \pi_M \\ \vdots & \vdots & \ddots & \vdots \\ \pi_0 & \pi_1 & \cdots & \pi_M \end{bmatrix} = \pi.$$

Recurrent, positive, aperiodic states are sufficiently important to have a name.

Definition 23.7.3. *A state j is ergodic iff it is recurrent, positive, aperiodic. A communicating class C is ergodic if any state is ergodic. A chain is ergodic iff it is irreducible and all states are ergodic.*

Remark: If any state in a communicating class C is ergodic, then all states must be ergodic.

Theorem 23.7.4 (Limit Laws for Ergodic States) *If j is an ergodic state in a homogeneous Markov chain with discrete state space, then*

1. $\pi_j = \lim_n p^n(j,j) = \dfrac{1}{m(j)}$ *where* $m(j) = E[T_1|X_0 = j] = \sum_{n=0}^{\infty} n F_n(j,j)$.

2. $\dfrac{1}{n+1} \sum_{m=0}^{n} I_j(X_m) \to \pi_j$ a.s. *as* $n \to \infty$.

— $\qquad\qquad\qquad\qquad\qquad\qquad\qquad\qquad\qquad\qquad\qquad\qquad$ $\square$

23.8.   Partitioning Finite Homogeneous Chains

If a homogeneous Markov chain has finite state space $\mathcal{E}$, the state space may be partitioned into the class of transient states and the class of recurrent states. The equivalence relation $\leftrightarrow$ ensures that the class of recurrent states is either irreducible or may be partitioned uniquely into closed, irreducible, communicating subclasses. If j is in such a closed subset C, then $\sum_{k \in C} p(j,k) = 1$. Therefore, such a subclass constitutes an irreducible, recurrent chain in itself. If the process reaches any of the recurrent states, it is absorbed or trapped in the corresponding communicating subclass.

Definition 23.8.1. *We refer to a closed, irreducible, communicating subclass of recurrent states in a homogeneous Markov as a* subchain.

By proper renumbering of the states, so that members of a subclass have consecutive numbers, the transition matrix is put into a form that displays the subchain structure. It is convenient to make all recurrent states precede the class of transient states. Then there is a finite number of subchains and a subclass of transient states. When partitioned, the matrix P, under the renumbering scheme, has the form

$$
P = \begin{bmatrix} P_1 & 0 & 0 & \cdots & 0 \\ 0 & P_2 & 0 & & 0 \\ \vdots & & & & \\ 0 & 0 & 0 & & P_s \\ Q_1 & Q_2 & Q_3 & \cdots & Q \end{bmatrix}.
$$

Each submatrix P_k is a stochastic matrix corresponding to the subchain C_k, which is an irreducible, recurrent Markov chain. The following example illustrates this pattern.

Example 23.8.1: A Matrix Displaying the Subchains and Transient States

Consider the transition matrix

$$
P = \begin{bmatrix}
0.2 & 0.5 & 0.3 & 0 & 0 & 0 & 0 \\
0.6 & 0.1 & 0.3 & 0 & 0 & 0 & 0 \\
0.2 & 0.7 & 0.1 & 0 & 0 & 0 & 0 \\
0 & 0 & 0 & 0.6 & 0.4 & 0 & 0 \\
0 & 0 & 0 & 0.5 & 0.5 & 0 & 0 \\
0.1 & 0.3 & 0 & 0.2 & 0.1 & 0.1 & 0.2 \\
0.1 & 0.2 & 0.1 & 0.2 & 0.2 & 0.2 & 0
\end{bmatrix}.
$$

It is apparent that the submatrices

$$
P_1 = \begin{bmatrix}
0.2 & 0.5 & 0.3 \\
0.6 & 0.1 & 0.3 \\
0.2 & 0.7 & 0.1
\end{bmatrix}
\quad \text{and} \quad
P_2 = \begin{bmatrix}
0.6 & 0.4 \\
0.5 & 0.5
\end{bmatrix}
$$

are stochastic matrices for recurrent, irreducible Markov chains, since each subclass is closed, communicating. Under the indicated numbering, 1 through 7, the transient class consists of $\{6, 7\}$. If the system is in state 6, it may remain in that state or go to state 7 or to one of the recurrent states. If it is in state 7, it may go to state 6 or one of the recurrent states. Hence, states 6 and 7 communicate. But they do not form a closed subclass, since other states may be reached from this subclass. If the system goes from either state 6 or state 7 into one of the recurrent states, then it is lost to the transient subclass, and is absorbed into whichever subchain it first reaches. If any state in the first subchain is reached from a transient state, then every state in that subchain is reached from that transient state. Thus, for example, $F(6, k)$ is the same for each $k = 1, 2, 3$.

———   □

The problem of identifying the transient class and the subchains by examining the transition matrix P may be tedious. If the number of states is not too large, the transition diagram is usually helpful in identifying clusters of communicating states.

We consider an alternate approach, which is useful if there is a computer program available for determining powers P^n of the transition matrix. To simplify exposition, we assume that the subchains are aperiodic, as they usually are.

1. The first task is to identify the transient states. By Theorem 23.6.6b and Corollary 23.7.2, state j is transient iff $p^n(i, j) \to 0$, for all $i \in \mathcal{E}$. This means that for large enough n, the jth column of P^n will be essentially zero.

Remove the rows of P^n corresponding to the transient states. The remaining rows correspond to recurrent states.

2. If there is only one subchain, consisting of the class of all recurrent states, then by Corollary 23.7.2, for each recurrent pair i, $j p^n(i,j) \to \pi_j > 0$. This means that for large enough n, all rows in P^n corresponding to recurrent states will be approximately the same. Thus, if rows for transient states are removed, all remaining rows must be essentially the same.

3. If there is more than one subchain, let C consist of the states in one of these subchains. Now

For i in C, $p^n(i,j) \to 0$ for j not in C and $p^n(i,j) \to \pi_j > 0$ for j in C.

Thus, for n large enough, the rows in P^n for states in a subchain must be alike, and the essentially nonzero elements must be in columns corresponding to states in the subchain.

Once the states of any recurrent subchain are identified, the corresponding rows may be removed and the remaining rows examined similarly to determine another subchain.

As an illustration of this procedure, consider the following.

Example 23.8.2: Identification of Transient States and Recurrent Classes

Consider the transition matrix P for a chain with seven states as follows:

$$P = \begin{bmatrix}
0.1 & 0.2 & 0.1 & 0.3 & 0.2 & 0 & 0.1 \\
0 & 0.6 & 0 & 0 & 0 & 0 & 0.4 \\
0 & 0 & 0.2 & 0.5 & 0 & 0.3 & 0 \\
0 & 0 & 0.6 & 0.1 & 0 & 0.3 & 0 \\
0.2 & 0.2 & 0.1 & 0.2 & 0 & 0.1 & 0.2 \\
0 & 0 & 0.2 & 0.7 & 0 & 0.1 & 0 \\
0 & 0.5 & 0 & 0 & 0 & 0 & 0.5
\end{bmatrix}.$$

The 16th power of P is

$$P^{16} = \begin{bmatrix}
0.0000 & 0.2455 & 0.1993 & 0.2193 & 0.0000 & 0.1395 & 0.1964 \\
0.0000 & 0.5556 & 0.0000 & 0.0000 & 0.0000 & 0.0000 & 0.4444 \\
0.0000 & 0.0000 & 0.3571 & 0.3929 & 0.0000 & 0.2500 & 0.0000 \\
0.0000 & 0.0000 & 0.3571 & 0.3929 & 0.0000 & 0.2500 & 0.0000 \\
0.0000 & 0.2713 & 0.1827 & 0.2010 & 0.0000 & 0.1297 & 0.2171 \\
0.0000 & 0.0000 & 0.3571 & 0.3929 & 0.0000 & 0.2500 & 0.0000 \\
0.0000 & 0.5556 & 0.0000 & 0.0000 & 0.0000 & 0.0000 & 0.4444
\end{bmatrix}.$$

It is apparent that states 1 and 5 are transient states, since columns 1 and 5 have negligibly small values. States 2, 3, 4, 6, and 7 are recurrent.

Consider row 2. It has nonzero entries in columns 2 and 7, so that it appears that the class of states $\{2, 7\}$ constitutes a subchain. This is confirmed by the fact that rows 2 and 7 are essentially identical, but no other row matches these.

This leaves states 3, 4, and 6. Examination of rows 3, 4, and 6 show them to be approximately identical, with nonzero entries in columns 3, 4, and 6. We conclude that class $\{3, 4, 6\}$ constitutes a subchain.

Since this exhausts the state space, we have two transient states and two subchains.

We effectively renumber the states by making the following rearrangements, first for all rows, then for the resulting columns.

$$
\begin{array}{lccccccc}
\text{Old position:} & 3 & 4 & 6 & 2 & 7 & 1 & 5 \\
\text{New position:} & 1 & 2 & 3 & 4 & 5 & 6 & 7
\end{array}
$$

The result is the same as the transition matrix presented in Example 23.8.1, which displays clearly the fact that the chain consists of two ergodic subchains and two transient states (new numbers 6, 7).

— ☐

Viewed from the transient states, each subchain appears as an absorbing "superstate," since once any state in the subchain is reached, the subsequent states are all in that subchain. The probability of transition from transient state i to a "superstate" is the sum of the transition probabilities $p(i, j)$ for all states j in the corresponding subchain. The behavior may be represented by a reduced matrix in which each subchain is considered an absorbing state and the transition probabilities from transient states to the absorbing "superstate" are suitably combined.

Example 23.8.3: Reduced Matrix for the System in Example 23.8.1

For the chain whose transition matrix is given in Example 23.8.1, the reduction of the subchains to absorbing states yields

$$
P_0 = \begin{bmatrix}
1 & 0 & 0 & 0 \\
0 & 1 & 0 & 0 \\
0.4 & 0.3 & 0.1 & 0.2 \\
0.4 & 0.4 & 0.2 & 0.0
\end{bmatrix}.
$$

— ☐

This reduced form, representing the chain as consisting of transient states and an absorbing state for each subchain allows us to address the following important question

> If the chain begins in a transient state, which subchain will the system reach?

This problem is dealt with by determining the $F(i,j)$ from transient state i to a state in each of the recurrent subchains. As noted in Theorem 23.6.13 (e), the probability $F(i,j)$ is the same for each state j in an irreducible, recurrent subclass. This must be the same as $F(i,j)$ in the reduced matrix, where i is a transient state and j refers to the appropriate absorbing state. By Theorem 23.6.7, if j is an aperiodic, recurrent state, $p^n(i,j) \to F(i,j)\pi_j$ as $n \to \infty$. Since for an absorbing state j the stationary probability $\pi_j = 1$, we have

$$F(i,j) = \lim_n p^n(i,j) \qquad \text{for an absorbing state } j.$$

Example 23.8.4: Probability of Reaching the Subchains (Examples 23.8.1 and 23.8.3)

$$P_0^8 = \begin{bmatrix} 1 & 0 & 0 & 0 \\ 0 & 1 & 0 & 0 \\ 0.56 & 0.44 & 0 & 0 \\ 0.51 & 0.49 & 0 & 0 \end{bmatrix}.$$

If the system is in transient state 6, it ends up in subchain 1 with probability 0.56 and in subchain 2 with probability 0.44. If it is in state 7, it ends in subchain 1 with probability 0.51 and in subchain 2 with probability 0.49.

— □

As a further example, consider the following.

Example 23.8.5: Second Numerical Example

A chain has eight states, three of which are transient. These are arranged, as discussed earlier, to give the following transition matrix.

$$P = \begin{bmatrix} 0.1 & 0.5 & 0.4 & 0 & 0 & 0 & 0 & 0 \\ 0.3 & 0.2 & 0.5 & 0 & 0 & 0 & 0 & 0 \\ 0.5 & 0.5 & 0 & 0 & 0 & 0 & 0 & 0 \\ 0 & 0 & 0 & 0.2 & 0.8 & 0 & 0 & 0 \\ 0 & 0 & 0 & 0.8 & 0.2 & 0 & 0 & 0 \\ 0 & 0 & 0 & 0 & 0 & 0.4 & 0.6 & 0 \\ 0.3 & 0.1 & 0 & 0.4 & 0 & 0 & 0 & 0.2 \\ 0.1 & 0 & 0.2 & 0 & 0.1 & 0.6 & 0 & 0 \end{bmatrix}$$

$$C_1 = \{1, 2, 3\} \qquad C_2 = \{4, 5\} \qquad D = \{6, 7, 8\}$$

where C_1 and C_2 are irreducible, recurrent classes, and D is the class of transient states. The reduced matrix is

$$\boldsymbol{P}_0 = \begin{bmatrix} 1 & 0 & 0 & 0 & 0 \\ 0 & 1 & 0 & 0 & 0 \\ 0 & 0 & 0.4 & 0.6 & 0 \\ 0.4 & 0.4 & 0 & 0 & 0.2 \\ 0.3 & 0.1 & 0.6 & 0 & 0 \end{bmatrix}.$$

We raise the reduced matrix to the 24th power to obtain negligible values in the transient columns. The result is

$$\boldsymbol{P}_0^{24} = \begin{bmatrix} 1.0000 & 0.0000 & 0.0000 & 0.0000 & 0.0000 \\ 0.0000 & 1.0000 & 0.0000 & 0.0000 & 0.0000 \\ 0.5227 & 0.4773 & 0.0000 & 0.0000 & 0.0000 \\ 0.5227 & 0.4773 & 0.0000 & 0.0000 & 0.0000 \\ 0.6163 & 0.3864 & 0.0000 & 0.0000 & 0.0000 \end{bmatrix}.$$

The probability of reaching the first subchain from transient state 6 is 0.5227, the probability of reaching the second subchain from transient state 8 is 0.3864, etc.

— $\qquad\qquad\qquad\qquad\qquad\qquad\qquad\qquad\qquad\qquad$ □

23.9. Evolution of Finite, Ergodic Chains

In Section 23.7, we consider several limit theorems for Markov chains with discrete state space. These show that in many important cases, the distribution of state probabilities converges to a long-run distribution as the stage number n becomes large. Corollary 23.7.2 states the facts for finite, irreducible, aperiodic chains, i.e., ergodic chains. The discussion in Section 23.8 shows that the state space for finite chains may be decomposed into a class of transient states and one or more irreducible, recurrent classes. The recurrent classes may be viewed as states for irreducible, recurrent subchains. If we require aperiodicity, which is the usual condition, then these subchains are ergodic.

In this section, we use techniques of generating functions for sequences of real numbers to examine the character of the convergence of finite, ergodic chains. In order to keep the details of the analysis simple, we impose certain restrictions, which are met in many practical chains. The resulting behavior is representative. Then we indicate how the behavior is modified if these restrictions are relaxed. Our argument provides a partial proof of Corollary 23.7.2. It does not show that the convergence must take place, but gives insight into the manner and rapidity of the convergence.

We consider an ergodic, homogeneous Markov chain X_N with finite state space

$$\mathcal{E} = \{0, 1, \ldots, M\}.$$

We let $p_j(n) = P(X_n = j)$ for all $j \in \mathcal{E}$ and all $n \in \mathcal{N} = \{0, 1, 2, \ldots\}$. Put

$$\boldsymbol{\pi}(n) = [p_0(n), p_1(n), \ldots, p_M(n)].$$

Then

$$
\begin{array}{rcl}
\boldsymbol{\pi}(0) & = & \text{the initial probability distribution} \\
\boldsymbol{\pi}(1) & = & \boldsymbol{\pi}(0)\boldsymbol{P} = \text{first-stage distribution} \\
\boldsymbol{\pi}(n) & = & \boldsymbol{\pi}(n-1)\boldsymbol{P} = \boldsymbol{\pi}(0)\boldsymbol{P}^n = n\text{th-stage distribution.}
\end{array}
$$

We wish to study the evolution of $\boldsymbol{\pi}(n)$ with increasing n.

GENERATING FUNCTIONS FOR SEQUENCES OF REAL NUMBERS

Generating functions for random variables are special cases of generating functions for sequences of real numbers. Such generating functions are used extensively in combinatorics and in the study of difference equations. In the latter case, they are often called z-transforms, since in the literature on these applications it is customary to use z rather than s for the parameter. We use the generating function for the sequence $\{p_j(n) : 0 \leq n\}$ of state probabilities in a Markov chain.

Definition 23.9.1. *For a sequence $\{a(n) : 0 \leq n\}$ of real numbers, the generating function $\hat{a}$ is given by*

$$\hat{a}(s) = \sum_{n=0}^{\infty} a(n)s^n \, |s| < s_0 \qquad \text{[usually s is taken to be real].}$$

Remarks

1. If the sequence is bounded, then $\hat{a}$ converges at least for $|s| < 1$.

2. If the sequence has only a finite number of nonzero terms, then $\hat{a}$ is a polynomial, defined for all real or complex values of the parameter s.

The usefulness of generating functions in the study of sequences rests on certain patterns, some of which are tabulated here.

I. Some General Patterns

Sequence	Generating function	Convergence region
$a(n)$	$\hat{a}(s)$	$\lvert s\rvert < s_0$
$ca(n) + db(n)$	$c\hat{a}(s) + d\hat{b}(s)$	Common region
$a(n-1)$	$s\hat{a}(s)$	$\lvert s\rvert < s_0$
$a(n+1)$	$\left(\hat{a}(s) - a(0)\right)/s$	$\lvert s\rvert < s_0$
$c^n a(n)$	$\hat{a}(cs)$	$\lvert cs\rvert < s_0$

The following properties of the geometric series, which converges for $\lvert s\rvert < 1$, yield several useful generating functions.

$$\sum_{n=0}^{\infty} s^n = \frac{1}{1-s} \qquad \sum_{n=0}^{\infty} n s^n = \frac{s}{(1-s)^2}$$

$$\sum_{n=0}^{\infty} n(n-1)s^n = \frac{2s^2}{(1-s)^3} \qquad \sum_{n=0}^{\infty} n^2 s^n = \frac{s^2+s}{(1-s)^3}.$$

II. Some Useful Generating Functions

Sequence	Generating function	Convergence region
$a(n) = a^n$, all n	$\dfrac{1}{1-as}$	$\lvert as\rvert < 1$
$a(n) = n$	$\dfrac{s}{(1-s)^2}$	$\lvert s\rvert < 1$
$a(n) = n^2$	$\dfrac{s^2+s}{(1-s)^3}$	$\lvert s\rvert < 1$

We state, without proof, one important limit theorem on generating functions.

Theorem 23.9.2 (Basic Limit Theorem for Generating Functions)
For a bounded sequence $\{a(n) : 0 \leq n\}$,

$$\sum_{k=0}^{n} a(k) \to A < \infty \quad \text{as } n \to \infty \qquad \textit{iff} \qquad \hat{a}(s) \to A \quad \text{as } s \to 1-0$$

$$\frac{1}{n}\sum_{k=0}^{n} a(k) \to L < \infty \quad \text{as } n \to \infty \qquad (1-s)\hat{a}(s) \to L \quad \text{as } s \to 1-0.$$

—
$\square$

APPLICATION TO THE DISTRIBUTIONS FOR A CHAIN

For each state j, we have a sequence $\{p_j(n) : 0 \leq n\}$ of state probabilities. Let $\hat{p}_j(s)$ be the generating function for that sequence of numbers. That is,

$$\hat{p}_j(s) = \sum_{n=0}^{\infty} p_j(n)s^n.$$

Now consider the sequence $\{\boldsymbol{\pi}(n) : 0 \leq n\}$ of matrices, where $\boldsymbol{\pi}(n) = [p_0(n), p_1(n), \ldots, p_M(n)]$. The generating function for this sequence of matrices is the row matrix, each of whose elements is a generating function

$$\hat{\boldsymbol{\pi}}(s) = [\hat{p}_0(s), \hat{p}_2(s), \ldots, \hat{p}_M(s)].$$

We wish to use $\hat{\boldsymbol{\pi}}(s)$ to find the corresponding sequence $\{\boldsymbol{\pi}(n) : 0 \leq n\}$ of distributions for X_N. As a first step, we note some facts about the matrix of generating functions.

1. If $\mathcal{A}$ is a square matrix and $\boldsymbol{\pi}^*(n) = \boldsymbol{\pi}(n)\mathcal{A}$, then by linearity of the generating function $\hat{\boldsymbol{\pi}}^*(s) = \hat{\boldsymbol{\pi}}(s)\mathcal{A}$.

2. Since $\boldsymbol{\pi}(n+1) = \boldsymbol{\pi}(n)\boldsymbol{P}$, the sequence $\{\boldsymbol{\pi}(n+1) : 0 \leq n\}$ has generating function $(\hat{\boldsymbol{\pi}}(s) - \boldsymbol{\pi}(0))/s$, which must be equal to $\hat{\boldsymbol{\pi}}(s)\boldsymbol{P}$.

To study the sequence $\{\boldsymbol{\pi}(n) : 0 \leq n\}$ of distributions, we exploit the relation

(A)
$$\frac{1}{s}[\hat{\boldsymbol{\pi}}(s) - \boldsymbol{\pi}(0)] = \hat{\boldsymbol{\pi}}(s)\boldsymbol{P}.$$

Now (A) reduces to $\hat{\boldsymbol{\pi}}(s)[\boldsymbol{I} - s\boldsymbol{P}] = \boldsymbol{\pi}(0)$, where $\boldsymbol{I}$ is the identity matrix. We solve for $\hat{\boldsymbol{\pi}}(s)$ by inverting $\boldsymbol{I} - s\boldsymbol{P}$ to get

(B)
$$\hat{\boldsymbol{\pi}}(s) = \boldsymbol{\pi}(0)[\boldsymbol{I} - s\boldsymbol{P}]^{-1} = \boldsymbol{\pi}(0)\boldsymbol{P}^*(s).$$

Again, we note several important facts.

1. $\Delta(s) = \text{Det}[\boldsymbol{I} - s\boldsymbol{P}]$ is a polynomial in s, as are all the cofactors of $[\boldsymbol{I} - s\boldsymbol{P}]$. Since

$$\boldsymbol{P}^*(s) = [\boldsymbol{I} - s\boldsymbol{P}]^{-1} = \frac{\mathcal{A}(s)}{\Delta(s)}$$

    where $\mathcal{A}(s)$ is the adjoint matrix for $[\boldsymbol{I} - s\boldsymbol{P}]$, it follows that $\boldsymbol{P}^*(s)$ is a matrix of rational functions in s, each with denominator $\Delta(s)$.

2. $\Delta(s) = \text{Det}[\boldsymbol{I} - s\boldsymbol{P}]$ is a polynomial of degree $M + 1$, with $\Delta(0) = 1$. Since both $\boldsymbol{P}$ and $\boldsymbol{I}$ are stochastic matrices, it follows that $\text{Det}[\boldsymbol{I} - \boldsymbol{P}] = \Delta(1) = 0$, so that $s_0 = 1$ is a zero of the polynomial $\Delta(s)$. The zeros s_i of $\Delta(s) = \text{Det}[\boldsymbol{I} - s\boldsymbol{P}]$ are the reciprocals of the eigenvalues λ_i of $\boldsymbol{P}$, since the latter are zeros of $\text{Det}[\lambda\boldsymbol{I} - \boldsymbol{P}]$.

3. We suppose that $\Delta(s)$ has distinct zeros $s_0 = 1, s_1, \ldots, s_M$, so that

$$\Delta(s) = (1-s)(1-s/s_1)\ldots(1-s/s_M) = (1-s)(1-\lambda_1 s)\ldots(1-\lambda_M s).$$

In this case, $P^*(s)$ has a partial fraction expansion

$$P^*(s) = \sum_{k\in\mathcal{E}} P_k \frac{1}{1-s/s_k} = \sum_{k\in\mathcal{E}} P_k \frac{1}{1-\lambda_k s}$$

where

$$P_k = \frac{\mathcal{A}(s_k)}{\prod_{j\neq k}(1-s_k/s_j)}.$$

4. The first entry in Table II, shows that $1/(1-as)$ is the generating function for the sequence with terms $a(n) = a^n$. Hence $1/(1-\lambda_k s)$ is the generating function for the sequence with terms $a(n) = \lambda_k^n$. Thus, the sequence corresponding to P^* is

$$P(n) = \sum_{k\in\mathcal{E}} P_k \lambda_k^n = P_0 + \sum_{k\neq 0} P_k \lambda_k^n.$$

5. From (B) we have $\pi(n) = \pi(0)P(n)$. Since $\pi(n) = \pi(0)P^n$, we have $P(n) = P^{(n)} = P^n$. There will be a long-run distribution iff each $|\lambda_k| < 1$ for $k \neq 0$. In this case, we have

$$\lim_{n\to\infty} P(n) = \lim_{n\to\infty} P^n = P_0.$$

6. It is known that if all $p(i,j)$ are positive, then $|\lambda_k| < 1$ for $k \neq 0$, as required for convergence. Theorems 23.6.7 and 23.7.1 show that this condition must hold more generally. What the generating function analysis shows is that the transient terms go to zero exponentially as n increases. The rate of decay is controlled by the largest $|\lambda_k|$ for $k \neq 0$, as the following argument shows.

From 4 and 5, we have $P^n - P_0 = \sum_{k\neq 0} P_k \lambda_k^n$. Let $|\lambda|$ be the largest of the $|\lambda_k|$, $k \neq 0$. Then on a term by term basis, since each term in $P_k \leq 1$,

$$|P^n - P_0| \leq |\lambda|^n \sum_{k\neq 0} |\lambda_k/\lambda|^n < a|\lambda|^n.$$

This gives an important estimate of the error of approximating P_0 by P^n.

7. Also, we note that

$$P(0) = P^{(0)} = I = \sum_{k\in\mathcal{E}} P_k.$$

Since P_0 is a stochastic matrix, $\sum_{k\neq 0} P_k$ must be a matrix whose rows add to zero.

Example 23.9.1: A Numerical Example

$$P = \begin{bmatrix} 1/2 & 1/2 \\ 2/5 & 3/5 \end{bmatrix} \qquad I - sP = \begin{bmatrix} 1 - s/2 & -s/2 \\ -2s/5 & 1 - 3s/5 \end{bmatrix}.$$

Direct expansion shows that

$$\Delta(s) = \text{Det}[I - sP] = (1 - s/2)(1 - 3s/5) - s^2/5 = (1 - s)(1 - s/10).$$

The zeros of $\Delta(s)$ are thus $s_0 = 1$ and $s_1 = 10$. Obtaining $\text{Adj}[I - sP]$ by the use of cofactors, we get

$$P^*(s) = \frac{1}{\Delta(s)} \begin{bmatrix} 1 - 3s/5 & s/2 \\ 2s/5 & 1 - s/2 \end{bmatrix}.$$

Use of the partial fraction expansion procedure gives

$$P_0 = \frac{1}{1 - 1/10} \begin{bmatrix} 1 - 3/5 & 1/2 \\ 2/5 & 1 - 1/2 \end{bmatrix} = \begin{bmatrix} 4/9 & 5/9 \\ 4/9 & 5/9 \end{bmatrix}$$

$$P_1 = \frac{1}{1 - 10} \begin{bmatrix} 1 - 30/5 & 10/2 \\ 20/5 & 1 - 10/2 \end{bmatrix} = \begin{bmatrix} 5/9 & -5/9 \\ -4/9 & 4/9 \end{bmatrix}.$$

Thus

$$P(n) = \begin{bmatrix} 4/9 & 5/9 \\ 4/9 & 5/9 \end{bmatrix} + \begin{bmatrix} 5/9 & -5/9 \\ -4/9 & 4/9 \end{bmatrix} 10^{-n}.$$

Suppose the system starts in state 1. Then $\pi(0) = [0, 1]$, so that

$$\pi(n) = [0, 1]P_0 + 10^{-n}[0, 1]P_1 = [4/9, 5/9] + 10^{-n}[-4/9, 4/9].$$

This converges rapidly to the long-run distribution. After n steps the state probabilities $p_i(n)$ differ by less than 10^{-n} from the long-run probabilities π_i, and the terms of P^n approach those of $P_0 = P^\infty$ in a similar way.

—   □

Direct determination of the partial fraction expansion in more complicated chains is tedious. However, with the use of standard computer packages, we can determine the eigenvalues of the transition matrix and the long-run distribution, quite easily. From this we can see how rapidly the long-run distribution is approached. The next example illustrates such calculations.

Example 23.9.2: The Chain of Example 23.4.1

The transition matrix is

$$P = \begin{bmatrix} 0 & 0.3 & 0.7 & 0 \\ 0.3 & 0.3 & 0.4 & 0 \\ 0 & 0.5 & 0.1 & 0.4 \\ 0.8 & 0 & 0.2 & 0 \end{bmatrix}.$$

Use of a numerical program for matrix operations shows that

$$P_0 \approx P^{32} = \begin{bmatrix} 0.2050 & 0.3268 & 0.3344 & 0.1338 \\ 0.2050 & 0.3268 & 0.3344 & 0.1338 \\ 0.2050 & 0.3268 & 0.3344 & 0.1338 \\ 0.2050 & 0.3268 & 0.3344 & 0.1338 \end{bmatrix}.$$

The eigenvalues are 1, $-0.3536 + 0.4591i$, $-0.3535 - 0.4591i$, and 0.1072. The complex eigenvalues need cause no difficulty, since it is the absolute values that determine the rapidity of convergence. For these, the absolute values are

$$|\lambda_1| = |\lambda_2| = (0.3536^2 + 0.4591^2)^{1/2} \approx 0.5795.$$

We have $|\lambda_1|^8 \approx 0.0127$ and $|\lambda_1|^{16} \approx 0.0002$. Thus, in fifteen to twenty periods, the system will reach equilibrium, in the sense that the long-run state probabilities prevail, regardless of the initial state probabilities.

———   □

Remark: If the zeros of $\Delta(s)$ are not distinct, but $|s_k| > 1$ for $k \neq 0$, the analysis is more complicated in detail, but gives transient terms that die out somewhat more slowly with increasing n. For example, if we have a double zero $s_k = s_{k+1}$ of $\Delta(s)$, the transient portion contains a term $B_k n s_k^{-n}$. Unless $|s_k|$ is very close to one, the convergence is still quite rapid.

23.10.   The Strong Markov Property for Sequences

We have used the Markov property to formulate the memoryless character of the system. Thus, probabilities of future states are conditioned by the present but not by the manner in which the present is reached. This suggests that a stronger form of the Markov property might hold. If the time of observation (i.e., the present) is determined by some rule based on the past, then probabilities for the future should depend only on this present. We wish to formulate this strong property in terms of the concept of Markov time introduced in Chapter 20.

Let $X_\mathcal{N}$ be Markov relative to $Z_\mathcal{N}$ and let T be a Markov time for the decision sequence $Z_\mathcal{N}$. The value of T represents the "present" as determined by some appropriate rule applied to the "past." Thus, the present state of the process is the value of X_T, and the past of the decision process is W_T, where

$$X_T = \sum_{n=0}^{\infty} I_{\{T=n\}} X_n + I_{\{T=\infty\}} X_\infty,$$

$$W_T = \sum_{n=0}^{\infty} I_{\{T=n\}} W_n + I_{\{T=\infty\}} W_\infty.$$

Suppose B is an event determined by the future, as specified by T. That is,

$$B = \bigcup_{n=0}^{\infty} \{T = n\} B_n \bigcup \{T = \infty\} B_\infty$$

where each B_n is an event determined by U^n, the future at n of the process X_N. We wish to show that in an appropriate sense

$$E[I_B|W_T] = E[I_B|X_T] \quad \text{a.s.}$$

Now the development of the concept of conditional expectation shows that $E[I_B|W_T]$ is almost surely equal to a random quantity whose events are determined by W_T. There is, however, a difficulty in expressing

$$E[I_B|W_T] = e(W_T)$$

since the dimension of $W_T(\omega)$ is determined by the value $T(\omega)$. Thus, $e(W_T)$ must be of the form

$$e(W_T) = \sum_{n=0}^{\infty} I_{\{T=n\}} e_n(W_n) + I_{\{T=\infty\}} e_\infty(W_\infty).$$

In the appendix to this chapter, we show that

$$E[I_B|W_T] = \sum_{n=0}^{\infty} I_{\{T=n\}} E[I_{B_n}|W_n] + I_{\{T=\infty\}} E[I_{B_\infty}|W_\infty] \quad \text{a.s.}$$

By the Markov property, this becomes

$$E[I_B|W_T] = \sum_{n=0}^{\infty} I_{\{T=n\}} E[I_{B_n}|X_n] + I_{\{T=\infty\}} E[I_{B_\infty}|W_\infty] \quad \text{a.s.}$$

For the usual case, $T < \infty$ a.s., so that we may neglect the ∞-term.

Now consider $E[I_B|X_T]$. Since there is no problem of dimension here, the notion of a function $g(X_T)$ is well-determined. Thus, if $E[I_B|X_T] = e(X_T)$, then it would seem plausible that

$$E[I_B|X_T] = \sum_{n=0}^{\infty} I_{\{T=n\}} e(X_n) + I_{\{T=\infty\}} e(X_\infty) \quad \text{a.s.}$$

Here we have the same function e for each n. It turns out that we need to relax this restriction and use the more general form

$$E[I_B|X_T] = \sum_{n=0}^{\infty} I_{\{T=n\}} e_n(X_n) + I_{\{T-\infty\}} e_\infty(X_\infty) \quad \text{a.s.}$$

That is, we allow a different e_n for each n. The motivation for this modification comes from a more general and abstract form of conditional expectation with respect to a sigma algebra of events. Our treatment, shows the appropriateness of this choice. Again, we show in the appendix that for a Markov sequence

$$E[I_B|X_T] = \sum_{n=0}^{\infty} I_{\{T=n\}} E[I_{B_n}|X_n] + I_{\{T=\infty\}} E[I_{B_\infty}|X_\infty] \quad \text{a.s.}$$

For $T < \infty$ a.s., we may neglect the X_∞ term. In the Markov case, we have

$$E[I_B|W_T] = \sum_{n=0}^{\infty} I_{\{T=n\}} E[I_{B_n}|X_n] = E[I_B|X_T] \quad \text{a.s.}$$

Examination of the arguments in the appendix shows that we could replace I_B with

$$Y = \sum_{n=0}^{\infty} I_{\{T=n\}} h_n(U^n)$$

where each h_n is a Borel function such that $E[h_n(U^n)]$ is finite for each $n \geq 0$. Thus, we have the following.

Theorem 23.10.1   (The Strong Markov Property for Sequences)
Suppose X_N is Markov relative to Z_N and T is an almost surely finite Markov time for Z_N. Let

$$Y = \sum_{n-0}^{\infty} I_{\{T=n\}} h_n(U^n)$$

where each h_n is a Borel function such that $E[h_n(U^n)]$ is finite. Then

$$E[Y|W_T] = \sum_{n=0}^{\infty} I_{\{T=n\}} E[h_n(U^n)|X_n] = E[Y|X_T] \quad \text{a.s.}$$

□

The functions h_n may take many forms. And they could all be the same function. For example, we could have $h(U^n) = X_{n+k}$ for a fixed k. In the time-homogeneous case, with all h_n the same, the theorem takes a very simple and useful form.

Corollary 23.10.2 (Strong Markov Property for Homogeneous Chains) *Suppose X_N is a time-homogeneous Markov process relative to Z_N and T is an almost surely finite Markov time for Z_N. Let*

$$Y = \sum_{n=0}^{\infty} I_{\{T=n\}} h(U^n).$$

Then

$$E[Y|W_T] = E[Y|X_T] = E[h(U^n)|X_n] \quad \text{a.s. invariant with } n \geq 0.$$

Corollary 23.10.3 (Transition Probabilities and Markov Times) *Suppose the transition probabilities for a time-homogeneous chain are*

$$P(X_{n+1} = k | X_n = j) = p(i, k) \qquad \text{invariant with } n.$$

Then the probabilities for transition at Markov time T are

$$P(X_{T+1} = k | X_T = j) = p(i, k).$$

A FUNDAMENTAL RESULT FOR HOMOGENEOUS CHAINS WITH DISCRETE STATE SPACE

We use Corollary 23.10.2 to obtain a fundamental result for time-homogeneous chains with discrete state space. We select any state j, and hold it fixed for the following analysis. We say the process visits state j at time n iff $X_n(\omega) = j$. We let $\{T_1, T_2, \ldots,\}$ be the succession of arrival times at state j, after the initial state. Thus

- If $X_1(\omega) \neq j, X_2(\omega) \neq j, \ldots, X_{n-1}(\omega) \neq j, X_n(\omega) = j$, then $T_1(\omega) = n$.

- If this is followed by a succession of visits to states other than j and $X_r = j$, then $T_2(\omega) = r$.

- In a similar manner we describe T_3, T_4, etc.

    1. If there is no arrival, set $T_1(\omega) = T_2(\omega) = \ldots = \infty$.

    2. If there is a finite number m of arrival times, set

$$T_{m+1}(\omega) - T_m(\omega) = T_{m+2}(\omega) - T_{m+1}(\omega) = \ldots = \infty.$$

Now $T_m(\omega) \leq n$ iff j appears in $\{X_1(\omega), X_2(\omega), \ldots, X_n(\omega)\}$ m or more times. Thus, the event $\{T_m = n\}$ is an event determined by W_n, so that each T_m is a Markov time for $Z_{\mathcal{N}}$.

If $T_m(\omega) < \infty$, then $T_m(\omega) < T_{m+1}(\omega)$ and $X_{T_m}(\omega) = j$.

First, we obtain a special case, which we use to obtain the more general result we need for subsequent analysis.

Theorem 23.10.4. *If $X_{\mathcal{N}}$ is a time-homogeneous Markov process relative to $Z_{\mathcal{N}}$ with discrete state space $\mathcal{E}$ and T_1, T_2, ... are times of successive visits to state j, then*

$$E[I_Q(T_{m+1} - T_m)|X_0, T_1, \ldots, T_m] = I_{\{T_m < \infty\}} E[I_Q(T_1)|X_0 = j].$$

Proof

1. Since $Q \subset \mathcal{N}$ is countable, by countable additivity of conditional expectation, it is sufficient to prove the theorem for $Q = \{k\}$.

2. To simplify writing, let $T = T_m$, $U = T_{m+1}$, and $A = \{T < \infty\}$. Then
$$I_{\{k\}}(U - T) = I_A I_M(U_{T,T+k})$$
where M is the subset of $\mathcal{E}^{k+1}$ such that the first and last coordinates are j and the other coordinates may be any elements of $\mathcal{E}$ except j. We note that $X_T = j$ a.s. and the event
$$\{U_k \in M\} = \{X_0 = j, T = k\}.$$

3. Let $V_T = (X_0, T_1, \ldots, T_{m-1}, T)$. Then $V_T = \phi(W_T)$ and

$$
\begin{aligned}
E[I_A I_M(U_{T,T+k})|V_T] & \\
= \; & I_A E\big[E[I_M(U_{T,T+k})|W_T] \,\big|\, V_T\big] \quad \text{a.s.} \quad \text{by (CE9)}^* \\
= \; & I_A E\big[E[I_M(U_{0,k})|X_0 = j] \,\big|\, V_T\big] \quad \text{a.s.} \quad \text{by Corollary 23.10.2} \\
= \; & I_A E[I_M(U_k)|X_0 = j] \quad \text{a.s.} \quad \text{by linearity} \\
= \; & I_A E[I_{\{k\}}(T)|X_0 = j] \quad \text{a.s.}
\end{aligned}
$$

— □

---

* Because the random quantities V_T and W_T are not ordinary random vectors, a more general form of the property (CE9) on repeated conditioning is used. It is possible to use arguments such as those employed for Theorem 23.10.1, but such a development involves tedious details. We simply assume the more general form, which is well-known in advanced treatments.

The next theorem provides the theoretical justification for a number of essential formulas used in the analysis of time-homogeneous chains.

Theorem 23.10.5. *If $X_{\mathcal{N}}$ is a time-homogeneous Markov process relative to $Z_{\mathcal{N}}$ with discrete state space $\mathcal{E}$ and T_1, T_2, ... are times of successive visits to state j, then*

$$E[I_{Q_m}(T_m - T_{m-1})I_{Q_{m-1}}(T_{m-1} - T_{m-2})\ldots I_{Q_1}(T_1)|X_0]$$
$$= E[I_{Q_m}(T_1)|X_0 = j]\ldots E[I_{Q_2}(T_1)|X_0 = j]$$
$$\times E[I_{\{T_m < \infty\}}I_{Q_1}(T_1)|X_0] \quad \text{a.s.}$$

Proof

On the set $\{T_m < \infty\}$, we have $0 < T_1 < T_2 < \cdots < T_m < \infty$.
  For $m = 2$,

$$E[I_{Q_2}(T_2 - T_1)I_{Q_1}(T_1)|X_0]$$
$$= E\big[E[I_{Q_2}(T_2 - T_1)I_{Q_1}(T_1)|X_0, T_1] \mid X_0\big] \quad \text{a.s. by (CE9)}$$
$$= E\big[I_{Q_1}(T_1)E[I_{Q_2}(T_2 - T_1)|X_0, T_1] \mid X_0\big] \quad \text{a.s. by (CE8)}$$
$$= E\big[I_{Q_1}(T_1)I_{\{T_2 < \infty\}}E[I_{Q_2}(T_1)|X_0 = j] \mid X_0\big] \quad \text{a.s. by Corollary 23.10.3}$$
$$= E[I_{Q_2}(T_1)|X_0 = j]E[I_{\{T_2 < \infty\}}I_{Q_1}(T_1)|X_0] \quad \text{a.s. by linearity}$$

For the general case, set up a proof by induction. Argue as before to show that if the theorem is true for $m = k$, then it must be true for $m = k + 1$.
—   □

We use this important property in the proof of Theorem 23.5.5, which plays a key role in establishing properties of homogeneous Markov chains.

Theorem 23.10.6. **[Theorem 23.5.5.] Suppose N_j is the number of visits to state j in a homogeneous Markov chain with discrete state space. Then*

$$P(N_j = m|X_0 = i) = \begin{cases} \big(1 - F(j,j)\big)F(j,j)^{m-1} & \text{for } m > 0, \ i = j \\ 1 - F(i,j) & \text{for } m = 0, \ i \neq j \\ F(i,j)\big(1 - F(j,j)\big)F(j,j)^{m-1} & \text{for } m > 0, \ i \neq j \end{cases}.$$

Proof

$$I_{\{\infty\}}(T_{m+1} - T_m) = 1 - I_{\mathcal{N}_1}(T_{m+1} - T_m)$$
$$P(N_j = 0|X_0 = i) = E[I_{\{\infty\}}(T_1)|X_0 = i] = 1 - F(i,j)$$
$$P(N_j = m|X_0 = i) =$$
$$E[I_{\{\infty\}}(T_{m+1} - T_m)I_{\mathcal{N}_1}(T_m - T_{m-1})\ldots I_{\mathcal{N}_1}(T_1)|X_0 = i]$$
$$= E[I_{\mathcal{N}_1}(T_m - T_{m-1})\ldots I_{\mathcal{N}_1}(T_1)|X_0 = i]$$
$$- E[I_{\mathcal{N}_1}(T_{m+1} - T_m)\ldots I_{\mathcal{N}_1}(T_1)|X_0 = i].$$

Applying Theorem 23.10.4 to the last expression, we get

$$P(N_j = m|X_0 = i) = \big(1 - E[I_{\mathcal{N}_1}(T_1)|X_0 = j]\big)E[I_{\mathcal{N}_1}(T_1)|X_0 = j]^{m-1}$$
$$\times E[I_{\mathcal{N}_1}(T_1)|X_0 = i]$$
$$= \big(1 - F(j,j)\big)F(j,j)^{m-1}F(i,j).$$

A similar argument holds for $P(N_j = m|X_0 = j)$.
—   □

23a

Some Theoretical Details

We provide proofs of several of the theorems stated and used in this chapter.

Theorem 23a.7   ([23.2.1] Extended Markov Property)
X_N is Markov relative to Z_N iff

(M) $$X_N \sim Z_N$$

and
(M*) $$\{U^n, W_m\} \text{ ci} | U_{m,n} \qquad \text{for all } 0 \le m \le n.$$

Proof

1. It is clear that (M) is a special case of (M*), with $m = n$ and $g(U^n) = X_{n+1}$.

2. Given $X_N \sim Z_N$ and $\{X_{n+1}, W_n\}$ ci$|$ X_n for all $n \ge 0$, we show

   (A) $$\{U_{n,r}, W_n\} \text{ ci} | X_n \qquad \text{for all } 0 \le n < r$$

   (B) $$\{U_{n,r}, W_m\} \text{ ci} | U_{m,n} \qquad \text{for all } 0 \le m \le n < r.$$

3. To establish (A), we use mathematical induction on r.

   (a) By (CI9), with $X = X_{n+1}$, $Y = W_n$, $Z = X_n$ and $U = g(X_n, X_{n+1}) = U_{n,n+1}$, we have

   $$\{U_{n,n+1}, W_n\} \text{ ci} | X_n \qquad \text{for all } 0 \le n \quad [r = n+1].$$

   (b) Suppose (A) is true for $r = n + k$. We wish to show that it must also be true for $r = n + k + 1$. Thus, we suppose

   $$\{U_{n,n+k}, W_n\} \text{ ci} | X_n \qquad \text{for all } 0 \le n.$$

   We make $n + k$ the new present, so that by (M)

   $$\{X_{n+k+1}, W_{n+k}\} \text{ ci} | X_{n+k}.$$

   By (CI10), with $X = X_{n+k+1}$, $Y = W_{n+k}$, $Z = X_{n+k}$, and $V = k(W_{n+k}) = U_{n,n+k-1}$, we have

   $$\{X_{n+k+1}, W_{n+k}\} \text{ ci} | U_{n,n+k}.$$

By (CI9), with $X = X_{n+k+1}$, $Y = W_{n+k}$, $Z = U_{n,n+k}$ and $V = h(W_{n+k}, U_{n,n+k}) = (X_n, W_n)$, we have

$$\{X_{n+k+1}, (X_n, W_n)\} \text{ ci} | \ U_{n,n+k}.$$

By (CI12), with $X = W_n$, $Y = X_n$, $Z = U_{n,n+k}$, and $W = X_{n+k+1}$, we have

$$\{(U_{n,n+k}, X_{n+k+1}), W_n\} \text{ ci} | \ X_n.$$

Since $(U_{n,n+k}, X_{n+k+1}) = U_{n,n+k+1}$, proposition (A) is true for $r = n + k + 1$.

(c) By mathematical induction, (A) is true for all $0 \le n < r$

4. To establish (B), we use (A) and the properties (CI9) and (CI10).

By (CI10), with $X = U_{n,r}$, $Y = W_n$, $Z = X_n$, and $V = h(W_n) = U_{m,n-1}$, we have

$$\{U_{n,r} W_n\} \text{ ci} | \ U_{m,n}.$$

By (CI9), with $X = U_{n,r}$, $Y = W_n$, $Z = U_{m,n}$ and $V = h(W_n, U_{m,n}) = W_m$, we have

$$\{U_{n,r}, W_m\} \text{ ci} | \ U_{m,n} \qquad \text{for all } 0 \le m \le n < r.$$

5. We assume a result from measure theory that allows us to replace $U_{n,r}$ for all $r > n$, with U^n.

— $\qquad\qquad\qquad\qquad\qquad\qquad\qquad\qquad\qquad\qquad\qquad\qquad$ □

Theorem 23a.8 (Nearest-Neighbor Property)

If X_N is Markov relative to Z_N, then

$$E[g(U_{m,n})|W_{m-1}, U^{n+1}] = E[g(U_{m,n})|X_{m-1}, X_{n+1}] \quad \text{a.s.} \qquad 1 \le m \le n$$
(N)

or, equivalently

(N*) $\qquad \{U_{m,n}, (W_{m-1}, U^{n+1})\} \text{ ci} | \ (X_{m-1}, X_{n+1}) \qquad 1 \le m \le n$

Proof

Consider any $1 \le m \le n$. By the extended Markov property (M*), we have

$$\{U^{m-1}, W_{m-1}\} \text{ ci} | \ X_{m-1}.$$

By (CI10), with $X = U^{m-1}$, $Y = W_{m-1}$, $Z = X_{m-1}$ and $U = g(U^{m-1}) = X_{n+1}$,

$$\{U^{m-1}, W_{m-1}\} \text{ ci} | \ (X_{m-1}, X_{n+1}).$$

By (CI9), with $X = U^{m-1}$, $Y = W_{m-1}$, $Z = (X_{m-1}, X_{n+1})$, and $U = g(U^{m-1}, X_{m-1}, X_{n+1}) = U_{m,n}$, we have

$$\{U_{m,n}, W_{m-1}\} \ \text{ci}| \ (X_{m-1}, X_{n+1}).$$

If we take the present to be $n+1$, the Markov property gives

$$\{U^{n+1}, W_{n+1}\} \ \text{ci}| \ X_{n+1}.$$

By (CI10), with $X = U^{n+1}$, $Y = W_{n+1}$, $Z = X_{n+1}$, and $V = k(W_{n+1}) = (X_{m-1}, W_{m-1})$,

$$\{U^{n+1}, W_{n+1}\} \ \text{ci}| \ (X_{m-1}, X_{n+1}, W_{m-1}).$$

By (CI9), with $h(W_{n+1}) = U_{m,n}$, we have

$$\{U^{n+1}, U_{m,n}\} \ \text{ci}| \ (X_{m-1}, X_{n+1}, W_{m-1}).$$

By (CI11), with $X = U_{m,n}$, $Y = (X_{m-1}, X_{n+1})$, $Z = W_{m-1}$, and $W = U^{n+1}$,

$$\{U_{m,n}, (W_{m-1}, U^{n+1})\} \ \text{ci}| \ (X_{m-1}, X_{n+1}).$$

— $\qquad\qquad\qquad\qquad\qquad\qquad\qquad\qquad\qquad\qquad\qquad$ $\square$

Next, we establish the expressions for $E[I_B|W_T]$ and $E[I_B|X_T]$ used in the proof of the strong Markov property, Theorem 23.10.1.

We suppose T is an almost surely finite Markov time for $Z_\mathcal{N}$, and let

$$B = \bigcup_{n=0}^{\infty} \{T = n\} B_n \qquad \text{where each } B_n \text{ is an event determined by } U^n.$$

1. By $E[I_B|W_T]$ we mean a random quantity such that

$$E[I_C I_B] = E\big[I_C E[I_B|W_T]\big]$$

for any event C of the form

$$C = \bigcup_{n=0}^{\infty} \{T = n\} C_n, \qquad \text{with } C_n \text{ an event determined by } W_n.$$

We note that

$$C\{T = n\} = C_n\{T = n\} \qquad \text{and} \qquad BC\{T = n\} = B_n C_n\{T = n\}$$

so that

$$I_C I_{\{T=n\}} = I_{C_n} I_{\{T=n\}} \qquad \text{and} \qquad I_B I_C I_{\{T=n\}} = I_{B_n} I_{C_n} I_{\{T=n\}}.$$

Also, we note that the event $C\{T = n\} = C_n\{T = n\}$ is an event determined by W_n, so that

$$E[I_{\{T=n\}} I_{C_n} I_{B_n}] = E\big[I_{\{T=n\}} I_{C_n} E[I_{B_n}|W_n]\big].$$

2. To show that $E[I_B|W_T] = \sum_{n=0}^{\infty} I_{\{T=n\}}E[I_{B_n}|W_n]$ a.s., we use the countable sum property (E8) and the defining property (CE1) for conditional expectation to assert

$$
\begin{aligned}
E[I_C I_B] &= E[\sum_n I_{\{T=n\}}I_{C_n}I_{B_n}] = \sum_n E[I_{\{T=n\}}I_{C_n}I_{B_n}] \\
&= \sum_n E\big[I_{\{T=n\}}I_{C_n}E[I_{B_n}|W_n]\big] \\
&= E\big[I_C \sum_n I_{\{T=n\}}E[I_{B_n}|W_n]\big].
\end{aligned}
$$

3. If $X_{\mathcal{N}}$ is Markov relative to $Z_{\mathcal{N}}$, then

$$
E[I_B|W_T] = \sum_n I_{\{T=n\}}E[I_B|X_n] \quad \text{a.s.}
$$

4. For the Markov case, we wish to show that

$$
E[I_B|W_T] = E[I_B|X_T] \quad \text{a.s.}
$$

Now $E[I_B|X_T]$ is a random quantity such that

$$
E[I_D I_B] = E\big[I_D E^*[I_B|X_T]\big]
$$

for any event D of the form

$$
D = \bigcup_{n=0}^{\infty} \{T = n\}D_n \quad \text{with } D_n \text{ an event determined by } X_n.
$$

5. We note that since D_n is an event determined by X_n and any event determined by X_n is an event determined by W_n, then $\{T = n\}D_n$ is an event determined by W_n. Thus

$$
E[I_{\{T=n\}}I_{D_n}I_{B_n}] = E\big[E[I_{\{T=n\}}I_{D_n}E[I_{B_n}|W_n]]\big] \quad \text{a.s.}
$$

Using this fact, we repeat the argument in 2 with event C replaced by D and C_n replaced by D_n to show that

$$
E[I_D I_B] = E\big[I_D \sum_n I_{\{T=n\}}E[I_{B_n}|X_n]\big].
$$

We thus have

$$
E[I_B|X_T] = \sum_n I_{\{T=n\}}E[I_{B_n}|X_n] \quad \text{a.s.}
$$

$\square$

PROBLEMS

23–1. For the inventory problem in Example 23.1.3, let $m = 2$, $M = 4$, and
$P(D = k) = 0.20, 0.45, 0.2, 0.1, 0.05$, for $k = 0, 1, 2, 3, 4$, respectively.

(a) Determine the transition matrix P.

(b) For the initial distribution $\pi(0) = [0\ 0\ 0\ 0\ 1]$, find the state
distributions $\pi(1)$, $\pi(2)$, and $\pi(3)$.

(c) Determine the long-run distribution.

23–2. A box has three balls, identical except for color: black or white. A
sequence of selections is made, with the following replacement policy:
If a black ball is drawn, replace it with a white ball. If a white ball is
drawn, flip a coin; if the coin turns up heads, replace it with a white
ball; otherwise, replace it with a black ball.

Let X_n be the number of white balls after the nth selection. X_0 is
the number at the start of the process.

(a) Explain why the sequence $\{X_n : 1 \le n\}$ is Markov, and determine the transition matrix.

(b) Suppose the system starts with one white ball and two black
balls. This means the initial distribution is $\pi(0) = [0\ 1\ 0\ 0]$.
Determine the state distributions $\pi(1)$ and $\pi(2)$.

(c) Determine the long-run distribution.

23–3. A gambler plays a game repeatedly, until he loses the money he
started with or until he runs his total to ten dollars. Let X_n be his
net worth after n plays. His strategy is as follows:

If $X_n \ge 4$, he bets two dollars, which returns zero, three, or four
dollars with respective probabilities 0.45, 0.30, 0.25, respectively. If
$1 \le X_n \le 3$, he bets one dollar, which returns zero or two dollars
with respective probabilities 0.55 and 0.45. If $X_n = 0$ or $X_n = 10$, he
stops. The maximum $X_n = 10$.

Let Y_n be the net gain on the nth play, so that $X_{n+1} = X_n + Y_{n+1}$.

(a) Determine $P(Y_{n+1} = k | X_n = i)$ for the various permissible val-
ues of i and k.

(b) Explain why the sequence $\{X_n : 0 \le n\}$ is Markov and determine
the transition matrix P.

(c) Classify the states.

23–4. An isolated research station depends upon an electric power generating unit. A second such unit is kept for standby. Not more than one
unit is in operation at any time. The unit in operation may break

down on any given day with probability p. There is a single repair facility, which takes two days to return a failed unit to operating condition. We consider the possible states of the system to be $(2, 0)$, $(1, 0)$, $(1, 1)$, and $(0, 1)$. The first component is the number of units in operating condition at the end of the day. The second component is one if one day's work has been done on a machine under repair, and is zero otherwise. Number these states 1, 2, 3, and 4.

(a) Determine the transition matrix P for the system in terms of p and $q = 1 - p$.

(b) Determine the long-run distribution for the system when $p = 0.1$. Repeat for $p = 0.05$.

23–5. A young woman visits a gambling casino for the first time. She plays three slot machines, which pay off with probabilities a, b, and c. If she wins on a machine, she plays again. If not, she selects one of the other two on an equally likely basis, without regard to the previous experience with any machine. Let X_n be the number of the machine she plays on her nth try. Then the sequence $\{X_n : 1 \leq n\}$ is a homogeneous Markov chain.

(a) Determine the transition probability matrix P in terms of a, b, and c.

(b) For $a = 1/3$, $b = 1/4$, $c = 1/4$, determine the long-run distribution that shows approximately what fraction of the time she plays each machine.

23–6. A white rat is placed in a box with compartments or cells, as shown in the following figure. It can move to adjacent cells, either to the left or to the right, but not diagonally. Let X_n be the number of the cell it occupies after n moves. Suppose it is equally likely to make any of the moves available to it in any cell, regardless of its previous moves; that is, there is no effective memory of previous moves. Thus, if it is in a corner cell, it will take one of two moves with probability $1/2$; if it is in the center cell 5, it will take one of the four possible moves with probability $1/4$.

1	2	3
4	5	6
7	8	9

(a) Write the transition matrix for the homogeneous Markov sequence $X_N = \{X_n : 0 \leq n\}$.

(b) Determine the long-run distribution.

23–7. For each of the following transition diagrams for homogeneous Markov sequence:

(a) Draw the transition diagram.

(b) Classify the states.

(1) $P = \begin{bmatrix} 0.2 & 0.7 & 0 & 0.1 & 0 & 0 \\ 0 & 0 & 1 & 0 & 0 & 0 \\ 0.2 & 0.3 & 0 & 0.1 & 0 & 0.4 \\ 0 & 0 & 0 & 0.5 & 0.5 & 0 \\ 0 & 0 & 0 & 0 & 0 & 1 \\ 0 & 0 & 0 & 0 & 1 & 0 \end{bmatrix}.$

(2) $P = \begin{bmatrix} 0.2 & 0.4 & 0.2 & 0.2 \\ 0.3 & 0 & 0.4 & 0.3 \\ 0.5 & 0 & 0 & 0.5 \\ 0.2 & 0.3 & 0.5 & 0 \end{bmatrix}.$

(3) $P = \begin{bmatrix} 0.1 & 0.1 & 0.3 & 0.5 & 0 \\ 0.3 & 0.2 & 0.2 & 0.2 & 0.1 \\ 0.1 & 0.2 & 0.4 & 0.2 & 0.1 \\ 0 & 0.1 & 0.2 & 0.4 & 0.3 \\ 0 & 0.1 & 0.1 & 0.4 & 0.4 \end{bmatrix}.$

23–8. For each of the following transition probability matrices:

(a) Draw the transition diagram.

(b) Show that the chain is irreducible and aperiodic.

(c) Determine the long-run distribution.

(d) Compare the three-step transition matrix to the limiting matrix (See Problem 23-9).

(1) $P = \begin{bmatrix} 0.1 & 0.5 & 0.4 \\ 0.6 & 0.1 & 0.3 \\ 0 & 0.8 & 0.2 \end{bmatrix}.$

(2) $P = \begin{bmatrix} 0.2 & 0.4 & 0.2 & 0.2 \\ 0.4 & 0 & 0.3 & 0.3 \\ 0.5 & 0 & 0 & 0.5 \\ 0.2 & 0.3 & 0.5 & 0 \end{bmatrix}.$

(3) $P = \begin{bmatrix} 0.1 & 0.4 & 0.2 & 0.3 \\ 0.5 & 0.1 & 0 & 0.4 \\ 0.4 & 0.2 & 0.1 & 0.3 \\ 0.1 & 0.2 & 0.4 & 0.3 \end{bmatrix}.$

23–9. The eigenvalues for the matrices in Problem 23-8 are

(1) $1, -0.3 \pm 0.141i.$

(2) $1, -0.5, -0.262, -0.038$.

(3) $1, -0.411, 0.0056 \pm 0.121i$.

For approximately what n will P^n approximate P^∞ to four decimal places?

23–10. Consider the homogeneous chain with state space $\mathcal{E} = \{1, 2, \ldots, 10\}$ and transition matrix

$$
P = \begin{bmatrix}
0.6 & 0 & 0.4 & 0 & 0 & 0 & 0 & 0 & 0 & 0 \\
0 & 0.3 & 0 & 0 & 0 & 0 & 0.7 & 0 & 0 & 0 \\
1 & 0 & 0 & 0 & 0 & 0 & 0 & 0 & 0 & 0 \\
0 & 0 & 0 & 0 & 1 & 0 & 0 & 0 & 0 & 0 \\
0 & 0 & 0 & 0.3 & 0.3 & 0 & 0 & 0 & 0.4 & 0 \\
0 & 0 & 0 & 0 & 0 & 0.7 & 0.3 & 0 & 0 & 0 \\
0 & 0 & 0 & 0 & 0 & 0.2 & 0.3 & 0 & 0.5 & 0 \\
0 & 0 & 0.2 & 0.2 & 0 & 0 & 0 & 0.3 & 0 & 0.3 \\
0 & 1 & 0 & 0 & 0 & 0 & 0 & 0 & 0 & 0 \\
0 & 0.4 & 0 & 0 & 0.3 & 0 & 0 & 0 & 0 & 0.3
\end{bmatrix}.
$$

(a) Identify the transient states and the recurrent classes.

(b) Renumber the states to display the subchains. Write the rearranged transition matrix.

(c) Produce the reduced matrix in which each subchain is treated as an absorbing state.

(d) Determine the probabilities of reaching the various subchains from each of the transient states.

23–11. For the system in Problem 23-3, determine the various probabilities $F(i, j)$ of reaching an absorbing state j from a transient state i. How rapidly does the system approach the steady-state condition? Interpret the results in terms of the player's chances of winning with various initial stakes. Is there any amount he could start with to make the game favorable? (If you do not have a suitable computer program or calculator, explain how you would solve the problem if you did have access to such a facility).

Appendix A

Some Mathematical Aids

Series

1. *Geometric series*: From $(1 - r)(1 + r + r^2 + \cdots + r^n) = 1 - r^{n+1}$, we obtain

$$\sum_{k=0}^{n} r^k = \frac{1 - r^{n+1}}{1 - r} \quad \text{for } r \neq 1.$$

   For $|r| < 1$, these sums converge to the geometric series $\sum_{k=0}^{\infty} r^k = 1/(1 - r)$. Differentiation yields the following two useful series:
   $\sum_{k=1}^{\infty} k r^{k-1} = \frac{1}{(1-r)^2}$ for $|r| < 1$ and $\sum_{k=2}^{\infty} k(k-1) r^{k-2} = \frac{2}{(1-r)^3}$ for $|r| < 1$.

   For the finite sum, differentiation and algebraic manipulation yields

$$\sum_{k=0}^{n} k r^{k-1} = \frac{1 - r^n (1 + n(1 - r))}{(1 - r)^2},$$

   which converges to

$$\frac{1}{(1 - r)^2} \quad \text{for } |r| < 1.$$

2. *Exponential series*: $e^x = \sum_{k=0}^{\infty} \frac{x^k}{k!}$ and $e^{-x} = \sum_{k=0}^{\infty} (-1)^k \frac{x^k}{k!}$ for any x.

   Simple algebraic manipulation yields the following equalities useful for the Poisson distribution:

$$\sum_{k=n}^{\infty} k \frac{x^k}{k!} = x \sum_{k=n-1}^{\infty} \frac{x^k}{k!} \quad \text{and} \quad \sum_{k=n}^{\infty} k(k-1) \frac{x^k}{k!} = x^2 \sum_{k=n-2}^{\infty} \frac{x^k}{k!}.$$

Some Useful Integrals

1. *The gamma function*: $\Gamma(r) = \int_0^{\infty} t^{r-1} e^{-t}\, dt$ for $r > 0$

   Integration by parts shows $\Gamma(r) = (r - 1)\Gamma(r - 1)$ for $r > 1$.
   By induction $\Gamma(r) = (r - 1)(r - 2)\ldots(r - k)\Gamma(r - k)$ for $r > k$.
   For a positive integer n, $\Gamma(n) = (n - 1)!$ with $\Gamma(1) = 0! = 1$.

2. By a change of variable in the gamma integral, we obtain

$$\int_0^\infty t^r e^{-\lambda t}\, dt = \frac{\Gamma(r+1)}{\lambda^{r+1}} \qquad r > -1,\ \lambda > 0.$$

3. A well-known indefinite integral gives

$$\int_a^\infty t e^{-\lambda t}\, dt = \frac{1}{\lambda^2} e^{-\lambda a}(1 + \lambda a)$$

and

$$\int_a^\infty t^2 e^{-\lambda a t}\, dt = \frac{1}{\lambda^3} e^{-\lambda a}(1 + \lambda a + (\lambda a)^2/2).$$

For any positive integer m

$$\int_a^\infty t^m e^{-\lambda t}\, dt = \frac{m!}{\lambda^{m+1}} e^{-\lambda a}\left(1 + \lambda a + \frac{(\lambda a)^2}{2!} + \cdots + \frac{(\lambda a)^m}{m!}\right).$$

4. The following integrals are important for the Beta distribution.

$$\int_0^1 u^r(1-u)^s\, du = \frac{\Gamma(r+1)\Gamma(s+1)}{\Gamma(r+s+2)} \qquad r > -1, s > -1.$$

For nonnegative integers m, n, $\int_0^1 u^m(1-u)^n\, du = m!\,n!/(m+n+1)!$

Extended Binomial Coefficients and the Binomial Series

The ordinary binomial coefficient is $C(n, k) = n!/(k!(n-k)!)$ for integers $n > 0,\ 0 \le k \le n$.

For any real x, any integer k, we extend the definition by

$$
\begin{aligned}
C(x, 0) &= 1. \\
C(x, k) &= 0 \quad \text{for } k < 0. \\
C(n, k) &= 0 \quad \text{for } n \text{ a positive integer, } k > n. \\
C(x, k) &= \frac{x(x-1)(x-2)\ldots(x-k+1)}{k!} \quad \text{otherwise.}
\end{aligned}
$$

Then Pascal's relation holds: $C(x, k) = C(x-1, k-1) + C(x-1, k)$
The power series expansion about $t = 0$ shows

$$(1+t)^x = 1 + C(x, 1)t + C(x, 2)t^2 + \cdots, \qquad \text{for all } x,\ -1 < t < 1.$$

For $x = n$, a positive integer, the series becomes a polynomial of degree n.

Cauchy's Equation

1. Let f be a real-valued function defined on $(0, \infty)$, such that

   (a) $f(t + u) = f(t) + f(u)$ for $t, u > 0$.

   (b) There is an open interval I on which f is bounded above (or is bounded below).

   Then $f(t) = f(1)t$ for $t > 0$.

2. Let f be a real-valued function defined on $(0, \infty)$ such that

   (a) $f(t + u) = f(t)f(u)$ for $t, u > 0$.

   (b) There is an interval on which f is bounded above.

   Then, either $f(t) = 0$ for $t > 0$, or there is a constant a such that $f(t) = e^{at}$ for $t > 0$
   (For a proof, see Billingsley, *Probability and measure*, 2., Appendix A20.)

Some Useful Analytical Lemmas

1. $\ln(1 + z) = z + R(z)$, where $|R(z)| < |z|^2$ for $|z| < 1/2$.

   Proof

$$\ln(1 + z) = z + \frac{z^2}{2} + \frac{z^3}{3} + \cdots = z + R(z) \qquad \text{for } |z| < 1$$

where

$$R(z) = \frac{z^2}{2}\left(1 + \frac{2}{3}z + \frac{2}{4}z^2 + \cdots\right).$$

For $|z| < 1/2$,

$$|R(z)| < \frac{|z|^2}{2}\sum_{k=0}^{\infty}\frac{1}{2^k} = |z|^2.$$

$\square$

2. Let z_i, w_i, $1 \le i \le m$, be complex numbers with $|z_i| \le 1$, $|w_i| \le 1$, $1 \le i \le m$. Then

$$\left|\prod_{i=1}^{m} z_i - \prod_{i=1}^{m} w_i\right| \le \sum_{i=1}^{m}|z_i - w_i|.$$

Proof

Consider

$$\prod_{i=1}^{k+1} z_i - \prod_{i=1}^{k+1} w_i = (z_{k+1} - w_{k+1}) \prod_{i=1}^{k} z_i + w_{k+1} \left[\prod_{i=1}^{k} z_i - \prod_{i=1}^{k} w_i \right].$$

By hypothesis,

$$\left| \prod_{i=1}^{k} z_i \right| = \prod_{i=1}^{k} |z_i| \le 1 \quad \text{and} \quad |w_{k+1}| \le 1.$$

From elementary properties of absolute value, we have

$$\left| \prod_{i=1}^{k+1} z_i - \prod_{i=1}^{k+1} w_i \right| \le |z_{k+1} - w_{k+1}| \le |z_{k+1} - w_{k+1}| + \left| \prod_{i=1}^{k} z_i - \prod_{i=1}^{k} w_i \right|.$$

Since the theorem is trivially true for $m = 1$, the last inequality provides the basis for a proof by mathematical induction.

—   □

3. As a corollary, we have

   If $|z| \le 1$ and $|w| \le 1$, then $|z^n - w^n| \le n|z - w|$.

—   □

The Little-o Notation:

1. Suppose $g(x) \ne 0$ for all x in some deleted neighborhood of a. We say

   $$f(x) = o\big(g(x)\big) \quad \text{as} \quad x \to a \quad \text{iff} \quad \lim_{x \to a} \frac{f(x)}{g(x)} = 0.$$

   The expression $f(x) = o\big(g(x)\big)$ is read "f is little-o of g at a" or "f is of smaller order than g at a."

2. A number of properties of the little-o symbol are worth noting.

   (a) If f_1 and f_2 are little-o of g at a, then so is $f_1 \pm f_2$.

   (b) If $c \ne 0$ and f is little-o of g at a, then so is cf.

   (c) If h is bounded and nonzero in a deleted neighborhood of a and f is little-o of g at a, then so is hf.

   (d) If h is little-o of f at a and f is little-o of g at a, then h is little-o of g at a.

   (e) If $g(x) \to 0$ as $x \to a$, then $1/\big(1 + g(x)\big) = 1 - g(x) + o\big(g(x)\big)$.

Appendix B

Some Basic Counting Problems

We examine three basic counting problems that are used repeatedly as components of more complex problems. Two of these problems are equivalent—that of arrangements and that of occupancy. The third is a basic matching problem.

I. *Arrangements* of r objects selected from among n distinguishable objects. Two cases are important:

   (a) The order is significant.
   (b) The order is irrelevant.

For each of these we consider two additional alternative conditions:

   (a) No element may be selected more than once.
   (b) Repetition is allowed.

II. *Occupancy* of n distinguishable cells by r objects. These objects are

   (a) Distinguishable.
   (b) Indistinguishable.

The occupancy may be

   (a) Exclusive.
   (b) Nonexclusive (i.e., more than one object per cell).

These counting problems can be shown to be equivalent. The results in the four cases may be summarized as follows:

(a) 1. Ordered arrangements, without repetition (permutations). Distinguishable objects, exclusive occupancy.

$$P(n,r) = \frac{n!}{(n-r)!}.$$

   2. Ordered arrangements, with repetition allowed. Distinguishable objects, nonexclusive occupancy.

$$U(n,r) = n^r.$$

(b) 1. Arrangements without repetition, order irrelevant (combinations).
Indistinguishable objects, exclusive occupancy.

$$C(n,r) = \frac{n!}{r!(n-r)!} = \frac{P(n,r)}{r!}.$$

2. Unordered arrangements, with repetition.
Indistinguishable objects, nonexclusive occupancy.

$$S(n,r) = C(n+r-1,r).$$

III. Matching n distinguishable elements to a fixed order. $M(n,k)$ is the number of permutations that give k matches.

Example B.1: $n = 5$

Natural order	1	2	3	4	5
Permutation	3	2	5	4	1

(Two matches—positions 2, 4).

—   □

We reduce the problem to determining $M(n,0)$, as follows:

(a) Select k places for matches in $C(n,k)$ ways.

(b) Order the $n-k$ remaining elements so that no matches occur in the other $n-k$ places, in $M(n-k,0)$ ways.
We conclude that $M(n,k) = C(n,k)M(n-k,0)$.

Some algebraic trickery shows that $M(n,0)$ is the integer nearest the number $n!/e$, where e is approximately 2.718.

Values for n up to 10 are as follows:

n	$M(n,0)$	n	$M(n,0)$
1	0	6	265
2	1	7	1,854
3	2	8	14,833
4	9	9	133,496
5	44	10	1,334,961

Index

Springer Texts in Statistics *(continued from p. ii)*

Pfeiffer	Probability for Applications
Santner and Duffy	The Statistical Analysis of Discrete Data